AF868715

The Topological Model of Genome and Evolution

Pradeep Chhaya

The Topological Model of Genome and Evolution

Understanding the Origin and Nature of Life

Springer

Pradeep Chhaya
Independent Technical Advisor
Mumbai, Maharashtra, India

ISBN 978-981-99-4317-3 ISBN 978-981-99-4318-0 (eBook)
https://doi.org/10.1007/978-981-99-4318-0

© The Editor(s) (if applicable) and The Author(s), under exclusive license to Springer Nature Singapore Pte Ltd. 2023
This work is subject to copyright. All rights are solely and exclusively licensed by the Publisher, whether the whole or part of the material is concerned, specifically the rights of translation, reprinting, reuse of illustrations, recitation, broadcasting, reproduction on microfilms or in any other physical way, and transmission or information storage and retrieval, electronic adaptation, computer software, or by similar or dissimilar methodology now known or hereafter developed.
The use of general descriptive names, registered names, trademarks, service marks, etc. in this publication does not imply, even in the absence of a specific statement, that such names are exempt from the relevant protective laws and regulations and therefore free for general use.
The publisher, the authors, and the editors are safe to assume that the advice and information in this book are believed to be true and accurate at the date of publication. Neither the publisher nor the authors or the editors give a warranty, expressed or implied, with respect to the material contained herein or for any errors or omissions that may have been made. The publisher remains neutral with regard to jurisdictional claims in published maps and institutional affiliations.

This Springer imprint is published by the registered company Springer Nature Singapore Pte Ltd.
The registered company address is: 152 Beach Road, #21-01/04 Gateway East, Singapore 189721, Singapore

This monograph is dedicated
To
The first living organism
who
still resides incognito within each one of us

Preface

This book marks a departure from my previous books, in more than one sense. With this book, I return to my roots in science. In my three previous books, I tried to understand the philosophical foundation of science. However, it was a scientist's way to understand philosophy and not a philosopher's way to look at science. However, with this book, I return to science, as I was trained to practice. As I begin this book, I realize that this journey is also returning to the roots of our own existence as a species. Therefore, it is a homecoming in more than one sense.

However, having said that, I realize that the book is still going to be occupied with the philosophical underpinnings of the evolution of life. Therefore, I admit that this book is also a manifestation of an old adage that the more things change, the more they remain the same. However, to the extent a scientist believes in a unified theory of science (and I do), such a recurring motif is bound to be present in any attempt to understand the nature of reality. Therefore, I accept the charge of being a single-minded individual who tries to understand Nature through the prism of one particular model. I personally see a merit in the criticism that if an individual has only a hammer, all the solutions that she/he can think of are the nailing solutions. However, in my defense, I submit that this hammer seems to be providing pretty good nailing solutions. At the same time, I am convinced that there exist some different perspectives of understanding Nature which are equally, if not more, useful in understanding Nature. It is just that I think that the model proposed here is still fertile enough to produce new insights into the nature of reality. Therefore, I propose to extend its use to deconstruct the nature of biological evolution and particularly the nature of genomic architecture and its role in biological evolution.

This book is going to focus on several aspects of genomics and evolution that have not been explored before. Therefore, it is relevant to articulate these aspects in this preface. It is all the more relevant because these aspects have some philosophical bearings and therefore, they have been ignored by and large. Largely under the Cartesian influence, we have chosen to overlook the unique nature of the evolution of sentience. Our tendency has been to treat our cognitive faculty as an accidental payoff, at best. While the conventional perspective of evolution explains the origins of sensory motor processes within the confines of the Darwinian paradigm, the evolution of epistemological processes is never explained from the Darwinian perspective. There are two aspects of our ability to comprehend reality which remain unanswered. Our scientific theories are remarkably congruent with the nature of reality. Secondly, these theories possess complex structural frameworks that are able to encapsulate the truth value. Both these features, viz. congruence with Nature and truth-bearing capacity, point towards some kind of ontological explanation of the evolution of our epistemological processes.

What is strange and what prompted me to deconstruct the Darwinian paradigm de novo is the fact that the Darwinian paradigm does not deal with the evolution of life, it merely deals with what happened after the first life forms appeared on Earth. It is all the more strange because the Darwinian paradigm, in spite of its probabilistic template, is essentially a causal theory. In fact, upon a little reflection, it is intuitively clear that the Darwinian paradigm has no explanation for the emergence of any form of complexity (and the abovementioned epistemological complexity is just one example of this). Conventionally, it has been argued that since first living organisms were simple due to probabilistic reasons, it is axiomatic that later species would turn out to be more complex than their predecessors. Thus, according to the conventional perspective, the emergence of complexity is a historical legacy of having started with simple organisms. Any improvement, by way of natural selection, will inevitably produce next generations with incrementally complex structuralism.

Such an explanation sounds reasonable, except for the fact that even those first living organisms had enough complexity in the form of genomes. Even if one were to disregard this argument of primitive organisms having complex genomes, the complexity of the molecules involved in creating the functionalities of replication and retention of biological information is of a very high order. Therefore, it is inevitable that sooner, rather than later, the Darwinian paradigm must incorporate the complexity argument in its semantics and its operative principles. Therefore, I have decided to take a look at the Darwinian theory from the genomic perspective. At the outset, there are two aspects which need to be deconstructed. Firstly, one needs to understand the role of complexity in natural selection. Secondly, one needs to understand whether the genome can be a unit of selection. We have historically dealt with the notion of the unit of selection from the perspective of group selection. Therefore, it is tempting to think whether the genome behaves as a group of genes, each of which undergoes natural selection and allows natural selection to operate without its own participation. Alternatively, one can think of the genome itself as a unit of selection. This is interesting because the first alternative postulates that genomic complexity is incidental to natural selection. On the other hand, the second possibility points towards a scenario wherein complexity plays an active role in natural selection.

However, before undertaking such a deconstruction, it is necessary to remodel our conventional perspectives of both Darwinian theory and genomic architecture. This is because the conventional perspective of both these domains has not yielded any insights into either the genomic architecture or the role of complexity in natural selection. The proposed model has an inherent advantage which makes it a good candidate for reinterpreting the role of complexity in natural selection and postulating a possible template for genomic architecture. This refers to an important feature of the proposed model of involuted manifold. According to the proposed model, changes in complexity can be defined by the changes in the dimensionality of a system. Therefore it should be possible to formulate natural selection in the terms of the changes in the dimensionality. Thus, emergence of complexity can be shown to have arisen during natural selection without invoking any teleological arguments. Moreover, if a complexity of any kind can be shown to be an outcome of natural selection, it is axiomatic that genomic architecture too must have arisen as a result of natural selection. This conjecture provides two interesting lines of investigation. Firstly, it suggests that the template of the genomic architecture of different genomes can be employed to define a phylogeny. Secondly, this conjecture provides a basis for formalizing a model of genomic architecture. Admittedly, I would focus on the second line of investigation viz. Formalization of genomic architecture.

There is another interesting possibility that arises when we choose to formalize the Darwinian paradigm using the changes in the complexity as a basis. It allows us to encompass the information theoretical perspective into the Darwinian paradigm. If genomes are viewed as repositories of information content, it is inevitable that natural selection must be defined using the rules of information transfers. This ought to enable us to formalize a domain agnostic description of the Darwinian paradigm capable of deconstructing all the other natural phenomena. Admittedly, the objectives outlined above are too large to be accommodated in a single volume. However, I would try to simplify some of the finer nuances of the topics discussed, if only to achieve semantic continuity.

Mumbai, Maharashtra, India

Pradeep Chhaya

Introduction

As mentioned in the preface, this book deals with evolutionary genomics from a philosopher's point of view. At the same time, the book focuses on the underlying principles of genomics. The reason for this dual emphasis lies in the fact that the book deals with the implicit semantics of the Darwinian paradigm and how it diverges from our current understanding of genomics. Thus, this book tries to deconstruct the semantics of the Darwinian paradigm from our experience in genomics, as well as its tries to deconstruct genomics by assuming that the Darwinian paradigm is valid. Surprisingly, as discussed in the following chapters, the incongruence between the two necessitates the reinterpretation of both these domains. The basic premise of this book is that the conventional perspective of the Darwinian paradigm shies away from explaining the emergence of complexity during biological evolution and natural selection. This reluctance historically arose from its aversion to any kind of teleology whatsoever, particularly the one implicit in the Lamarckian interpretation.

One of the most remarkable features of the Darwinian paradigm has been to rejuvenate itself after every paradigm shift in biology. Leaving aside the possibility of Darwin's own awareness of Mendel's pioneering work, genetics was the first of such paradigm shifts that Darwin's theory faced. Not surprisingly, genetic principles provided a foundation for Darwin's theory. Next came the statistical framework of population genetics. Here also, the Darwinian paradigm prospered with its enriched semantics. Similarly, the advent of molecular biology provided a new impetus for the revival of the Darwinian paradigm. Therefore, it is intuitively clear that the Darwinian paradigm must progress further by the inclusion of the semantics of genomics. Admittedly, this is not a new concept. However, there is one aspect of such a symbiosis of these two domains that has not been articulated so far. This refers to the mismatch between these two domains. On one hand, the Darwinian paradigm possesses, perhaps due to its historical legacy, a rich semantics and even a philosophy of its own. The genomics, on the other hand, has rich experimental details, but without any theoretical or semantic underpinnings. This is not to suggest that both these domains are unidimensional, far from it. However, there exists a definitive bias in each of these two domains.

For instance, the Darwinian paradigm lacks any predictive methodologies to map the possible course of evolution. Of course, as discussed in the first chapter, there is a valid reason for this lack. However, if the Darwinian paradigm is a causal explanation (and it is indeed to a large extent), it must possess a methodology for predictions. Of course, we can revert back to concepts like the neutral (or near neutral) rate of mutations or to the plurality of phylogenetic trees to justify the absence of any predictive methodologies. However, it is a category mistake. The Darwinian paradigm is much more fundamental than these individual examples. The absence of any predictive methodologies in the Darwinian paradigm arises because the paradigm lacks a theoretical framework. Thus, the Darwinian paradigm is a domain rich in its semantics and philosophy, but without any native structural template.

Similarly, genomics, in spite of its humongous experimental details, lacks a definitive theoretical perspective of genomic architecture. Our progress in this domain has been amazing. However, it has come about because of the small incremental increases in the details of how a genome operates. Our progress in genomics has not come about due to any theoretical modeling. It is tempting to think that this absence of any models of genomic architecture based on theory is due to the historical aversion to teleological arguments that we have carried forward from the Darwinian paradigm. We have assumed that any theoretical model of genomic architecture would be antithetical to the Darwinian paradigm, particularly the nondeterminism implicit in Darwin's theory.

When viewed from this perspective, it is intuitively clear that our present conception of genomics is dichotomous. On one hand, we wish to adhere to the Darwinian paradigm and its inherent nondeterminism. On the other hand, we intuitively know that genomics is actually a branch of information theory. It is important to keep in mind that according to information theory, be it a biological system like a genome or a cognitive system like the human brain, there must exist a certain number of rules defining the information transfers and transformations. These rules provide not just a theoretical framework, but they also provide a certain degree of predictivity. In fact, it is tempting to think that genomic processes involving information transfers and transformations are actually epistemic processes. Thus, just as the notion of self (and its attendant self-reference) defines the rules of epistemology, a genome too must manifest a notion of self and its consequent self-reference. Once we accept this analogy, it is intuitively clear that the Darwinian nondeterminism is not due to any absence of a framework, but due to inherent incompleteness of any system containing information content.

With this perspective in mind, the book tries to explore various issues concerning evolutionary genomics. The first chapter tries to deconstruct the nature of Life. It is important to understand why Life, in spite of being a natural phenomenon, is unlike any other natural phenomena. It outlines the arguments why it is difficult to formalize Life. The key point is that in other natural phenomena, there exists a definitive relationship between structuralism and functionalities of these phenomena. It is suggested that the reason why we can't formalize Life lies in the fact that we don't have any formal description of the relationship between structuralism and functionalities of living organisms. In order to understand this lack of description, the chapter outlines the relationship between structuralism and functionalities of natural phenomena using the model of modified involuted manifold. Using this approach, this chapter deconstructs the ambiguity about the relationship between structuralism and functionalities of Life. This deconstruction enables us to understand why we need to reinterpret Darwinian theory.

The second chapter tries to apply the reasoning developed in the first chapter to the typical relationship between structuralism and functionalities to the specific topic of the relationship between genotype and phenotype. Having done that, this chapter outlines the broader case of dualities in the Darwinian paradigm. It seeks to articulate the semantic ambiguities of the relationship between genotype and phenotype. These ambiguities can be traced back to the ambiguity about the emergence of complexity during biological evolution. It is suggested that if we could define and correlate the notion of complexity with the relationship between genotype and phenotype, it is possible to understand the evolutionary relevance of dualities. This new perspective offers a valid foundation for reinterpreting the Darwinian paradigm. This new perspective on the Darwinian paradigm offers a new explanation for the emergence of complexity during biological evolution and natural selection.

The third chapter carries forward the complexity argument to justify the belief that genomes must have a good architectural template which itself must be subject to the Darwinian principles of natural selection. The chapter begins with the conventional perspective of natural selection and tries to redefine it in the context of this new explanation of complexity. For this purpose, it employs the modified involuted manifold model. Having outlined a simplistic description of such a model of natural selection, a new template of genomic architecture is outlined wherein functional and structural templates of genomes are linked to one another. This linkage is defined using topological dimensions. This reorganization of the genome into a topological arrangement between functional genome and structural genome gives rise to the definition of a topological modularity. This topological unit of genomic architecture is named and formalized as a Genotope. A tentative scenario of genomic evolution using this definition of genotope is outlined. This chapter ends with a discussion on the semantic congruence and incongruence between the conventional perspective of natural selection and the perspective arising from the genotopic architecture.

The fourth chapter tries to define biological evolution and natural selection as a typical natural phenomenon by defining it as a natural algorithm. In order to do that, this chapter borrows the notion of singularity from cosmology and tries to define an entity named genomic singularity. Just as the nonstructural entity of the cosmic singularity gives rise to different types of complexities, it is postulated that the corresponding genomic singularity gives rise to different types of complexities during the course of evolution. This chapter postulates that there exists a single mechanism of involution which gives rise to different types of pluralities. Thus, it is postulated that the genomic singularity gives rise to modularity. During the first step, functionalities separate from structuralism. Then during successive steps different types of functionalities get separated out. Similarly the structural template of the genome undergoes successive separation of different structural modules. A topological model of different types of modularities each linked with the remaining modularities is outlined. This hierarchy of modularity enables us to distinguish between genomic expressions and genetic expressions. This, in turn, allows us to define a new class of long range influences of the genome. This postulate of the unitary mechanism of genomic and genetic expressions can be viewed as the mechanism of natural selection. More importantly, it allows us to include the evolution of Life into the general class of natural phenomena. Having given an outline of the proposed model, the remaining chapters focus on its specific application to different aspects of genomics.

The fifth chapter applies the proposed model to the specific case of the mammalian developmental biology. In particular, the chapter discusses the concepts behind HOX and Homeobox genes. To begin with, it briefly summarizes the conventional perspective of these two entities. The chapter outlines the need to separate the functional and structural templates of Homeobox. It outlines the semantic imperative for this separation. Having done that, the chapter discusses the need to separate the static and the dynamic templates of Homeobox. This chapter discusses the semantic ambiguities of the conventional perspective of Homeobox. This chapter discusses how the proposed model helps us to remove these semantic ambiguities of the conventional perspective. This chapter ends with a brief discussion on the therapeutic possibilities of the proposed model.

The sixth chapter seeks to apply the proposed model to the genetics of aging. The chapter begins with an outline of our present understanding of the molecular biology of aging. This chapter tries to unravel the effects of individual genes from the effects of the genome as a whole, on the process of aging. Having done that, this chapter deconstructs the evolutionary perspective of aging. It tries to demonstrate that aging is an integral element of the evolution of genomic architecture. Using the proposed model, this chapter tries to articulate the influence of genomic architecture on the individual genes conventionally associated with aging. This new perspective offers new therapeutic approaches to control geriatric pathologies.

The seventh chapter deals with the evolutionary perspective of cancer. While it is generally conceded that cancer is an inevitable consequence of the complexity of genomic architecture, our conventional perspective of genomic architecture fails to formalize this widely held belief. This is partly because we do not have any formal description of genomic architecture. Secondly, even from the evolutionary perspective, we do not know how complexity emerged during the course of evolution. The proposed model offers insights into both these topics. Therefore, this chapter tries to deconstruct the evolutionary perspective of cancer, particularly its relationship with genomic architecture. Conventionally, we think of cancer as an outcome of failure of control mechanisms that decide on the individual gene expressions. However, the conventional perspective restricts itself to our current knowledge of gene expression initiators and inhibitors. However, the proposed model offers a systemic description of long range influences arising from genomic architecture. Therefore, it is intuitively clear that this model of genomic architecture is inherently capable of explaining the origin of cancer pathologies. A brief description of how the proposed model offers a mechanistic perspective of the onset of cancer pathologies is presented in this chapter. This chapter ends with several therapeutic possibilities for treating cancer pathologies.

The eighth chapter returns to the details of the proposed model and tries to articulate two important themes that are central to this book. Firstly, this chapter describes how not just the genome but the entire ecosystem can be formalized using the proposed modified involuted manifold model. Having outlined a generic description of an ecosystem as an involuted manifold, the proposed model explains how a genome itself could be viewed as an ecosystem of its own. Having established the analogy between the two systems, this chapter defines how the regulatory genome can bring about natural selection of individual genes. This chapter refines the notion of genotope and tries to provide its formal description. Using this description, this chapter seeks to define the regulatory genome. It seeks to differentiate between a DNA sequence from the genome. It demonstrates that the DNA sequence of a genome is a three-dimensional representation and the regulatory genome is a higher-dimensional representation of the genome. This chapter seeks to define all the types of gene expressions using the operator of involution and explains how the local effects of gene initiators and suppressors are similar to the long range influences of genomic architecture. Both these types of influences can be formalized using the operator of involution. This chapter ends with the semantic perspective of the proposed model.

The ninth chapter focuses on the principle of natural selection. It revisits the debate on the units of selection. It also deconstructs the evolutionary imperative of having dualities, viz. the duality of the unit of selection versus the units of inheritance, the duality of genotype and phenotype, and the duality of structuralism and functionalities. While the conventional perspective is ambiguous on these dualities, the proposed model offers a new insight into this relationship. Having done that, this chapter outlines general principles of genomic evolution. This chapter ends with an information theoretical perspective of genomic evolution and the reasons why we need to reinterpret the Darwinian paradigm.

Contents

About the Author

Pradeep Chhaya has acted as a technical advisor to the pharmaceutical industry for 30 years. He has postdoctoral research experience. Prior to starting a consultancy, he was a chief executive and a partner of an information content company specialized in value addition to research published globally. He has spent the last 15 years writing research monographs on the foundation of science. At present, seven of his monographs are under various stages of publication. His domains of expertise include spacetime theories, quantum mechanics, mathematics, computations theory, cognitive architecture, molecular biology, medicinal chemistry, and linguistics.

Nature of Life: Structuralism and Functionalities

1

Abstract

Purely from the naturalistic perspective, the phenomenon of life offers a challenge. There is no formal way to define what life is. In a generic sense, it is possible to think of life as a phenomenon having certain structuralism and/or having certain functionalities. This has always been implicit in the Darwinian paradigm and even in genomics. However, it is not possible, at least so far, to formalize both these notions of structural templates and functionalities of life. More importantly, it is not clear whether one can formalize the relationship between the structuralism per se and their biological functionalities. This lacuna has handicapped not just the Darwinian paradigm, but it has also stymied genomics. In this chapter, one would try to formalize the biological structuralism and its functionalities and their relationship using the formalism of the involuted manifold. This approach offers a paradigm shift in understanding both these disciplines.

1.1 Introduction

From a scientific point of view, it is difficult to define life in a formal sense. Though, at an intuitive level, we know what life means, we are never able to explicate in a cogent manner. It is tempting to think that this is due to inadequacies of our formal and natural languages. However, there is a far deeper reason for this inexplicability of life. Upon a little reflection on this inexplicability of life, it is intuitively clear that when we think of life as a natural phenomenon, there are two aspects of life which are generic in nature. Incidentally, these features have also been the subject matter of scientific inquiry in the last couple of centuries. We have realized, over a period of time, that there are certain structural templates which are definitive of life. Similarly, there are certain functionalities which are defining features of life.

© The Author(s), under exclusive license to Springer Nature Singapore Pte Ltd. 2023
P. Chhaya, *The Topological Model of Genome and Evolution*,
https://doi.org/10.1007/978-981-99-4318-0_1

The trouble with our current understanding of these features is that we have no clear understanding of the relationship between these features. The nature of the relationship between structural complexity and functionalities of living systems has remained nebulous. There is no formal theory of linking structuralism of living systems and their functionalities. Although the same can be said about some other natural phenomena like the quantum field, in the case of the phenomenon of Life, this is problematic. This is because the prevailing paradigm of origin and evolution of life is essentially a causal process. For instance, in quantum field theory, we have certain structural templates of a quantum field. Similarly, we have quantum phenomenology in the form of particles and their properties. However, both these perspectives can coexist without much difficulty. This is because quantum theory is a noncausal theory (Bell and Gao 2016, see p. 91).

However, the same is not true for structuralism and the functionalities of life because these two templates are causally connected. This shortcoming is also reflected in the very edifice of Darwinian theory (Hodge and Redick 2009). The process of natural selection rests on two disjointed pillars of genotype and phenotype. This shortcoming also manifests genomics. While we have very detailed information on the structural template of the genome and its replication and translation, there is no detailed knowledge of the functionalities of the genome as a whole (Pevsner 2015). Therefore, it is imperative that we look at what life is in a formal sense and seek to explicate the relationship between structural templates of life and their functions. Therefore, in this chapter, we would try to address this problem of formalizing the phenomenon of life. Using a topological model discussed previously (Chhaya 2022a, see Chapter 3), we will try to deconstruct the nature of living organisms, their characteristic structuralism, and their biological functionalities. We will try to define the relationship between structuralism and functionalities in the biological context. In the following chapter, we will extend the model to define the Darwinian connotations of genotype and phenotype.

Admittedly, these issues are intricately connected to one another. However, for the sake of linearity, this chapter has been further divided into 17 sections. Section 1.2: Life as a Natural Phenomenon, Sect. 1.3: What Separates Life from Other Natural Phenomena?, Sect. 1.4: Why Does Life Defy Formal Definition?, Sect. 1.5: Involuted Manifold Model of a Typical Natural Phenomenon, Sect. 1.6: Involuted Manifold Model of Life, Sect. 1.7: The Reason Why Life Is Difficult to be Formalized, Sect. 1.8: Is There a Common Structure of Life?, Sect. 1.9: Are There Any Definitive Functionalities of Life?, Sect. 1.10: The Semantics of the Relationship Between Structuralism and Functionalities, Sect. 1.11: Why Does Life Defy Common Semantics?, Sect. 1.12: Involuted Model of the Relationship Between Structuralism and Functionalities, Sect. 1.13: Involuted Model of Biological Structuralism and Functionalities, Sect. 1.14: Involuted Model of Natural Selection, Sect. 1.15: Conventional Perspective of Natural Selection, Sect. 1.16: Why Natural Selection Cannot Be Rationalized by the Conventional Model, Sect. 1.17: Need to Reinterpret Darwinian Paradigm, Sect. 1.18: Conclusion.

1.2 Life as a Natural Phenomenon

Modern science is ambivalent about the nature of life. On the one hand, we have a well-defined scientific framework in the form of biological sciences, to study the nature of life. On the other hand, we are moving toward the realization that there is something elusive about life that escapes our scientific inquiry through our framework of biology. Ironically, the origin of this nagging doubt that our present scientific inquiry is inadequate comes from that very inquiry itself. Therefore, we need to take an outsider's perspective of biological sciences to understand what is lacking. Unfortunately, there is no such Archimedean point from which we can observe our process of scientific inquiry, much less that of biological sciences. Therefore, we need to deconstruct our biological framework from within and try to understand what changes one can make.

While this deconstruction has been discussed in preceding monographs (Chhaya 2020, 2022b), here we will confine ourselves to deconstruct our ambivalence about the nature of life that biological sciences provide. To begin with, it is intuitively clear that biological sciences are compromised by the phenomenon of self-reference. Admittedly, this is true for all scientific theories, but it is all the more pertinent in the case of biology. Purely from an abstract perspective, when we study biology, it amounts to a biological organism studying its own nature. This problem becomes more acute when it comes to investigating our cognitive faculty. If our cognitive faculty is a biological phenomenon, then it can't investigate itself, at least not completely. This is best exemplified by one of the features of our cognitive faculty. Our cognitive faculty is structured in such a way that it can observe its own outputs but it can't observe the mechanisms by which these outputs are produced. Therefore, we can comprehend the nature of our thoughts (and this includes all our scientific theories), but we can't comprehend the process by which our cognitive faculty arrives at these thoughts. Therefore, if our cognitive faculty were to be imperfect or biased, we will find it difficult to discover these shortcomings.

There are two glaring examples of this situation. One belongs to sensory perceptions, and the other belongs to epistemological conception. We know from our experience in neurology that our vision is limited to only a certain range of visible light. However, this is not the case with other organisms. As a result of this limited cognitive capabilities, our world view is limited to Nature as it manifests itself in the visible light. However, we know that there exists a whole spectrum of electromagnetic radiation which also reflects myriads of images of Nature, though they are unavailable to us. This argument can be extended to all types of sensory perceptions. Similarly, we have realized during the development of mathematics that there are types of algebras which will remain beyond our cognitive capabilities (cf. The axiom of choice; Herrlich 2006, see Chapter 2). Thus, if due to evolutionary compulsions, our cognitive faculty is compromised, our attempts to understand the biological perspective of Life will always be ambivalent. We will be able to formalize some aspects of Life, but not in its entirety. As mentioned above, this kind of incompleteness plagues every scientific discipline. However, in the case of biological sciences, this problem is more acute.

For instance, if one chooses to investigate the nature of the universe, one can investigate the distribution and spread of stars and galaxies and arrive at a reasonable understanding of the universe. Admittedly, it will be incomplete due to the inherent shortcomings of the instruments used or due to shortcomings of the mathematical models employed in formalizing cosmological perspective. Our inability to formalize the cosmic singularity is the clear example of this (Montani et al. 2011, see Chapter 1). However, such a perspective is not directly influenced by our biological nature. In other words, a comparable view of the universe can be built using a computer, except that a human being is required to make such a computer. Otherwise, there is not much of a difference between a model created by a human being or a computer. However, the same cannot be said about Life. A computer generated definition of Life can never define what Life is (at least not as yet). Therefore, before we begin to try to define Life, we must concede that our present scientific framework is inadequate to carry out this task. Our problem is that we know that Life is like any other natural phenomena, and yet, we can't formalize its definition. Having accepted this shortcoming, it is time to understand what separates Life from other natural phenomena.

1.3 What Separates Life from Other Natural Phenomena

We mentioned above that we are ambivalent about the nature of Life. Let us understand the origin of this ambivalence. There are two grounds on which our ambivalence rests. Firstly, Life, as a natural phenomenon, has some features which are absent from all the other natural phenomena. Therefore, it is difficult to define Life using our conventional definition of a natural phenomenon. Secondly, there are some features of Life which have remained beyond our cognitive capabilities of formalizing. Therefore, we find it difficult to formalize Life. In other words, our ambivalence about the nature of Life derives itself from our ontological and epistemological understanding of Life. Therefore, in this section, we will look at both these perspectives.

Let us begin with the ontology of life and why it is different from any other natural phenomena. It might sound strange in the days of genomics and systems biology, but the fact remains that we do not know clearly how Life originated. It is true that we have some well-founded conjectures about the origins of Life. However, there is no formal scientific theory of the origin of Life. If there was such a theory, it would have enabled us to start Life de novo, if not on the Earth then, somewhere else. However, there is no such theory. The question is why don't we have any such theory? There have been innumerable cases when we encountered a natural phenomenon, we investigated it and eventually developed a formal theory of that phenomenon. However, this has not been possible in the case of Life. Although we know a great deal about Life, both formally and informally, we have not been able to convert that knowledge into a formal description of Life. The reason behind our inability to formalize Life is that there are some inherent features of Life which we cannot deconstruct. Of course, we will discuss these features in more detail in the later sections, but let us enumerate them in this section. These features are intelligence, sentience, and self-reference. Of these three features, we, as a species, seem to exclusively possess the feature of self-reference, while the features of intelligence and sentience seem to manifest in all forms of life to varying degrees.

It must be admitted at the outset, that these three features are not easily defined and generations of scientists have spent their lives trying to formalize these three features. Therefore, we will take a simplistic definition of these concepts and instead try to deconstruct them in a nontechnical language. For the purpose of this chapter, we will define intelligence as an ability to respond to the environment. Similarly, we will define sentience as an awareness of the reality and self-reference as an ability to define one's own self vis a vis the reality. When viewed from these definitions, it is intuitively clear that the reason why we can't formalize these features of life is that we can't pinpoint the underlying structural template in our brains. This is in contrast to any other natural phenomena wherein it is possible to define the features of a natural phenomenon in terms of its structural template. For instance, think about a storm. It is characterized by two features, high speed of winds on the periphery and relatively calm at its center. It is possible to connect both these features of a storm to its structuralism which meteorologists create on their computers. There exists a set of mathematical equations which describes a storm. Of course, it is a different matter that the nature of these equations is such that due to their sensitivity to the initial parameters, they give rise to chaotic outcomes (Gribbin 2009, see Chapter 1). However, in principle, this set of equations is capable of defining and predicting (to a reasonable accuracy) the generation and after effects of any storm.

Similarly, one can think about the Sun. Its enormous size, its extreme volatility of temperature, and its variable outputs of electromagnetic radiation are all the features that define the Sun. Moreover, since we know more about the Sun, we can add several more features like its internal structure, its elemental composition, and the type of nuclear reactions responsible for its extremely high temperature. The pertinent point is that all these features are accounted for by the structural template of the Sun. One can extend this argument to the largest or the smallest natural phenomena, from the farthest galaxies to tiniest subnuclear particles. In each of these cases, there exists a direct relationship between their structuralism and their features. Admittedly, due to our limited cognitive capabilities, these natural phenomena are not completely known. However, we, as a community of scientists, are convinced that in each of these cases there exists a definitive relationship between the structuralism of any given natural phenomenon and its manifest features. However, the same cannot be said about Life. The three features mentioned above defy such direct correlation with the underlying structuralism.

It is possible to argue, particularly by theoretical physicists, that there are aspects of reality like quantum phenomena, which also defy such a relationship between its structuralism and its features. Such an argument is valid. However, as discussed in the preceding monographs (Chhaya 2022b, see Chapter 5; Chhaya 2022c, see Chapter 8), it is possible to ascribe counterintuitive nature of quantum phenomena to inherent mismatch between quantum reality and our cognitive faculty which has evolved from the features of the classical reality as formalized in the Newtonian paradigm. More importantly, quantum phenomena do happen to share some structural congruence with that of our cognitive faculty as evident from the phenomenon of quantum entanglement. However, the key point here is that the characteristics of features of Life do not seem to correlate with the inherent structuralism of Life. Therefore, in spite of being a natural phenomenon, Life is different from any other natural phenomena.

In addition to the three features mentioned above, there is one more feature of Life which distinguishes life from other natural phenomena. This refers to the complexity of the structuralism of living organisms. However, this feature has been deliberately kept out of the discussion here because we will return to it in the later sections. This exclusion is necessary because it involves a fundamental semantic proposition of Life and we will discuss it not only in this chapter, but also in the remaining chapters of this monograph. Having discussed the reasons why Life is different from any other natural phenomena, it is time to discuss why it is difficult to formalize Life.

1.4 Why Does Life Defy Formal Definition?

Even if we were to concede that Life as a natural phenomenon is different from the rest of the natural phenomena, the question of why it is difficult to formalize Life remains to be answered. After all, there are several natural phenomena that are counterintuitive if not weird in nature. Quantum phenomenon is the most obvious example of it. However, even in the case of quantum phenomena, we have been able to formalize its structuralism and employ it to predict the outcomes of quantum computation with considerable success (Wieder 2012, see Chapter 3). Though, it must be admitted that we do not understand the nature of quantum phenomena, but at least we have a formal description in place. This is not the case with the phenomenon of Life.

Today, it is possible to define a DNA sequence of any living organism. More importantly, it is also possible to "manufacture" a living cell containing the entire subcellular machinery, including a genome (Prakash 2007, see Part I). Moreover, such a "manufactured" cell would behave like a normal living cell. However, it is still not possible to define what constitutes life in such a "manufactured" cell. The reason why we can't define Life has something to do with how we try to understand complexity. Our analytical skills and the reductionist paradigm of modern science has resulted in a Janus faced structural template of all our scientific theories. On the one hand, we have very detailed deconstruction of any given natural phenomenon into its smallest constituents. On the other hand, we have certain functionalities which arise only when these smallest constituents are assembled together. Therefore, there is a fundamental semantic lacuna in our scientific theories. In order to explain the emergence of complexity, particularly the functional complexities, we resort to what can be called the emergence principle (Smith and Morowitz 2016, see Chapter 4). We are convinced, and perhaps rightfully, that as more and more constituents are linked to one another, the resulting functionalities emerge as a natural consequence.

This emergence principle is best exemplified in the case of scale free networks (Arbib 2003). A network acquires functionalities due to the number of nodes present in the network. However, beyond a certain number of nodes, there is no linear relationship between the number of nodes and the nature of functionalities of the network. There is no explanation as to why such a scenario arises. Apparently, there is some relationship between the number of nodes present in a network and its functionalities. However, at some point, after a certain number of nodes are included in the network, this relationship ceases to exist. Thus, the semantics behind this relationship between the number of nodes and the nature of functionalities remains unarticulated. It might sound harsh, but the fact remains that we treat this semantics as a black box. We don't know what is inside the box, but we use it whenever we fail to explain the origins of these functionalities. It must be kept in mind that this strategy is not confined to just machine learning. It is prevalent in every scientific discipline.

For instance, in theoretical physics, we can think of quarks as building blocks of the universe. These quarks give rise to fundamental particles. These particles, in turn, give rise to the nucleus. We can go further and further and arrive at how galaxies and even the cosmos can be constructed from these elementary building blocks of quarks. However, what is missing in this approach is how the functionalities as each level of structural complexity arise. Fundamental particles have different functionalities. Nuclei have their own functionalities. We can go on ad infinitum in this manner. Similarly, we can think of genomics. At the most fundamental level, we have nucleotides. When these are assembled together, we obtain genes. When we connect these genes together, we arrive at a genome. Here again, the functionalities of nucleotides are different from those of genes and the functionalities of genes are different from those of genome.

In all such cases, what we do is to assign different functionalities to different levels of organizations of the smallest constituents. We feel comfortable with this correlation between the level of organizations and the corresponding functionalities at each of these levels. It must be kept in mind that this approach has a lot of benefits and has given us excellent results. For instance, this approach has enabled us to discover newer and newer materials. It has helped us to develop newer therapies for complicated diseases like cancer. Therefore, we have become complacent. However, it has to be admitted that unless science gives up this "black box" model of complexity, it will remain stagnant. It is time for us to admit that there exists some relationship between structural complexity and functionalities. More importantly, this relationship manifests itself in each of these natural phenomena. In fact, this is implicit in our conception of a unified theory of science. However, this relationship between structuralism and functionalities has eluded formalization. Before we look at any such attempts to formalize this relationship, let us look at the reasons why we have not been able to formalize it so far. This is important because we have necessary wherewithals, but we have chosen not to employ them.

The prime reason why we have not sought to formalize this relationship is that we are not sure about the nature of mathematical constructs and their origins. While this aspect is discussed in the preceding monograph (Chhaya 2022a), we will briefly outline it here. Since Plato (Panza and Sereni 2013, see Introduction), we have realized that mathematics in general and mathematical constructs in particular, are independent of our cognition of them. They seem to exist whether we are capable of perceiving them or not. In consonance with the then prevalent philosophy, Plato categorized them as transcendental entities and absolute. In congruence with the then practiced theology, they were assigned a divine status. It was only when Descartes (Cottingham 2008, see Part III) found a way to separate divine from mundane that we stopped worrying about the origins of mathematical objects. Therefore, in modern terminology, we call them a priori. While this designation saves modern science from the embarrassment of employing transcendental (and therefore divine) entities, it doesn't alter the basic dilemma of mathematics. The question of why mathematics should be absolute remains unresolved.

More importantly, the relationship among different mathematical constructs remains obscure. With the passage of time, we have discovered more and more such relationships between different mathematical constructs. With each such discovery, we realize that there exist some aspects of the structure of mathematics that are beyond our cognitive capabilities. On the one hand, it reaffirms the a priori status of mathematics; however, on the other hand, it leaves us with a nagging doubt that this structural complexity among the mathematical constructs has certain ontology. It is this ontological ambiguity, or rather our ignorance of how different mathematical constructs are connected to one another, that is at the heart of the above mentioned relationship between structuralism and functionalities. Let us see why these two aspects are related to one another.

When we think of the relationship among different mathematical constructs, we think in terms of category theory (Riehl 2017, see Chapter 1). This theory formalizes the relationship between different mathematical constructs. It formalizes the rules of transformations of one category into another category. This is achieved by formalizing the notion of functors. A functor defines the rules of transforming one category of mathematical objects into another category of objects. This formal description of category theory has given us tremendous insights into the nature of mathematics. However, this edifice stands on two troublesome premises. Firstly, it undermines the immutability of mathematical objects which is a necessary prerequisite for defining the transcendental status of mathematics. Secondly, and perhaps to avoid this problem, it places functors as a mathematical category itself. Thus, category theory tries to define ontology of mathematical constructs without compromising its transcendental nature. However, in the process, it creates a self-reference in the form of functors being a category by themselves. (This is because it leads to infinite regression by asking a rhetorical question about who transforms the functors? Is there any special category of functors which transforms one functor into another? etc.)

However, things would be different if we knew the origin of mathematical objects. However, the trouble with this possibility that mathematical objects have any origins is that it leads to certain infirmity to their absolute nature. If mathematical objects were subject to mutations, the world that we know would collapse. Therefore, we have accepted the ontological ambiguity of mathematics as a lesser evil.

However, the key point is that if mathematical constructs had certain ontology, then the changes in their complexity could be used to justify what was called the emergence principle above. In that case, one can argue that different structural complexity gives rise to different functionalities because it is inherently due to the underlying mathematical constructs. One could argue that the phenomenon of scale free networks arises because the underlying mathematical structuralism demands it to be so (Caldarelli 2007, see Chapters 3 and 4). However, we cannot do it at present because under the Cartesian influence, we have kept mathematics out of our spatio-temporal universe. Therefore, it would be very convenient if mathematics was an integral part of the spatiotemporal universe. In that case, we could always point toward the underlying mathematical formalism to explain the origins of the functionalities of any given natural phenomenon.

One such model of defining origins of mathematical objects was proposed in the preceding monograph (Chhaya 2022a). In the later sections, we will employ that model to define the relationship between structuralism and the functional complexities of Life. However, what is germane to the present discussion is that even with the proposed model, it is difficult to formalize Life. Therefore, let us see why it is so. Like many other natural phenomena, Life seems to manifest different functionalities, with different degrees of complexity. Therefore, there is no uniform way to formalize different functionalities, much less that their different degrees of complexity. Normally, it is possible (or it ought to be possible) to link different functionalities of a given natural phenomenon to a common framework. For instance, we have different types of fundamental particles, which we ascribe to being a member of different generations of particles. Each of these generations of particles and their individual members, are vastly different from one another. Therefore, it would have been difficult to explain their origins and their different features in the absence of any common framework. However, we have such a framework available in the form of the standard model (Montani et al. 2011). It is this common framework of standard model that provides a necessary backdrop and a semantic content to justify the emergence of different functionalities and their complexities.

In the case of Life, there is no such common structuralism that connects different functionalities of Life. Admittedly, we have different frameworks for each type of functionalities of Life. For instance, we have a detailed description of neurological templates of signal transduction. We also have a fairly good knowledge of neural networks (or to be more specific, neuronal networks). Therefore, one would think that we are within the reach of defining cognitive processes. However, the neuronal architecture, with its detailed description of synaptic junctions, does not extend itself to cognition per se. There is a structural gap between cognition and its underlying neurology (Alon 2006, see Chapter 2). Similarly, in genomics, we have two parallel frameworks. We have detailed theory genes and their expressions. Similarly, we have a detailed description of the DNA sequence of a large number of genomes. However, we do not have any (at least as yet) a framework which links genomic architecture to each and every gene expression. Admittedly, with the advent of systems biology (Alon 2006) or functional genomics (Roy and Kundu 2021, see

Part III), we are trying to fill this gap. However, the fact remains that we do not have any common framework.

As mentioned above, there are historical reasons for this lack of a common framework. To begin with, under the Cartesian influence, we have chosen to keep our cognitive faculty, in the form of consciousness, out of the scientific pursuits. Though this lacuna has been sought to be filled in the last hundred years or so, the theoretical backlog needs to be cleared. Similarly, Darwinian theory (Hodge and Redick 2009) has focused on natural selection by sidestepping the problem of evolution of Life. Thus, we have very nuanced Darwinian semantics of natural selection, and our scientific theories of the origin of Life are on less firmer grounds. Even here, we are trying to develop the theory of biological evolution using the postulates of RNA or RNA proteins being a common ancestor to all living organisms (Yarus 2010, see Chapter 2). However, even here, one is treading on thin ice of speculative arguments.

Both these instances are manifestations of a deeper malaise of the Cartesian split. Without being aware of it, we adhere to Cartesian dualism and concede that Life is something transcendental beyond the reach of our scientific inquiry. If we could find a way to convince ourselves that Life and the emergence of consciousness in living organisms is a natural phenomenon like any other natural phenomena, then we can think of defining Life like any other natural phenomena. Thus, we need two tools as prerequisites to define a common structuralism of all the natural phenomena and to prove that Life is just another example of natural phenomena. These tools are a model to describe the origins of mathematical objects and their functionalities. This will enable us to justify the emergence principle. Secondly, we need a framework which links our cognitive faculty to its underlying biological template. Once we have these two tools, we can hope to formalize a universal theory of natural phenomena, including Life and its cognitive faculty. Therefore, in the next section, we will look at the structural template of one such model.

1.5 Involuted Manifold Model of a Typical Natural Phenomenon

In the preceding sections, we looked at the reasons why Life is unlike any other natural phenomena and why it is difficult to formalize Life. During the discussion, we found several aspects of Life, which if included in its formalization, would enable us to formalize Life. Therefore, in this section, we will take a first step by defining a natural phenomenon in an abstract sense. Purely from the theoretical perspective and particularly from the mathematical modeling perspective, it has to be admitted that there are several excellent models available in the literature (Laudal 2021, see Introduction). In fact, it is possible to think of mathematics itself, as a model for this purpose. However, as discussed above, the problem with the conventional perspective is that it remains ambiguous about the origins of mathematical objects. Therefore, if we were to employ the conventional perspective of mathematical modeling, we would fall into the trap of failing to accommodate the unique features of Life which anyway, is one's objective. Therefore, in this section, we will take a look at a novel approach outlined in the preceding monograph (Chhaya 2022a).

We will begin with a postulate that mathematical constructs are embedded in the spatiotemporal universe itself. If these entities are not observed in our experiments, it is only because they are located in the higher dimensions of spacetime. It must be kept in mind that the notion of spacetime being a higher dimensional entity (higher than four dimensional perspective), is an integral part of the theories of physics (Randall 2006, see Part I). Therefore, our postulate that spacetime is a higher dimensional entity is not far-fetched. However, there are a couple of points that must be clarified here before we begin to formalize a typical natural phenomenon. Firstly, though the theories of physics regularly invoke higher dimensional models, not all of them assume that spacetime is indeed higher dimensional. These theories merely resort to higher dimensional models because they want to accommodate a greater number of parameters into the proposed formalism. Therefore, we will not use this type of employment of higher dimensional modeling as a justification for the proposed model.

This is because the model proposed here employs a higher dimensional model of spacetime in a physical sense. According to this model, the higher dimensional model of spacetime is not a hermeneutic device to help us to formalize a given natural phenomenon. Rather, according to this model, spacetime is indeed higher dimensional. In that sense, the proposed model is similar to the theories of spacetime like string theories (Conlon 2016, see Chapter 14) or unified quantum field theory (Lawrie 2013, see Chapter 5). These theories also postulate that spacetime is indeed a higher dimensional entity in a physical sense. Using the higher dimensional models of spacetime, these theories try to explain the phenomenology of the quantum field. In this chapter, we will employ this underlying semantic proposition that spacetime is a higher dimensional physical entity.

Of course, as discussed in the preceding monograph (Chhaya 2022a), there is not much difference between these two connotations of a higher dimensional model. This is because of two fundamental beliefs of modern science. Firstly, we are convinced due to quantum computation theory (Nielsen and Chuang 2010) that information, irrespective of its form, is a physical entity. Therefore, when we employ a higher dimensional model for a hermeneutic purpose, we are simply acknowledging the greater information content of the natural phenomena under observation. Therefore, if any natural phenomenon possesses a greater information content, it would per force be occupying a larger span of spacetime. Therefore, this can only be explained by using a higher dimensional model because we operate from the four dimensional spacetime. Thus, the material status of the information content and our initial conception of the natural phenomena under observation having more number of parameters, leads to higher dimensional spacetime by implication.

The second reason why both these connotations of higher dimensional models are synonymous, is that we are convinced due to our experience in quantum phenomena, that spacetime is not empty. It contains a whole lot of structural details. Therefore, it is less problematic to assume that the parameters that we choose to define any natural phenomena are already embedded in the spacetime itself. Therefore, every natural phenomenon must be viewed as a manifestation of this hidden information content of spacetime. A natural phenomenon merely explicates the implicit structuralism of spacetime. Therefore, it is legitimate to employ a higher dimensional model of spacetime in defining a universal template of a typical natural phenomenon. With these caveats in place, let us outline the proposed higher dimensional model of a typical natural phenomenon.

The notion of a parent system having more than one subsystems embedded in it, is a universal template of our scientific disciplines. In cosmology, we conceptualize cosmic singularity (Montani et al. 2011) in which the manifest universe, with all its natural phenomena, can be placed. Similarly, in biology, one can think of an Ecosystem (Raffaelli and Frid 2010) in which all the living organisms, with their complex genomes, can be placed. Similar examples can be found in all scientific theories. Perhaps, the origin of this nested hierarchy of a parent system with its numerous subsystems owes its origins in our cognitive faculty, particularly the processes responsible for formalizing scientific theories. However, since our cognitive faculty is incapable of perceiving any nonanthropic perspective of reality, it seems sensible to accept this limited cognitive capability of perceiving some types of hierarchy, and try to improve upon it. In fact, as discussed in the preceding monograph (Chhaya 2020), it is possible that our ability to comprehend reality owes its origin to the universe itself. Therefore, such an approach, even if it is incomplete, seems to be the best strategy to understand what a natural phenomenon should be. However, it must be kept in mind that since our cognitive faculty itself is a natural phenomenon, its conception of a typical natural phenomenon would be incomplete in more than one sense.

Once we accept this hierarchical perspective, it is possible to define the interactions between the parent system and its various subsystems (and the interactions between various subsystems) as transactions. Therefore, according to the proposed model, these transactions can be defined as transactions between different dimensionalities. Thus, every influence of the parent system on its various subsystems can be visualized as an inward folding of one of the dimensions of the parent system onto its subsystems. This is formalized as a class of mathematical operators of involution. Conversely, the influence of any of its subsystems on the parent system can be defined as the inverse of the operator of involution. Temporarily, we will assume that the interactions between various subsystems of any given parent system passes through the parent system and not directly between the participating subsystems. This assumption is admittedly ad hoc and its necessity would become self-evident in the following chapters. Presently, we will accept this ad hoc assumption as a priori.

The sketch outlined above is very abstract and perhaps inadequate. However, it will serve the purpose of deconstructing our conventional perspective of theoretical modeling. We are primarily interested in finding out why we find it difficult to formalize Life. Therefore, in this section, we will use this strategy to deconstruct our conventional perspective and demonstrate why Life cannot be defined using this approach. Having done that, we will suggest a modification of this nested architecture to demonstrate that this modification enables us to formalize Life.

One of the key features of this way of defining a natural phenomenon is that it allows us to understand the nature of interactions between the parent system and its subsystems. Suppose, we wish to understand the phenomenon of a planet orbiting a star. Then, according to this model, the parent system is the planetary system consisting of a star and the planets circling around it. Therefore, according to this model, one can explain the reasons why a planet orbits around the star. It can be defined in terms of the entire planetary system, which would be taken as the parent system in this model. Now, in this model, it is possible to define gravity as a property of the parent system. However, its influence on a planet and even the star can be defined as an involutive algebra. Of course, as mentioned above, in this model, we will have to define the gravitational field between a star and its planet not in a direct manner, but only in the context of the entire planetary system.

Admittedly, the suggestion that gravitational fields can be defined as an involutive algebra is not a part of the conventional perspective. However, the rest of the scenario outlined here is obviously part of the conventional perspective. For instance, the Newtonian theory of gravitation (Capuzzo-Dolcetta 2019, see Chapter 2) was not based on the perspective of the parent system, in this case, the solar planetary system. It was defined in terms of individual planets and the Sun. However, that is the reason why it failed to explain how gravity can act at a distance. However, the relativistic perspective of gravity exemplified the proposed model. By investing gravity in spacetime itself, rather than to either the Sun or any of its planets, the relativistic paradigm (Schutz 2009, see Chapters 7 and 8) defines it as a property of the parent system. Therefore, the problem of force of gravity acting at a distance doesn't arise. Moreover, by defining gravity as a property of spacetime, it obeys the above placed restriction of not defining the interactions between subsystems directly, but only through the parent system. The success of the general theory of relativity in explaining the movements of Mercury owes its origin to this decision to define gravity as a systemic property and not as a sum of all binary influences. (Incidentally, the mathematical difficulties in what is conventionally defined as three body problems also played a part in this paradigm shift.)

Now let us look at a different example from a totally dissimilar scientific theory of natural selection (Delisle 2021, see Part II). Even here it is possible to define the parent system as the Ecosystem of the Earth. All the species can be defined as subsystems of this ecosystem. Naturally, one can define natural selection as a competitive survival amidst finite resources (Incidentally, it must be kept in mind that Darwin himself was influenced by the Malthusian economic model of limited resources (Flew 2017, see Part III)). Once again, it makes sense to think of resources of an ecosystem as a feature of the entire system rather than being localized as a feature. The original idea of natural selection was based on the interactions between the competing species in the food cycle. This is an example of binary interactions which, as mentioned above, should be avoided. However, this conventional perspective left quite a few things unexplained. However, when a new paradigm of population genetics (Provine 2001, see Chapter 5) arrived, natural selection became a systemic feature. Instead of competing species, scientists started looking for competing genes and their spread in a given population. Thus, once again, it was this paradigm of hierarchy of subsystems that has refined our understanding of a given natural phenomenon. However, it must be admitted that while this hierarchical perspective is an integral part of our scientific theories, the approach of defining the relationship within a hierarchy subsystems as an involutive algebra is not a part of the conventional perspective. As the chapters unfold, the justifications for the insistence on the use of involutive algebras will become self-evident.

Prima facie, let us accept that this conventional perspective of understanding any natural phenomenon using this hierarchy of subsystems (sans the involutive algebra) is consistent with both the nature of reality and the nature of our cognitive abilities. Therefore, it is imperative that we look for the reasons why Life can't be defined in this model. As mentioned above, there are three features of Life that seem to defy formalization using this hierarchical perspective. These features are intelligence, sentience, and self-reference. Therefore, let us see why it is difficult to define these features. Once we do that, we will realize that there exists a far more fundamental problem in defining Life. It consists of a semantic ambiguity about the nature of the relationship between structuralism and functionalities. However, let us begin with these three features of Life and see why we can't formalize them, at least not without assuming something transcendental.

The biological foundations of intelligence (Heyes and Huber 2000, see Chapter 2) have been researched extensively. There exists a fairly detailed description of the neurological mechanisms of how an intelligent behavior arises in living organisms (Wolstenholme and O'Connor 2009). Irrespective of the modalities of intelligence, it is possible to explain its biological template. In addition, there exists a definitive rationale to explain the evolutionary relevance of origin and selection of intelligent behavior. Therefore, it might appear, at least prima facie, that the emergence of intelligence in the evolution of Life is inevitable in some sense (Heyes and Huber 2000). Therefore, why should it be difficult to formalize Life using intelligent behavior? After all, thanks to recent advances in artificial intelligence, we have a good formal description of intelligence (Ramsay 1991).

The reason why we can't formalize Life on the basis of intelligence is twofold. Firstly, intelligence manifested in living organisms is distinctly different from the one manifested in machines. It must be kept in mind that though computers do perform the tasks carried out by human beings, they do so in different manners. It is possible to assert today that computers give comparable outputs to the tasks assigned vis a vis the outputs produced by human beings. However, these outputs are not arrived at by mechanisms that operate in our cognitive faculty. It is just that the outputs are the same in both these cases, but the methods are not. Therefore, to seek to define Life on the basis of intelligence, using the methodology of computer intelligence is a category mistake.

The second reason why we can't define Life using intelligence as a basis, lies in the fact that we do not know the mechanisms by which living organisms act intelligently. For instance, in the case of human beings, it would have been possible to employ our understanding of artificial intelligence to understand human intelligence, if we knew how human intelligence operates. This is because we could have transferred formalism of artificial intelligence to the domain of human intelligence with suitable modifications. However, we have no idea how the human mind operates, at least in its mechanistic details. All we know about human intelligence is in the normative form. We can describe human intelligence but cannot replicate it. As mentioned above, we can devise machines which can give us comparable intelligent responses, but not by the same mechanisms that operate in the human mind. Therefore, it is legitimate and even imperative that we employ computers to replace human beings in performing tasks. However, this paradigm cannot be applied to define Life on the basis of biological intelligence. If we wish to define Life using intelligence, we need a structural template of biological foundation of intelligence which is not available. It is important to keep in mind that we know why intelligence has evolved during biological evolution, but we don't know how it has evolved.

In addition, there exists another problem about the choice of intelligence to define Life as a typical natural phenomenon. No other natural phenomena possess intelligence. All the other natural phenomena unfold as a consequence of the laws of Nature. There is a certain degree of inevitability about manifestation of the other natural phenomena. It is true that the world we live in (and to a large extent, the spatiotemporal universe itself) is governed by probabilistic edifice. There is no definitive way to predict the future. More importantly, the choice of what kind of future unfolds is not in the hands of the natural phenomena itself. The choices, if we can call them choices, are decided by the parameters of the initial conditions. We have learnt from chaos theory (Gribbin 2009) that there are infinitely large numbers of outcomes of any given natural phenomena and outcome is decided by the initial conditions. In contrast, when we think of intelligence in the biological context, we realize that intelligence essentially means that we make conscious choices and therefore influence the possible outcomes. Therefore, if we wish to prove that Life is like any other natural phenomena, we need to demonstrate that biological intelligence is a natural consequence of the structural template of living organisms. More importantly, we need to define the relationship between biological intelligence and the underlying structuralism of the living organisms. Before we make any such attempts, let us look at the remaining two features that separate Life from other natural phenomena, viz., sentience and self-reference.

Let us begin with the feature of sentience. Sentience can be best defined as an awareness of the surroundings and its relationship with the self. Thus, there are two prerequisites of sentience, ability to grasp the nature of reality and grasp its context with the self. The trouble with the feature of sentience is that it cannot be observed from outside. This is quite unlike the features of intelligence. It is possible to observe a living organism in its natural habitat and infer the motivation behind its behavior. Of course, this approach has its own shortcomings. The biggest problem with this strategy is that we, as a species, unknowingly impose our own definition of motivation and interpret the organism's behavior through our anthropocentric perspective. However, with a careful construction of experiments, one can evaluate a species agnostic measure of intelligence. However, such a luxury is not available in investigating sentience. Sentience, by definition, is an internal feature. Therefore, while discussing it, we will have to restrict ourselves to the anthropic interpretation of sentience. However, it is not unreasonable to assume that different species would have a comparable degree of sentience. As mentioned above, the capacity of sentience consists of two features, viz., an ability to grasp the nature of reality and an ability to grasp the nature of reality in the context of the self. Therefore, sentience requires two functionalities. Firstly, it requires an ability to create a representation of reality in the human mind and an ability to define self. Both these functionalities are essentially cognitive functionalities without any corresponding biological structural template. Upon a little reflection, it is intuitively clear that the functionality of the representation of the self necessarily gives rise to self-reference. Thus, sentience and self-reference are conjoined functionalities. We can't have one in the absence of another.

When one thinks of sentience from an evolutionary perspective, it is easy to understand why it should be favorably selected. However, the problem is the functionality of sentience (and therefore that of self-reference) doesn't have any underlying biological template. Therefore, it is not possible to apply Darwinian theory of natural selection to its evolution. This is because when we apply the principle of natural selection to any other biological functionalities, there exists a corresponding biological structuralism. This structuralism, in turn, depends on one or more genes. Therefore, in all such cases, we can clearly demarcate genotype and its corresponding phenotype. Once we do that, it is easy to construct its phylogeny. This phylogenetic relationship, spanning a diverse range of species, can be used to define the course of biological evolution and its attendant natural selection.

The classic example of this strategy is that of biological evolution and natural selection of eyes (Glaeser and Paulus 2015, see Chapter 1). There exists a definite structural template of vision. It has also been possible to identify the number of genes and even the products of these genes. Therefore, it is possible to trace the evolution and natural selection of functionality of vision. Thus, a well-defined relationship between a functionality and its underlying biological template is a prerequisite for defining the course of its evolution. However, such a possibility doesn't exist in the case of sentience and self-reference. Admittedly, this problem also manifests itself in the case of the functionality of intelligence, but to a lesser extent. This is primarily because even though there is no underlying biological template for intelligence, it is possible to observe its manifestation in almost all living organisms. Therefore, there exists an external behaviorist paradigm to compare and evaluate the degree of intelligence across the species.

It is possible to argue that this lack of knowledge about the biological underpinnings of these functionalities is temporary and maybe in the future, we will be able to deconstruct the biological context of these functionalities and then we will be able to outline a similar evolutionary perspective of these functionalities. However, such an optimistic perspective is misplaced. This is because we are complacent about the nature of biological evolution and its attendant natural selection. The Darwinian paradigm is ridden with fundamental semantic ambiguities. While these semantic ambiguities have been discussed in the preceding monograph (Chhaya 2020, see Chapter 8), it is necessary to summarize these semantic ambiguities here. In discussing the gap between a functionality and its corresponding structural template in the preceding paragraphs, it is possible to mistakenly assume that the gap between cognitive functionalities and the neurological templates is something specific to the cognitive functionalities. However, this gap between a functionality and structural template is manifest at all levels in the Darwinian paradigm.

For instance, if one were to ask a rhetorical question about the evolutionary relevance of having different units of inheritance and natural selection. If natural selection operates on phenotypic variation, why is it necessary to have genotype to begin with? Why can't natural selection operate on genotype directly? Apparently, it doesn't. The reason behind the need to have two distinct units of inheritance and natural selection eludes us. This duality hides behind itself some of the deeper semantic propositions which need to be articulated. It must be kept in mind that duality of genotype and phenotype is not the only instance. There exist several such dualities. DNA and RNA is another such duality. Brain and Mind are also dualities. In fact, upon a little reflection, it is intuitively clear that these dualities refer to the duality of structuralism and functionalities. Nature, in its wisdom, has kept these two features separate from one another. When viewed from this perspective, now we can understand the origins of the debate on the units of selection (Okasha 2010, see Chapter 2).

Contemporary debates on Darwinian theory are centered around the units of selection. It began with the postulate of group selection (Borrello 2010) and culminated in the postulate of Gaia (Lovelock 2000). The fact that we can perceive units of selection at every level of complexity suggests that the Darwinian paradigm of natural selection operates at every conceivable level, making it a universal mechanism. It is possible to argue that these debates are much ado about nothing. It is our innate compulsion to seek unitary framework that has forced us to stretch the analogy of Darwinian selection to unacceptable dilution. Maybe, such cynicism is justified. However, if we were able to decode the semantic necessity of this duality of structuralism and functionalities, we would know where to stop. Maybe the Darwinian paradigm is confined to biological evolution and nothing else. However, the onus is on us to articulate the semantics behind the duality of structuralism and functionalities in living organisms.

It is the contention of this monograph that this duality exists to avoid the problems of self-reference. This might appear to be self-contradictory as it was pointed out above that Life is characterized by its feature of sentience and its prerequisite of self-reference. Therefore, if Life is characterized by such a self-reference, then Nature couldn't have used these dualities as a mechanism for avoiding self-reference. This paradox of evolution of self-reference by the mechanisms that avoid self-reference, can be solved if we were to assume that Life manifests itself at more than one level. At every level of Life, there is a process of natural selection. However, whenever a functionality appears during natural selection which also manifests itself at more than one level, we observe the phenomenology of self-reference. Thus, sentience, which characterizes Human species, arises from natural causes, but arises inevitably whenever the functionalities of different levels overlap over one another.

We will try to demonstrate this rationale in greater detail and different evolutionary contexts in the following chapters. Presently, we will provisionally accept it as a scientific hypothesis and see how we can create a model which can help us to understand why Life is unlike any other natural phenomena. Therefore, in this section, we will outline a brief sketch of a topological model wherein the overlap of different levels can be represented. The details of this model are discussed in the preceding monographs (Chhaya 2020, see Chapter 3; Chhaya 2022a, see Chapter 3). Therefore, its justification and validity will be taken as a priori. What we will try to do is to see whether this model can be used to define natural phenomena, particularly Life. It must be kept in mind that this model has been applied to some of the fundamental natural domains of cosmology, particle physics and quantum field theory. Therefore, it is possible to argue that if this model can deconstruct the Darwinian paradigm, then it is possible that the Darwinian paradigm can be extended to all natural phenomena, thereby elevating Darwinian theory to the status of a law of Nature and not just a law of biology. However, we will take a skeptical view of such a possibility, at least until we redefine the Darwinian paradigm in the language of the proposed model. The proposed model is essentially a mathematical formalism. However, we shall avoid the mathematical nuances here and outline the proposed model in a domain agnostic perspective. For the sake of simplicity and linearity, the model is described in a point-wise manner.

1. It is possible to formalize any phenomena using this model provided it manifests itself in more than one dimensionality.
2. Any phenomena that contain more than one type of information content can be formalized using this model.
3. This model seeks to connect different dimensionalities with one another through a mathematical formalism of involuted manifolds.
4. The fundamental postulate of this model is that it is possible to reduce dimensionality within a system by applying the operator of involution. Similarly, it is possible to increase the dimensionality of a system by applying the inverse of the operator of involution.
5. Therefore, for any system operating in more than one dimensionality, it is possible to shift from one dimensionality to another by using the operator of involution or its inverse.
6. This formalism can also be applied to any system possessing different types of information content.
7. This is possible because according to this model, each dimensionality possesses its own metric and this metric is represented by its information content. Therefore, if a system possesses more than one type of information content, it can be represented by assigning different dimensionalities to different types of information content.
8. Therefore, it is possible to define structuralism and functionalities as different kinds of information content and therefore, they can be assigned different dimensionalities.

9. The proposed model postulates that every functionality is connected to its underlying structuralism by a single involutive algebra.

It must be admitted that this outline is simplistic and needs to be elaborated separately for each scientific theory. Moreover, the proposed model rests on a modified definition of involution. As discussed in the preceding monograph (Chhaya 2022a, see Chapters 1 and 2), there are enough semantic and epistemological reasons why the conventional definition of involution needs to be modified. However, it must be kept in mind that the proposed modification is not formally derived from any first principles. Rather, it has been designed to accommodate epistemological compulsions. Therefore, the validity of the proposed model rests only on its successful application to standard theories of science. In the preceding monographs (Chhaya 2022a, b, c), this model has been used to understand the nature of spacetime, quantum field and the origin of mathematical objects. Therefore, in this monograph, we will employ this model to deconstruct the genomic functionalities and their Darwinian perspective. Having looked at the brief description of the proposed model, in the next section, we will try to define Life using this model.

1.6 Involuted Manifold Model of Life

In the preceding sections, we looked at the reasons why Life is unlike any other natural phenomena. Our dilemma has been that though we are inclined to think that Life is a natural phenomenon, it cannot be defined using formalisms that we employ in formalizing the remaining natural phenomena. It was pointed out that Life is different from any other natural phenomena on account of three characteristics of intelligence, sentience, and self-reference. No other natural phenomenon behaves with a goal-directed behavior which characterizes intelligence in living organisms. It was suggested that in order to manifest intelligence, any system must possess sentience. The notion of sentience can be defined as an ability to grasp the nature of reality and an ability to comprehend it in the context of one's own self. Therefore, self-reference is a prerequisite for sentience.

It was pointed out that the reason why we find it difficult to define intelligence, sentience, and self-reference in the biological context, is that there is no specific biological template for these cognitive features. It was mentioned above that this disconnect between structuralism and functionalities is not confined to cognitive functionalities only. It manifests itself at every level of biology. Therefore, it was proposed that we need a model which connects structuralism and functionalities of every natural phenomenon, including Life. One such model was outlined in the previous section. Therefore, in this section, we will try to deconstruct the relationship between structuralism and functionalities of living organisms using the proposed model.

Before going further into the details of this approach, it is necessary to understand the role of information content in these unique functionalities of Life, viz., intelligence, sentience, and self-reference. This is necessary for two reasons. Firstly, the proposed model rests on the notion of metric as a type of information content. Therefore, according to this model, each of these three functionalities must employ the nature of information content to manifest themselves. Secondly, if we were to think that Life is unlike any other natural phenomena, then this distinction between Life and other natural phenomena must also reflect in their respective information content. Therefore, it is the contention of this model that origins of intelligence, sentience, and self-reference must lie in the way living organisms process information content. Moreover, it is this unique way to process information that separates Life from other natural phenomena. Therefore, let us try to understand the features of intelligence, sentience, and self-reference in the language of information content.

Intelligence, in the biological context, requires two features. Firstly, living organisms must have an ability to gather and store information. While in the context of the cognitive functions, this is self-evidently true, in living organisms, this ability manifests itself in various ways. More often than not, an organism responds to external stimuli without being cognizant of it. Therefore, intelligent behavior in the form of goal-directed behavior, doesn't necessarily involve cognitive faculty. It can happen at a cellular and even at subcellular levels. Chemotaxis and transport of molecules from one organelle to another are prime examples of this. To broaden the scope of this definition, we can cite the example of genomic and immunological retention of information. It is possible to argue that information content of any natural phenomenon contains the historical legacy of that phenomenon. Therefore, it is not a unique feature of Life. However, it must be kept in mind that no other natural phenomenon utilizes the information stored in itself for any particular objectives.

For instance, it is possible to argue that the spatiotemporal universe contains within itself information about its origin from the cosmic singularity. In fact, because of the presence of this legacy information, we have been able to understand the cosmological perspective of our universe. Therefore, information of the past is retained in every natural phenomenon. So why should Life be singled out? The explanation is simple. No other natural phenomenon employs its information content to decide its future course. Living organisms do. Our Milky Way galaxy evolves in accordance with the laws of Nature. It doesn't opt for any particular type of future for itself. Living organisms do precisely that. When an amoeba moves itself through pseudopods toward a source of food, it is deciding for itself a particular future.

This ability to utilize the information available toward a particular objective requires not just storage of information, but it also requires an ability to define the context. This is where the significance of sentience lies. More importantly, this ability to define context requires an ability to define the notion of self. Once an organism defines a notion of self, it can contextualize the information content available with reference to that notion of self. Admittedly, this distinction between Life from other natural phenomena is not that simple. This is because while it is possible to reject the concept that any nonliving natural phenomena can possibly have an ability to comprehend the notion of self, there is no clarity about the level at which living organisms manifest the notion of self. Admittedly, at a cognitive level, organisms do seem to possess the notion of self, but in the noncognitive processes of living organisms, there is no clarity whether such a notion of self and its subsequent goal-directed behavior is evident.

However, there is another yardstick by which one can locate the notion of self. Admittedly, conventionally, we associate the notion of self with cognitive faculty. However, if we replace this conventional perspective with the notion of systemic features, it is possible to expand the notion of self. For instance, if you take a genome of a typical organism as a system possessing a certain amount of information content, then it is possible to think of evolutionary changes as the changes in the information content of the genome. This has been a conventional perspective. However, now, we know that these evolutionary changes are not confined to DNA sequence alone. There are a host of structural changes in the genome which also bring about functional changes without necessarily bringing out any changes in its DNA sequence. Therefore, it is possible to conceive the genome as a unit, by itself and therefore as an independent entity which transcends its DNA sequence. Thus, we have a DNA sequence at one level and its higher dimensional organization as a genome at another level. When viewed from this perspective, it is intuitively clear that the genome operates as an independent entity. Therefore, we can think of the total information content of the genome as an entity with its own sense of self.

By expanding the notion of self from cognitive to systemic perspective, it is intuitively clear that all living organisms have a certain sense of self, albeit of varying degrees. This might appear to be a pedagogical fine hair splitting exercise. However, it is not so. We have to look at the debates on the units of selection in the Darwinian paradigm and its invocation in the discussion on the idea of group selection (Borrello 2010). Right from Dawkins's selfish genes (Dawkins 1999) to Wilson's social biology (Alcock 2001, see Chapter 3), all these debates center around the notion of units of selection. What is often overlooked in these debates is the fact that the real debate is about which of these entities are autonomous and therefore eligible to be units of selection. However, if we were to use the above mentioned scenario, it is clear that if we can define a sense of autonomy of a unit of selection, then it can be considered as a unit of selection in the Darwinian context. At a certain level of organization of the information content, a system can acquire a certain degree of autonomy. Therefore, it makes sense to assume that beyond that level of organizational complexity of information content of a system, it must be considered as an autonomous entity and therefore subject to Darwinian selection. Therefore, it is axiomatic that any system manifesting this degree of complexity of its information content will undergo competitive survival, provided it faces the Malthusian conditions of survival (Flew 2017).

One might object to such a generalization by pointing out any number of natural phenomena having comparable complexity and yet not subject to Darwinian selection. After all, as mentioned above, Life is unlike any other natural phenomena. Therefore, there must be something else, other than the informational complexity, that separates Life from other natural phenomena. It is the contention of this monograph that it is only when this informational complexity is accompanied by the feature of self-reference that one ends up with a definition of a living organism. Thus, there exists a certain structuralism which endows the informational complexity of the living organisms with the feature of self-reference, and it is only then that such a complex system with self-reference undergoes Darwinian selection. Therefore, let us see how the feature of self-reference arises, and more importantly, why it is limited to living organisms and not any other nonliving complex systems.

In the preceding monograph (Chhaya 2022a, see Chapter 2), a topological model of naturalistic epistemology has been described. The domain of epistemology presupposes the notion of self. Classical epistemology (BonJour 2010, see Part one) was concerned about the problem of epistemological access because it was based on the Cartesian paradigm (Cottingham 2008, see Part III) which assumed that knowledge, in its transcendental form, exists and our cognitive faculty must get access to that Platonic realm. Admittedly, Descartes's idea of splitting the universe into two separate realms, laid the foundation of modern science by segregating scientific inquiry from the prevailing theological beliefs. However, this also resulted in restricting the scope of epistemology. Since the transcendental knowledge also included mathematics, the origin of mathematical objects remained outside the scope of classical epistemology. This lacuna has plagued modern science till today. Given the naturalistic paradigm of modern science, it has chosen to remain agnostic about the origin of mathematical constructs. On the one hand, modern science cannot subscribe to platonic absoluteness because it implies some denomination of deism. On the other hand, modern science has no explanation of the origin of mathematical objects. Therefore, as a compromise, modern science has labeled mathematical constructs as a priori.

As a result, while formalizing our scientific theories, we invoke mathematical constructs to define a structural template for a given natural phenomenon under investigation. When we do so, we are literally importing the mathematical structural template and imposing it on the natural phenomenon under investigation. If we were to classify this method of formalizing a scientific theory as an information processing activity, it is clear that it is akin to the manner in which a computer operates. The only difference between these two instances is that the necessary structural template has to be fed to the computer before it can go ahead. On the other hand, when we try to formalize a scientific theory, our cognitive faculty already "knows" the details of the information content of the mathematical constructs employed in formalizing that theory. What is strange about this analogy is that in spite of the fact that both these instances operate by different mechanisms, we think that they are synonymous just because their outputs are similar. This is a category mistake. In the days when artificial intelligence and machine learning are hotly pursued topics of research, it is difficult to accept that our cognitive faculty operates using entirely different mechanisms. However, under the Cartesian influence and particularly after the articulation of the Church Turing hypothesis, we have led ourselves to complacency. Our cognitive computation (which is what our attempts to formalize a scientific theory are) is different from the template of the universal Turing machine. If at all, our cognitive computation bears any resemblance to any other computation, it is the quantum computation and not the Turing machine paradigm. However, this is a very important topic to be discussed parenthetically in a monograph dealing with genomics. It will be addressed in a separate monograph (Chhaya 2022e).

Presently, let us leave aside the cognitive perspective of information processing and let us look at the genome and how it processes information. This is important because it would demonstrate that unlike computers, the genome isn't required to be provided with a program. Genome carries its program in its own architecture. Therefore, the genome must be viewed as a computing device that uses its own data as its own program. Rather, genome's data is a program! Genome can use its own data as a program because its architecture enables it to possess what we have called self-reference earlier. A genome can operate its own self and give rise to the structuralism of the organism it resides in. This ability of self-governance is characteristic of not only all the living organisms, but also of their organs and organelles. Incidentally, this also includes our cognitive faculty. It is just that since we can't perceive how our cognitive faculty operates, that leads us to believe that mathematics comes from outside and we use this imported knowledge in formalizing scientific theories. Our cognitive faculty possesses all the knowledge it needs to survive and thrive. Our acquisition of knowledge, including the mathematical constructs, is triggered by the books that we read or the lectures that we attend. However, understanding of these mathematical constructs is inbuilt. This is something very similar to us looking at a photograph and grasping the details of that photograph. The visual sensory perceptions merely trigger our cognitive faculty. The understanding of the content of the photograph is built into our cognitive faculty. It is tempting to think that the structural template of our cognitive faculty is a program by itself.

Of course, we will discuss in detail how our genome employs its own architecture as a program in the following chapters. Presently, let us see how this distinguishes Life from other natural phenomena. Therefore, let us see how this functionality of self-reference manifests itself in the living organisms and more importantly, why no other natural phenomenon manifests it. It is important to understand that this functionality of self-reference is not something mystical or an occult feature. It arises from a particular structural template of living organisms, particularly that of the genome. This structural template has two prerequisites. Firstly, a natural phenomenon must possess multiple dimensionalities within itself. Secondly, it must possess a feature wherein different dimensionalities of a given natural phenomenon must be able to exchange information among themselves. These two prerequisites appear to be simple and perhaps inadequate to explain the complex functionalities of living organisms. However, as discussed in the following sections, these apparently simple conditions are powerful enough to give rise to the complex features of life. More importantly, no other natural phenomena, except perhaps quantum phenomena, possess these two features. Therefore, its combination of these two features is adequate to define life. Incidentally, it is this very combination of these two features that makes it difficult to formalize the definition of Life. More importantly, it also gives rise to separation of structuralism and functionalities in living organisms, thereby making them unique among other natural phenomena.

To illustrate this model, we will try to deconstruct two characteristic features of Life. These features are the relationship between neurological templates and cognitive functionalities of the human mind and the relationship between genotype and phenotype. Admittedly, both these features are extensively covered in the literature and even in the present monograph. Therefore, we will merely outline a brief sketch of these two dualities. Incidentally, both these dualities also represent the duality of structuralism and functionalities. Therefore, they are a natural choice. Of course, prima facie, it is possible to argue that the relationship between neurological templates and cognitive functionalities is not analogous to the relationship between genotype and phenotype. Upon a little reflection, it is clear that they are in fact analogous. Just as a genotype gives expression to various phenotypes, a single neural network can give rise to different types of concepts. Moreover, just as different phenotypes compete with one another to occupy a given environmental niche, several concepts struggle to occupy a given semantic content. Most importantly, just as the environment acts as a final arbiter, semantics acts as a final arbiter too. The analogy is remarkable because in a given environmental niche several phenotypes could coexist, a given semantic content can accommodate more than one type of concept. Thus, both these dualities not only represent the duality of structuralism and functionalities, but they also exemplify the operative natural selection.

Let us begin with the relationship between a neurological template and its corresponding functionalities. Since a more detailed description of this duality will appear in the following monograph (Chhaya 2022d, see Chapter 3), presently, we will look at a general schema. For that purpose, we will designate the underlying neural networks as the brain and the corresponding cognitive functionalities as mind. Therefore, in the next few paragraphs, we will use the terms brain and mind in a generic sense without any reference to any particular neurological templates or any particular cognitive functionalities. The arguments presented may also sound like generalizations, but are adequate for this introductory essay. More detailed descriptions would be made available later.

Let us see how the classical brain mind duality gets formalized using this model. It is apparent to anyone studying cognitive science that there exists a definite semantic and structural gap between neurology and cognitive theory (DeVos and Pluth 2015, see Part I). In spite of tremendous advances in cognitive science (Minski 1988, see Chapters 2 and 3), including the computational paradigm of cognition, it is not possible to translate a cognitive feature into its neurological counterpart. Admittedly, this lacuna is simply a carry forward from the earlier paradigm of psychology and its relationship with neurology. When it was realized that the subjective experience of qualia cannot be translated into its synaptic counterpart, a paradigm of network was conceptualized. The logic being that the individual synapses may not offer any explanation about the different types of quale, a neural network will have sufficient structural depth to offer an explanation of how different quale could arise. Admittedly, our experience in analyzing and parsing sensory perceptions (Barth et al. 2012, see Parts II and IV), particularly the visual signals has justified our choice of employing neural networks in understanding complex task vision.

However, while it is possible to map different regions of the brain involved in different tasks of visual processing, the corresponding cognitive processes of mind remain beyond formalization. This is primarily due to the fact that we do not know how the brain /mind stores and processes information content. Of course, we have been given a convenient label to hide our ignorance about how information content is stored in the brain. We call it a virtual memory (Parker et al. 2002). Admittedly, this notion of virtual memory is a placeholder for our ignorance. There are several reasons why we should not take this notion of virtual memory seriously. For instance, our experience with quantum computation (Nielsen and Chuang 2010) has demonstrated that every type of information is a material entity. There is nothing abstract about any information. Therefore, there is no reason why our cognitive memory should be virtual. It has to be like those intangible properties of the fundamental particles, and must be a physical entity.

Second reason why we should reject the notion of virtual memory is that if it is really virtual how the human brain can get access to it. Take the analogy of our present day usage of the Internet. We normally refer to keeping our data on the cloud. However, in reality our data is physically stored in some server operating some remote corner of the Earth. The usage of cloud computing is simply a metaphor and shouldn't be taken literally. Similarly, the notion of virtual memory should not be taken at face value. At best, we can think that the usage of the term virtual memory is a placeholder for our ignorance. The reason why we have used this notion of virtual memory lies in the philosophical foundation of our modern science. It can be traced back to the Cartesian paradigm (Cottingham 2008). By placing knowledge, or rather, the knowledge in Platonic sense, in the transcendental realm of *Res cogitans*, Descartes could separate divine from mundane and thereby providing an impetus to the advent of a secular version of knowledge in the form of modern science.

Therefore, when we accept the notion of virtual memory, we are subconsciously regressing to the Cartesian legacy. Therefore, it is time to accept that our present understanding of our cognitive faculty and particularly that of its structuralism and functionalities needs to be revised. As we will discuss later in this chapter, this duality needs to be reconfigured and more importantly, the notion of virtual memory needs to be redefined. However, before doing so, let us look at the second duality of genotype and phenotype. It is intuitively clear that the duality of genotype and phenotype is a typical example of the duality of structuralism and functionalities. However, it is not clear, at least at first sight, where to draw the boundary between the two. Conventionally, we have believed that DNA sequence is the structural template, and therefore, it represents genotype. However, as we discover more and more features of long range influences of other genes present in the genome (Shmulevich and Dougherty 2014), we are forced to concede that structuralism of the genome is not confined to its DNA sequence. It actually resides in genomic architecture.

Secondly, we need to ponder over one of the most overlooked features of genomics. It refers to the gene expressions. When a transcriptor initiates a translation of DNA sequence, is this activity of transcription a genotype or phenotype? Apparently, for the transcriptor, its own sequence is its genotype. However, its ability to bring about transcription is apparently a phenotypic expression. Similarly, if the genome has some higher level architecture, should it be taken as a genotype or phenotype? These rhetorical questions may appear superfluous and pedantic. However, they point toward the semantic ambiguities of the Darwinian paradigm. Why does natural selection operate on phenotype when the information details are preserved in genotype? Apparently, there is some reason behind the need for duality of genotype and phenotype in natural selection. However, we have yet to articulate it. It must be admitted that the reason why this semantic ambiguity about the evolutionary relevance of this duality arises is because we don't know how information is stored and utilized.

This is not to suggest that the information content present in DNA sequence is not important. Of course, it is important. However, we also need to accept that there exists a different kind of information coded in the form of genomic architecture. There are two important semantic implications of this possibility of the genome being a repository of information content which is other than that preserved in its DNA sequence. Firstly, it too must be taken as virtual memory. Secondly, we must accept that there exists a possibility that Life is the interactions between chemical memory present in its structuralism and the virtual memory present in its functionality. Once we accept this duality of information content, it is easy to see why Life cannot be defined as a natural phenomenon. It can't be defined as a natural phenomenon because our definition of a natural phenomenon is based on the material status of the information content. Our definition of a natural phenomenon doesn't accommodate any such information content which is virtual. It is the contention of this monograph that what is referred to as virtual information above is in fact, real information at higher dimensionalities. It appears to be virtual because we are looking at it from the four dimensional spacetime.

Of course, it is possible to argue that this reasoning implies that no other natural phenomena possess any virtual information, and therefore, Life must be treated not as a natural phenomenon. However, this is not true. Most natural phenomena possess explicit and implicit information content. It is just that all the other natural phenomena have different types of interactions between its material and virtual information content. Life, on the other hand, has a different kind of interaction between its material and virtual information content. It is this difference between Life and the remaining natural phenomena that makes it difficult to define Life using the formal description of natural phenomena. Therefore, in the next section, we will discuss the reason why Life is difficult to be formalized.

1.7 The Reason Why Life Is Difficult to Be Formalized

In the previous section, it was suggested that there exists a marked separation between structuralism and functionalities of living organisms. Moreover, this separation manifests in all the characteristic features of Life. It was pointed out that this separation arises because living organisms have two sets of information content. Firstly, there is explicit information content in the form of the structuralism of life forms. Secondly, there is an implicit or virtual information content in the form of the functionalities of living organisms. More importantly, it was pointed out that while every natural phenomenon possesses both these types of information content, it is the type of interaction between this implicit and explicit information content in the living organisms that separates it from the other natural phenomena. Therefore, in this section, we will try to deconstruct the nature of interactions between the material and virtual information content of the living organisms. Having done that, we will compare it with the interactions between material and virtual information content with other natural phenomena. An attempt would be made to demonstrate that both these types of information, material and virtual, are essentially the same. What separates them is our ability to access them. Possibly, this should provide us with a new approach to define Life.

Let us begin to understand what exactly is the distinction between these two types of information content of living organisms. More importantly, why this aspect has been overlooked in our scientific theories? In order to understand the distinction between these two types of information, we will restrict ourselves to the two above mentioned examples of cognitive processing and genomic functionalities. These two domains offer themselves as obvious choices because both these functionalities are definitive characteristics of Life. Moreover, from the historical perspective, they together represent the fundamental shortcomings of the Darwinian paradigm. The evolution of cognitive faculty, particularly the human mind, has always defied any explanation from the conventional Darwinian theory. Therefore, if we can demonstrate certain structural congruence between these two functionalities, then it will pave the way for reinterpreting the Darwinian paradigm in the context of our cognitive faculty. Presently, let us begin with the cognitive processing and how material and virtual information appear to be different and to what extent our conventional perspective of science has contributed to this distinction. Having done that, we will look at the genomic functionalities.

When we think of material information in the context of cognitive processing, we end up with the scenario of signal processing of our neural networks (It must be kept in mind that we will employ the term neural networks to denote neuronal networks. This clarification is necessary because conventionally one uses this term to indicate an architecture of multiple processing units linked to one another. Moreover, these processing units need not be neurons, but they could be any other computing units including computer switches, quantum spin states or even biological switches like RNA proteins. Conventionally, the term neural network has been employed to denote any ensemble of multiple processors linked to one another (Donahue and Dorsel 1997, see Part I). However, in this chapter we will use this term to refer to various arrangements of neurons necessary for cognition.). Therefore, the material information being processed consists of electric potentials that pass through synaptic junctions and the chemical messengers that are exchanged at the synaptic junctions. We have a very detailed description of the types of electric potentials and the chemical messengers that are exchanged at the level of an individual synaptic junction (Albers and Price 2012). This detailed information has laid the foundations of several neurological templates and some of them have found therapeutic use.

However, there is no theoretical framework which could integrate these details into an architectural model. In fact, one can divide the research reported in the neurological domain into two broad categories. Firstly, we have research focused at the level of an individual synaptic junction that has produced an awesome picture of how a synaptic junction operates. Secondly, we have a humongous amount of research focused on the systemic behavior of the human brain. However, the second approach rests on our perception of cognition derived from psychology and our conception of computation theory. We have, rightly or wrongly, assumed that the human brain is essentially a computing device. This is a valid assumption because it is possible to think of any natural phenomenon as a computation. The computation paradigm of the universe is a very alluring notion and it is legitimate to extend it to cognition. However, while thinking of the human brain as a computing device, we have made a category mistake. We have, under the influence of our successful paradigm of the universal Turing machine, assumed that the human brain is also a Turing machine (Chhaya 2022e, see Chapter 1).

The key mistake is not that we have assumed that the human brain is a Turing machine. Rather, it is that we have assumed that it is nothing but the Turing machine. In the following monograph (Chhaya 2022e), we will deconstruct the significance of the universal Turing machine paradigm. Presently, it will suffice to add that a typical Turing machine merely produces the identical outputs that one obtains from cognitive faculty. However, the mechanism by which the Turing machine computes is different from the mechanisms by which our cognitive faculty computes. In fact, it is possible to define a universal computation which subsumes both, the Turing machine and the cognitive computation protocols.

Returning to the present discussion, this dichotomy between the research at the level of individual synaptic junctions and the systemic level based on computational paradigm, has resulted in creating a semantic gap. More importantly, it is this semantic gap that has given rise to the distinction between the material and virtual information content. Let us see how. Before we begin to deconstruct this distinction between the material and virtual information, it must be kept in mind that these terms have a legacy of conventional perspective, and therefore, they need to be taken as mere tokens without any semantic nuances attached to them. In other words, when we refer to virtual information, we are still referring to material information. It is just that since we don't know its location, we call it virtual information. It appears to be a self-contradictory nomenclature. However, it finds its justification from our implicit belief in the Cartesian paradigm (Cottingham 2008). According to this paradigm, knowledge or information resides in a transcendental realm of *Res cogitans*. Moreover, since we live in a material realm called *Res extensa*, we can't locate any entity that is present in the transcendental realm of *Res cogitans*. Though, how we, the residents of the material realm, can get access to the things present in the transcendental realm has been a topic of an unending debate, we will overlook this aspect. What is relevant to the present discussion is that because any form of knowledge belongs to the transcendental realm, we can't naturally locate it. Therefore, to us, this information appears to be virtual because it can't be located in any specific location.

While this historical perspective explains why we have named this type of information as virtual information, it also finds its echo in our cognitive science. On the one hand, we can pinpoint, with a great deal of accuracy, the type and magnitude of material information, in the form of electrical signals and chemical messengers. On the other hand, we have not been able to define the exact nature and location of the outputs of our cognitive processes. Therefore, in continuation with the Cartesian legacy, we have continued to label the outputs of our cognitive processes as virtual information. It is a matter of debate whether we have carried forward the Cartesian legacy or we have come around a full circle to the same conclusion as that of the Cartesian paradigm. The key point is that in spite of its nomenclature, virtual information present in our cognitive faculty must also be material information. It is just that we don't know where it is located. This insistence on material status of the so-called virtual information rests on the increasing evidence that information is indestructible. Even in the extreme scenario like the one present in the Black hole, the information survives (Susskind and Lindesay 2005, see Chapter 9). Therefore, there is no reason why this so-called virtual information vanishes once our cognitive faculty is finished dealing with it.

In the context of the present discussion, it is intuitively clear that we can't formalize Life because we can't assign any specific location to this virtual information that our cognitive faculty employs during cognition. If we can't locate such information or find out how our cognitive faculty gets access to it, there is no way we can formalize our cognitive faculty. Therefore, as a logical extension, we can't formalize Life itself. Now, let us look at the second characteristic feature of Life. This refers to the information that is passed from one generation to the next. With the rapid advances in the field of genomics, we have been overwhelmed by its dramatic success. It is indeed awesome to realize that we have the necessary expertise in mapping the entire genomes of all the living organisms, including ourselves. Its importance to the future of the human race can hardly be overemphasized. Genomics has a potential to solve the problems of mortality, incurable diseases, our food resources and even the longevity of human life. Therefore, it is inevitable that genomics will remain our primary focus in research.

However, the domain of genomics is plagued by a fundamental semantic shortcoming. Genomics too, has chosen to neglect the study of the nature and mechanisms of information transfer. This statement might appear to be scandalous, if not slanderous. However, the fact remains that our current paradigm of genomics is focused on the material information content that is available in the form of the DNA sequence. Admittedly, DNA sequence does and must enjoy primacy in genomic research. However, it is a category mistake to think that genomics is solely concerned about DNA sequence. Genomics is concerned about how the information is stored, transmitted and expressed during the life cycle of a living organism. Obviously, DNA sequence is the primary source of this information. However, it is not the only source of information. To extend the analogy, one can say that genomics also deals with two types of information, material and virtual. The material information consists of DNA sequence and the virtual information consists of the sequence of gene expressions. This virtual information is stored somewhere in the genomic architecture, but we can't locate it, at least not completely.

Let us see why we have not been able to locate the virtual information present in the genome. As mentioned above, this virtual information consists of, among other things, the sequence of gene expressions. Our current understanding of the factors controlling the sequence of gene expressions is restricted to long range influences within and across the different chromosomes (which are named as cis and trans effects) and the presence and absence of gene expression initiators (Shmulevich and Dougherty 2014). However, since there are thousands of different genes present in any given genome, we find it difficult to isolate these factors in each of these genes and their expressions. In addition, in the case of Eukaryotes, the things are further complicated by the fact that even a single gene can give rise to several different proteins depending on the intracellular signals present (Dragini 1998). Therefore, our conventional perspective of the virtual information present in the genome is partial. Of course, the details of the large number of genes and their multiple products are being mapped out by a large number of research groups. However, our insights into the nature of the virtual information present in the form of genomic architecture are based on this incremental increase in the data available. Therefore, our current understanding of the exact nature of this virtual information present in the genome remains partial.

One of the reasons for this situation is that we don't have any theoretical model of genomic architecture. What we are doing at present, is to find out this genomic architecture by removing bricks after bricks from the edifice of the genome and to find out the consequences. However, if we had a theoretical model of genomic architecture, we would find it easier to understand the nature of the virtual information present in the genome. Surprisingly, the reason why we don't have any theoretical model of genomic architecture lies in the very conception of the Darwinian paradigm (Grene 1986). The Darwinian paradigm, in spite of its ability to reinvent itself with every paradigm shift in biology and in spite of its semantic primacy among the scientific theories, is not a proper scientific theory. Prima facie, we can define a scientific theory as a proposition that serves two important purposes. Firstly, it explains the nature of reality in a logically consistent manner. In other words, it has a certain explanatory power. Secondly, a scientific theory must have certain predictive power. A scientific theory must be able to predict the possible outcomes in a given starting conditions. Thus, a scientific theory must possess certain semantic propositions capable of explaining the reality and at the same time, it must possess certain structuralism capable of predicting the future of the system under investigation.

The Darwinian paradigm fails on the second feature of providing predictions. Admittedly, its ability to explain the process of natural selection is so exceptional that we have overlooked its lack of predictive power. Conventionally, we have justified this lack of predictive power of the Darwinian paradigm by pointing out the random nature of the underlying process of mutation and the resulting changes in genotypes (Bonner 2013). However, to the extent a process is causal, it must be predictable. However, in this case, it is possible to argue and perhaps justifiably, that the reason why the process of natural selection operates not on the genotype which is subject to causal processes of mutations, but it operates on phenotype which are not governed by mutational causality. This presence of duality of genotype and phenotype is a semantically loaded proposition, and we have yet to unravel its significance. More importantly, it is not confined to the duality of genotype and phenotype. In biological evolution and therefore in the process of natural selection, there are two more instances of duality. They are the duality of DNA and RNA and the duality of structuralism and functionalities.

Upon a little reflection, it is intuitively clear that these dualities, in fact, roughly correspond to the distinction between the material information and virtual information. Interestingly, this distinction between these two types of information is an outcome of the process of natural selection. In the most primitive forms of Life, say archaebacteria, RNA served the dual function of information transfer and information translation (Garett and Klenk 2007, see Chapter 3). Therefore, it is tempting to think that perhaps biological evolution was like any other natural phenomena. It is only when the process of natural selection began operating that the separation of functions began. In other words, the origin of the above-mentioned dualities lies not in biological evolution per se, but in the subsequent natural selection. Therefore, it is possible to view this separation of material information from the virtual information, in the form of these three dualities, as definitive features of Life. Perhaps this separation of the information is what makes it difficult to formalize Life as a natural phenomenon.

It is not a coincidence that there exists another natural phenomenon which defies our attempts to formalize it. More importantly, this phenomenon is also characterized by two types of information, material and virtual. This phenomenon is quantum phenomena. When we think of quantum superposition states (Wieder 2012), we are faced with a similar situation. Prima facie, a quantum superposition state appears to be nonstructural, and yet, it embodies within itself a whole lot of information in the form of its constituents, viz., spin states of quantum field and spacetime. Yet, a quantum superposition state appears to be nonstructural, a state having virtual information content. It is tempting to think that perhaps both these phenomena share a common structural template capable of holding virtual information content, and therefore, they are capable of entangling with one another. Leaving aside this parallelism, let's focus on the nature of separation between the virtual and material information present in the living organisms. In this section, we talked about two instances of separation of material and virtual information, viz., cognitive information and genomic information. However attractive this postulate may appear to be, it still needs to be fleshed out with details. Therefore, in the next section, we will discuss whether these dualities have any common framework or not. Admittedly, we will include the cognitive perspective within the broader perspective of the separation of structuralism and functionalities. It is true that the cognitive duality of structuralism and functionalities is too important a topic to be treated marginally. Therefore, a separate account of cognitive architecture and its functionalities will be presented in the following monograph (Chhaya 2022d, see Chapter 1). Presently, we combine the distinction between cognitive structuralism and cognitive functionalities with the rest of biological structuralism and functionalities. Of course, wherever possible, an attempt will be made to exemplify the arguments with the examples from the cognitive domain.

1.8 Is There a Common Structure of Life?

If at all, the postulate of the separation of material and virtual information being a characteristic of Life is true, then it is axiomatic that all the structural templates of the biological processes must have a common framework which results in the separation of these two types of information. Therefore, in this section, we will look at the conventional perspective of biological structuralism and try to find out any commonality among different biological processes. For this purpose, we will once again focus on our cognitive faculty and genomics. Using the conventional perspective, we will try to find out how the structural templates of both these domains are connected to their respective functionalities. Having done that, we will try to find any common framework that connects the structural templates of both these domains with their respective functionalities. It is possible that while the structural templates of both these domains are different, the relationship between structuralism and functionalities, in each of these cases, could still be common. Therefore, we will focus on both these aspects, viz., the structuralism and functionalities of both these domains and the relationship between structuralism and functionalities in each of these domains.

In case we succeed in establishing a common structuralism of both these domains, it seems reasonable to think that this parallelism extends to all the biological processes as well as to all the living organisms. This extrapolation is justified on the grounds that these two domains are so different from one another that if they happen to share structural commonality, it must be taken as a characteristic feature of all the living organisms. Therefore, let us begin with our cognitive faculty. Before we begin with the details, it must be mentioned that it is possible to argue that the structural template of our cognitive faculty could be seen to be determined by the human genome. Admittedly, we don't know much about how the human genome gives rise to the complexity of neuronal architecture. However, it can still be argued that the human genome by hitherto undefined mechanisms sculpts our complex organs like the brain. Therefore, in principle, even the possibility of genome and cognitive faculty sharing a common framework cannot be taken as a definitive proof of the suggestion that Life has certain unique structural templates.

However, there is one observed fact about human developmental biology that denies any such ontological perspective. This refers to the phenomenon of neuronal migration (Rubinstein and Rakic 2013, see Chapter 13). During embryogenesis, the human brain is not manufactured like a machine. There are no specific numbers of genes giving rise to certain fixed types of neurons which are then assembled into a complete brain. Even as more and more neurons are produced at a frenetic pace, they tend to migrate. More importantly, their migration is not defined beforehand. During this migration, the Darwinian model of the survival of the fittest ensures that only the selected neurons give rise to structural complexity of the brain. Moreover, this migration is brought about by the process of chemotaxis. Each neuron follows a chemical trail by " sniffing" out the gradient of the chemical it is programmed to follow. Therefore, these two factors, viz., the random natural selection among the neurons and the mass migration of neurons, ensure that the genome cannot directly influence the cognitive architecture.

It is possible to argue that the signaling molecules which map out the migration path of these neurons are programmed to be expressed at a particular location in the embryo. Therefore, the human genome could still plan the neuronal architecture. However, there are two arguments which refute this argument. Firstly, the scenario of natural selection deciding which neuron should succeed in reaching its target area eliminates any preplanned or teleological arrangement. Since the genome has no control over which neuron would eventually survive and where it would enter the neuronal architecture, the structuralism of our cognitive faculty cannot be dictated by our genome, at least not completely. Secondly, it is not the genome as a whole or even a small group of genes that decides the map of the resulting architecture of the brain. It is a combination of genomic architecture and the random nature of the Darwinian natural selection among the competing neurons together that lay the foundation of neuronal architecture. Therefore, if our cognitive faculty and our genome were shown to possess a common structuralism, it must be taken as a characteristic feature of Life.

Finally, this possibility that the genome partially contributes to the structural template of our cognitive faculty is precisely what we are interested in. This is because this partial contribution arises from what was earlier described as a virtual information content of the genome. The location in the embryo where the genes responsible for synthesizing the signaling molecules are expressed is not decided by the material information in the form of DNA sequence. It is decided by the virtual information present in the genome in the form of long range influences within the genome. Therefore, prima facie, we will assume that if our cognitive faculty and genome possess a common framework of the relationship between structuralism and functionalities, then it is justified to think that this relationship between structuralism and functionalities is a characteristic of Life.

Therefore, let us look at the structural template of our cognitive faculty. The conventional perspective (Minski 1988) on this topic can be summarized as follows. Our cognitive faculty possesses different heterogeneous cognitive processes. Some of these cognitive processes get integrated into several modules. In other words, our cognitive faculty can be best described as a modular faculty rather than as a simple collection of different cognitive processes (Carruthers 2006, see Chapter 3). Admittedly, the nature and the origin of modularity are not well defined. However, the existence of modularity as such, has been generally conceded. It is possible to justify the notion of cognitive modularity from the Darwinian perspective because such a scenario offers better survival chances. At the same time, it must be admitted that barring this semantic congruence between the modular design of our cognitive faculty and the Darwinian semantics, nothing much is known about the mechanisms by which modularization occurs or how natural selection selects any such modular design.

This structural ambiguity can be traced back to the structural agnosticism about the nature of our cognitive functionalities. For instance, it is possible to conceptualize distinct cognitive functionalities and their formalization as separate cognitive processes. However, it is not possible to define their neurological templates. This lacuna can be ascribed to the fact that our present understanding of cognitive science rests on twin foundations of psychology and neurology. Our definition of different cognitive functionalities is based on psychology, whereas our definition of structuralism of our cognitive faculty is based on neurology. Therefore, to the extent the domains of psychology and neurology are disjointed, the structural agnosticism about our cognitive functionalities would remain.

This dichotomy between cognitive functionalities and their neurological templates is best exemplified by the fact that our neurological templates are configured on the premise that each synaptic junction is identical to another (Albers and Price 2012). True, there are different types of chemical messenger molecules that are exchanged at these synaptic junctions. Similarly, it is also true that there exists a limited variation in the magnitude of signal transduction across the synaptic junctions. However, none of these variations can be used to explain different cognitive functionalities. Therefore, there exists a definitive semantic and structural gap between the cognitive functionalities and their underlying neurology. This lacuna has resulted in two problems. Firstly, we have not been able to establish a structural template for formalizing our cognitive functionalities. Secondly, it frustrates any attempt to accommodate cognitive functionalities into the Darwinian model of natural selection. In fact, this dichotomy points toward a larger class dualities of the Darwinian paradigm. If our neurological templates are subject to Darwinian selection, then there is no way to claim that our cognitive functionalities are products of natural selection. To make such an assertion it is imperative that we must redefine the relationship between cognitive functionalities and their neurological templates. This problem is similar to a much broader problem of why natural selection operates on phenotype when it is the genotype which is capable of undergoing structural changes via mutations. In other words, the duality of cognitive functionalities with the corresponding neurological templates is a member of the larger class of dualities consisting of genotype and phenotype; DNA and RNA, and the units of selection and the units of inheritance.

The key point is that we must accept that the reason why we can't formalize Life lies in our inability to formalize the relationship between these above-mentioned dualities. Now, let us turn to the genome and its separation of functionality and structuralism. As modern science moved from genetics to genomics, it was apparent that the genome cannot be viewed as a compilation of genes. Genomes have their own separate identity and functionalities. The functionalities of a genome are not an arithmetic sum of the functionalities of its constituent genes. While dealing with individual genes, it is self-evidently true that the functionality of a gene arises from the structuralism of genes in the form of its DNA sequence. However, this is not true for Genome. The genomic functionalities are not coded in its DNA sequence, at least not completely. Most of the genomic functionalities arise not from its underlying DNA sequence, but from the order of expressions of the DNA sequences of different genes. If we accept the proposition that the genome is a repository of all the instructions needed to develop an organism, then it is axiomatic that this sequence of expressions of different genes must also be encoded in the genome. However, we don't know much about it, except that there exists some long range, cis and trans, effects (Shmulevich and Dougherty 2014). Therefore, by implication, this information must be stored in a higher level of genomic architecture. Thus, even in the case of the genome, we find a similar paradoxical situation. There exists a set of functionalities and there exists a definitive structural template. However, we do not know how they are connected to one another.

It is possible to argue that this semantic gap is transient and will be eliminated as more and more information about the structural details of both these domains becomes available. Admittedly, such a possibility exists. However, till such time, it is sensible to think about alternative strategies to solve this dichotomy between functionalities and structuralism. Therefore, in this monograph, we will make one such attempt to define a model for the relationship between functionalities and structuralism which can be applied to any biological processes and their outputs. Admittedly, one such model has been outlined in the preceding monograph (Chhaya 2022a, see Chapter 3). Therefore, we will try to extend the same approach here. As a first step, in this section, we will take a minimalist approach and define what we expect from such a relationship between functionalities and structuralism. Having done that, we will demonstrate that the proposed model fulfills just such expectations.

Prima facie, we will concede that there exists a definitive relationship between biological functionalities and their underlying structuralism. Secondly, it is intuitively clear that in spite of this belief, such a relationship is not discernible. Had it been otherwise, we would have formalized this relationship and reinterpreted the Darwinian paradigm. Therefore, if any such relationship between functionalities and structuralism of the biological systems were to exist, it would have to be defined by some indirect means. It is intuitively clear that whatever the formal description of this relationship may turn out to be, it must be amenable to formalization using the information theoretical perspective. This is because if there is anything common between the functionalities and their structural details, it is their information content. When one thinks of structuralism of any system (and particularly a biological system), it can only be perceived as spatiotemporal details of its constituents. Therefore, these spatiotemporal details are nothing but the information content of each of these constituents and their relationship with one another. In the present context, DNA sequence exemplifies this assertion. We can think of the genome as a collection of chromosomes, chromosomes as a linear assembly of genes, genes as a linear assembly of nucleotides and nucleotides as a stereochemical assembly of individual atoms. Thus, the structuralism of the genome can be formalized using information theoretic perspective, at least in principle.

Similarly, it is possible to argue that cognitive functionalities can also be configured on the basis of the information content present in the outputs of these processes. Let us think of a simple functionality of vision (Wechsler 1992). On the one hand, we have sensory signals being sent out by the retina to different parts of the brain. On the other hand, we have a neuronal response to these signals sent by the retina. Prima facie, both these features are constituents of structuralism of the cognitive functionality of vision. On the other side, we have perception of objects that comes about when the cognitive functionality of vision operates on the sensory signals present in the neural networks responsible for vision. This perception, which is an output of our cognitive functionality of vision, also consists of information content. Thus, structural template and output of the functionality of vision are both amenable to formalization using the information theoretical perspective. Therefore, it is axiomatic that cognitive functionality of vision too must have an information theoretical perspective. Incidentally, this scenario is congruent with the two vision hypothesis which distinguishes between cognitive and sensory processing. Historically, we have confused the details of biological functionalities with the details of the underlying structuralism. Therefore, we have conflated the semantics of both these otherwise incompatible information content.

In a more abstract language, we can conceive that there exists a set of information elements (say the signals sent by retina and processed by neural networks) present in structural templates. Similarly, one can conceive that there exists another set of information elements present in the outputs of the functionality of vision, say in the form of the details of our perception of the object being observed. Therefore, it seems logical if at all we wish to formalize the functionality of vision, it must be formalized as a mathematical construct which transforms one set of information elements (of neural signals) into another set of information elements (of perceiving the image). Once we accept that functionalities, biological or otherwise, are in essence, like mathematical operators, it is intuitively clear that we can assign certain structural details to these mathematical operators.

The reason why we have not been able to formalize functionalities in this manner is that under the Cartesian influence, we believe that mathematical operators are abstract, if not transcendental entities. Therefore, these functionalities, like any other mathematical operators, cannot be invested with any information content or semantic propositions. As a result of this Cartesian legacy, we have committed a category mistake of conflating the structural template with the functional template. Therefore, once we accept that functionalities can have their own templates, it is intuitively clear that all we need to do is to define the rules of composition between the templates of structuralism and functionalities. This is precisely what we will do in this monograph. Before going further into this approach, it is legitimate to ask if the structuralism and functionalities have their own templates, then it must be manifest in every natural phenomena. However, as we have been arguing so far, Life is unlike any other natural phenomena. Therefore, we need to understand how the relationship between structuralism and functionalities of Life is different from that of the remaining natural phenomena. Therefore, using the proposed arguments, we will try to define characteristic functionalities of Life using the arguments presented above, in the next section.

1.9 Are There Any Definitive Functionalities of Life?

In the previous section, it was suggested that functionalities can have their own templates which can be different from those of structural templates. Even if we accept this scenario provisionally, the key point that is germane to the present discussion is that this scenario must be valid for all natural phenomena and not just to Life. Therefore, if it can be demonstrated that the separation between functionalities and structuralism exists only in the living organisms and not anywhere else, then it provides us with a new paradigm for formalizing Life. Therefore, in this section, we will be concerned with two questions, viz., whether the separation between functionalities and structuralism is restricted to Life? Secondly, if it is so, whether it is possible to define any definitive functionalities of Life in the language of this separation? For this purpose, we will select three functionalities that we discussed earlier, viz., intelligence, sentience, and self-reference.

Let us try to find an answer to the first question about whether the separation between functionalities and structuralism exists only for the living organisms and not anywhere else. When one thinks of intelligence as a functionality, it is self-evident that no other natural phenomenon seems to possess it. However, this could be due to our anthropic interpretation of intelligence. For instance, if Nature obeys some of the fundamental laws, then it can appear to anyone watching from outside as an intelligent behavior. This is because there will always be predictable consequences of fundamental laws. Therefore, to an outside observer, our spatiotemporal universe will appear to manifest a regular pattern of behavior. Therefore, to that observer, particularly whose intelligence consists of recognizing patterns (just as ours is), the universe would appear to be an intelligent entity just because it is predictable.

Therefore, we will expand the notion of intelligence to include sentience. As discussed above, this extension forces us to include self-reference into the ambit of the definition of intelligence. Admittedly, it is not possible to verify the presence of sentience and self-reference from outside. Therefore, the above mentioned outside observer would find it difficult to conclude either presence or absence of sentience and self-reference. However, if that observer were to look for goal-directed behavior, she would realize sooner or later that nothing in the universe possesses such goal-directed behavior, except the living organisms.

Therefore, it makes sense to look for any goal-directed behavior in all the natural phenomena to find out whether Life is the only natural phenomenon that manifests such goal-directed behavior. Once we accept that the goal-directed behavior is a good yardstick to evaluate a given natural phenomenon, it is intuitively clear that no other natural phenomenon manifests it. In fact, our conception of Nature is governed by nondeterministic and nonteleological foundations. Therefore, if we accept that the feature of goal-directed behavior arising from the simultaneous manifestation of intelligence, sentience, and self-reference is a characteristic of Life, then no other natural phenomenon possesses it.

Now let us look at the second question of whether it is possible to define this goal-directed behavior in the language of the separation between functionalities and structuralism. Prima facie, such a yardstick of goal-directed behavior doesn't fit into our conventional perspective of biology which is based on the Darwinian paradigm. In fact, the only possible biological theory that accepted goal-directed behavior, viz., Lamarck's theory has been discarded. Therefore, it is counterintuitive to employ a yardstick of goal-directed behavior as a definitive feature of Life. Before we look at the details of these arguments, few comments are necessary. Lamarckian model may have been discarded (and justifiably so), its modern version in the form of epigenetics has reemerged. Therefore, one needs to evaluate a scientific hypothesis on the basis of its implicit semantics and not on the basis of its structuralism. In Lamarck's theory, its semantic implications were and still are invalid. However, its structuralism is present in the form of epigenetic mechanisms. There is another similar example in the history of biology. Before the advent of molecular biology, or even before the advent of genetics, it was suggested that living organisms were governed by "Vital Force." Admittedly, its implicit deism is incongruent with our present Darwinian perspective of biology. However, if we were to eliminate the implicit deism from the conception of this so-called vital force, what we have is long range genomic influences which are still not formalized. The vitalism implicit in that hypothesis is a placeholder for our ignorance about structuralism and not that of any deism, implicit or otherwise.

Having looked at the historical perspective of why a scientific theory gets rejected on the basis of its semantics rather than on its structuralism, let us return to the question whether any goal-directed behavior can be thought of as a characteristic feature of Life and if so, how does it fit into the Darwinian paradigm which rests on probabilistic edifice of population genetics (Provine 2001; Bonner 2013). The reason why we find the notion of goal-directed behavior as being incongruent with the Darwinian paradigm, is that we think in terms of organisms manifesting such goal-directed behavior. While it is true that individual organisms do behave in such a manner, the correct way to look at the goal-directed behavior is to look for it at the level of biological systems rather than at the individual organisms. When viewed from such a systemic perspective, it is intuitively clear that most of the biological systems, be it a genome or be it our cognitive faculty, each one of them, manifests such a goal-directed behavior. A genome has a goal of allowing expressions of each of its constituent genes. Our cognitive faculty has a goal of processing every stimulus it is fed. This becomes easily discernible with the onset of pathologies. A random mutation may trigger a wrong sequence of gene expressions resulting in the death of the organism concerned. However, the genome has a fixed goal of allowing its constituents to express themselves. Admittedly, it is a closed subroutine which is followed blindly by the genome. The key point is that unlike machines, genomes don't stop functioning when its constituents malfunction. The genome continues with its goal-directed behavior, albeit blindly and much against the survival of the parent organism.

Similar situation occurs in our cognitive faculty. The signal processing continues even if one of the cognitive processes malfunctions. In biology, this is referred to as a graceful degradation (Wechsler 1992). Every biological system continues operating even after some of its constituents are nonfunctional. This is in contrast to machines which stop functioning when its components malfunction. This is because every biological system has a hierarchy of controls and it is this distributed architecture that gives rise to goal-directed behavior. It is possible to argue that since these goal-directed behaviors are blind, obviously there are no features of sentience or self-reference. Prima facie, this is a valid argument. However, upon a little reflection, it is intuitively clear that this is not so. In these biological systems, what is missing in the characteristics of sentience? The feature of self-reference is present because when a biological system operates even when its constituents malfunction, it is obviously functioning from a higher level of control which bypasses the malfunctioning constituents. However, it must be admitted that biological systems do not manifest sentience.

This dilemma can be resolved by placing biological systems in a hierarchy within an organism. Therefore, self-reference operates at the level of an individual biological system, but the sentience manifests at the level of organisms. Thus, if an organism is formalized as a hierarchy of biological systems, it is possible to arrive at self-reference, goal-directed behavior, and sentience at different levels of that hierarchy. It is legitimate to be skeptical about this approach by pointing out that there exists a host of natural phenomena which also manifest some kind of a hierarchy. Therefore, a mere presence of a hierarchy cannot, by itself, distinguish Life from other natural phenomena. This is where the second characteristic of hierarchy comes into the picture. What distinguishes Life from other natural phenomena is not a mere hierarchy of subsystems, but it is the separation of structuralism and functionalities within this hierarchy that separates Life from other natural phenomena. Let us see how. In the next section, we will look at the semantics of the relationship between structuralism and functionalities of natural phenomena in general and see how the relationship between structuralism and functionalities is different in the case of Life from that relationship in other natural phenomena.

1.10 The Semantics of This Relationship Between Structuralism and Functionalities

In continuation with the arguments presented above, let us see how structuralism and functionalities are related to one another in natural phenomena. Having done that, we will find out how Life differs from other natural phenomena. It must be admitted at the outset that under the Cartesian influence (Cottingham 2008), we have chosen to distinguish between structuralism and functionalities of all natural phenomena. While this structuralism is linked to some mathematical templates that we can perceive, the functionalities do not have any template as such. Instead, we define functionalities from the outcomes of the natural processes. Thus, in the Cartesian paradigm, structuralism is rooted into the transcendental realm of *Res cogitans* and functionalities are rooted into the nature of Natural laws. Admittedly, prior to and even during the Cartesian paradigm, functionalities were rooted into the implicit deism. It was on later, with the advent of modern science that this implicit deism was replaced by Nature. Therefore, the functionalities which were earlier ascribed to divine wish or the "First Cause" were later assigned to the laws of Nature. Modern science, being rooted in the naturalistic paradigm, finds it difficult to accept causal explanations because it leads to now rejected deism. Therefore, it is axiomatic that modern science has to abandon the semantic antonym of causality, viz., teleology as well. If the first cause can't be accepted, as a logical corollary, teleological doctrine too cannot be accepted.

The trouble with this reasoning is that modern science looks for causal explanations of all the natural phenomena without any commitment to the first cause. This is reflected in our conception of the cosmic singularity (Eden et al. 2013). There is no way our scientific theories can explain the origins of structural complexity of the spatiotemporal universe from a nonstructural entity like the cosmic singularity. At the other extreme lies the second law of thermodynamics (Haynie 2008). It tells us what kind of future awaits any given natural phenomena, but it can't be expressed in any teleological terms. Therefore, we have a statistical foundation for the second law of thermodynamics. Honestly, this is not a serious problem. However, the trouble with this formalization of the second law of thermodynamics is that it shapes our notion of functionalities. It is not really difficult to realize that our notion of functionality is determined by the final outcomes of any given natural phenomena. However, these outcomes are, in the final analysis, governed by the second law of thermodynamics. Therefore, the notion of functionality, per se, doesn't exist in our scientific theories except in the form of possible outcomes.

The problem with this reasoning is that it denies any template of functionalities per se. This is best exemplified by quantum phenomena. According to the conventional perspective (Gao 2017, see Chapter 1), there exists a boundary between classical reality and quantum reality. Moreover, the laws governing both these realms are distinctly different. However, there is no way to draw a line between these two realms. Therefore, in order to circumvent this ambiguity, we have formalized the notion of quantum decoherence or objective reduction (Schlosshauer 2007, see Chapter 2). This process of quantum decoherence essentially formalizes how a noncausal form of reality can be transformed into the causal form of reality which is normally called classical reality. Apparently, the intermediate state between these two forms of reality, quantum superposition state, must have some functionality which allows a noncausal form of reality into a causal form of reality. However, under the influence of the Cartesian paradigm, modern science cannot assign any such functionalities to the quantum superposition state. Therefore, instead of accepting this semantic lacuna, we have come up with many world interpretations of quantum reality (DeWitt and Graham 1973). It would have been simpler if we had conceded that functionalities can have their own templates which give rise to multiple outcomes and the second law of thermodynamics merely informs us which outcomes are more likely to occur. As discussed in the preceding monograph (Chhaya 2022c, see Chapter 3), the topic of Einstein Podolski Rosen experiments and its formalization in the form of Bell's inequalities, points toward a certain structural template of quantum superposition state. Therefore, it should be possible, at least in principle, to define functionalities of quantum superposition states. However, the key point here is that modern science, under the Cartesian influence, doesn't allow us to assume that functionalities, per se, can have their own templates of possible outcomes.

It is possible to argue that it is unfair to cite quantum nondeterminism as an example of a typical natural phenomenon. After all, quantum phenomena are counterintuitive. Therefore, one cannot cite an exception to a rule to invalidate it. Therefore, we will look at a typical natural phenomenon and see how there is inherent mismatch between our conception of its structuralism and functionalities. For this purpose, we will look at the phenomenon which is not a biological phenomenon and which is not counterintuitive like quantum phenomena. Therefore, we will select the phenomenon of catalysis. There are several reasons why this choice is justified. Firstly, catalysis is a typical natural phenomenon which can be and is explained by the laws of Nature. It occurs abundantly in Nature and more importantly, we can exploit it for our own benefits. Its prevalence suggests that it is a typical phenomenon and not an exception. Secondly, since we have been able to exploit it, it suggests that its mechanisms are intuitive. Moreover, catalysis is also observed in biological processes, thereby providing a link between Life and other natural phenomena.

Having justified our choice, let us try to understand what constitutes catalysis. More importantly, let us understand the nature of its structuralism and functionalities vis a vis its formal concepts. The most intuitive way to define catalysis is to classify it as any process that facilitates, accelerates and enhances the outcome of any process which is otherwise inefficient, slow and unproductive. In terms of its structuralism, we have a clear thermodynamic picture of how catalysis is brought about. It is intuitively clear that catalysis is a process whereby the activation energy necessary for the reaction to proceed is reduced. Normally, we think of any chemical reactions in terms of energies of the starting materials and that of the products. Therefore, the most intuitive way to enable a given reaction to go through is to provide sufficient energy from outside. The need for this external supply of energy arises because the starting materials align themselves in such a manner that their mutual orientations facilitate the product formation. Since every molecule can be deemed to be a cloud of electrons clothing the skeleton of nuclei, it is intuitively clear that we cannot push different molecules close enough to achieve the orientation necessary to bring about the reaction. Therefore, as a logical corollary, we provide an external source of energy which enables the participating molecules to approach one another with sufficient velocity. It is this short lived proximity between the reacting molecules that gives rise to a chemical reaction.

This simplistic scenario is universally valid. Therefore, very few reactions occur at ambient temperatures. However, whenever there is a rise in temperature, we witness more and more reactions going through. It is in this temperature controlled scenario that catalysis finds its relevance. The process of catalysis reduces the amount of external energy required for the reactions which are otherwise difficult to carry out or do not even occur. These reactions can now be brought about by catalysis. Having understood the basic concept of a catalysis, let us look at its structuralism. Before the articulation of the theory of thermodynamics, particularly the parameters of free energy, energy of activation, enthalpy and entropy, the catalysts were thought to possess some mystical powers. They seemed to bring about changes which were otherwise unthinkable. However, with the advent of thermodynamics, the phenomenon of catalysis became routinely acceptable and its power was harnessed. Today, the theory of catalysis (Teoh et al. 2021, see Section I) is formalized and we keep on looking for newer and newer catalysts for commercially beneficial reactions. Modern theory of catalysis encompasses, not just thermodynamics, but it also includes molecular structuralism and stereochemistry. Therefore, it is reasonable to assume that the structuralism of catalysis is very much established.

What is germane to the present discussion is whether it is possible to define the functionality of catalysis and how it is related to, by now, its well-established structuralism. There are two points that are relevant to the present discussion. Firstly, is it possible to define functionality of catalysis only on the basis of its structural template or does the functionality of catalysis have its own structural template which we have overlooked so far? Secondly, are both these structural templates of catalysis synonymous with another? The answer to the first question will enable us to define the relationship between structuralism and functionalities in a typical natural phenomenon which we believe the catalysis is. The answer to the second question will enable us to understand how the corresponding relationship between structuralism and functionalities in living organisms is different from any other natural phenomena. Therefore, let us begin with the first question about the functional template of catalysis.

As mentioned above, structuralism of catalysis is by now well defined. However, that is not the case with the contours of the functionality of catalysis. If one were to pick up a putative candidate catalyst and decide to find its exact catalytic functions, it is not possible to do so, at least not completely. The reason why it is not possible to completely predict the catalytic functions of any given catalyst lies in the fact that it is possible to define the surface structural details including the map of electron density at every point on its surface. It is also possible to define the fine structure of the surface of the catalyst using modern spectroscopic techniques. Therefore, given this data, one can guess what kind of reactions this putative candidate of catalysis can catalyze. Using the Internet or any other specialized databases, it is possible to guess which types of reaction the candidate catalyst can catalyze. Admittedly, the researchers today employ far more sophisticated approaches in studying new catalysts. However, the scenario outlined above sums up the basic plan of these researchers.

However, if one were to ask a question about any given catalyst, whether it is possible to predict its entire catalytic profile? The answer is no. Using a library-based methodology, it is possible to make a reasonably good guess about possible reactions that a given catalyst may be useful. However, there is no theory of catalysis which can formally predict the entire catalytic functions of any given catalyst. Conventionally, we don't expect to develop such a formalism for the simple reason that it is not possible to do so because our present approaches require the details of the reactants undergoing the reaction sought to be catalyzed. Therefore, as exemplified by the above mentioned library based methodology, we need to know the structural details of the reactants as well to predict whether the given catalyst can be effective or not. This sums up our current understanding of the theory of catalysis.

Before we answer the second question posed earlier, let us look at the semantics behind this conventional wisdom. The reason why we have chosen to ignore the possibility that functionalities, by themselves, can have their own templates, lies in the belief that any such functional templates would be synonymous with some kind of teleological theory which can predict in advance about the catalytic functions of any given catalyst. To be frank, this stance is eminently sensible. Moreover, it is congruent with the founding principle of modern science of logical positivism (Ayer 1959, see Chapter 3). This is essentially a minimalist view which discourages any additional postulates. However, there are two scientific domains, which are incidentally well established, that challenge this complacency. These domains of quantum chemistry and immunochemistry basically challenge this conventional wisdom. More importantly, they also tell us why we need to re-examine the theory of catalysis. Admittedly, we will not be trying to reformulate the theory of catalysis here. Therefore, we will merely present the gist of the rationale why these domains compel us to leave our comfort zone and try to seek a needed theory of catalysis. More importantly, these two domains also provide a clue to distinguish between Life and other natural phenomena.

Quantum chemistry (Szabo and Ostlund 1989, see Chapters 2 and 3) rests on a single fundamental assertion that the wave function is a repository of all the chemical reactions that a molecule may be capable of undergoing. Without going into the details of this semantic proposition, it is intuitively clear that if the functionality of catalysis is included in the wave function, then it is axiomatic that all the potential catalytic functions must be already present in the wave function of any putative catalyst. Therefore, the wave function contains the information about the total catalytic functions of any given catalyst. As a logical corollary, such a catalytic description of wave function must be amenable to formalization, thereby providing a separate template of the functionality of catalysis. Therefore, by implication, such a theory of catalysis can be articulated, if only we have a template for catalytic functionalities. However, it is our belief that the structural template of catalysis is identical to the functionalities of catalysis, that prevents us from exploring the template for catalytic functionalities.

Now, let us look at the second domain of immunochemistry (Inman 2012). Admittedly, we can think of immune response as a type of catalytic functionality. This is because, at a fundamental level, immune response arises when antibodies catalyze the degradation of cellular components of the invading organisms. However, our emphasis here is to demonstrate that the template of catalytic functionalities can be synchronized with the structural template of catalysis. The basic scenario runs like this. All living organisms have arisen from some, as yet unidentified ancestor. Therefore, they carry common structural and functional ontologies. Therefore, during an infection, the immune response of the host is confronted with the structural template of the vector of infection. However, due to shared ontology, the host is capable of decoding and identifying the parasite's structural template in the form of its molecular surface. The conventional wisdom suggests that this is possible because due to the very effective tendency to produce a library of epitopes, the host is able to create a structural template which is complementary to the surface structural template of the parasites. There is no reason to doubt this scenario because it has been verified under a diverse range of infections.

However, what is missing from this perspective is that this library of epitopes is produced by higher level genomic architecture which decides the sequence of gene expressions of the genes responsible for producing antibodies. While it is conventionally understood that this ability to produce a right mix of antibodies is governed by rapid generation of epitopes, followed by the Darwinian selection of the epitopes as defined by the structural template of the parasite. However, upon a little reflection, it is intuitively clear that this enormous diversity of epitopes is a product of the higher level control arising from the peculiar genomic architecture. The randomness of the structural template of the resulting epitopes is due to a particular type of long range influences within the genomic architecture. What the proposed model described in the following chapters suggests that the higher level details of genomic architecture contains the template for different functionalities which is passed on to the structural template of immune response encoded in the form DNA sequence of the concerned genes.

The key point of the above discussion is that in the case of chemical catalysis, we have no access to the template for catalytic functionalities and therefore, we are required to employ a method based on educated guesswork. On the other hand, the same catalytic functionalities are encoded in living organisms and these functionalities are available for translation into corresponding structural templates. Thus, there are two possible inferences available to us from this discussion. Firstly, there exists a separate and perhaps distinct template of functionalities and structuralism. However, the templates of functionalities are not easily accessible to our cognitive faculty. Therefore, we are forced to base our scientific theories on structural templates available to us and employ educated guesswork to discern the corresponding functional template. However, if a natural phenomenon is sufficiently complex, like quantum superposition state or living organisms, then those phenomena contain both these types of templates. This brings us to the second inference from this discussion. Living organisms not only contain within themselves both these structural templates, but more importantly, in the living organisms, both templates operate at different levels.

This separation of structuralism and functionalities in the living organisms is the reason why Life is unlike any other natural phenomena. It is different because the structuralism and functionalities are present in living organisms but they are separated from one another. In most of the other natural phenomena, the templates for structuralism and functionalities are not separate and therefore not amenable to separate formalizations. Interestingly, quantum phenomena occupy the boundary between the separation of the template for structuralism and functionalities and their unification. If we were to think of other natural phenomena as spread over the four dimensional spacetime, then one can think of the quantum superposition state whenever it arises in other natural phenomena, as some kind of inward folding of the underlying spacetime. This higher dimensional version of spacetime gives rise to the separation of the templates for structuralism and functionalities of other natural phenomena. Therefore, quantum chemistry can formulate a theory of catalysis, but classical theory of catalysis cannot express it. However, Life due to its inherent complexity is able to separate these two types of templates and practice immunological "omniscience" envisaged in quantum theory of catalysis. Admittedly, this is a too radical theory to be accepted uncritically. Therefore, in the following chapters, we will examine it in some detail. Presently, let us confine ourselves to the distinction between Life and other natural phenomena. Therefore, in the next section, we will examine why Life defies the common feature of the unification of structuralism and functionalities. More importantly, what are the semantic implications of this separation?

1.11 Why Does Life Defy Common Semantics?

In the previous section, we looked at the idea that the functionality, per se, could have its own templates. Moreover, it is not necessary for this functional template to coincide with the structural template of a given natural phenomenon. To justify this assertion, we discussed two phenomena, viz., quantum superposition and cognitive faculty. As mentioned above, this idea needs to be scrutinized before one can think of it as a serious scientific hypothesis. However, prior to doing such a scrutiny, it is necessary to deconstruct the semantics of having two separate templates for structuralism and functionalities. More importantly, it is necessary to deconstruct why Life defies the norm of having indistinguishable templates for structuralism and functionalities, which seems to be the case with most of the natural phenomena.

Therefore, let us look at the semantics of the necessity of having separate templates for structuralism and functionalities. Admittedly, such a concept of having two templates for structuralism and functionalities is not accepted by modern science. Therefore, we cannot deconstruct this concept from the conventional perspective. However, there is an analogous concept accepted by our contemporary theoretical framework. Therefore, let us look at it. This refers to the belief in the emergence of new functionalities with the increase in the complexity of a given system. Of course, when we think of an example of such a principle of emergence (Smith and Morowitz 2016), we immediately think of network theory (Arbib 2003). Given the predominance of the Internet and computers, it is inevitable that one thinks of such an example. However, it is a mistake to think that this idea is confined to network theory only. It is present in almost all the scientific domains. We have to think of molecular biology to realize that such an emergence of functionalities as a consequence of structural complexity is a norm in that domain. Similar examples are available in the domains of composite materials, semiconductors, medicinal chemistry. In fact, the list of such examples is endless.

Therefore, let us take this emergence principle as an analogous concept and see what are its semantic implications. Firstly, this concept is devoid of any causal explanation. It is more of a generalization obtained by observation. Therefore, if we were to ask a rhetorical question why should structural complexity give rise to additional functionalities? There are no answers available. In that sense, this concept is empirical evidence rather than a theoretical construct. Of course, those familiar with network theory would point out several mathematical constructs which quantify the relationship between structuralism and functionalities, say in the form of the relationship between number of nodes and the parallel signal processing. However, it must be kept in mind that good mathematical equations merely correlate different parameters, they don't explain the correlations. The most obvious example of this semantic agnosticism of mathematical expressions is available in the second law of thermodynamics (Haynie 2008). The equations do not explain why entropy increases or why the arrow of time is always in one direction. These questions are addressed by scientific theories and not by mathematical expressions.

It is not surprising that our belief in the emergence of functionalities with the increase in structural complexities, hasn't gone beyond empirical observations. In fact, in the later chapters, we will realize that the model proposed here not only explains why Life is unlike any other natural phenomena, but it also explains why functionalities should be dependent on the structural complexities. The key point is that our conventional belief that new functionalities emerge from an increase in the structural complexities is an incipient acknowledgment of some kind of relationships between structuralism and functionalities. In that sense, the proposed model presents the next step. It helps us understand the exact relationship between structuralism and functionalities.

Now, let us return to the need for separate templates for structuralism and functionalities and why they happen to be identical in most of the natural phenomena. The proposed model has something radical to offer on this issue. As discussed in the preceding monograph (Chhaya 2020), every natural phenomenon, beginning with the cosmic singularity itself, possesses different templates for structuralism and functionalities. However, what prevents us from perceiving the separation between structuralism and functionalities in our limited access to the nature of reality. As pointed out above in the case of chemical catalysis, functionality of catalysis can have its own template, but it is not available to our cognitive faculty for formalizing. However, it is possible to define a wave function representing the entire catalytic profile of a given catalyst. However, since our access to such a wave function is limited only to orthogonal solutions of the wave function, we cannot deconstruct the catalytic profile in its entirety. Therefore, the functional template of a catalysis exists, but it is not available to us, at least not completely. On the other hand, the structural template of any given catalyst is available to us.

As discussed in the preceding monograph (Chhaya 2022a), this happens because our cognitive faculty can operate from a narrow range of dimensionalities whereas the functionalities of any natural phenomenon may or may not exist in the range of dimensionalities from which our cognitive faculty operates. Therefore, in such cases, we mistake structural templates to be the functional template. However, in the case of Life (and also in the case of quantum phenomena), this doesn't happen. Let us understand why it doesn't happen. Firstly, Life manifests itself at more than one level of organization. As we know from our experience in molecular biology, genes and their expressions operate at a molecular level. However, genomic controls over these gene expressions operate from a higher level. Similarly, each organ operates on their own as well as in coordination with other organs. Therefore, there exists a hierarchy of controls that ranges from the entire organisms to different organs, to different cells of each organ, to different organelles of each cell and to each molecule of each organelle. Therefore, it is inevitable that there is an overlap between these different levels.

It is possible to argue that this is not the case because in such a scenario, even complex machines like computers would manifest this feature. This is because every complex machine has overlapping controls and commands. Therefore, mere presence of hierarchy of subsystems, by itself, cannot be the reason why Life behaves differently. However, there is one feature of the hierarchy present in biological systems which is missing in the case of machines. In living organisms, both structural and functional hierarchies have evolved together. Therefore, these functional and structural hierarchies have a common ancestral hierarchy from which they have differentiated. This is not the case with machines. In computers, functional hierarchy is created separately and structural hierarchy is manufactured afterward. Therefore, both hierarchies remain distinctly different from one another. However, because both these hierarchies are designed by human beings, they are congruent, but not unified.

It is possible to be skeptical about such a rationale and dismiss it as another type of "Vital Force" argument. However, there is another feature of these biological hierarchies that is absent from the hierarchies present in machines. This is the feature of self-reference. Admittedly, the notion of self-reference is nebulous and more often than not, used as an alibi. However, in the present discussion, it has a direct relevance. Both these hierarchies, structural and functional, in the living organisms can refer to their internal states and respond accordingly. To illustrate this feature of self-reference, let us take an example of something as simple as body temperature. In multicellular organisms, there exists a mechanism for regulating body temperature. There exists a parallel feature of temperature regulation in machines, say, computers. Now, let us see how both these features operate differently and why. In the case of computers, just like in the living organisms, there is a sensor, which helps to define the range of temperature in which systems can operate safely. Although the nature of sensors is different in both these systems, the objective is the same. Now, let us see how this parallel feature differs in both these systems.

In the case of computers, we need to feed in the numerical values of the safe range of temperatures in which the computer can operate. Having done that, the sensor sends the signal to the hardware whenever it senses that the actual temperature of the processor has breached the given range of safe temperature. The key point is that this is achieved only after the safe range of temperature is fed from outside. The sensor, by itself, cannot define the safe range of temperature. However, in the case of living organisms, this functionality of temperature control is achieved differently. Admittedly, just as in the case of computers, there exists a sensor (albeit of a different kind) to observe the actual temperature of the body. However, it doesn't require any external standard, in the form of a predetermined safe range of temperature, to send a signal to different organs of the body to activate the process of cooling.

This is possible because the structure of the sensor is such that its functionality itself, is the definition of the range of safe temperature. Whenever this range of safe temperature is breached, a new functionality sets in. Therefore, the functionalities of the sensor itself are the limits of the safe range of temperature. Since functionalities themselves define parameters like temperature, they don't need any external input of parameters. This is a correct perspective of self-reference. Admittedly, this is definitely missing in computers, or any machine for that matter.

Downside of this mechanism is that it lacks precision. However, the upside of this mechanism is that it allows the system to operate under a diverse range of conditions. Moreover, this flexibility ensures that the living organisms undergo what is called "Graceful degradation" of their functionalities (Wechsler 1992). This enables living organisms to operate at a suboptimal level in the face of a hostile environment. The machines, on the other hand, simply fail to operate under such adverse conditions. It is important to keep in mind that this self-reference has occurred only because biological evolution began with a single framework for both structuralism and functionalities. It was during the course of biological evolution that the separation of structuralism and functionalities occurred, leading to separation of templates of both and emergence of new functionalities. Thus, one can sum up this discussion by saying that Life is unlike any other natural phenomena because it has different templates for structuralism and functionalities and more importantly, because both these templates arose from a common ancestral framework. Till now, we have looked at the arguments presented above in the terms of theoretical postulates. Therefore, in the next section, we will try to define a structural template that corresponds to these postulates.

1.12 Involuted Model of the Relationship Between Structuralism and Functionalities

In the previous section, we discussed the reasons why Life is unlike any other natural phenomena. It was suggested that the main reason for this distinction is the separation of structuralism and functionalities in living organisms. It was suggested that every natural phenomenon has two templates, one each for structuralism and functionalities. However, in most cases, the separation between these two templates is not perceptible because the level at which the template for functionalities exists is beyond our cognitive capabilities. Therefore, we resort to mathematical templates to define natural phenomena. However, we have no explanation for the fact that mathematical templates are "unreasonably effective" in formalizing natural phenomena.

However, there are two natural phenomena which seem to be exceptions to this general feature. It was suggested above that quantum phenomena and living organisms have two separate templates for structuralism and functionalities that have a unique feature. In the case of quantum phenomena, the template for functionalities, which brings about collapse of the wave function, is within the reach of the template for functionalities of our cognitive faculty. This has resulted in two important consequences. Firstly, it allows our cognitive faculty to get entangled with the quantum superposition state, thereby initiating the collapse of the wave function. Secondly, it enables our cognitive processes, particularly those responsible for semantic processing, to conceptualize wave function as a repository of all the semantic propositions. Admittedly, both these consequences have enormous importance and are discussed separately (Chhaya 2022c).

In this chapter, we will focus on the separation of templates for structuralism and functionalities in living organisms. The unique feature of this separation in the living organisms is that they have a functionality of self-reference. In other words, functional template and structural template are not only connected to one another, but the functional template possesses a functionality of self-reference. Thus, the separation of the templates for structuralism and functionalities exists in every natural phenomenon, including Life. However, the functional template of Life, exceptionally possesses a property of self-reference which is absent from the functional templates of all the other natural phenomena. Irrespective of the validity of this postulate, it is necessary to formalize it. This is because only when we have a formal description of this postulate that we can generate verifiable predictions. Therefore, in this section, we will look at one such model. The mathematical foundation of this model has been articulated in the preceding monograph (Chhaya 2022a). Therefore, we will look at a summary of this model that is relevant to the present discussion. Having done that, in the following sections, we will try to apply it to biological systems.

One of the original motivations to look for an alternative model was the ambiguity about the relationship between the mathematical constructs and natural phenomena. Our conventional wisdom about the relationship between mathematics and the nature of reality is ambivalent. On the one hand, modern science has consciously sought a naturalistic paradigm for itself. However, it cannot hope to do so without employing mathematics. The trouble with this strategy is that the origins of mathematics are anything but naturalistic. Admittedly, modern science cannot subscribe to Platonic absoluteness because it implies some denomination of deism. This transcendental status of mathematics has survived the paradigm shift from Platonism to the Cartesian paradigm. Therefore, modern science has forsaken any ontological perspective of mathematics. Instead, it has taken mathematics as a priori. However, as science approaches its logical climax of explaining the cosmic singularity (Eden et al. 2013), this stop gap assignment of a priori status to mathematics needs to be revisited and possibly revised.

With this perspective in mind a new topological model of ontology and epistemology of mathematical objects was outlined in the preceding monographs. As a logical progression, we will apply this model to Life and try to describe it in purely naturalistic idioms in this monograph. For the sake of simplicity, the proposed model has been described here in a point-wise manner. Moreover, only those details which are relevant to the present discussion have been included here. For more information, the reader is requested to refer to the preceding monographs.

1. The proposed model is built on a modified definition of involution. It seeks to define every natural phenomena, including the cosmic singularity, in the form of an involuted manifold.
2. Conventional definition of involution refers to the influence of the parent manifold being passed on to any of its submanifolds via the operator of involution.
3. Thus, in the conventional involutive algebras, the parent manifold and its submanifold retain their identities, but the structural template of the parent manifold is passed to any of its submanifolds via the operator of involution.
4. As discussed in the preceding monograph, the proposed model modifies this conventional definition of involution. It creates a special case of involution in which the operator of involution acts on the parent manifold itself. Since every manifold is its own submanifold, the resulting modified operator remains the operator of involution.
5. The modified operator of involution results in two important consequences. Firstly, at the end of the application of the operator of involution, the dimensionality of the manifold is decreased by one. This operation can be visualized as an inward folding of one of the dimensions of the manifold onto the remaining dimensions of the manifold.
6. Secondly, the application of the modified operator of involution results in the increase in the complexity of the metric of the manifold. This can be visualized as integration of the information content of the dimension undergoing involution, with the information content of the remaining dimensions of the manifold.
7. The proposed model postulates that natural phenomena exist in multiple dimensionalities simultaneously.
8. Depending on the dimensionality from which we make an observation, we find different types of metric. We can think of metric as a type of "granularity" of the observation being made.
9. However, this granularity remains fixed for a given dimensionality.
10. At the same time, if the observations of a given natural phenomenon are made from different dimensionalities, the granularity of each observation will be different.
11. More importantly, the different granularities observed in the case of a single natural phenomenon from different dimensionalities, are connected to one another by the involutive algebras.
12. Thus, the proposed model offers a framework for unifying different scientific theories into a single framework of unified theory of science.
13. Moreover, since the proposed model is applicable to all natural phenomena, it can be applied to the cosmic singularity as well as to our cognitive faculty. This provides a common structuralism of ontology and epistemology, thereby eliminating the Cartesian split.
14. In fact, the proposed model suggests that the Cartesian paradigm is a phenomenology of the four dimensional spacetime.

15. The proposed model, by assigning different mathematical constructs at the granularity of the spatiotemporal universe, converts the classical problem of epistemological access to that of topological access.

Admittedly, the proposed model leads to several heterodoxical consequences. These consequences and their semantic justifications are discussed in the preceding monographs. For the present discussion, we will take this model as a priori and try to understand how the relationship between structuralism and functionalities gets represented in this model. In the following chapters, we will look at a novel approach to define genomic architecture and its Darwinian interpretation using this model. Presently, let us look at the relationship between structuralism and functionalities according to this model.

1.13 Involuted Model of Biological Structuralism and Functionalities

In this section, we will begin with the relationship between structuralism and functionalities of natural phenomena and its description according to this model. Having done that, we will try to find how Life differs from other natural phenomena in the language of the formal description of the relationship between structuralism and functionalities in the proposed model. Therefore, let us begin with the relationship between structuralism and functionalities. As discussed in the preceding sections, it is due to our limited cognitive capabilities that we have not been able to discern the concept that functionalities can have their own templates which need not coincide with the structural template. Our inability to do so is further complicated by the fact that these two templates themselves could be related to one another by some kind of algebra. Therefore, when we try to understand functionalities of any given natural phenomenon, we, of course under the mistaken belief, try to think in terms of the structural template of that phenomenon. However, because these two templates are interconvertible through a proper involutive algebra, we achieve a reasonable degree of success in profiling the functionalities of that natural phenomenon by defining its structural template. This partial success has led us to intellectual complacency. It is only when we are faced with the inadequacies of this indirect approach to define functionalities through structuralism that we are forced to revisit the paradigm. There are two such examples of the inadequacies of the conventional perspective of mathematical formalization of natural phenomena. They are quantum phenomena and cognitive functionalities. While the nature of quantum phenomena is discussed in the preceding monograph (Chhaya 2022c), in this and the next monograph (Chhaya 2022d), we will look at biological and cognitive functionalities respectively.

It is intuitively clear from the brief description of the proposed model given above that according to this model, structuralism and functionalities must occupy different dimensionalities. Moreover, as mentioned above, it is possible that the dimensionality of functionalities of several natural phenomena could be beyond the dimensionalities from which our cognitive faculty operates. However, such a possibility, even if it is true, offers no way to confirm it. Therefore, it is reasonable to assume that this possibility is a hypothesis and try to seek its falsifiability (Popper 1963). Therefore, the real question is what kind of verifiable consequences this possibility offers? Apparently, we can always argue that this has always been implicit in the conventional perspective. For instance, in the case of medicinal properties of different molecules, we normally try to predict the medicinal properties of candidate molecules based on their structural template. Literature is replete with a variety of methods based on different mathematical descriptions of molecular structure of candidate molecules (Roy 2015, see Chapter 4). This includes molecular mechanics, molecular orbitals and even the combinatorial approach to different molecular fragments. Admittedly, each of these approaches is focused on the structuralism of molecules.

However, even within these approaches, the most insightful methods are those that try to deconstruct the medicinal property of a given set of molecules using topological models (DeVillers and Balaban 1999) rather than those who use other combinatorial paradigms. This is possible because there exists a definitive relationship between algebraic operations allowed in different types of geometries and topologies. Conventionally, we intuitively feel that there exists a definitive relationship between geometry and topology. However, the exact nature of this relationship has eluded us (Brendon 1993, see Chapter 2). The proposed model fills this gap between geometry and topology. Therefore, our conventional perspective of mathematical formalisms which tacitly accepts the efficacy of topological methods over the geometric paradigm is congruent with the proposed model.

However, even if we were to accept the reasoning given above as valid, we still need to flesh out the exact relationship between structuralism and functionalities. Therefore, without appealing to the implicit congruence between the conventional wisdom and the proposed model and without appealing to our intuitive belief about the relationship between geometry and topology, let us define the relationship between structuralism and functionalities as defined in the proposed model. For this purpose, we will delve upon two features of the proposed model. Firstly, according to this model, every dimensionality of a given natural phenomenon has its own characteristic granularity in the form of its metric. Therefore, according to this model, if structuralism and functionalities were to occupy different dimensionalities, then both these features are similar to one another. Therefore, the question arises: What is the difference between structuralism and functionalities? Apparently, according to this model, functionalities are nothing but structuralism at a different dimensionality. Therefore, how functionalities are different from structuralism? There are two possible inferences available to us. Let us begin with the assertion that functionalities and structuralism are different in some basic fundamental sense. Therefore, the proposed model must define the distinction between these two domains in purely topological terms. In such a scenario, it is intuitively clear that the relationship between a higher dimensional entity and the corresponding lower dimensional entity must be asymmetric. In other words, functionality can influence structuralism but not vice versa. Alternatively, structuralism influences functionalities but not vice versa. Therefore, according to this model, the relationship between structuralism and functionalities must be one sided, though at present, we can't decide what is the direction of this one sidedness. However, such a proposition is unacceptable from the topological perspective. The operator of involution, either as conventionally defined or as modified here, possesses a group symmetry. Therefore, every operator of involution must have its own inverse. However, the operators of involution, individually, can be symmetry breaking if its inverse operator is not possible. This is precisely what is manifest in the case of cosmic singularity devolving into the asymmetric manifest universe.

The second inference available from the proposed model is that functionalities and structuralism are in fact analogous, but our perception of their different nature is a cognitive artifact. Let us examine this possibility. Purely from the topological perspective, the only difference between any two different dimensionalities is their granularity. Therefore, let us assume that functionalities, as a rule, occupy higher dimensionality vis a vis its corresponding structuralism. In that case, it is intuitively clear that according to this model, the granularity of functionalities will be coarser than that of the corresponding structuralism. This is because according to this model, granularity becomes finer and finer as the number of involutions increase, leading to lower and lower dimensionality. Therefore, any distinction between structuralism and functionalities must rest on the degree of coarseness of their granularities.

Let us examine the consequences of this proposition. Without going into the semantics of functionalities, it is intuitively clear that whenever an element of coarser granularity devolves onto lower dimensionality, it will reshape a larger number of elements of granularity of the lower dimensionality. This is axiomatically true, irrespective of the dimensionalities involved and their corresponding granularities. Therefore, let us say that there exists an element of functionality which devolves into a lower dimensionality and the number of elements of granularity of the lower dimensionality influenced by this involution is finite. From the perspective of the lower dimensionality, this influence will appear to be unanalyzable because it will appear to be more than the sum of all the elements of granularity influenced by the element of granularity which has devolved from the higher dimensionality.

It must be kept in mind that this phenomenology is not specific to any particular dimensionality. It is generic in the sense that it will manifest at all but the highest dimensionality. This may sound outlandish, but this is precisely what we normally observe. Take the case of a medicinal property. Presently, we arrive at the conception of such property by trying out different fragments of molecules and creating a library of candidate drug molecules. The assumption behind this approach is that a single medicinal property can arise from multiple pathways of combining different molecular fragments. This is precisely what the above discussion suggests. There is another example of this proposition, the one that is closer to the theme of this monograph. Conventionally, we agree that the genome is more than the sum of its constituent genes. Therefore, according to this model, the genomic properties appear to us as functionalities because our genomic paradigm is founded upon the primacy of individual genes.

It is possible to counter this perspective by arguing that according to this model, what we perceive to be structuralism, must also be functionalities for the dimensionalities lower than that of this structuralism. Apparently, this is not the case. However, upon a little reflection, it is intuitively clear that this argument is fallacious. What we conventionally categorize as structuralism actually is a functionality arising from still lower dimensionality. Take the example of genes. It is true that genes constitute a structural template for genomic functionalities. However, genes themselves have their own functionalities independent of their place in the genome. It is true that it is the genome that controls expressions of individual genes. However, genes can and do express themselves on their own under suitable laboratory conditions.

We can cite examples of this all the way down to individual atoms and even beyond. Nucleotides have their own functionalities which arise from their own constituent molecular fragments of pyrimidines, sugars, and phosphates. They in turn acquire their functionalities from their respective constituent atoms. One can go on ad infinitum. The point is that the distinction between structuralism and functionalities is essentially phenomenological. However, because our cognitive faculty operates from a narrow range of dimensionalities, we can't perceive this duality of structuralism and functionalities. It depends on the dimensionality from which one makes an assessment. What is asymmetric and therefore definitive about this perspective is that at highest dimensionality, there is no granularity and therefore no structuralism. This is a true definition of the cosmic singularity. The rest of the manifest universe is nothing but the interplay between different dimensionalities and their granularities.

Let us temporarily grant this model a provisional status of being a serious scientific hypothesis. The question that is pertinent here is can it explain why Life is unlike any other natural phenomena in the language of this separation? Apparently, the separation between structuralism and functionalities is a matter of perspective of the dimensionality from which one makes an observation. Therefore, Life, like any other natural phenomena ought to have a similar dimensionality based definition of structuralism and functionalities. The separation of structuralism and functionalities, per se, can help us to explain why Life is unlike any other natural phenomena. There has to be something more specific about the relationship between structuralism and functionalities in the case of Life, to make it unique among natural phenomena. This unique feature consists of the functionality of self-reference. Therefore, if this model can justify why self-reference arises in the case of Life and in no other natural phenomena and link this explanation to the separation of structuralism and functionalities, then we can accept the proposed model as a serious scientific hypothesis. Let us know what the proposed model offers on this feature of self-reference.

To begin with, let us see how self-reference can be formalized in this model. Having done that, we will look at why it manifests only in the case of Life and nowhere else. Purely from the mathematical perspective, the operator of involution constitutes a self-reference. This is because the operator of involution consists of passing on the information content from the parent manifold to any of its submanifolds. Therefore, if one were to take the parent manifold as a frame of reference, naturally, every operator of involution is an operator of self-reference. However, if one were to define the frame of reference by taking a submanifold which is a recipient of involution, then the operator of involution is not an operator of self-reference because from the frame of reference of the submanifold, the information content is imported from outside the frame of reference.

As an aside, it must be kept in mind that the conventional definition of involution which uses the frame of reference of the parent manifold is congruent with the Cartesian paradigm because according to the Cartesian paradigm, mathematical operators are part of the transcendental realm of *Res cogitans*. Therefore, they need to be "imported" into our cognitive faculty. Therefore, the eventual formalism would again be part of the transcendental realm of *Res cogitans*. Therefore, the frame of reference remains, either before or after in *Res cogitans*. Therefore, it makes sense to use the parent manifold as a frame of reference. However, as discussed in the preceding monograph (Chhaya 2022a), scientific theories based on this approach cannot be truly naturalistic. Therefore, it was suggested in that monograph that in order to avoid this implicit transcendentalism, we need to modify the conventional definition of involution.

Returning to the present discussion, the proposed model modifies the operator of involution where the feature of self-reference is built into the very definition of involution. Therefore, every time we succeed in formalizing a given natural phenomenon using this model, we are introducing the feature of self-reference into the formal description of the natural phenomena. Therefore, if a natural phenomenon possesses the feature of self-reference, it would find its formal expression. Thus, our task is now reduced to demonstrate that only Life, among all the natural phenomena, possesses a formal expression of this feature. Let us see how this comes about. It is intuitively clear that this feature of self-reference must consist of two-way information transfer, or more precisely, two-way simultaneous information transfer.

Conventionally, any mathematical operator can have its own inverse. In fact, this is a prerequisite for any operators belonging to a mathematical group which is the case with the modified operator of involution. Therefore, two-way information transfer by the invocation of an operator and its inverse is a routine protocol. However, it is not possible, at least not normally, to employ an operator and its inverse simultaneously. We normally employ them sequentially but not simultaneously. However, in the present case, since the frame of reference of the "donor" manifold and the frame of reference of the "recipient" manifold are identical because the involution happens on the same manifold, simultaneity is possible. The only precondition of this simultaneity of the information transfer is the nature of information being transferred. However, according to this model, the nature of information content depends on the dimensionality of the manifold. Thus, simultaneous information transfer is decided by the dimensionality of the manifold in which this simultaneity of information transfer can be brought about. Therefore, our task is reduced to demonstrate that the dimensionality of the living organisms is such that it allows this simultaneous information transfer leading to the feature of self-reference.

More importantly, our task is to demonstrate that this dimensionality is exclusively manifest in Life and nowhere else.

This reasoning might appear to be contradictory to our earlier inference that there exists a marked separation between structuralism and functionalities in living organisms, whereas in all the other natural phenomena structural template can adequately represent its functional template. However, this contradiction arises due to misconception. Let us revisit the proposition that every natural phenomenon has two templates for structuralism and functionalities. However, we are not able to perceive functional templates of most of the natural phenomena, and therefore, we try to understand them by trying to interpret their structural templates which are available to our cognitive faculty. However, according to this model, these two templates, since they are different from one another, must occupy different dimensionalities. This is because each dimensionality possesses its own metric. Therefore, if any two templates possess different features, they must occupy different dimensionalities.

Therefore, when we try to understand functionalities of any given natural phenomenon, particularly the one whose functional template is not discernible, we use that phenomenon's structural template and try to deconstruct its functional template. This is possible because according to this model, different types of metric manifest in different dimensionalities are themselves interconvertible through the involutive algebras. However, in congruence with the axiom of choice (Herrlich 2006), some of the metrics, say those representing functional templates, are beyond our cognitive capabilities. Therefore, we are forced to base our understanding of these functional templates (whose metrics are unavailable for cognition) on the structural templates which are available to us. This could be visualized as the templates for functionalities occupying dimensionalities beyond those accessible to our cognitive faculty. It is possible to argue that in such a scenario, how are we able to understand the functionalities of these natural phenomena when their templates of functionalities are not available? The answer to this question lies in the functionality of semantic processing. Our cognitive faculty seems to possess an access to the highest dimensionality which connects all the lower dimensionalities simultaneously. Since our cognitive faculty has an access to this highest dimensionality representing the semantics of a given natural phenomenon through its structural template, we obtain an intuitive understanding of functionalities of that phenomenon through noesis. However, since the functional template and its inherent metric are not accessible to our cognitive faculty, we cannot formalize it though we have an intuitive understanding of its functionalities. Quantum phenomenon is the classic example of this simultaneous intuitive understanding and incomplete formalization by our cognitive faculty.

With this rationale in place, let us see how Life differs from other natural phenomena. Apparently, the explanation must lie in the magnitude of difference between these two dimensionalities. As mentioned above, in most of the natural phenomena, the separation between structuralism and functionalities, as defined by their respective dimensionalities, must be so large that the difference between the metrics of these two dimensionalities must be too radical. This results in two important consequences. Firstly, since the gap between these dimensionalities is large, it cannot be observed from our cognitive faculty, thereby rendering functional templates incomprehensible. Secondly, this large gap between these two dimensionalities of structuralism and functionalities ensures that simultaneous information transfer, a prerequisite for self-reference, doesn't occur.

Therefore, according to this model Life must have two prerequisites. Firstly, the gap between the dimensionality of structuralism and functionalities must be narrow to allow simultaneous information transfer. Secondly, both these dimensionalities of structuralism and functionalities of Life must occupy lower dimensionalities among the other natural phenomena. This is because of the inherent complexities both structural and functional which according to the proposed model have lower dimensionalities. The advantage of having a narrow gap between the dimensionality of structuralism and functionalities is that their respective complexities must be comparable. Therefore, their respective metrics too must be comparable. It is this similarity between the metrics of structuralism and functionalities of Life that ensures the simultaneous information transfer leading to self-reference.

This brings us to the last point of the present discussion. As mentioned above, let us provisionally accept that Life has two templates representing structuralism and functionalities which are very close to one another, and therefore, Life manifests a feature of self-reference. Let us also accept, albeit provisionally, that it is this feature of self-reference that distinguishes Life from other natural phenomena. However, there is still one unresolved dilemma. This refers to the notion of complexity and the emergence principle mentioned above. There is a basic contradiction between the conventional perspective of Life which concedes that Life is perhaps the most complex natural phenomena, and therefore, it acquires functionalities due to this increased complexity as formalized in the emergence principle and the proposed model. According to this model, the more complex metrics will manifest at lower dimensionality. In other words, lower the dimensionality, more complex the metric. If this is true, then according to this model, Life, being the most complex phenomenon, must manifest at the lowest of the dimensionality. This is apparently counterintuitive, if not absurd.

Therefore, let us see how this contradiction can be resolved. This contradiction arises because we assume that our cognitive faculty creates models for explaining natural phenomena from an external perspective. In other words, when we conceive a transformation of a circle to a sphere, we create a perspective which views this transformation from outside. This external view is actually a Cartesian legacy (Cottingham 2008). The Cartesian paradigm postulates that our cognitive faculty (which was equated with consciousness in those days) is outside the spatiotemporal universe. It is a part of the transcendental realm of *Res cogitans*, along with all the mathematical constructs, including the conception of a circle and a sphere. Since our cognitive faculty (in the form of consciousness) devolves into the human brain from outside (outside the spatiotemporal universe in which the human brain is present) our scientific theories are essentially an external projection embedded into the human brain.

However, as we know from our current understanding of cosmology, there is no such external point outside the spatiotemporal universe. Therefore, what was earlier described as a transcendental realm, which includes mathematical objects, must be present inside the spatiotemporal universe itself and not outside it. This scenario is discussed in the preceding monograph (Chhaya 2022a). From the perspective of the present discussion, let us see how this model resolves this contradiction between the emergence principle, which ascribes the unique functionalities of Life to its exceptionally complexity on the one hand and the inference available from the proposed model which asserts that more complex a natural phenomenon, lower must be its dimensionality.

The resolution of this paradox lies in the nested hierarchy of dimensionalities. Our conventional perspective of our cognitive faculty assumes that our epistemological capabilities are absolute and therefore independent of the dimensionality. However, according to this model, our cognitive faculty operates from multiple dimensionalities simultaneously. Thus, every type of cognitive processes has their own dimensionality. Therefore, our conception, as a rule, is a product of composite perceptions blended together. It must be kept in mind that while our conventional perspective of cognition doesn't accept this postulate of each cognitive process having its own dimensionality, it accepts the composite nature of our perception of reality (Carruthers 2006).

Therefore, even within the conventional perspective, the composite nature of conception is implicit. However, the proposed model embellishes this composite nature of conception with the argument that its origin lies in different dimensionalities of the participating cognitive processes. Once we accept this additional argument, it is possible to resolve this paradox. Once we accept that our conception of reality is a view from within, it is intuitively clear that our conception of dimensionality and its concomitant complexity is an artifact of the dimensionalities from which our cognitive processes operate simultaneously. Thus, in addition to the fact that our conception of reality is composite, the fact remains that our conception is dependent on the dimensionality of our cognitive faculty.

This is where the concept of nested hierarchy comes into the picture. When we conceptualize, we create a projection by borrowing details from different dimensionalities. Therefore, let us assume that our sense of comprehension operates from a particular dimensionality. Therefore, whenever we conceptualize, we transfer different perceptions from different cognitive processes on the cognitive process responsible for comprehension. However, according to this model, this process of transferring different perceptions available in different cognitive processes to the cognitive process responsible for comprehension, involves changes in the dimensionalities from different cognitive processes sending their perceptions to the cognitive process responsible for comprehension. This change in the dimensionalities brings about the corresponding changes in the nature of perceptions being transferred.

Thus, while comprehending, we create a hierarchy of dimensionalities within the dimensionality of the cognitive process responsible for comprehension. Thus, within the cognitive process responsible for comprehension, now there exists a nested hierarchy. To borrow from topological vocabulary, there is a nested submanifold representing the composite conception of reality within the dimensionality of the cognitive process responsible for comprehension. It is the presence of dimensionalities within the dimensionality that gives rise to this paradox. Thus, when viewed from inside our cognitive faculty, we perceive the increased complexity and the resulting emergence principle. However, when viewed from outside an Archimedean point (which never exists) the proposed model offers a cogent way to define complexity.

Let us look at a nested hierarchy and how it gives rise to these two mutually incompatible propositions. Let us assume, at least temporarily, that our cognitive processes give rise to a composite perception of any given natural phenomenon. Therefore, according to this model, this composite perception would manifest itself in a particular dimensionality, say a dimensionality wherein our comprehension happens. Now, at this dimensionality of comprehension, our cognitive faculty will conjure up a multidimensional image of the natural phenomena under observation. Therefore, our conception of that natural phenomenon doesn't really reflect the dimensionality of the natural phenomena under observation, but it reflects the dimensionality which is a composite of the outputs of different cognitive processes that have contributed in creating this composite perception. Thus, the dimensionality which we perceive during our epistemological processing refers to the number and the type of cognitive processes involved. Thus, during the formalization of theories, our cognitive faculty creates a nested hierarchy of dimensionalities in the dimensionality wherein the process of comprehension operates.

There are two indirect proofs for this scenario. Firstly, while formalizing a scientific theory, we assign a certain dimensionality to the model proposed. Conventionally, we decide the dimensionality on two considerations. Firstly, it represents the physical dimensions of the natural phenomena under observation. Secondly, we decide the dimensionality of the model based on the number of parameters we wish to investigate. In reality, what we are doing is that we assign a dimensionality to a model based on the number of cognitive processes involved. This is because if each cognitive process were capable of handling one type of parameters, then the number of parameters that we choose to employ in formalizing, must refer to the number of cognitive processes involved. The second indirect evidence comes from our incorporation of time in our formal theories. Admittedly, we perceive time differently from the way we perceive space. However, we incorporate time as a parameter, not because we perceive time directly, but because we may have a separate cognitive process responsible for our perception of time. Thus, our ascription of dimensionality to different scientific models actually refers to the number and types of cognitive processes that give rise to the composite perception of a given natural phenomenon.

Let us take an example of spacetime to see how we arrive at our spacetime theories (Friedman 1983, see Chapters VI and VII). More importantly, how our perception of the dimensionality of these theories is inconsistent. Admittedly, since we perceive three spatial dimensions and one temporal dimension, our intuitive theories formalize four dimensional models. However, as we came to know more about the fundamental particles, we realized that there exist more parameters to be incorporated into the model of spacetime. Therefore, we arrived at the ten or eleven dimensionality model of spacetime in string theory (Conlon 2016). Thus, our higher dimensional models are decided by the number of parameters. Therefore, according to this model, it merely reflects the number of cognitive processes involved in the conception of spacetime.

One might wonder if such a scenario is at work, why should our scientific theories be so good in depicting the nature of reality? The answer according to this model is that our scientific theories really depict the nature of reality because the structuralism of the universe and that of our cognitive faculty are identical. Nature too places parameters in different dimensions and then blends them in different manners to give rise to different natural phenomena. Thus, an unanswerable problem of the conventional perspective about why we should be able to comprehend the nature of reality can now be answered by the proposed model. We can comprehend the nature of reality because our cognitive faculty shares the same structuralism with the spatio-temporal universe itself. Finally, let us look at the problem of complexity of life and why it leads to two diametrically opposite inferences.

As mentioned above, our conventional perspective leads us to the emergence principle which states that as the complexity of a system increases, newer functionalities emerge. Therefore, higher complexity is associated with higher dimensionality. However, the proposed model takes an opposite view. According to this model, higher complexity arises as more and more involutions are carried out. However, this results in the decrease in the dimensionality. Admittedly, both these positions appear to be irreconcilable. The resolution comes in the form of distinction between two definitions of dimensions. Our conventional perspective rests on the notion of dimensions as an embodiment of information content. This is because we are looking at the epistemological perspective of reality. However, according to the proposed model, the notion of dimensions must be taken in a physical sense. Therefore, when the spacetime dimensions get embedded into one another, that dimensionality gets reduced. Thus according to this model, as more and more dimensions of spacetime get involuted, the information content of the involuted manifold will become more and more complex. Therefore, according to this model, Life is complex because of the large number of dimensions that have been folded inward during biological evolution. Thus, it is possible to define the emergence principle in the proposed model. The increase in complexity gives rise to new functionalities, but this is due to the compression of the information content of spacetime into a lower dimensionality.

Now, using this reasoning, we can explain the contradiction between the conventional perspective and the proposed model about the relationship between complexity and dimensionality. Since our cognitive faculty operates from a particular range of dimensionalities, what we perceive is projections from different dimensionalities which are otherwise beyond the range of dimensionalities of our cognitive faculty. Therefore, structuralism of these otherwise inaccessible dimensionalities gets blended into the structuralism of the dimensionalities from which our cognitive faculty operates. Moreover, since our cognitive faculty is capable of epistemology, it converts the structural details of these projections from other dimensionalities into additional dimensions. This is primarily because our cognitive faculty converts every bit of information into a separate dimension. Thus, it is the projections from the dimensionalities beyond the range of dimensionalities of our cognitive faculty that increase the information content, thereby giving rise to perception of additional dimensionalities. The key point is that our cognitive faculty cannot perceive dimensions or different dimensionalities. It is capable of perceiving information content present in different dimensionalities. Moreover, since the range of dimensionalities from which our cognitive faculty operates is small, it tries to compress the information content of other dimensionalities into its own range of dimensionalities. It is this feature of information processing that gives rise to the perception that more information is equivalent to higher dimensionality.

Admittedly, what is described in the preceding paragraphs is difficult to comprehend. However, we will return to this paradox and its resolution in the following chapters. Presently, we will accept the scenario presented by the proposed model as provisionally valid. If the scenario outlined above is valid, then all that is required to prove the proposed model is to demonstrate that Life can be defined by using the proposed model. Therefore, in the following chapters, we will look at the genomic functionalities, structuralism and their relationships using this model. Presently, let us assume that the proposed model is provisionally true. Then, the best way to verify it will be to test it against the most enduring paradigm of biology, viz., natural selection. Therefore, in the remaining sections, we will try to deconstruct natural selection as implicit in the Darwinian paradigm and see how they compare.

1.14 Involuted Model of Natural Selection

As mentioned above, the best way to evaluate any new model is to test it against the principle of natural selection as implicit in the Darwinian paradigm. While the semantics of the Darwinian paradigm and the ontological primacy it enjoys among all the scientific theories is discussed elsewhere in this monograph, in this section, we will focus on three aspects. Firstly, we will look at the reasons why it is imperative to evaluate any new model against this paradigm. Secondly, we will discuss, albeit briefly, what are the semantic propositions of natural selection as implicit in the Darwinian paradigm. Finally, we will try to translate natural selection in the language of the proposed model. Having done that, in the following sections, we will look at the shortcomings of the conventional perspective of natural selection and how the proposed model offers a way to overcome these shortcomings.

Let us begin with the reason why we should use the Darwinian paradigm as a touchstone to evaluate a new model. Our current understanding of the Darwinian paradigm is that it seems to be the ultimate explanation of the origin and evolution of Life. This perception has arisen from the historical perspective. It is the ability of the Darwinian paradigm to withstand and to find new justifications for every subsequent paradigm shift that has convinced most of us that the Darwinian paradigm contains some fundamental insights into the nature of Life (Grene 1986). It is possible that in future, this paradigm might evolve into a better paradigm. However, the paradigm, as it is understood today, by itself, is an expression of semantics of Life. The subsequent paradigm shifts like genetics, population genetics, molecular biology and genomics, each of them, have enriched the Darwinian semantics. While we will discuss these semantic changes in later chapters, what is germane to the present discussion is that this ability to rejuvenate itself places the Darwinian paradigm in a preeminent position. Therefore, it makes sense to test every new hypothesis using the Darwinian paradigm as a yardstick. This predilection toward the Darwinian paradigm is universal and best summarized by Dobzanski (1982) that "Nothing makes sense in Biology, unless it is viewed from the Darwinian perspective."

Now let us look at the semantic propositions of the conventional perspective of the Darwinian paradigm. Once again, as mentioned above, we will look at this topic in a brief manner. This is because this topic is highly debated in literature and has acquired a very nuanced semantics. Therefore, here we will confine ourselves to the mainstream or conservative interpretation of the Darwinian semantics. To maintain brevity and simplicity, these semantic propositions are given below in a point-wise manner. Admittedly, it is possible to expand the list of these semantic propositions given below. However, we will take a minimalist approach and list only those propositions that are germane to the present discussion.

1. Natural selection, according to this paradigm, is essentially a process of competitive survival amidst finite resources.
2. The environment is a repository of all the resources necessary for survival of living organisms.
3. The ability to survive is passed on to subsequent generations by some fixed mechanisms.
4. Every organism exploits its inherited ability to survive, to outsmart its competitors and use the limited resources available in the environment for its individual survival and proliferation.
5. Inherited abilities of every organism find their expressions in accordance with the nature of the environment.
6. Environment enables and modifies this expression of inherited abilities.

These semantic propositions adequately describe the core concepts of natural selection as implicit in the Darwinian paradigm. How these propositions add up to give us semantics of the Darwinian paradigm will be discussed in the next section. Presently, let us see how these propositions can be formalized using the proposed model. Even without going into the details, it is intuitively clear that the environment with its various resources can be formalized as a hierarchy of different resources connected to one another. Similarly, one can think of different organisms, occupying the same ecological niche and competing for the resources available, as occupying different levels of this hierarchy.

For the present discussion, we will focus on three aspects of this hierarchical perspective. These aspects are the relationship between the inherited abilities and their expressions; the role of environment in these expressions and the role of environment in natural selection. Admittedly, one can also include the nature of competition among the competing species and the variable nature of the environment. Both these aspects are important in their own right, however, they have been extensively discussed and formalized in literature in the domain of population genetics (Provine 2001). More importantly, these aspects contribute significantly to the semantics of nondeterminism or randomness that characterizes the Darwinian paradigm. Therefore, they would be discussed separately. Presently, we are concerned about the mechanisms by which functionalities arise from the structural templates. Therefore, the three aspects mentioned above seem to be good topics to be deconstructed.

Therefore, let us begin with the first aspect of the relationship between the inherited abilities and their expressions. In fact, this relationship is synonymous with the relationship between structuralism and functionalities discussed above. Therefore, this relationship is of immediate concern for this discussion. Conventionally, it is intuitive to think of this relationship as a flow chart depicting structuralism giving rise to functionalities. Therefore, it is possible to visualize this relationship as a relationship between two levels of the hierarchy. Therefore, to an extent, we can construct a corresponding topological model linking structuralism with functionalities, each occupying a different dimensionality. In addition, the conventional perspective puts an additional restriction on the relationship between these two different dimensionalities. The conventional perspective insists that whatever the nature of this relationship may be, it must be one sided. Functionalities cannot shape structuralism. This restriction is to eliminate any Lamarckian model being surreptitiously incorporated. (Of course, since the discovery of epigenetic mechanisms, it is a matter of debate whether Lamarckian perspective must be rejected in toto. However, we will adhere to the classical interpretation of Darwinism in the present discussion (Grene 1986).

Therefore, prima facie, it is possible to visualize this relationship using the proposed model, by assigning two different dimensionalities to genotypes and phenotypes. However, as discussed in the following sections, there are some unexpected semantic consequences that force us to seek the reinterpretation of the conventional perspective of natural selection. Let us look at the second aspect of the role of the environment in defining this relationship between structuralism and functionalities. When viewed from the perspective of the proposed model, it is intuitively clear that these different dimensionalities of structuralism and functionalities must be embedded within the topological model of the environment. In other words, we can think of the relationship between structuralism and functionalities of any given organism as a submanifold present within the parent manifold of the environment. Here again, the conventional Darwinian theory imposes a restriction that the parent manifold itself doesn't participate in the interaction between different dimensionalities of any of its submanifolds. This is necessary because according to conventional wisdom, the environment doesn't actively influence the relationship between genotype and phenotype. Thus, prima facie, the conventional perspective of the role of environment in the relationship between genotype and phenotype can be represented in the proposed model.

However, here again, our current understanding of genotype and phenotype has changed considerably. As our understanding of the units of selection expanded (Okasha 2010), we have been forced to include genes, genomes, and groups into our conception of the units of selection. Therefore, the boundary between the conceptions of genotype and phenotype is getting blurred. Therefore, we require a flexible demarcation between these two categories. This is precisely what this model offers. A topological surface separating two domains allows us to be flexible. However, classical Darwinism, which insists on the defined distinction between genotype and phenotype, can also be adequately represented in the proposed model. Now, let us look at the third aspect of the role of the environment in natural selection. As mentioned above, the environment plays a passive role in natural selection. Of course, it provides resources for survival to all the competing species, it doesn't select any particular species by itself. The survival is a result of the availability of resources and the ability of the competing species to exploit these resources. Of course, since the environment itself changes, the process of natural selection also changes its direction. This phenomenon adds an additional degree of randomness to the inherent randomness arising due to the random nature of mutation of the genotype. From the topological perspective, it is possible to formalize the passive role of the environment in natural selection. However, it is not easy to formalize the random changes in either the environment or in genotype using a topological model.

Having looked at the brief description of the conventional perspective of natural selection as implicit in the Darwinian paradigm, let us try to conceptualize it using the proposed model. As mentioned above, it is intuitively clear that the environment itself can be formalized as a topological manifold with each of its resources being defined as a dimension. Similarly, it is possible to define individual members of the competing species as submanifolds present within the manifold representing the environment. The consumption of natural resources can be formalized as involutions arising from the parent manifold of the environment into submanifolds representing individuals.

Now let us see how the relationship between genotype and phenotype can be represented by this model. Admittedly, the conventional perspective on this relationship is that genotype giving rise to phenotype is a one way relationship. Therefore, this relationship can be defined as a mathematical operator connecting genotype with phenotype. However, using the arguments presented above, we can add another feature. We can place genotype and phenotype in different dimensionalities. Then, it is possible to define the relationship between genotype and phenotype as an operator of involution. Now let us define the role of the environment in shaping the relationship between genotype and phenotype. Admittedly, according to the conventional perspective, the environment doesn't play any active role in this. Therefore, prima facie, there is no need to define any operators to represent this. However, even in the conventional perspective, there is an indirect influence of the environment in triggering gene expressions. This feature can be defined as a pair of operators of involution, one representing the triggering mechanisms and the other representing the emergence of phenotype.

This brings us to the third aspect of the role of the environment in natural selection. Since the environment does directly play any role in natural selection, there is no need to define a separate operator to represent it. However, according to the conventional perspective, depletion of resources will alter the course of natural selection. This can be represented by this model. Since the proposed model employs a separate dimension for every resource present in the environment, the depletion of resources would alter the metric of the manifold representing the environment. In the worst case scenario, say, if one of the resources present in the environment is completely consumed, this can be represented by the reduction in dimensionality of the manifold representing the environment. This too can be formalized using the proposed model. In summary, the conventional perspective of natural selection as implicit in the Darwinian paradigm can be formalized using the proposed model. Therefore, in the next section, we will look at the conventional perspective of natural selection that is generated by this model.

1.15 Conventional Perspective of Natural Selection

The Darwinian paradigm, as mentioned above, has survived several paradigm shifts. During each of these transitions, semantics of the Darwinian paradigm has acquired newer nuances. Accordingly, literature on this subject is voluminous. It is therefore well-nigh impossible to summarize it here. Instead, we will select a few semantic ambiguities which have remained unresolved in the classical as well-modified interpretations of the Darwinian paradigm. Even here, we will focus on three aspects. They are a role of natural selection in the evolution of Life; the need for separate units of inheritance and selection and the origin of complexity. Admittedly, each of these topics has been extensively discussed in literature. Therefore, we will take them as having been read and take a fresh start.

Let us begin with the role of natural selection in biological evolution itself. Classical perspective of the Darwinian paradigm takes the evolution of Life as a priori and tries to explain morphological parallelism on the basis of natural selection. However, in more recent times, particularly when a hypothesis of RNA being progenitor of Life is gaining acceptance (Yarus 2010), it is logically consistent to extend the principle of natural selection to biological evolution itself. While this idea has a certain romantic appeal, the semantic ambiguities of natural selection raise their heads. This is primarily because according to the conventional perspective of natural selection, there are two prerequisites for the process of natural selection to manifest. First is the limited availability of resources and the second prerequisite is competition among different entities, each having different capabilities. However, in the case of emergence of the first living organisms, these prerequisites are not fulfilled. Even if one were to assume that RNA or rather different types of RNA proteins are competing species, the prerequisite of limited resources still remains unmet. Moreover, even the syntheses of different RNA proteins must also be governed by the same principle because each of these RNA proteins has certain functionalities and structuralism of their own. Such an argument leads us to logical regression. If we were to extend the principle of natural selection to include even the catalytic functions of molecules, it would amount to ascribing unwarranted properties to molecules. Therefore, it is imperative that we must resolve the semantic ambiguity about the role of natural selection in the evolution of Life.

At a more fundamental level, this problem of extending the principle of natural selection to the evolution of Life itself, points toward the second semantic ambiguity about the Darwinian paradigm. It points toward the ambiguity about the need to have separate units of inheritance and selection. From the semantic perspective, the Darwinian paradigm would have been simpler if Nature had employed a single unit of inheritance and selection. However, the fact remains that Nature has opted for this duality and therein lies its semantic ambiguity. For instance, if there was some definitive relationship between structuralism and functionalities even at the level of molecules, say, RNA proteins, then one would know why the Darwinian model works only on the living organisms. The real problem, as discussed in the preceding monograph (Chhaya 2020), is that there are reasons for us to believe that there exists a definitive relationship between structuralism and functionalities at every level of complexity. However, this relationship eludes formalization. Therefore, we have to assume that this relationship between structuralism and functionalities is different in the case of Life, and therefore, Life is unlike any other natural phenomena. Given this scenario, the conventional perspective of the Darwinian paradigm remains ambiguous about the relationship between genotype and phenotype on the one hand and between structuralism and functionalities on the other hand. More importantly, our current understanding of the semantics of the Darwinian paradigm seems incapable of offering any resolution of this ambiguity. However, the proposed model offers a way to remove this ambiguity.

Let us now look at the third semantic ambiguity of the origin of complexity during the course of biological evolution. Prima facie, there are two problems with this emergence of complexity. Firstly, this is contrary to our thermodynamic perspective (Haynie 2008) of natural phenomena. Every natural phenomenon, barring Life, tends to increase disorder and thereby entropy. However, Life is perhaps the only natural phenomenon which results in increasing order in Nature. Admittedly, the thermodynamics of the open system has been worked out elegantly by Prigorgine (1968), but the problem of complexity remains unresolved. This is particularly true about the second problem with biological complexity. This refers to the problem of the source of this complexity. To put it differently, the problem of the emergence of complexity during biological evolution is which kind of complexity is selected by natural selection. This is a far deeper problem because according to the conventional wisdom, natural selection employs functionalities to decide survival whereas the biological complexity rests on the structural template of the living organisms and not on their functionalities. Thus, the problem of the emergence of complexity during biological evolution is twofold. Firstly, why does a natural phenomenon work against the law of thermodynamics? Secondly, is the selection of complexity based on structuralism or on functionalities? Apparently, the conventional perspective cannot resolve this semantic ambiguity. Therefore, we need to reinterpret the Darwinian paradigm. However, before going into any such attempts, it is necessary to find out why the conventional perspective of the Darwinian paradigm cannot resolve these semantic ambiguities. Therefore, in the next section, we will look at the conventional perspective as represented in the proposed model and find out the reasons why the conventional perspective is inadequate by itself and needs to be replaced.

1.16 Why Natural Selection Cannot Be Rationalized by the Conventional Model

In the previous section, we looked at three semantic ambiguities of the Darwinian perspective of natural selection viz. The role of environment in biological evolution of Life; the need for separate units of inheritance and selection and the origin of complexity. It is possible to argue that these ambiguities are temporary and as our understanding of biological evolution increases, these ambiguities will be resolved. After all, that is how science progresses. However, it is the contention of this monograph that these ambiguities do not arise just from the lack of information about the mechanisms by which biological evolution has occurred. These ambiguities arise because of the inherent semantic propositions of the conventional perspective of the Darwinian paradigm. More importantly, these ambiguities represent a certain semantic lacuna in the conventional perspective. Therefore, in this section, we will try to deconstruct these ambiguities from the conventional perspective and demonstrate that there is something lacking in the conventional perspective.

Let us begin with the first ambiguity about the role of the environment in biological evolution as implicit in the Darwinian paradigm. As mentioned in the previous section, Darwinian theory of natural selection insists on two prerequisites of limited resources and inheritable capabilities to exploit these resources. Although we aren't sure where Life originated, it is possible to accept that such an environmental niche could possess limited resources necessary for biological evolution. Even if these resources were to be abundant initially, at least during and after biological evolution, these resources would eventually be in short supply. Therefore, the first prerequisite of Darwin's theory of natural selection would always be fulfilled in any scenario wherein evolution of Life occurs. However, the second prerequisite of inheritable capabilities of exploiting these limited resources for proliferation is problematic from the semantic perspective. Let us assume, albeit temporarily, that these competing species (either RNA proteins or any other progenitors) have the necessary structural template which makes them capable of exploiting these limited resources for their own proliferation.

The real problem is the relationship between structuralism and functionalities of these progenitors. To be precise, the problem is the exact nature of this relationship. On the one hand, our naturalistic paradigm of modern science suggests that functionalities cannot exist on their own even in the absence of any structural details. This is because it leads us to postulate some magical, if not divine explanation of these functionalities. Therefore, we are convinced that functionalities are consequences of structuralism and cannot exist without any structural foundations. On the other hand, we also know from our experience in catalysis and from immunochemistry that a single structural template can manifest multiple functionalities under the right environment. Therefore, the relationship between structuralism and functionalities is many to many. The same functionality can be manifested by different structural templates and a single structural template can manifest multiple functionalities.

Therefore, it is intuitively clear that the role of the environment in biological evolution is to select a particular type of relationships among these many to many types of relationships between structuralism and functionalities. In order to select one of these many relationships between structuralism and functionalities, the environment cannot play a passive role. This is because it can't select either structuralism or functionality in isolation. Since the environment selects the relationship and not the entities manifesting this relationship, it has to operate from a higher level. Moreover, since this type of natural selection cannot be passive since the details of these relationships at a higher level (from which environment carries out natural selection) will be more abstract than physical. It seems sensible to think of the role of the environment as some kind of mathematical operator which acts on abstract entities like mathematical objects. Since this kind of process demands congruence between the operator and the operand, it can never be thought of as a passive role. Thus, it is not possible to justify biological evolution on the basis of the Darwinian perspective of natural selection. The only way to do so would require a more active role of the environment in biological evolution. Thus, it is the semantic propositions of the Darwinian paradigm that prevents us from explaining biological evolution using the Darwinian paradigm.

Of course, it is possible to argue that maybe biological evolution is outside the purview of the Darwinian paradigm. However, this will push us back to pre-Darwinian times when theological beliefs were employed in formalizing the theories of evolution of Life. Therefore, it is better to modify the conventional perspective of natural selection than going back pre Darwinian biology. Incidentally, this reasoning leads us to the second semantic ambiguity about the need to have separate units of inheritance and selection. The conventional perspective on this topic is nebulous. While Darwin's own writings (Hodge and Redick 2009) show no distinction between genotype and phenotype, the later paradigms acknowledge the duality of genotype and phenotype, there is no semantic justifications available in these paradigms. In fact, the tendency is to employ the inherent random nature of mutations of genotype to explain the randomness necessary for natural selection. For instance, in population genetics (Provine 2001), we employ tools like genetic drift and near neutral rate of mutations to quantify the frequency of gene expressions. Similarly, in molecular biology, one employs mechanisms like changes in promoter or suppressor genes to explain mutations and the subsequent changes in gene expressions. However, these arguments never explain why the environment requires both genotype and phenotype, to bring about natural selection.

Given this scenario, it is axiomatic that the Darwinian perspective of natural selection cannot explain the need for separate units of inheritance and selection.

However, it has to be admitted that by accepting the possibility of variable boundary between genotype and phenotype, the conventional perspective does admit that there are some semantic justifications for this duality. However, it fails to deconstruct this semantic imperative. The question therefore arises whether the conventional perspective, by itself, can provide the necessary semantic propositions to explain the need for this duality or not? The answer to this question is no, at least not without changing the semantics of the conventional perspective.

Let us now look at the third ambiguity about the origin of complexity during the course of natural selection. There are two aspects of the problem of complexity. Firstly, whether the kind of complexity obtained by natural selection is different from the complexity obtained by different natural processes or not? Secondly, are there any mechanisms by which natural selection leads to complexity? Let us begin with the first question. It is possible to argue that almost all natural phenomena have different types of complexity. Therefore, why should we single out biological evolution and its attendant natural selection for investigating the origins of complexity of its outcomes? Admittedly, this point is valid. Given the right degree of precision, it is possible to demonstrate that the simplest of the natural phenomena is very complex. Therefore, why should we treat biological evolution as an exception?

The answer to this question lies in the fact that the kind of complexity that biological evolution has given rise to is different from the nature of complexity manifest in any other natural phenomena. This difference between biological complexity and any other natural complexity is both qualitative and quantitative. The biological complexity gives rise to functionalities (particularly the functionality of self-reference) that is not manifest in any other natural phenomena. Secondly, the degree of complexity in living organisms is unmatched by any other type of complexity present in all the natural phenomena. Take an example of the architecture of the brain. The number of neurons present in the brain is humongous. Moreover, the number of synaptic junctions connecting these neurons is larger by several powers of ten than the number of neurons. We can compare this number with the number of atoms present in the spatiotemporal universe. The magnitude of these numbers is comparable. However, the atoms present in the spatiotemporal universe are not connected to one another, but the neurons are. This gives us an idea of the degree of complexity evolved during biological evolution.

It is possible to argue that maybe the degree of complexity of living organisms is exceptional, but qualitatively biological complexity could still be similar to the complexity present in different natural phenomena. However, this argument is fallacious because the functionalities of Life are not present in any other natural phenomena. As mentioned above, the cognitive functionality of self-reference is an obvious example of it. In addition, the functionality of reproduction with retention of complexity is not manifest anywhere else. There are natural phenomena which can recur, but none can pass on the information from one generation to the next. It is as if Life has a unique feature of information retention and transfer that is not manifest in any other natural phenomena. More importantly, this feature arises from the way the information is organized in the living organisms. Therefore, it seems reasonable to think that the nature of complexity, by itself, distinguishes Life from other natural phenomena. Therefore, the emergence of complexity has to be explained by the theories of biological evolution and since the Darwinian paradigm is the only theory that we have, it is imperative that this explanation must be an integral part of Darwinian theory and its semantics.

This brings us to the second question whether any such expectations are available in the Darwinian paradigm. The only explanation that is available from the Darwinian paradigm is an argument of historicism. The emergence of complexity during biological evolution is often explained by pointing out that the earliest living organisms were simplest in the structural terms. Therefore, it is inevitable that during natural selection, the successive generations will be more and more complex, if only to endow better survival mechanisms. Admittedly, this argument is very intuitive and can hardly be refuted. However, the problem with this reasoning is that it fails to explain why only certain functionalities have evolved and not any of the other functionalities. Nature, for instance, could have evolved living organisms which never age or which never die. Apparently, this hasn't happened. More importantly, there must be some reason why only some of all the possible functionalities have evolved. The reason why only some functionalities and not others have evolved must have something to do with the mechanism by which natural selection operates. However, the Darwinian paradigm refuses to permit us to formalize any such mechanisms. This is primarily because the Darwinian paradigm is founded upon nondeterminism and randomness (Bonner 2013). Therefore, by allowing any such mechanisms, it would undermine its own foundation.

Historically, this emphasis on randomness in explaining natural selection has arisen from the tendency to avoid any theist or teleological arguments. Therefore, it is quite natural, in fact, inevitable that the Darwinian paradigm must adhere to nondeterministic doctrine. The question however, is whether there is any way by which one can postulate a mechanism that explains the emergence of complexity during biological evolution, without appealing to implicit or explicit theism or teleology? In the next section, we will look at this possibility.

1.17 Need to Reinterpret Darwinian Paradigm

As discussed in the previous section, the conventional perspective of natural selection as implicit in the Darwinian paradigm cannot explain three features of Life. These are the role of the environment in the relationship between genotype and phenotype, the need for separate units of inheritance and selection and the origin of complexity during biological evolution and its attendant natural selection. It was pointed out that these three features of Life distinguish Life from other natural phenomena. Therefore, Life as a natural phenomenon is unique, and therefore, the Darwinian paradigm must provide a naturalistic explanation for these three features. However, since the Darwinian paradigm is founded upon nondeterministic arguments, it cannot invoke any mechanisms to explain these features. Therefore, it is imperative that we must redefine the Darwinian paradigm de novo.

The solution to the dilemma of creating a mechanistic perspective of these features while maintaining nondeterminism lies in redefining what constitutes a naturalistic explanation. Let us assume, albeit temporarily, that there exists a mechanism which requires the duality of genotype and phenotype and which gives rise to complexity in the living organisms. Such a putative mechanism will naturally be amenable to formalization using the mathematical constructs. However, the problem with this reasoning is that if such a mechanism and its formal description exist, then it must be transcendental in nature. Therefore, such a mechanism cannot be a naturalistic explanation!

In fact, this kind of fallacy exists in every formalism of science. Even when one employs a statistical model for population genetics (Provine 2001), the use of a mathematical formalism doesn't compromise the underlying naturalism of biological evolution. As discussed in the preceding monograph (Chhaya 2022a), it is intuitively clear that modern science has forsaken, albeit for pragmatic reasons, the troublesome question of why should any mathematical constructs be able to formalize a given natural phenomenon. This is because on the one hand we concede that the mathematical constructs are immutable, if not transcendental entities. Therefore, every time we invoke mathematical constructs to define a natural phenomenon, we are accepting the nebulous ontology of mathematical objects. Of course, we conveniently place mathematical objects in the category of a priori status, thereby eliminating any ontological explanation. Therefore, if a putative mechanism mentioned above were to exist, we could still use it to explain these three features.

This preamble is necessary because there exists a natural phenomenon which manifests the same degree of duality and complexity. This refers to mathematics itself. However, the problem is that we can't accept mathematics as a natural phenomenon. It is considered to be a priori. However, as described in the preceding monograph, it is possible to argue that mathematics can be taken as a natural phenomenon provided we redefine the Cartesian split. A topological model of the cosmic singularity was outlined in the preceding monograph. Using that model, it can be demonstrated that the mathematical objects reside within the spatiotemporal universe itself, albeit in higher dimensions. In fact, the proposed model postulates that spacetime has a unique metric at each of its dimensionality. Moreover, spacetime exists in multiple dimensionalities simultaneously. Therefore, the metric of any given natural phenomenon depends on the dimensionality in which it manifests. Moreover, our conception of the metric of any given natural phenomenon depends on two factors, viz., the dimensionality of the natural phenomena and the dimensionality from which we make an observation. This is because not only the natural phenomena, but our cognitive faculty too possesses different heterogeneous cognitive processes, with each such cognitive process operating from its characteristic dimensionality.

Therefore, whenever there is a mismatch between the dimensionality of the natural phenomena under observation and the dimensionality of our cognitive processes, what we perceive is the projections from one dimensionality to another. More importantly, the proposed model offers a formalism that connects different metrics of different dimensionalities using a single class of operators of involution.

The details of this model are discussed in the preceding monograph. In this section, we will merely demonstrate how this model offers a mechanism for explaining these three features. More importantly, it doesn't undermine the nondeterministic foundation of the Darwinian paradigm. Therefore, let us look at how these three features can be explained by the proposed model. Admittedly, we will briefly summarize these arguments. A more nuanced explanation will be discussed in the final chapter.

Let us begin with the formal description of an ecosystem. According to this model, since structuralism and functionalities must have different templates, their respective metrics too must be different. Therefore, structuralism and functionalities must be assigned different dimensionalities. Thus, according to this model, genotype and phenotype must occupy different dimensionalities. Moreover, as mentioned above, the genotype and phenotype of any given organism must be represented as a submanifold within the parent manifold representing the environment. Now, according to this model, any influence of the environment must be represented by an operator of involution. Thus, in principle, the influence of the environment must be passed on to the submanifold representing individual organisms and then to both genotype and phenotype. Moreover, the relationship between genotype and phenotype too must be represented by an operator of involution. Between genotype and phenotype, it is intuitively clear that the template of functionalities must possess relatively coarser "granularity" than that of genotype. This is because from the four dimensional perspective, structuralism of genotype appears to consist of molecules and atoms. On the other hand, the template for functionalities appears to be a continuum (cf. Almost infinite capacity for our immune response (Inman 2012).). Therefore, according to this model, phenotype will occupy a higher dimensionality as compared to its corresponding genotype. This is a basic sketch of the formal description of the relationship between the environment, genotype and phenotype.

Within this framework, let us see how we can represent these three questions. Let us begin with the influence of the environment on the relationship between genotype and phenotype. Admittedly, this model doesn't deny the possibility of the environment influencing both these domains of genotype and phenotype. More controversially, it doesn't prevent the possibility of a phenotype influencing its own genotype. Admittedly, while the first possibility is consistent with the conventional perspective of natural selection as implicit in the Darwinian paradigm, it is yet to be understood in its totality. On the other hand, the second possibility of phenotype influencing genotype is antithetical to Darwinian semantics. It points toward the Lamarckian model which has been rejected long ago. Therefore, let us look at this representation in some more detail.

Let us begin with the influence of the environment on genotype and phenotype. The fact that genotype is getting altered by the environmental impact is a mainstream notion. Of course, our current understanding of the Darwinian paradigm, postulates that this influence is indirectly incorporated in the form of mutations. Therefore, prima facie, this is not a problem. However, the environmental impact of the phenotype is slightly problematic. There are two possible types of environmental impacts. Firstly, it could in the form of the impact of the environment on the process of gene expressions. Once again, this is a mainstream idea. More so if we include influence of the remaining genes of the parent genome influencing a given gene expression. In this case, the boundary between genotype and the environment shifts toward the boundary between genome and the individual genes. In other words, the genome becomes the environment and an individual gene becomes a genotype. While this is consistent with our current understanding of the Darwinian paradigm, the influence of the environment on the phenotype is indirect.

However, there is a distinct possibility, according to this model, that the environment directly influences the phenotype. Admittedly, this is a problem. If one were to think of phenotype as a physically manifest feature, the environmental impact is a routine process which comes about due to wear and tear during the lifespan of an individual organism. However, if phenotype were to be thought of as a functionality, then it is problematic. This is because functionalities are abstract and therefore immutable. However, if we distinguish between phenotype and its underlying functionality, it is possible to reconcile the proposed model with the conventional perspective of the Darwinian paradigm. However, according to this model, we can think of functionalities as phenotypes in an abstract sense. In that case, we need to think of the environment in its abstract sense and replace it with the nature of reality. Thus, the proposed model is consistent with the conventional perspective of the Darwinian paradigm. However, the proposed model seeks to extend the Darwinian paradigm to abstract entities like functionalities and the nature of reality. Whether this extension is valid or not needs to be justified. We will attempt to do so in later chapters.

Now let us look at the second question of the need for separate units of inheritance and selection. As mentioned above, the conventional perspective of the Darwinian paradigm is silent on the need for separate units of inheritance and selection. Of course, it is possible to argue that since Darwin himself didn't include genetics explicitly in his discourse, it wasn't necessary for him to address this problem. However, it is also possible to construe his opposition to Lamarckian model as his implicit endorsement of the duality of the units of inheritance and selection. This is because Lamarckian theory rests on the direct influence from the environment to the phenotype and then to genotype. Thus, it seems reasonable to think that Darwin himself was aware of this duality of the units of inheritance and selection, but didn't dwell on it.

Irrespective of the reasons why the classical Darwinian theory is silent on this issue, what is germane to the present discussion is whether there is any implicit explanation for this duality. Apparently, as mentioned above, the separation of units of selection from the units of inheritance provides a ground to reject any teleological explanations for natural selection. However, apart from this predilection toward naturalism, Darwin's theory must possess a structural justification for the need for separate units of inheritance and selection. It ought to provide some mechanisms which necessarily require both these units as prerequisites. Apparently, this is not the case with the Darwinian paradigm. In fact, it is one of the peculiarities of the Darwinian paradigm, as a scientific hypothesis, that it lacks any mechanistic details. As a result, Darwin's theory has almost no capacity to offer any predictions. A scientific theory must possess two features, viz., an ability to explain and an ability to predict. However, Darwin's theory is remarkable in the sense that it has an exceptional power to explain, but hardly any theoretical tools to predict the future course of biological evolution.

Conventionally, it has been argued that this lack of mechanisms is due to the inherent randomness of biological evolution. Therefore, the only predictive method that can be used in biological evolution is that of statistical modeling. This belief is justified by the success of population genetics. However, this is a category mistake. To the extent the influence of the environment on the genotype is a causal influence, there ought to be some mechanistic details of natural selection. As a corollary, if there are some causal mechanisms, there ought to be some predictive power available in the theory. Thc statistical analysis may obviate the need to study causal explanations, but it doesn't replace causal explanations. It is in this context that we must examine the structural imperative of this duality of units of inheritance and selection. Moreover, once we accept this scenario, it is intuitively clear that this structural imperative will find its echo in the very semantics of the Darwinian paradigm. Therefore, even if we don't know the structural details of natural selection, its semantics ought to provide clues for these structural details.

Therefore, let us see whether the proposed model offers anything new on this topic. As mentioned above, according to this model, all the three entities, viz., the environment, genotype, and phenotype, occupy different dimensionalities. Therefore, any justification for the need for separate units of inheritance and selection must come from the differences among the dimensionalities of these three entities. Let us see how this happens. If there was no need to have separate units of inheritance and selection, natural selection should have occurred by the direct interactions between the units of Life (which would possess the features of genotype and phenotype) and the environment. This is essentially a teleological theory wherein the environment, as some kind of superior force, directs or orchestrates the evolution of Life and its subsequent natural selection. In the classical theories of Life, prior to the advent of the Darwinian paradigm, this scenario was articulated, depending on the individual's own religious beliefs.

This scenario has one fatal flaw. It can't explain the morphological diversity of living organisms. If the environment were to decide the course of biological evolution, there would be no need to select among different forms of living organisms. This is because the environment would have all the details of living organisms because these organisms are integral to the environment. Thus, in such a scenario, whatever form of living organisms are produced would be in harmony with the environment. Thus, in such an approach, the sense of identity between the environment and the living organisms is complete. There is no separation between Life and Nature. Upon a little reflection, it is intuitively clear that this is nothing but a religious doctrine! Nature replaces omniscient and omnipresent God.

It is in this context of quasi-religious explanation of the origin of Life that we can see why Darwin's theory is unique. The Darwinian paradigm doesn't merely separate Divine from mundane, but it also separates Nature from individual members of Life. Nature is not the sum of all the living organisms. It has something more to contribute to biological evolution. If one thinks of Life as more organized arrangements of diverse and different molecules, then the role of the environment is not just to provide these constituent molecules, but it is to provide the rules of organizing these molecules. This is the true meaning of the ecosystem. All forms of biota, together, form the environment. More importantly, this hierarchy of different resources gives rise to different features of the environment. Thus, the conventional perspective of the Darwinian paradigm can be reconfigured from the information theoretical perspective. When we do that, it becomes clear that the need for separate units of inheritance and selection arises because the information content of the units of inheritance is different from the information content of the units of selection. The units of selection (phenotype) contain the kind of information that is amenable to direct influence from the environment. As we will discuss in later chapters, it is possible to define the interactions among the environment, genotype and phenotype using a hierarchy of dimensionalities within which each of these three entities can have different types of interactions among themselves. More importantly, this model explains the need for separate units of inheritance and selection.

Now let us look at the third question about the origins of complexity during biological evolution. In this context, there are two aspects on which the conventional perspective of the Darwinian paradigm is silent. Firstly, there is a question of the emergence of complexity during biological evolution. Secondly, there is a question of the kind of complexity that has been selected during the course of biological evolution. In the context of the present discussion, the question arises whether natural selection, per se, has a certain preference for any particular type of complexity or the choice of the types of complexities selected is random, just like the emergence of complexity itself? As mentioned above, the conventional perspective justifies its agnosticism about the emergence of complexity by citing the inherent randomness of the process of natural selection. It suggests that since the earliest living organisms possessed simple structure templates by default, it is axiomatic that the process of natural selection will lead to more and more complex living organisms. Moreover, statistically, the chances of the earliest living organisms being complex are extremely rare. Therefore, there is no need to explain the emergence of complexity during biological evolution.

However, this argument is fallacious. The probabilistic models that one normally employs are based on thermodynamics. Surprisingly, in thermodynamics, complexity doesn't arise naturally. It requires a certain amount of work or energy expenditures and increase in entropy. Therefore, no natural phenomena, except Life, lead to the emergence of complexity. Therefore, the real problem of explaining the emergence of complexity is not how simple living organisms give rise to complex living organisms during the course of biological evolution. The real problem is how did the earliest living organisms evolve from less complex raw materials. The conventional perspective of the Darwinian paradigm justifies the emergence of living organisms out of nonliving raw materials on the basis of catalysis. The employment of catalysis eliminates or minimizes the energy expenditure in creating more complex structures, living as well as nonliving.

This is where the second question of which types of complexities are selected comes into the picture. If Life originated purely as an outcome of some catalysis, then we would have witnessed different types of living organisms having different structuralism. However, with the advent of molecular biology techniques, we are certain that all living organisms, extinct or not, have a common ancestor, and therefore, they have one structural template from which a diverse range of morphology arises. If the catalytic explanation of the origin of Life is true, then we should have witnessed different types of living organisms having different types of genes, different types of genomic architecture and most importantly, different genetic code. However, we have never come across any organisms having different genetic code or different types of biochemistry. Therefore, it seems reasonable to think that biological evolution has come about because natural selection has certain inbuilt preferences for certain structural complexity. Therefore, our present agnosticism about the origins of complexity during biological evolution is merely a placeholder for our ignorance.

The reason why we adhere to this ontological agnosticism about complexity is our mistaken belief that any structural explanation of the origin of complexity during biological evolution will compromise the fundamental nondeterminism of the Darwinian paradigm. This is nothing but a category mistake. It is possible to formalize ontology of complexity in biological evolution and its attendant natural selection without any fear of the resurrection of teleological interpretations. The ontology of complexity in biological evolution can be naturally explained if we invoke mathematical complexity. If we could demonstrate that mathematical complexity has a certain ontological perspective, then we don't have to worry about teleology, theological or otherwise. However, the problem with this argument is that we don't have any naturalistic ontology of mathematical objects. However, as discussed in the preceding monographs, a new topological model of naturalistic ontology of mathematical objects is now available. Therefore, in the following chapters, we try to understand the origins of genomic complexity in the form of its architecture using this model. We will try to deconstruct the genomic architecture and demonstrate that it is the inherent mathematical template that distinguishes Life from other natural phenomena. We have covered a wide range of topics in these sections. Therefore, we will summarize these arguments in the concluding section.

1.18 Conclusion

In this concluding section, we will try to summarize different arguments presented above. Since the preceding sections covered diverse topics, we will summarize these arguments in a point-wise manner.

1. In spite of our belief that Life is a natural phenomenon, we have not been able to formalize it. Therefore, it is necessary to deconstruct the reason why Life is unlike any other natural phenomena.
2. Every natural phenomenon possesses its own structuralism and functionalities. Moreover, there is a definitive relationship between structuralism and functionalities of each of these natural phenomena. In fact, this common

relationship between structuralism and functionalities provides a foundation for the formal scientific theories. Since we employ mathematical constructs to formalize scientific theories, the underlying mathematical formalism paves the way for the unified theory of science.

3. Therefore, if Life seems unlike any other natural phenomena, its explanation must lie in the nature of the relationship between structuralism and functionalities of Life. Therefore, it is necessary to deconstruct the relationship between structuralism and functionalities of Life vis a vis that of other natural phenomena.
4. Therefore, it was suggested that we should postulate that functionalities have their own templates which are different from those of the corresponding structuralism. In such a scenario, the relationship between structuralism and functionalities can be defined as a mathematical operation connecting two different metrics, one each for structuralism and functionalities.
5. It was suggested that since our cognitive faculty operates from several heterogeneous cognitive processes simultaneously, it is possible that some of the metrics remain beyond the capabilities of our cognitive processing.
6. However, if we can formalize the mathematical operators defining the relationship between structuralism and functionalities, as a special class of operators, then we can obtain a glimpse of metrics which are otherwise beyond our cognitive capabilities.
7. This approach enables us to expand our formal description of natural phenomena which have defied our attempts to formalize them.
8. Using this rationale, a distinction between Life and other natural phenomena was sought to be articulated in this chapter. It was argued that barring two natural phenomena of quantum phenomena and Life, the remaining natural phenomena seem to possess structuralism and functionalities whose templates fall within the range of our cognitive processing capabilities.
9. In such cases, the metric of functionalities is adequately represented by the mathematical constructs amenable to cognition; we are able to formalize these natural phenomena.
10. Since the templates of functionalities of Life remain beyond our cognition, we are not able to formalize Life. However, the approach outlined, which defines a new class of mathematical operators, provides a way to formalize functionalities of Life from its structuralism, thereby helping us to understand hitherto unknown features of the origins of Life, its natural selection and the origin of complexity.
11. This opens up a new opportunity to reinterpret the Darwinian paradigm. As a first step, it was demonstrated in the preceding sections that some of the fundamental semantic ambiguities of the Darwinian paradigm can't be resolved without making any additional postulates. The proposed approach provides these additional semantic propositions necessary for explaining these ambiguities.
12. Therefore, this approach provides a tool to deconstruct and reinterpret the Darwinian paradigm.

13. In following chapters, we will try to articulate functionalities and structuralism of genomes using this model.
14. We will try to deconstruct structuralism of genomes using the modified involuted manifold model. Since this model is already successfully used in formalizing other natural phenomena, any reasonable success of defining the structuralism of genomes would demonstrate that Life is like any other natural phenomena.
15. Using the special class of mathematical operators, we will try to predict hitherto unknown templates of functionalities.

References

Albers RW, Price DL (2012) Basic neurochemistry principles of molecular, cellular and medical neurobiology. Elsevier, Amsterdam

Alcock J (2001) The triumph of sociobiology. Oxford University Press, Oxford

Alon U (2006) An introduction to systems biology: design principles of biological circuits. CRC Press, Boca Raton

Arbib MA (2003) The handbook of brain theory and neural networks. MIT Press, Cambridge

Ayer AJ (ed) (1959) Logical positivism. Simon and Schuster, New York

Barth FG, Giampieri-Deutsch P, Klein H (eds) (2012) Sensory perception: mind and matter. Springer, Dordrecht

Bell M, Gao S (2016) Quantum nonlocality and reality: fifty years of Bell's theorem. Cambridge University Press, Cambridge

BonJour L (2010) Epistemology: classic problems and contemporary responses. Rowman and Littlefield, Lanham

Bonner JT (2013) Randomness in evolution. Princeton University Press, Princeton

Borrello ME (2010) Evolutionary restraints: the contentious history of group selection. University of Chicago, Chicago

Brendon GE (1993) Topology and geometry. Springer, Dordrecht

Caldarelli G (2007) Scale free networks: complex webs in nature and technology. Oxford University Press, Oxford

Capuzzo-Dolcetta RA (2019) Classical Newtonian gravity: a comprehensive introduction. Springer, Dordrecht

Carruthers P (2006) The architecture of mind. Oxford University Press, Oxford

Chhaya P (2020) The universe within: and a view from that within. Universal Press, Irvine

Chhaya P (2022a) The origin and nature of mathematics. (Submitted)

Chhaya P (2022b) The origin and nature of spacetime. (Submitted)

Chhaya P (2022c) The origin of the counterintuitive nature of quantum phenomena. (Submitted)

Chhaya P (2022d) The origin and nature of cognition. (To be Submitted)

Chhaya P (2022e) A new paradigm for therapeutic genomics. (To be Submitted)

Conlon J (2016) Why string theory? CRC Press, Boca Raton

Cottingham J (2008) Cartesian reflections. Oxford University Press, Oxford

Dawkins R (1999) The selfish genes. Oxford University Press, Oxford

Delisle RG (ed) (2021) Natural selection: revisiting explanatory role in evolutionary biology. Springer, Dordrecht

DeVillers J, Balaban AT (eds) (1999) Topological indices and related descriptors in QSAR and QSPR. CRC Press, Boca Raton

DeVos J, Pluth E (eds) (2015) Neuroscience and critique: exploring the limits of neurological Thursday. Routledge, London
DeWitt BS, Graham N (eds) (1973) The many-worlds interpretation of quantum mechanics. Princeton University Press, Princeton
Dobzanski T (1982) Genetics and origin of species. Columbia University Press, New York
Donahue JW, Dorsel JW (1997) Neural network model of cognition: behavioral foundations. North Holland Publications, Amsterdam
Dragini TA (ed) (1998) Human polygenic diseases. Harwood Academic Press, Amsterdam
Eden H, Moor JH, Soraken JH, Steinhart E (2013) Singularity hypothesis: a scientific and philosophical assessment. Springer, Dordrecht
Flew A (2017) Darwinian evolution. Routledge, London
Friedman M (1983) Foundation of spacetime theories: relativistic physics and philosophy of science. Princeton University Press, Princeton
Gao S (2017) The meaning of the wave function: in search of the ontology of quantum mechanics. Cambridge University Press, Cambridge
Garett RA, Klenk HP (eds) (2007) Archaea: evolution, physiology and molecular biology. Blackwell Science, Oxford
Glaeser G, Paulus HF (2015) The evolution of eye. Springer, Dordrecht
Grene M (1986) Dimensions of Darwinism: themes and counter themes in the twentieth century evolutionary theory. Cambridge University Press, Cambridge
Gribbin J (2009) Deep simplicity: chaos, complexity and the emergence of life. Penguin Press, London
Haynie DT (2008) Biological thermodynamics. Cambridge University Press, Cambridge
Herrlich H (2006) Axiom of choice. Springer, Dordrecht
Heyes C, Huber L (eds) (2000) The evolution of cognition. MIT Press, Cambridge
Hodge J, Redick G (eds) (2009) Cambridge companion to Darwin. Cambridge University Press, Cambridge
Inman F (2012) Contemporary topics in immunochemistry, vol 1. Springer, Dordrecht
Laudal OA (2021) Mathematical models in science. World Scientific, Singapore
Lawrie ID (2013) A unified grand tour of theoretical physics. CRC Press, Boca Raton
Lovelock J (2000) Gaia: a new look at life on earth. Oxford University Press, Oxford
Minski M (1988) The society of mind. Simon and Schuster, New York
Montani G, Battisti MV, Benini R, Imponente G (2011) Primordial cosmology. World Scientific, Singapore
Nielsen MA, Chuang IL (2010) Quantum computations and quantum information. Cambridge University Press, Cambridge
Okasha S (2010) Evolution and the levels of selection. Oxford University Press, Oxford
Panza M, Sereni A (2013) Plato's problem: an introduction to mathematical Platonism. Springer, Dordrecht
Parker A, Wilding EL, Bussey TJ (eds) (2002) The cognitive neuroscience of memory: encoding and retrieval. Psychology Press, Sussex
Pevsner J (2015) Bioinformatics and functional genomics. Wiley, Hoboken
Popper K (1963) Conjectures and refutations: the growth of scientific knowledge. Routledge, London
Prakash S (ed) (2007) Artificial cells: cell engineering and theory. CRC Press, Boca Raton
Prigorgine I (1968) Introduction to thermodynamics of irreversible processes. John Wiley and Sons, Hoboken
Provine WB (2001) The origins of theoretical population genetics. University of Chicago, Chicago
Raffaelli DG, Frid CLJ (eds) (2010) Ecosystem ecology: a new synthesis. Cambridge University Press, Cambridge
Ramsay A (1991) Formal methods in artificial intelligence. Cambridge University Press, Cambridge

Randall L (2006) Warped passages: unraveling the universe's hidden dimensions. Ecco Press, New York
Riehl E (2017) Category theory in context. Dover Publications, Mineola
Roy K (2015) Quantitative structure activity relationships in drug design, predictive toxicology and risk assessment. IGI Global Publications, Hershey
Roy S, Kundu T (2021) Chemical biology of the genome. Academic Press, London
Rubinstein J, Rakic P (eds) (2013) Cellular migration and formation of neuronal connections. Academic Press, San Diego
Schlosshauer MA (2007) Decoherence and the quantum to classical transition. Springer, Dordrecht
Schutz BF (2009) A first course in general relativity. Cambridge University Press, Cambridge
Shmulevich I, Dougherty ER (2014) Genomic signal processing. Princeton University Press, Princeton
Smith E, Morowitz HJ (2016) The origin and nature of life on earth: the emergence of the fourth geosphere. Cambridge University Press, Cambridge
Susskind L, Lindesay J (2005) String theory revolution. World Scientific, Singapore
Szabo A, Ostlund NS (1989) Modern quantum chemistry: an introduction to advanced electronic structure theory. Dover Publications, Mineola
Teoh WY, Urakawa A, Ng YH, Sit P (2021) Heterogeneous catalysts: advanced design, characterization, vol I. Wiley VCH, Weinheim
Wechsler H (ed) (1992) Neural networks. Academic Press, San Diego
Wieder S (2012) The foundations of quantum theory. Elsevier, Amsterdam
Wolstenholme GEW, O'Connor CM (2009) Neurological basis of behavior. Wiley, Hoboken
Yarus M (2010) Life from an RNA world: the ancestor within. Harvard University Press, Cambridge

2 Nature of Relationship Between a Genotype and Phenotype

Abstract

In the previous chapter, a new formal representation of life based on the duality of structuralism and the functionalities was described. In continuation with that approach, this chapter will explore the exact relationships between the biological structures and their functionalities, in the context of the Darwinian paradigm. It is suggested that the corresponding duality of genotype and phenotype holds the secret of biological evolution. The Darwinian paradigm is silent on the need to have different sets of units of inheritance and the units of natural selection. Using the formalism of the involuted manifold, this chapter would seek to reinterpret the notions of a genotype and a phenotype. It is argued that by redefining these notions, it is possible to explain the emergence of complexity during biological evolution. More importantly, this model ensures that this emergence of complexity does not introduce any form of determinism or teleology.

2.1 Introduction

In the preceding chapter, a formal description of life was developed using a topological model of involuted manifolds. One of the objectives of that model was to understand the relationship between structural templates of living organisms and functionalities of these templates. It was suggested that it is the lack of clarity about this relationship between structuralism and the functionalities that characterizes the biological sciences, including the Darwinian paradigm. In this chapter, we will explore this formalization of the relationship between structuralism and the functionalities to deconstruct the Darwinian paradigm (Grene 1986), particularly the relationship between genotype and phenotype. Purely from the naturalistic perspective, there is no rationale for the need for separate entities of phenotypes and genotypes. The question of why the units of inheritance must be distinct from the

© The Author(s), under exclusive license to Springer Nature Singapore Pte Ltd. 2023
P. Chhaya, *The Topological Model of Genome and Evolution*,
https://doi.org/10.1007/978-981-99-4318-0_2

units of selection remains unanswered (Okasha 2010, see Chapter 2). In fact, Darwinian theory takes this distinction as a priori. In addition, the emergence of complexity during the course of evolution has not been satisfactorily settled. It is the contention of this reinterpretation of the Darwinian paradigm that the emergence of complexity (Smith and Morowitz 2016, see Chapter 4) can only be explained by deconstruction of the relationship between genotype and phenotype. Therefore, we need to qualify this assertion by valid inferences from the explication of the exact nature of the relationship between genotype and phenotype.

Before we look at the nature of relationships between genotype and phenotype, it is necessary to understand the semantic implications of the need to have units of inheritance separate from the units of selection. This duality encompasses within itself some hidden meaning of the Darwinian paradigm. This is evident from the fact that there are several such dualities within the Darwinian paradigm. It is reasonable to think that all such dualities, viz., the dualities of genotype and phenotype, DNA and RNA, and structuralism and functionalities of living organisms, embody this hitherto unknown semantic content which needs to be articulated. In order to tease out this dormant semantics, we will employ a topological model described in the previous chapter. Thus, we will try to define genotype and phenotype using the proposed model. Having done that, we will try to deconstruct the nature of relationships between the two. Using this as a basis, we will try to generalize the nature of all such dualities. Using this generalized template of duality, we will look at the nature of natural selection. Having defined natural selection, we will look at possible explanations for the emergence of complexity during natural selection.

For the sake of linearity, this chapter is further divided into 17 sections, even though the issues involved in natural selection are intricately connected with one another. Section 2.2: Formal Description of the Relationship Between Structuralism and Functionalities, Sect. 2.3: Comparison with the Relationship Between Genotype and Phenotype, Sect. 2.4: Generic Description of Dualities, Sect. 2.5: Significance of Genotype and Phenotype, Sect. 2.6: Conventional Perspective of the Relationship Between Genotype and Phenotype, Sect. 2.7: Semantic Ambiguities of the Conventional Perspective, Sect. 2.8: Origins of Semantic Ambiguities, Sect. 2.9: Can We Redefine Genotype and Phenotype Within the Darwinian Paradigm?, Sect. 2.10: Problem of Complexity in the Darwinian Paradigm, Sect. 2.11: Semantics of Complexity in Biological Evolution, Sect. 2.12: Why Do We Need to Reinterpret Darwinian Theory?, Sect. 2.13: Emergence of Complexity in the Involuted Model, Sect. 2.14: Redefining Genotype and Phenotype, Sect. 2.15: New Explanation of the Emergence of Complexity, Sect. 2.16: Role of Genotype and Phenotype in the Emergence of Complexity, Sect. 2.17: Semantics of Dualities, Sect. 2.18: Conclusion.

2.2 Formal Description of the Relationship Between Structuralism and Functionalities

Modern science is primarily concerned with formal descriptions of natural phenomena which can help us to understand the nature of these phenomena and to predict their outcomes. Thus, formal description of a natural phenomenon has a certain mathematical template which has inbuilt semantic propositions and certain rules of transformations. Thus, it is these inherent features of a given mathematical template that endows the formal description of a natural phenomenon with its explanatory and predictive powers. Admittedly, this protocol demands that the semantic propositions and the rules of transformations of any mathematical template are available to us prior to our formalization of a given natural phenomenon. However, in reality, this is not the case. When we choose one mathematical template over another, we do so not because we are fully aware of the semantics and structuralism of the mathematical formalisms under consideration. We decide on one particular mathematical template because we intuitively feel it to be suitable for the given natural phenomenon. In other words, we make our choice not because we know, but we choose because we intuitively feel the correctness of our choice.

This preamble is necessary because there exists a parallelism between the relationship of semantic propositions and the rules of transformations of a mathematical formalism on the one hand and the relationship between structuralism and functionalities of any given natural phenomenon. Therefore, when we formalize a theory to explain a given natural phenomenon, we intuitively feel a certain degree of congruence between these two patterns of the mathematical constructs and the semantics of their transformations on one the hand and the structuralism and functionalities of the given natural phenomenon on the other hand. Moreover, since this congruence is partial sometimes we find that the resulting theory is falsifiable under certain conditions (Popper 1963).

As discussed in the previous chapter, Life is unlike any other natural phenomena because there is no way to formalize the relationship between the functionalities and structuralism of Life (Quantum phenomenon is another such example.) Therefore, in the previous chapter, a new topological model for formalizing the relationship between structuralism and functionalities was briefly described. In this section, we will summarize the gist of the proposed model. Having done that, in the following sections, we employ this model to understand the nature of the relationship between genotype and phenotype. For the sake of brevity, we will describe the proposed model in a point-wise manner here.

1. Historically, we have assumed that the functionalities of any natural phenomenon are predictable from the structural template of that phenomenon. In the process, we have overlooked the possibility that the functionalities of any given natural phenomenon can have their own template.
2. In most cases, it is possible to overlook the distinction between the template of functionalities and structuralism because there exists a direct correspondence between these two templates.

3. However, in the case of Life and quantum phenomena, there exists an unbridgeable gap between the structuralism and functionalities of these two phenomena.
4. Due to our historical bias of equating the templates of functionalities with the template of structuralism, we have never sought to define the template of functionalities as distinct from the template of structuralism of Life and quantum phenomena.
5. This has led to a general consensus that Life remains difficult to formalize. While this problem in the case of quantum phenomena has been discussed in the preceding monograph (Chhaya 2022c, see Chapter 8), this monograph is focused on the difficulties in formalizing Life as a natural phenomenon.
6. In order to distinguish the template of functionalities from the template of structuralism of all the natural phenomena, a new paradigm has been discussed in the preceding monographs.
7. The proposed model postulates that the metric of any template can be defined in a topological formalism. Moreover, according to this model, every dimensionality has its own metric. Therefore, different types of metrics can be connected to one another by formalizing them in a single involuted manifold.
8. In such an involuted manifold, each dimensionality is connected to the remaining dimensionalities by the operator of involution.
9. Thus, it is possible to convert different metrics into one another by defining a set of operators of involution. Therefore, it is possible to formalize any given natural phenomenon having different templates using this model. For instance, our cognitive faculty possesses different types of cognitive processes, each having different functionalities. These different functionalities can be represented by the proposed model. Moreover, as discussed elsewhere (Chhaya 2022d, see Chapter 4), it is possible to integrate different cognitive functionalities operating in different dimensionalities to create composite perception.
10. As a logical corollary, the proposed model can be used to define the templates of functionalities and structuralism of any given natural phenomenon in a single structure. In this model, we can demonstrate that the template of functionalities of a given natural phenomenon can be defined separately from its structural template in the same formal structure by assigning different dimensionalities to the structuralism and functionalities of the given natural phenomenon.
11. Once this model is formalized, it is possible to define the exact nature of the relationship between the structuralism and functionalities of a given natural phenomenon by defining a proper set of operators of involution.
12. Using this model, it was suggested that when the operators of involution defining the relationship between structuralism and functionalities are not available, we find a mismatch between the structuralism and functionalities of those natural phenomena. Life and quantum phenomena are the prime examples of this.
13. This mismatch arises because the difference between the dimensionality of structuralism and the dimensionality of functionalities is large, thereby requiring

more than one set of operators to formalize the relationship between structuralism and functionalities of these natural phenomena.
14. It was suggested that the relationship between the structuralism and functionalities of Life belong to this category.

Using this rationale, we will try to deconstruct the relationship between genotype and phenotype in the next section.

2.3 Comparison with the Relationship Between Genotype and Phenotype

In the previous section, we looked at the relationship between structuralism and functionalities of a typical natural phenomenon. Therefore, it is natural to expect that the relationship between the unit of structuralism, viz., genotype, and the unit of functionality, viz., phenotype, must conform to this typical description of the relationship between structuralism and functionalities. However, as discussed in the first chapter, Life, as a natural phenomenon, doesn't seem to obey this universal description of the relationship between structuralism and functionalities. Therefore, in this section, we will try to deconstruct the conventional perspective of the relationship between genotype and phenotype. We will try to understand how this relationship is different from other instances found in Nature. In the following sections, we will try to reinterpret the relationship between genotype and phenotype based on this deconstruction.

Let us begin with the conventional perspective of the Darwinian paradigm about this relationship. Historically, it is not clear whether Darwin was aware of Mendel's pioneering work on genetics. However, the key point is that Darwin anticipated it. Therefore, what emerged in later decades in this domain must be taken as a correct interpretation of Darwin's theory. Therefore, we will assume that the conventional perspective of the relationship between genotype and phenotype embodies Darwinian paradigm's implicit semantics. As mentioned in the first chapter, Darwin's theory (Hodge and Radick 2009) accepts the need for genotype and phenotype as separate units of inheritance and selection as a priori. Therefore, Darwin's theory doesn't need to justify this duality. However, the theory does contain some implicit semantic propositions which seem to justify the need for the separate units of inheritance and selection. Therefore, let us deconstruct these semantic propositions.

The notion of descent is central to the Darwinian paradigm and so is the notion of descent with modifications. Similarly, the idea of natural selection occupies a center of the Darwinian paradigm. At the same time, the principle of natural selection seems to be applicable to the phenotypes in the classical interpretation of the Darwinian paradigm. Admittedly, there is no mention of the relationship between genotype and phenotype, at least explicitly, in Darwin's own writings. However, in order to avoid the teleological arguments, Darwin's theory is categorically explicit that natural selection operates only on the phenotype. As the Darwinian paradigm progressed from Darwin's own interpretation of natural selection to the new synthesis (Delisle 2021, see Part II) which integrated Mendelian genetics with Darwin's theory, the relationship between genotype and phenotype was clearly shown to be unidirectional. The genotype would give rise to the phenotype, but the phenotype will have no influence on the genotype. This unidirectional influence was necessary to exclude any Lamarckian interpretation of natural selection (Steele et al. 1998).

It is important to note that the fact that Mendelian genetics rests on the discrete units of inheritance, the semantic ambiguities about what constitutes a genotype do not arise. Subsequent development of population genetics successfully provided a mathematical foundation for the randomness implicit in Darwin's theory (Provine 2001, see Chapter 5). In fact, it appears retrospectively that the triad of classical Darwinian theory, Mendelian genetics, and population genetics, collectively, articulated what is now described as the conventional perspective of the Darwinian paradigm. We will sidestep the enormous impact that this perspective has on modern science and instead focus on the specific topic of the relationship between genotype and phenotype. In the next paradigm shift that molecular biology ushered (Nei 2013, see Chapter 4), the rigidity of definition of what constitutes a genotype and phenotype started eroding. Moreover, in parallel, population genetics too moved toward the articulation of normal and near normal rates of mutations (Kimura 1983, see Chapter 3). As our concepts of genetic drift began acquiring more sophistication, the relationship between genotype and phenotype started becoming more and more nuanced. As a result, we are forced to question the need for having separate units of inheritance and selection. Surprisingly, any explanation of this duality lies in the nature of natural selection. Therefore, it has become imperative that we must redefine the Darwinian paradigm.

At first sight, it is not obvious that the explanation for the need for separate units of inheritance and selection lies in natural selection. This is because there exists a semantic gap between the random nature of natural selection (as manifest in the statistical formulation of population genetics) and the causal processes of conversion of genotype into phenotype (as exemplified by molecular biology). The fear that any such causal explanation will take us back to some form of teleology has prevented us from exploring the semantic gap between randomness implicit in Darwin's theory (Bonner 2013) and causality implicit in molecular biology (Nei 2013). The conventional defense of our reluctance to explore this semantic lacuna is that molecular events operate at the level of individual nucleotides, whereas there are billions of such nucleotides in an average-sized genome. Therefore, overall changes in genotype (and therefore in phenotype) are bound to be indeterminate. In addition, the changes in the environment are also random. Therefore, it is inevitable that the final outcomes of natural selection are bound to be random and therefore beyond the scope of any causal explanation. It has to be admitted that this rationale is eminently justified and backed by rigorous statistical analysis.

However, there is another side of natural selection which has been overlooked. The edifice of the statistical analysis rests on the plurality of discrete units. Therefore, when applied to individual nucleotides this scenario is justified. However, in the post genomic era, we have come to realize that the genome is not a simple aggregation of individual genes. Rather, the genome has deeper linkages among its constituent genes. More importantly, it is due to the inherent nature of genomic architecture that gives rise to some functionalities (Pevsner 2015). Thus, as we move away from discrete units in the form of the individual genes to larger and larger structural units of genomes, it is intuitively clear that we will have to define a less random paradigm.

To understand this, let us take an example of the common usage of languages. In English, there are 26 alphabets. Therefore, according to any standard statistical model, there are an infinite number of words that can be generated from these 26 alphabets. However, as we know, semantic and phonetic rules ensure that only a small fraction of all the possible words are permissible. Therefore, we abandon purely statistical models and instead incorporate deeper level structural templates of phonetics and semantics to introduce a certain minimal level of predictivity. In fact, we do the same thing in evolutionary biology when we employ phylogenetic models (Bromham 2008, see Chapter 5). We accept some deeper level connectivities to reduce the degree of randomness.

Therefore, mere introduction of causality into the Darwinian paradigm need not undermine the essential semantic proposition of randomness implicit in Darwin's theory. In the postgenomic era, we have realized that the distinction between genotype and phenotype are not ironclad. The boundary between them shifts, depending on the genomic context. Therefore, there must be a generic template of the relationship between genotype and phenotype which remains unchanged even when we shift the boundary between the two. It is this context free template of the relationship between genotype and phenotype that will help us formalize Life as a natural phenomenon. Before we articulate this relationship, let us see how this relationship is unlike any other relationship between structuralism and functionalities in the rest of the natural phenomenon.

2.4 Generic Description of Dualities

As discussed in the first chapter, the relationship between structuralism and functionalities of Life, as manifest in the form of genotype and phenotype respectively, is unlike the corresponding relationships in other natural phenomena. Therefore, in this section we will look at the generic description of the relationship between structuralism and functionalities of natural phenomena other than Life. Prima facie, in the sense that in the absence of any knowledge of the relationship between genotype and phenotype, it is intuitively clear that this relationship is different from other such relationships between structuralism and functionalities, because of the role played by the environment. In any other natural phenomena, structuralism and functionalities are the instantiations of the environment itself (with the notable exception of quantum phenomena). It is as if the environment undergoes structural changes, which, in turn, gives rise to different functionalities of the environment. The environment, itself, plays the role of structuralism and functionalities. The environment as a separate entity does not influence the relationship between structuralism and functionalities. This is because structuralism and functionalities of a natural phenomenon are the constituents of the environment itself.

Therefore, we define the relationship between structuralism and functionalities of a typical natural phenomenon, and we are merely describing the changes in the environment itself. This is obviously true in the case of terrestrial natural phenomena that we closely monitor for weather. For instance, when we try to monitor melting of polar ice caps or the size of the ozone hole, there exists a structuralism behind these phenomena which we try to define by using different models. In these cases, the environment, as a separate entity, does not influence the size of ice caps or ozone holes. Rather these features are integral facets of the environment. More importantly, this is also true for man-made innovations like sending a rocket to the Moon. There is a certain structuralism of spacetime in the form of gravitational force and there is a powerful thrust to escape the effects of the gravitational force. In this case, gravity and the combustion of fuel in the rocket are integral parts of the environment. The environment cannot be separated from natural phenomena.

However, this is not the case with living organisms. Life seems to possess a greater degree of autonomy from the environment it operates in. Once generated, the structuralism of Life, acts on its own and sometimes against the Laws of Nature. It is this sense of autonomy that separates Life from other natural phenomena. It is intuitively clear that this very conception of autonomy is a kind of functionality that the peculiar structuralism endows on the living organisms. Therefore, prima facie, there is something unique about the structuralism of Life that gives rise to the distinction between Life and other natural phenomena.

In addition to this unique structuralism of Life, there is another feature of natural phenomena that we need to investigate. As mentioned above, living organisms seem to operate in a manner that is contrary to the Laws of Nature. This unique feature points toward another important facet of natural phenomena. All the natural phenomena (except Life and quantum phenomena) are the instantiations of the Laws of Nature. In other words, the changes in the environment are governed by the Laws of Nature and different natural phenomena are nothing but different algorithms embodying different Laws of Nature. This feature is in congruence with our earlier assertion that structuralism and functionalities of natural phenomena are an integral part of the environment. Therefore, there is no demarcation between structuralism and functionalities of natural phenomena on the one hand and the environment on the other hand. They both are constituents of the environment.

As discussed in the first chapter, it is possible to argue that if this is true then there is no distinction between structuralism and functionalities of any natural phenomenon. They are simply different configurations of the underlying environment. This is self-evidently true. However, when we think from an epistemological perspective, the trouble begins. When we try to formalize structuralism of any given phenomena, we employ mathematics. Obviously, our understanding of mathematics depends on our epistemological capabilities. As we know from our experience in the development of modern science, we use a trial and error method to arrive at the structuralism of the natural phenomena under investigation. On the other hand, we arrive at the functionalities of the natural phenomena under investigation by using the properties of the mathematical constructs we have deployed in formalizing the structuralism of that natural phenomenon. Therefore, there exists a hitherto unknown relationship between mathematics and natural phenomena which enables us to formalize different functionalities from a given structuralism of the natural phenomena. It is as if mathematics connects the structural template and the corresponding functional template of every natural phenomenon. In view of our earlier assertion that structuralism and functionalities of every natural phenomenon are integral part of the environment itself, it is intuitively clear that the environment is nothing more than a repository of all the mathematical constructs. Therefore, different types of algorithms instantiate different natural phenomena by picking up different types of mathematical constructs and applying different rules of transforming these mathematical constructs.

In addition, the fact that we can use mathematical templates of the structuralism of natural phenomena to infer the functionalities of any given natural phenomenon points toward a certain definitive relationship between the structuralism and functionalities. It is true that, as mentioned above, that mathematical formalism connects with the natural phenomena, and therefore, it is possible to infer the functionalities from the structuralism of the natural phenomena. However, this will be possible only if the structuralism and functionalities are also directly connected to one another. Moreover, as discussed in the first chapter, we often don't perceive the template of functionalities, and therefore, we base our estimate of functionalities from the template of structuralism which is available to us through our epistemological processes.

Therefore, the same method should work in the case of the functionalities of Life. However, it is intuitively clear that this is not the case. We cannot formalize our cognitive functionalities on the basis of our neurological model of the brain (DeVos and Pluth 2015, see Part I). Similarly, we cannot define functionalities of the genome from its DNA sequence (Pevsner 2015). Therefore, the relationship between structuralism and functionalities is not a straightforward equation. It must be more nuanced than simple rules of transformations. Therefore, it seems reasonable to think that the kind of mathematical operators required to connect the structuralism of Life to its functionalities must be of a different type. Thus, we will be justified in claiming that the relationship between structuralism and functionalities in a typical natural phenomenon can be formalized using the standard mathematical operators. However, in the case of Life, these operators are inadequate and therefore, we need a different type of mathematical operator to connect the functionalities of Life with its structuralism. With this reasoning, it is time to evaluate the relationship between genotype and phenotype. This is important because perhaps by redefining this relationship, one can eliminate the semantic ambiguities that are manifest in the Darwinian paradigm.

2.5 Significance of Genotype and Phenotype

Before we look at the conventional perspective of the relationship between genotype and phenotype and its inherent semantic ambiguities, it is important to understand why this relationship is important to the Darwinian paradigm. As mentioned above, there is no clarity on whether Darwin was aware of Mendel's pioneering work. However, irrespective of this, what is significant is that Darwin's writings anticipated genetics. Therefore, it is interesting to find out what kind of relationships between genotype and phenotype were implicit in the classical interpretation of Darwin's theory (Hodge and Radick 2009). The central theme of Darwin's theory is best summarized in the phrase "descent with modifications." Moreover, since Darwin was categorically against any kind of teleology, including the Lamarckian theory (Steele et al. 1998), he implicitly believed in the random nature of modifications. Moreover, his theory essentially argues on the basis of comparative morphology, it is intuitively clear that he implicitly endorsed genetic inheritance of morphological features. The key point of Darwin's theory (which is actually a stroke of genius) is competitive survival. Nature would produce a variety of morphological features. However, natural selection ensures that only those features that help an organism to survive would continue to manifest. Thus, Darwin separated the morphological variations from the process of natural selection. These variations occur even in the absence of natural selection. Natural selection merely sifts through these morphological variations to allow only those variations which help the organisms to survive. Thus, in the classical interpretation of Darwin's theory, the process of generating morphological variations is separated from the process of natural selection.

In the contemporary jargon, according to classical interpretation of Darwin's theory, genotype and its variations are causally separate from natural selection. Therefore, by implication, natural selection operates only on phenotypes and not on the factors giving rise to phenotypes, i.e., genotype. Admittedly, as discussed in the first chapter, this has led to semantic ambiguities about the need for separate units of inheritance and selection. If we were to ask a rhetorical question about why can't natural selection operate directly on genotype? There is no way to answer it. It is tempting to think that this duality of genotype and phenotype could be one of Nature's quirks. However, this feature of operating through such dualities is a structural prerequisite for natural selection. The duality is not confined to genotype and phenotype. It extends to DNA and RNA and structuralism and functionalities of Life. Therefore, it seems reasonable to think that this prerequisite of operating through such dualities hides some deeper semantics of natural selection. Before we delve further, let us see how this relationship between genotype and phenotype is conceptualized in the later paradigms of genetics, population genetics, and molecular biology. Of course this duality also manifests in the domain of genomics. However, we will discuss this aspect in later chapters.

Admittedly, the genetic paradigm was the first articulation of the discrete units of inheritance. Of course, it is true that it was the fortuitous choice of the plant and its inherently easily identifiable characteristics by Mendel that laid the foundation of genetics. However, from the semantic perspective, this choice reinforced the discrete nature of the units of inheritance in the form of genes. However, this obvious inference of the discrete nature of genes was in congruence with the reductionist approach of science and therefore occupied the center stage in the semantics of genetics. It was only when our knowledge of genomics and the associated long-range influences increased that we had to abandon this structuralism definition of genes as the discrete units of inheritance. It is important to understand that genetics, as a scientific discipline, is silent on the mechanism by which genotype gives rise to phenotype. Moreover, genetics also influenced our notion of phenotypes also being discrete entities.

Both these imputations of discreteness on genotype and phenotype continued in study of population genetics (Provine 2001). It is important to note that the problem with this inference of discreteness in genotype and phenotype is not that it is wrong. In fact, the success of population genetics lies in quantifying these notions. The problem is that it obviates the need for having any alternative explanation. There are several long-range influences of the genome which suggest that discreteness of either inheritance or selection is not fundamental to either biological evolution or to natural selection. However, due to inherent structuralism of statistical analysis, population genetics could not deconstruct the nondiscrete nature of inheritance. It was only with the advent of molecular biology that we could understand the finer nuances of gene expressions (Weinzierl 1999, see Chapter 2) and the importance of large scale structuralism of the genome (Pevsner 2015).

Surprisingly, the notion of discreteness is overhauled in molecular biology. There are discrete units, but the boundaries of this discreteness are variable. A single gene can give rise to multiple proteins by splicing the RNA molecule in more than one way (Jeanteur 2003). Similarly, during the course of evolution, a gene might undergo structural changes and acquire different phenotypic expressions. Still more strangely, a redundant copy of a gene may acquire a different expression in the form of exaptation (Gould 2002, see Part II). Thus, in molecular biology, the very notion of discreteness turns out to be a phenomenology of the genome. Therefore, it is legitimate to question the conventional perspective of the relationship between genotype and phenotype. More importantly, it is legitimate to question the need for separate units of inheritance and selection.

It is intuitively clear from the above discussion that the semantics of genotype and phenotype have undergone a paradigm shift. However, this is not exactly reflected in our conception of natural selection. As mentioned in the first chapter, the most striking feature of Darwin's theory is its ability to endure paradigm shifts and rediscover its semantics. There is something inherently fundamentally true about Darwinian semantics. However, it is yet to be articulated. As discussed above, the notions of genotype and phenotype have changed considerably since their original conception. However, there has never been any need to abandon these notions. Semantics of dualities, of which genotype and phenotype is just one example, has a certain resilience and that needs to be articulated. There is something more to biological evolution and natural selection that has escaped our attention. It is implicit semantics that needs to be articulated. Before we attempt such an articulation, let us look at the conventional perspective of the relationship between genotype and phenotype. In the next section, we will articulate the relationship between genotype and phenotype as implicit in the conventional perspective. Having done that, in the following sections we try to deconstruct the semantic ambiguities behind this conventional perspective.

2.6 Conventional Perspective of the Relationship Between Genotype and Phenotype

In the preceding sections, we discussed the historical perspective of the relationship between genotype and phenotype. We looked at some of the semantic content of the conventional perspective of this relationship. In this section, we will look at the structural template of the relationship between genotype and phenotype. This is necessary because this conventional perspective of the structural template of this relationship is at variance with the semantic perspective outlined above as well as the one outlined in the first chapter. Once we have outlined the conventional structuralism of this relationship, we will be in a position to realize the extent of mismatch between semantics and structuralism of the relationship between genotype and phenotype. While outlining this relationship, we will take an integrated approach as it has emerged over these transitions from Darwin's own interpretation of natural selection to the genetic perspective of "New Synthesis" (Delisle 2021); to molecular biology (Nei 2013); to genomics (Pevsner 2015). The reason for this approach is that while each paradigm has given an increasingly more nuanced description of the relationship between genotype and phenotype, there is a certain semantic continuity in all these paradigms. Therefore, there exists a definitive structural congruence among these paradigms. Therefore, if we wish to reinterpret the Darwinian paradigm, it is necessary to deconstruct the common structuralism of the relationship between genotype and phenotype. Having done that, in the following sections, we will look at the mismatch between this common structuralism and the conventional semantics of the Darwinian paradigm.

Let us begin with the processes that enable genotype to give rise to phenotype. It is intuitively clear, at least in the contemporary context, that these processes must be the processes directly responsible for the gene expression or indirectly responsible for initiating the gene expression. More importantly, these processes must be chemical and catalytic in nature. This insistence on these processes being chemical reactions is significant because it implicitly provides a semantic proposition that both genotype and phenotype are essentially chemical entities. This equivalence between genotype and phenotype of being chemical entities is important because of two reasons. Firstly, it eliminates any transcendental entities like "Vital Force" from biological evolution and natural selection. Biological evolution and natural selection are essentially chemical processes only. Therefore, the parameters that influence chemical reactions will also influence the relationship between genotype and phenotype. Admittedly, we need to be cautious about this chemical paradigm because it amounts to reductionism. At the same time, the chemical paradigm allows us to understand the nature of randomness implicit in Darwin's theory. If a given chemical reaction has a certain thermodynamic profile, its kinetics, its free energy and energy barrier between genotype and phenotype, would give rise to a certain degree of randomness in the manifestation of a phenotype from a given genotype.

At the next level of the structural template of the relationship between genotype and phenotype, there is a cascade of such processes. This is true not only because multiple genes undergo simultaneous expressions, but also because there are multiple agents operating on a given single gene expression. Therefore, this nested architecture of gene expressions adds another degree of randomness. The voluminous literature on polygeny and pleiotropy is a testimony of this structural and functional parallelism. It is tempting to think that this description of the relationship between genotype and phenotype (Dragini 1998) adheres to the classical Darwinian belief that the relationship between genotype and phenotype is unidirectional. Phenotypes, as a rule, don't influence genotypes. However, our experience of molecular biology indicates that this nested hierarchy of gene expressions also contains a chain of gene expressions. This creates a situation wherein a genotype in one case of gene expression becomes a phenotype from the perspective of another gene expression. Thus, the boundary between genotype and phenotype gets blurred. Moreover, the very definition of the unit of selection can include from a single gene expression product to a network of several genes, genome, individual members of a species and even a group of individuals from a single species. More importantly, the semantics behind each of these definitions of the unit of selection is different. Therefore, there exist a considerable amount of ambiguities about the distinction between genotype and phenotype and the exact nature of the relationship between genotype and phenotype.

Even the earlier belief that the relationship between genotype and phenotype is unidirectional, which was the cornerstone of Darwin's theory, has been diluted. Admittedly, the Lamarckian interpretation of natural selection still remains an anathema to the principle of natural selection. However, epigenetic phenomena, including genetic imprinting (Robert 2004), do require that we re-examine the relationship between genotype and phenotype. Earlier, the arrow of causality was straightforward, from genotype to phenotype and from the environment to phenotype. Thanks to recent advances in genomics and molecular biology, the causal relationships seem to exist between all the three entities, genotype, phenotype and the environment. Therefore, we need to redraw the boundaries between genotype, phenotype and the environment. More importantly, these boundaries are not static, but rather context dependent. Therefore, instead of adhering to the conventional definition of what constitutes a genotype or phenotype or the environment, it is time to define the semantics of genotype as a source of information, phenotype as a reorganized information content and the environment as a transformer of information content. Once we develop such a semantic definition of genotype, phenotype, and the environment, we will be able to formalize Darwin's theory in a domain-independent manner, thereby enlarging the scope of the Darwinian paradigm. More importantly, we will be able to articulate the semantic ambiguities of the conventional Darwinian paradigm which have remained beyond resolution. Before we develop such information theoretical semantics of natural selection, let us look at the semantic ambiguities of the Darwinian paradigm.

2.7 Semantic Ambiguities of the Conventional Perspective

Without going into the historical perspective of the Darwinian paradigm and the emergence of the semantic ambiguities during several paradigm shifts that Darwin's theory has undergone, we will directly enumerate and summarize them. Prima facie, there are three main types of ambiguities of the Darwinian paradigm. They are the need for the dualities in the form of genotype/phenotype, DNA/RNA and structuralism/functionalities; the origin of complexity during biological evolution and emergence of design in fundamentally random phenomena. In this section, we will briefly outline the gist of these ambiguities because each of these ambiguities has been the subject matter of long standing academic debates. Therefore, instead of summarizing the literature of these debates, we will define these ambiguities in a simple nontechnical language. At the outset, it must be admitted that these three types of ambiguities are linked to one another and perhaps they are manifestations of a more fundamental single semantic proposition that has not been articulated. However, for the moment, we will treat these three ambiguities as separate and discuss them individually.

Let us begin with the first semantic ambiguity about the need for having dualities in the form of genotype/phenotype, DNA /RNA and structuralism/functionalities. It is somewhat easier to deconstruct the notion of dualities because we intuitively feel that they are connected with one another. More importantly, the relationship between each of these pairs must be similar. Admittedly, this is all the more true in the case of dualities of genotype/phenotype and structuralism/functionalities because there is semantic parallelism in both these pairs. Our conventional perspective of genotype is that of a structural unit and that of phenotype as a functional unit. Therefore, we can substitute one pair with another. We can conceptualize structuralism as a structure of a gene and a functionality as a phenotypic feature. Thus, these two types of dualities look synonymous, except for the scope of their definitions. The notion of structuralism is not confined to the structure of genotype. Rather, genotype is a type of structuralism. There could be structuralism beyond the conception of genotype. The same is true for the relationship between functionalities and phenotype. There are functionalities other than those defined as phenotypes. In other words, the duality of genotype/phenotype is just one example of a more general definition of the duality of structuralism/functionalities.

If this reasoning is true, then why should we treat the duality of genotype/phenotype as a separate category different from the category of structuralism/functionalities? The reason why we should treat genotype/phenotype separately is that this duality behaves differently from other instances of the duality of structuralism/functionalities. There are instances of the duality of structuralism/functionalities within the living organisms and outside, which do not behave like the duality of genotype/phenotype. The relationship between genotype and phenotype is characterized by one unique feature not manifest in any other instances of the duality of structuralism/functionalities. This feature is that of encapsulation. Phenotype is encapsulated within genotype. Purely from the perspective of information theory, the information content of phenotype is embedded in genotype. What is provided from outside is material that constitutes phenotype. The information about how to organize this material into phenotype resides within genotype.

It is possible to argue that after all, both genotype and phenotype are made up of molecules, so how could genotype encapsulate phenotype because the material necessary to construct phenotype comes from outside. Moreover, as we know from our experience in developmental biology that the environment, in the form of physiological conditions, influences gene expression (Kurzfield 2009). Actually, these are two different arguments. Therefore, we will look at them separately. The answer to the first objection is simple. We are making a category mistake when we think of genotype as nothing more than DNA sequence. In reality, genotype is the information hidden in the form of the sequence of nucleotides and the sequence of nucleotides itself. The process of gene expressions is not a simple chemical reaction of transcription. The gene expression also requires the information of where and when to carry out transcription. Therefore, the correct way to conceptualize genotype is not to think in terms of individual genes, but in terms of genomic architecture. This feature of encapsulation of phenotype in genotype is not manifest in any other instances of the duality of structuralism/functionalities. To confirm this, just think of any other instances of the duality of structuralism/functionalities, say, a computer chip. There is a structural template of circuits, albeit very complicated. However, that template does not encapsulate the functionalities of processing. It is true that the functionalities of a computer chip arise from the structural template of the circuit, but it is not encapsulated in the circuit. The functionality manifests itself when an electric current is passed through the circuit. In the absence of an electric current, the functionalities of a computer chip do not manifest. It doesn't manifest because it is not encapsulated in the circuit.

Now, let us look at the second objection of physiological conditions influencing the gene expressions. It is a valid objection because if the physiological conditions influence the relationship between genotype and phenotype (and we know from our experience that it does (Kurzfield 2009)), then to claim that genotype encapsulates phenotype, cannot be sustained. In fact, if one includes the physiological conditions as a part of the environment, then one can expand the classical interpretation of natural selection to include the causal role of the environment in shaping the course of evolution. The contemporary interpretation of natural selection has assimilated this scenario. It defines four types of such influences, viz., heterotopy, heterometry, heterotypy, and heterochrony. However, there is a subtle distinction between these two approaches. As mentioned above, the information content of genotype doesn't simply consist of a sequence of nucleotides. The information content of genotype also consists of its placement within the genome. Therefore, the physiological conditions and the four types of influences mentioned above are still part of the information content of genotype. However, in this case, the genotype consists of a genome and not an individual gene.

This points toward two important inferences. Firstly, it is possible to stick to the classical interpretation of Darwin's theory if one is willing to alter the definition of genotype and phenotype. In other words, the semantics of Darwin's theory can be retained by changing the semantics of the terms genotype and phenotype. Alternatively, we can retain the semantics of the terms genotype and phenotype, but only if we are willing to change the semantics of Darwin's theory. The second inference available from the above discussion is that there exists a more fundamental context-independent semantics of natural selection and the classical interpretation of Darwin's theory and our current semantics of natural selection are variants of that fundamental principle. Thus, the semantic ambiguities about what constitutes a genotype and phenotype and what constitutes natural selection are contextual. However, to arrive at the context-free semantics of the Darwinian paradigm, it is imperative that we develop a new paradigm of semantics. The proposed model offers one such approach which relies on the information theoretical semantics and a topological representation.

Now, let us turn to the second ambiguity of emergence of complexity during biological evolution. It must be admitted that the third ambiguity about the emergence of design in an essentially random phenomenon of natural selection is synonymous with the second ambiguity about complexity. However, this synonymy is only partial. Therefore, for the sake of simplicity, we will treat these two ambiguities as separate phenomena. The emergence of complexity during biological evolution remains enigmatic. As mentioned above and in the first chapter, it is possible to argue that since the first living organisms were simple, the process of natural selection would have produced more and more complex living organisms by default. This is essentially a phase space argument. If the earliest living organisms occupied the region of the phase space representing least complex structuralism, then axiomatically, the successive species will be forced to occupy the region of the phase space representing more and more complex structuralism. Thus, the emergence of complexity during biological evolution is inevitable.

However, the trouble with this reasoning is that it is never manifest in any other natural phenomena. In fact, the second law of thermodynamics excludes such a possibility. It is true that when the theory of thermodynamics of the open systems was eventually formalized, it was possible to demonstrate that biological evolution leading to more and more complex organisms doesn't violate the second law of thermodynamics (Prigorgine 1968). However, while this justifies the evolution of Life, it doesn't justify the emergence of complexity during natural selection. Of course, there are two nuances of complexity involved here. Firstly, we need to justify the kind of complexity that emerged during natural selection. Secondly, we need to articulate the principles which allow only a certain kind of complexity to be favored during natural selection. Moreover, even if we try to articulate these principles of natural selection, the formalization would elude us. This is because if the selection is carried out on some structural template, then the underlying principle too must be structural. This may be contradictory to the randomness implicit in Darwin's theory (Bonner 2013). Moreover, such a protocol for formalizing demands mathematical operators which do not have any structuralism whatsoever. This is contradictory to our conception of algebraic mathematical operators. Therefore, we will have to define these principles in the language of topology. However, this technique is not available in the Darwinian paradigm.

Now let us look at the third ambiguity about the type of complexity that gets selected. Conventionally, we invoke the physiological conditions mentioned above as being responsible for the types of complexities that emerge during the course of evolution. Even in the classical interpretation of Darwin's theory, this issue is addressed by invoking the legacy of organisms. These kinds of phylogenetic models have been extensively verified, and therefore, they must be considered as valid (Bromham 2008). However, the real problem with this reasoning is that at every stage of the evolution, there are multiple structural options available for selection. In fact, this is true for the first living organisms as well. When Life evolved for the first time on Earth, it could have emerged in more than one type of complexity. Therefore, even at that time, the principle of natural selection must have used some structural template to evaluate the fitness of these earliest living organisms. Thereafter, during the process of natural selection, as more and more types of living organisms evolved, the selection process must have wherewithals to evaluate different types of complexity for their biological fitness. In other words, the selection principles must have a repertoire of all the possible types of complexities to evaluate their fitness.

It is possible to argue that such a foreknowledge is not a prerequisite because these different types of complexities can be evaluated as and when they emerge. Natural selection per se or the principles behind it need not have such foreknowledge. However, this is a fallacious argument. Let us for the moment, think of natural selection as a kind of computation. Therefore, it must have, like any other algorithm, a definition of complexity and its fitness, encoded beforehand. For an analogy, think that this algorithm of natural selection is like an algorithm to find out whether a given number is a prime number or not. Apparently, the algorithm for finding out prime numbers must possess a definition of what constitutes a prime number and how to decide whether a given number is a prime number or not. Thus, foreknowledge is a prerequisite for any algorithm. Therefore, if natural selection operates like an algorithm, it must have a definition of complexity and a method of computing its fitness, encoded within itself.

Of course, it is possible to argue that natural selection is not an algorithm. However, if every other natural phenomenon can be visualized as algorithms, it is imperative that we must treat the process of natural selection as a type of algorithm. Otherwise, there is no point in claiming that natural selection is a natural phenomenon. Thus, the choice is between accepting that natural selection operates on a structural template of complexity or that natural selection is not a typical natural phenomenon. The second option leads us back to vitalism or some other transcendental explanations, including teleology. Therefore, it is less problematic to think of natural selection as a kind of computation that decides the biological fitness of different types of complexities. Therefore, it is axiomatic that the principles of natural selection must possess a structural template of complexity as a priori. Purely from the semantic perspective, this shouldn't be a problem because we can posit this prior knowledge of complexity in the nature of the environment which anyway plays an important part in natural selection. However, we will have to redefine the notion of the environment and its role in natural selection. However, the bigger problem with any such attempts to redefine the role of the environment in natural selection is that it should not be antithetical to the semantic primacy of randomness of the Darwinian paradigm.

Conventionally, we rely on two types of randomness implicit in Darwin's theory. Firstly, we know that the changes in genotype via mutations are random. Secondly, we also know from our paleobiological models that the nature of the environment is also subject to random changes. Therefore, conventionally we have employed these twin sources of randomness to justify the emergence of design during natural selection. Therefore, we need to reinterpret the role of the environment in such a manner that foreknowledge doesn't lead to any form of teleology. Thus, it is intuitively clear from the above discussion that these ambiguities of the conventional interpretation of Darwin's theory arise from its semantics. Moreover, it is possible to resolve these ambiguities if we redefine the semantics of natural selection. We need to redefine the definition of genotype, phenotype and the environment for resolving these ambiguities. Before we look at any such attempts, let us go back to the classical interpretation of the semantics of Darwin's theory and find out the origins of these ambiguities.

2.8 Origins of Semantic Ambiguities

Let us, at least temporarily, assume that we need to redefine the notions of genotype, phenotype, and the environment to remove the ambiguities of the need for dualities, emergence of complexity during biological evolution, and emergence of design in a fundamentally random phenomenon. However, it is possible to argue that these semantic ambiguities have arisen from the lack of experimental evidence and not due to any inherent deficiency of Darwinian semantics. However, this argument is not valid, at least in the case of the Darwinian paradigm. As mentioned above, Darwin's own writings do not acknowledge having been informed of Mendel's research. However, that lack of experimental evidence didn't prevent the semantics of Darwin's theory to almost anticipate Mendelian genetics. Even if we were to think that Darwin might have been indirectly influenced by Mendel's work, there are later paradigm shifts which happened after Darwin's demise. Even these paradigms find semantic congruence with Darwin's theory. Therefore, it is erroneous to think that semantic ambiguities arise because of the lack of experimental evidence. They arise because of a lack of clarity of thought. Darwin's genius lies in the fact that his theory foresaw these paradigm shifts prior to the later day paradigm shifts, albeit approximately.

This belief that semantic ambiguities arise from the lack of factual knowledge is deep rooted and it is not confined to Darwin's theory. This belief owes its origin to our current understanding of the relationship between syntax and semantics (Jackendoff 2002, see Chapter 8). While the topic of the relationship between syntax and semantics is a contentious one, its resolution using the proposed model has been discussed in the preceding monograph, and it will be further articulated in the forthcoming monograph (Chhaya 2022e, see Chapter 13). In the present context, it would suffice to say that the relationship between syntax and semantics bears an uncanny resemblance with the relationship between structuralism and functionalities in biology, a topic that is at the core of this monograph. To sum up the interpretation of this relationship between syntax and semantics, as implicit in the proposed model, semantics enjoys an ontological primacy over syntactic template. It is just because our cognitive faculty is blind toward semantic processing that we perceive syntactic templates as of primary importance.

Returning to the present discussion, the semantic ambiguities discussed above arise not because of any lack of experimental evidence, but they arise from the inherent drawback of our conception of what a scientific theory is and how it is formalized. In other words, the above mentioned semantic ambiguities of the Darwinian paradigm arise from the fact that Darwin's theory has not been deconstructed to trace its epistemology. There are certain peculiarities of our cognitive faculty which shapes every scientific theory that we have been able to formalize. To borrow Shakespeare's phrase, "The fault does not lie in our scientific theories, but in the way we formalize them," sums up the gist of this section. Leaving aside such a linguistic flourish, let us get back to the nitty-gritty.

According to the proposed model, any given semantics can be devolved into multiple syntactic templates. However, every such template will, by definition, leave out some semantic propositions from its conception. This "fractionation" of semantic singularity into different mathematical formalisms leads to two of the most fundamental features of scientific theories. Firstly, it gives rise to the Popperian notion of falsifiability (Popper 1963). There are no absolutely correct or wrong theories. Every scientific theory is valid for a certain semantic content. Beyond this native semantic content, every scientific theory is falsifiable under certain conditions. Secondly, because of this partial "fractionation" semantics into mathematical formalisms, every formal system of knowledge remains incomplete in the Godelian sense (Smullyan 1992, see Part I). Therefore, it is axiomatic that Darwin's theory too will have its own set of semantic ambiguities. However, at the same time, it must be kept in mind that according to this model, it is possible to reformulate the semantics of every scientific theory in more than one way. Therefore, there exists a novel way to reinterpret every natural phenomenon. Admittedly, the new way of interpreting too would be incomplete, but it would offer us new insights into how different semantic propositions can be composed into a new interpretation. While the semantic singularity remains unchanged its compositional syntax can be redefined. Unfortunately, it remains beyond our cognitive capabilities of formalization.

When viewed from this perspective, let us see how these three semantic ambiguities could have arisen from the conventional perspective of the Darwinian paradigm. Let us begin with the need for separate units of inheritance and selection. As mentioned above, what is fundamental to the Darwinian paradigm is that it requires some form of dualities to bring about natural selection. Let us temporarily concede that the conventional definition of genotype and phenotype is inadequate to explain the semantics of natural selection. However, as discussed above, one could always shift the boundary between genotype and phenotype to expand our understanding of natural selection. Thus, the definition of what constitutes a genotype and phenotype can be varied, but the semantic need for the duality is constant. We can extend similar arguments for the remaining dualities of DNA/RNA and structuralism/functionalities. Therefore, what we need to reinterpret is why any form of dualities is a prerequisite. Of course, it is possible to argue that there is no reason to assume that these different dualities are mere instantiations of a more fundamental form of duality. Therefore, such a supposition could, by itself, be a new and perhaps unwarranted semantic proposition.

However, there are two reasons behind this unified perspective of these dualities. Firstly, with the passage of time, we have realized that the Darwinian paradigm seems to embody a general and perhaps more fundamental aspect of reality which is domain neutral. If we can extend the Darwinian model of competitive survival to as diverse domains as neuronal architecture, stock market behavior, and programming (Dennet 1995), apparently, the Darwinian paradigm must possess a semantic template that is generic in nature. Therefore, the need for dualities also must be context independent. Secondly, the principle of parsimony or the Ockham's razor (Sober 2015, see Chapter 1) suggests that there exists a unified template for all the forms of dualities. Thus, the semantic ambiguity about the need for separate units of inheritance and selection must originate from the conventional perspective of the Darwinian paradigm which assumes the existence of these dualities as a priori.

Now let us look at the ambiguity about the origin of complexity during biological evolution. Upon a little reflection, it is intuitively clear that this emergence of complexity could be better understood if we can define the relationship between structuralism and functionalities. Admittedly, both these features, viz., structuralism and functionalities, have their own definition of complexity. Therefore, it might appear that it is incorrect to think that the emergence of complexity during biological evolution can be explained by deconstructing the relationship between structuralism and functionalities. However, if we were to assign a more active participation of the environment itself in biological evolution and natural selection, it is intuitively clear that what appears to be an emergence of complexity, either in structuralism or in the functionalities of living organisms, could be transfer of complexity from the environment to the successive generations of living organisms during the process of natural selection. More importantly, in such a scenario, it is intuitively clear that the emergence of Life itself could have arisen from the active transfer of complexity from the environment to prebiotic chemicals. In other words, biological evolution and natural selection share a common framework for ontology in the form of active participation of the environment itself in the then prevalent chemical reactions in the primordial soup from which Life evolved.

Admittedly, to suggest that the environment plays an active role in biological evolution and natural selection amounts to intellectual sophistry, unless it is articulated in a testable model. However, if such a model is available, then it is intuitively clear that the best way to explain the emergence of complexity during biological evolution and natural selection is to deconstruct the relationship between structuralism and functionalities in the light of this active participation of the environment itself. This is because in such a scenario, the perception of the emergence of complexity is a mere phenomenology. In reality, it is merely a transfer of complexity from the environment to Life. Therefore, in that case, it makes sense to deconstruct the relationship between structuralism and functionalities to define the role of the environment and thereby explain the origin of complexity.

Now, let us look at the third ambiguity about the design principle arising from a random process of natural selection. Admittedly this enigma of design predates the Darwinian paradigm. More importantly, it still eludes resolution. The distinction between the second ambiguity of emergence of complexity during biological evolution and the ambiguity of design is rather subtle. Design, per se, refers to some degree of complexity. However, the reverse is not true. Complex systems need not obey design principles. The basis of this distinction between design and complexity lies in their semantic propositions. The notion of complexity, as discussed in the first chapter, deals with the plurality of components and their relationships among themselves. Incidentally, a similar template is also employed in formalizing the notion of semantics. Therefore, complex systems can possess semantics of their complexity. The notion of design, on the other hand, is basically an anthropic notion. This is because design requires two perspectives. Firstly, the notion of design requires some kind of teleological argument. A design is meant to fulfill some objective or intended use. However, this argument depends on the anthropic view of end use. Thus, the notion of design rests on the purpose for which a given system is meant to perform. The second perspective of design rests on human aesthetics. Our sense of aesthetics decides what is a good design and what is not a good design. Thus, the notion of design is essentially an anthropic notion.

However, there is an overlap between these two notions when we replace the perspective of end use or purpose with functionality. It is this overlap between the notions of complexity and design that manifests as a semantic ambiguity in the Darwinian paradigm. For instance, we concede that the genome is a complex machine because it is assembled from different genes and different regulatory elements. However, we can assert that the genome is a designed entity only if we think of design as a synonym of functionalities. However, if we were to think of design as a perspective from end use, the genome cannot be called a designed entity. No one, not even Nature, started to create a genome with the purpose of creating an intelligent organism like a human being. Therefore, to the extent the notion of design implies some teleological arguments, then evolution doesn't give rise to designed entities.

The trouble however, with this distinction between design for end use and design for functionalities is that there is no well-defined boundary between the two. The semantic ambiguity about the emergence of design in biological evolution arises from our conviction that natural selection is an outcome of several random mechanisms operating in tandem. The notion of design, as mentioned above, is an anthropic notion because it requires an entity of self which can define a framework for the possible end use or purpose. It is common to overlook the fact that all the judgments of what is useful require a framework from which we can define the notion of utility. This framework is precisely what the agency of self provides. Therefore, when we try to evaluate whether the genome has some design principles manifest in its architecture, we would have to decide whether such a notion of self is available for the genome or not.

Surprisingly, it is not impossible to assign the status of an independent entity to the genome. This is because it has several long-range influences built within itself. Therefore, it is intuitively clear that a genome acts as an independent entity and therefore, it must have a framework necessary for defining selfhood. However, if one were to assign intentionality to this selfhood, it is problematic. It was this implicit intentionality that Darwin consistently rejected because it would sooner, rather than later, lead to some form of deistic teleology. However, the notion of design based on functionalities is integral to the architecture of a genome and needs to be deconstructed with a modified notion of selfhood. Therefore, it is imperative that we must redefine the relationship between structuralism and functionalities and still retain the semantics of randomness implicit in Darwin's theory. Therefore, in the next section, we will discuss whether it is possible to redefine the relationship between genotype and phenotype within the Darwinian paradigm.

2.9 Can We Redefine Genotype and Phenotype Within the Darwinian Paradigm?

As mentioned in the introduction, the Darwinian paradigm has undergone quite a few paradigm shifts, and yet, it has survived. In fact, every paradigm shift has enriched the semantics of the Darwinian paradigm. Therefore, it is legitimate to seek another revision of Darwin's theory. However, the question is to what extent we can modify the original conception of Darwin's theory and yet claim that the new version is still Darwinian in nature. In other words, there must be some core semantic propositions of Darwin's theory which cannot be reinterpreted during any such reinterpretation. Of course, this is not to suggest that we can never question Darwin's theory or its semantic core. The point is that if we do it then, the resulting formulation of natural selection cannot be called a Darwinian theory. In the preceding sections, we have realized that there are several semantic ambiguities about what constitutes a Darwinian paradigm. It was also mentioned that there are possible ways to resolve these semantic ambiguities provided we reinterpret the key concepts of Darwin's theory. Therefore, in this section, we will try to find a way to redefine a few key concepts of the Darwinian paradigm and see whether the core semantic content of the Darwinian paradigm remains valid.

Upon a little reflection, it is clear from the above discussion that the origin of these semantic ambiguities lies in the way we define three concepts of genotype, phenotype and the environment. There were historical reasons why Darwin emphasized on a unidirectional influence from genotype to phenotype and the passive role of the environment in natural selection. In order to appreciate the genius of Charles Darwin, we must realize that this emphasis was not based on his own predilection but it was based on his prescience. There were concepts and mechanisms which were yet to be articulated and still they find their echoes in Darwin's writings. This is possible only because Darwin, through his prescience, could grasp the semantic core of natural selection. Therefore, one must accept that there exists a definitive semantics of natural selection and more importantly, it finds its expression in Darwin's writings. Therefore, we will try to reconfigure this semantic core in the language of contemporary science and look for the possible ways to improvise upon the conventional perspective.

In view of the fact that successive paradigms of biological evolution and natural selection have perpetuated the semantic ambiguities of Darwin's theory, it is necessary to develop a new approach starting ab initio. It must be admitted that every paradigm shift has enriched the semantics of the Darwinian paradigm. However, the fact remains that the original semantic ambiguities have remained unaddressed. This is a good enough justification for developing a new paradigm. Moreover, since the objective of this new approach is to resolve the semantic ambiguities present in these paradigms, it is imperative that it must be based on semantic arguments. Admittedly, semantics of Darwin's theory and its successive interpretations have been articulated extensively in literature. However, they have succeeded in creating more nuanced versions of the semantics of Darwin's theory. Therefore, there is a need for such ab initio attempts to take an outsider's perspective of the history of these paradigm shifts. Once we accept this need for a new semantic framework, it is intuitively obvious that this new approach must be based on semantic considerations and not on biological considerations.

In the preceding monographs (Chhaya 2022c, see Chapter 8), one such information theoretical model of semantics has been outlined. Therefore, we will try to employ that model to arrive at a new understanding of the Darwinian semantics. For this purpose, we will reconstruct the key concepts of genotype, phenotype and the environment, in the context of their information content. Having done that, we will try to deconstruct the semantic ambiguities discussed above, using these redefined concepts. Let us begin with the concept of genotype. As conventionally understood, genotype refers to inherited information content of a living organism. While Darwin himself refers to genotype as an inherited trait, with the advent of modern genetics, we refer to genotype as a sum of all the inherited information in the form of genes of an organism. In molecular biology, genotype can be defined as a DNA sequence that an organism inherits. Similarly, in genomics, we can refer to the entire genome as a genotype. Thus, irrespective of the paradigm that we employ, genotype refers to the inherited information content of a living organism.

There are two features of genotype that remained unchanged during these successive paradigm shifts. These are inheritability and mutability. Of course, the mechanisms by which these features manifest changes from paradigm to paradigm, these features remain invariant. Since we wish to develop an information theoretical perspective, we will sidestep these mechanistic details and instead, focus on the nature of the information content during inheritance and mutations. Prima facie, it is intuitively clear that inheritance refers to information transfer and mutations refer to information transformation. Thus, it is possible to formalize biological evolution and natural selection in the language of information transfers and transformations. Upon a little reflection, it is apparent that the origin of semantic ambiguities arises from the fact that genotype contains the inherited information that also contains mechanisms of information transformation. Thus, genotype has two types of information content, information content per se and information about how to bring about information transformation. Thus, from the information theoretical perspective, genotype has a functionality of self-reference. It is as if genotype knows what is to be transferred and what is to be transformed. This inference leads to two important consequences. Firstly, biological evolution must consist of the creation of self-reference. Secondly, natural selection must operate on this functionality of biological self-reference.

Now, let us look at phenotype from the perspective of information theory. Apparently, the information content of phenotype arises from genotype. In the classical interpretation of Darwin's theory, this process of generation of phenotype is solely governed by genotype. As mentioned above, genotype contains instructions on how to give rise to phenotype. The role of the environment in this interpretation is passive. The environment doesn't shape the nature of phenotype. Rather, once phenotype is ready, the environment decides on its survival fitness and depending on its fitness, phenotype survives or perishes. Thus, in the classical interpretation of Darwin's theory, the information content of phenotype is solely decided by the information content of genotype. It must be kept in mind that this interpretation implicitly suggests that the information content of phenotype has certain rigidity and this rigidity is programmed in the information content of genotype. This implication was in congruence with our earlier belief in the discrete nature of genotype and phenotype (cf. Mendelian genetics (Darden 1991)).

However, our current understanding of the role of the environment in influencing the nature of phenotype is slightly more nuanced. Since we know that the boundary between genotype and phenotype is not sharp, the role of the environment itself takes on active overtones. The environment, in the form of genomic long-range influences, actively shapes the relationship between genotype and phenotype. Purely from the information theoretical perspective, there are multiple information flows that converge during the expression of phenotype. Firstly, there is an information transfer in the form of gene expressions of the DNA sequence of genotype. Secondly, there is an information transfer from the environment in the form of pre-existing initiators and facilitators toward the gene expression. Thirdly, there is an information transfer in the form of cis and trans influences of the genome on the gene expression. Finally, there is an information transfer in the form of formation of translation complex consisting of transcriptase and its associated enzymes.

Without getting overwhelmed by the structural details of these processes, it is intuitively clear that these different information flows can be divided into three categories of information transfers. For the time being, we will omit the process of information transformation in the form of translation of DNA sequences of genes into proteins that give rise to phenotype. This omission is necessary because this information transformation is like any other chemical transformation and can be overlooked in the present discussion. These three categories of information transfer are information transfer from genome to the individual genes, information transfer from DNA sequence of gene to phenotype in the form of the tertiary structures of the protein and the information transfer from the environment to phenotype in the form of the challenges to its survival.

Admittedly, this selection of categories is a little peculiar and its logic will be clear as the discussion progresses. However, there are two points that must be made about this categorization. Firstly, it consciously abstains from the chemical paradigm. As mentioned above, the process of translation of DNA sequence to protein has been kept out of this categorization. This is because this aspect has been focused on molecular biology extensively (Weinzierl 1999). Therefore, there is no point in repeating it here. More importantly, it must be apparent to everyone who studies evolution that molecular biology can provide the nitty-gritty of how the genome operates, but it is inherently incapable of defining the semantics of the Darwinian paradigm. Natural selection as formalized in Darwin's theory is not just about how chemical reactions progress. Admittedly, the chemical transformations are also information transformations, but it is a category mistake to equate synthesis of proteins from a given DNA sequence through RNA as information transformations that define natural selection.

It is possible to argue that every chemical reaction, including the translation of DNA sequence to proteins, is governed by a kind of selection governed by thermodynamics, and therefore, they too are subject to a different type of natural selection, and therefore, there is no need to look beyond this chemical perspective while defining natural selection in biology. After all, thermodynamic principles also play a role in biological evolution (Prigorgine 1968). There is no denying the fact that biological evolution and natural selection are not governed by thermodynamic principles. However, it is important to keep in mind that the living organisms are something more than chemical ensembles. Even without resorting to some arguments belonging to the "Vital Force" variety, it is intuitively clear that living organisms have a few additional properties that are manifest due to higher levels of organization of molecules (Smith and Morowitz 2016). This is best exemplified by the functionalities of the genome. These functionalities are not the sum of the functionalities of the individual genes. The genome has certain functionalities that have arisen from the way these individual genes have been assembled. Therefore, we will overlook the chemical paradigm from the present discussion.

The second reason for this choice of categories is that even the conventional perspective is veering toward the view that different types of information, other than that of DNA sequence, are critical to natural selection. Our contemporary trends in functional genomics exemplify this realization. We are increasingly looking for protein domains as units of selection. The secondary structure of a protein is distinctly different from its primary structure of the sequence of amino acids. Yet, it is possible to show that it is the secondary structure of proteins and not their amino acid sequences that are selected and conserved. By implication, natural selection must operate on topological constructs, rather than on the sequence of nucleotides or amino acids. Therefore, we will sidestep the chemical paradigm and instead think in terms of abstract information content of genotype, phenotype and the environment. Admittedly, this choice is further reinforced by the fact that we wish to develop a topological model of natural selection.

Therefore, let us elaborate a little on these three categories of information transfers. Of course, a more detailed description of them will be discussed in the following chapters. Let us begin with the first category of information transfer, viz., the information transfers from the genome to individual genes. As the paradigm of molecular biology progressed, it was evident that there exist long-range influences of the genome on the individual gene expressions. Conventionally, these influences have been classified as cis and trans effects depending on the fact whether the long-range influence arises from the same chromosome or from the other chromosomes. This is essentially a topological classification. Surprisingly, there is no theoretical model available to formalize these effects. This is primarily because the stereochemical orientations of individual chromosomes change during the life cycle of a cell. Therefore, although the underlying topological perspective is accepted, it is not formalized.

The second tacit acknowledgment of genomic influence on the individual gene expressions comes in the form of pre-existing initiators and facilitators. The conventional wisdom suggests that the order in which genes are expressed, by itself, is governed by genomic architecture and therefore, it constitutes a long-range influence. If the cis and trans influences were spatial long-range influences, then the sequence of gene expressions itself must be taken as a temporal long-range influence. Thus, the conventional perspective of genomic architecture accepts the notion that genomic influences on the individual gene expressions are essentially spatiotemporal in nature. However, functional genomics stops at this generalization and doesn't venture into formalizing any such spatiotemporal topological model of the genome. As discussed in the following chapters, it is possible to develop such a model of the genome.

Now, let us look at the second category of information transfers, viz., the information transfers from genotype to phenotype. As discussed above, we will include the pre-existing initiators, facilitators and even the translational machinery in the form of protein complexes as the environment for these information transfers. Moreover, we will overlook the chemical paradigm implicit in the process of translation. It is intuitively clear that these information transfers from genotype to phenotype are essentially processes of unfolding of latent information into explicit information content. Let us take an example of protein synthesis which is the result of the translation. Apparently, this protein will have some functionality, say, in the form of enzymatic catalysis; otherwise, its genotype would not have been selected. (It must be kept in mind that even the noncoding DNA sequences are also translated and used in controlling gene expressions in an indirect manner.) Therefore, the emergence of functionality of enzymatic catalysis is essentially an unfolding of the latent information through the gene expression. It is because we have conditioned ourselves to think in terms of chemical paradigms that we miss this distinction. This is because from the chemical perspective, the synthesis of protein and its catalytic functionality belong to the same category. However, from the information theoretical perspective, information present in the sequence of amino acids and its catalytic functionality belong to two different categories of information. Thus, the process of translation of the sequence of nucleotides into the sequence of amino acids is not confined to chemical details (as we have been conditioned to think), rather it is the process of information transformation from structuralism to its functionalities.

This brings us to the third category of information transfers, viz., the information transfers from the environment to phenotype in deciding its survival fitness. In continuation with our earlier discussion, this category will include only the classical influence of the environment in natural selection as implicit in Darwin's theory. As mentioned above, in the classical interpretation of Darwin's theory, the role of the environment in natural selection is passive. Therefore, we will focus on this aspect of the environment. From this limited perspective, there are two passive forces that influence the survival of phenotype. Firstly, it is in the form of food, as in a food cycle wherein one species is a food for another. Secondly, the environment could provide hostile elements which make survival difficult, like the changes in temperature, ecological niche in which the given phenotype thrives etc. These influences can also be viewed as information transfers wherein certain elements of the environment impinge on phenotypes as a different type of information.

In both these cases, phenotype is not able to respond to the changed information content of the environment and therefore, cannot bring about information transformation within itself. Therefore, phenotype can't survive long enough to produce the next generation of phenotype. This is a nondescript theoretical perspective of the classical interpretation of Darwin's theory. What is often overlooked is that this arises because of the incompatibility between the information content of the environment itself and that of phenotype. For instance, take the example of a food cycle. It is possible to visualize that a given species feeds on another species, and therefore, the survival of the given species would be threatened whenever the species on which the given species is feeding is depleted. In such a scenario, the real threat to the given species is not really the lack of food, but rather the given species' inability to consume different species as food. In any ecosystem, there are a large number of species. Therefore, food habits of different species change under difficult conditions like the lack of availability of normal food sources. The man eating wild cats is the classic example of species being forced to seek new sources of food in the absence of regular sources of food.

The key point is that it is only when the given species cannot find alternative sources of food that its survival is threatened. Thus, it is the lack of compatibility of the information content, in the form of thinking of other forms of Life as a source of food, that causes the threat to the survival of the given species. Purely from the physiological perspective, this situation arises because of the particular metabolism of the given species. Therefore, in a way, this can be ascribed to the genotype of the given species. However, this limitation of the given species' metabolism is nothing but the incompatibility between the information content of the digestive capabilities and the information content of the food in the form of its digestibility. Thus, it is the particular type of complexity of information content of phenotype that gives rise to threat to its survival.

The same explanation works for the hostile elements of the environment. These hostile elements, say, in the form of low temperature during the ice age or in the form of pollutants in our water resources, cause threat to flora and fauna present on Earth.

Even here the threat to different species comes in the form of changed information content of the environment and its inherent incompatibility with the information content of phenotype. Moreover, this incompatibility arises because of the inherent complexity of the information content of phenotype. Those phenotypes whose informational complexity was compatible with the changed information content of the environment, like mammals during the extinction of dinosaurs, would always survive. Thus, purely from the information theoretical perspective, the passive role of the environment consists of changes in the information content of the environment and its incompatibility with the existing informational complexity of phenotypes.

The reason why this perspective is important is because it highlights the often overlooked aspect of complexity of phenotypes. Phenotypic survival is threatened not primarily because of the changes in the environment, rather it arises from the legacy of the informational complexity that is pre-existing in phenotype. Biological evolution's primary purpose is to fix a certain kind of informational complexity. Thus, the process of natural selection is actually a conflict between the plasticity of the information in the ever-changing environment and the rigidity of the informational complexity of phenotype. This leads us to the most fundamental semantic enigma of biological evolution. Why should biological evolution lead to the emergence of rigidity of complexity when the environment is essentially nothing more than an ever-changing mixture of simple elements. Therefore, in the following sections, we will delve deeper into this enigma of complexity.

2.10 Problem of Complexity in the Darwinian Paradigm

The problem of complexity in the Darwinian paradigm is twofold. Firstly, there is a question about whether the feature of complexity is selected during natural selection or not? This is important because if complexity is one of the parameters of natural selection then, natural selection must possess an ability to decide what complexity is and which type of complexity is good for survival. This amounts to ascribing the ability to "know" to Nature. The second problem of the emergence of complexity during biological evolution and natural selection is that is there any design principle (Fodor and Piatelli-Palmarini 2011) embedded in the process of biological evolution and natural selection? This is an uncomfortable question in view of the fact that the Darwinian paradigm rests on the notion of randomness (Bonner 2013). Darwin himself fought against any imputations of teleology (which is what the design principle would lead to). More importantly, the later paradigms have provided very sound semantic justifications for adhering to the nonteleological arguments of Darwin. Thus, on either count, it is difficult to justify the emergence of complexity during biological evolution and natural selection. The conventional defense for this emergence of complexity rests on the phase space type of argument. It is often argued that by the principle of parsimony, it is self-evident that the first living organisms were of the simplest possible types. Therefore, any natural selection from these organisms would inevitably lead to more complex organisms. Moreover, once this effect is set into motion, it would be unstoppable. Therefore, we witness the arrival of more and more complex organisms with the passage of time. In order to justify this argument, the feature of loss of function is cited (Alberts 2004). There are innumerable examples of species losing several phenotypic features during the course of evolution. Therefore, it is argued that natural selection, per se, is neutral toward the feature of complexity. Natural selection can work both ways, it can increase or decrease complexity depending on the environmental impact.

However, there are two reasons why this is not a correct interpretation. Firstly, the number of cases of loss of function is miniscule in comparison with the number of cases of gain of function (Alberts 2004). It amounts to citing an exception to prove that the rule is wrong. Secondly, if there exists a mechanism which defines when there should be a loss of function and when there is a gain of function, it amounts to admission to design principles which are anathema to the semantics of the Darwinian paradigm. Either which way, the problem of complexity is unresolved. Moreover, if we were to think of biological evolution and natural selection, as natural phenomena, then it is the only natural phenomenon that seems to defy the second law of thermodynamics. This law, in its simplest form, suggests that all the natural phenomena tend to produce more disorder than before. However, biological evolution and natural selection seem to produce more order or less disorder in the form of more and more complex organisms. Admittedly, there exists a theory of the thermodynamics of the open systems (Prigorgine 1968). This theory demonstrated that even biological evolution and natural selection produce more disorder. It must be kept in mind that this theory refers to a parameter of entropy to formalize the notion of disorder. The semantics of entropy has been contentious. However, even if we were to accept that biological evolution and natural selection do produce more disorder, like any other natural phenomena, the problem of explaining the emergence of complexity still remains unanswered. This is because the theory of thermodynamics of the open systems merely states that while the complexity does indeed emerge during these processes, there is a corresponding entropy increase in the environment, thereby resulting in the net increase in entropy. Thus, even this theory concedes that biological evolution and natural selection do indeed give rise to complexity with the concomitant local decrease in entropy. Therefore, the problem of the emergence of complexity during biological evolution and natural selection remains unresolved. Moreover, if the notion of complexity were to be formalized into different types of complexities, it is inevitable that we need to incorporate this formalization into the formal description of the Darwinian paradigm, which given the random nature of natural selection, is difficult to conceptualize. Therefore, in this section and the next section, we will try to understand how the proposed model offers a way out of this semantic impasse.

The proposed model, which we will discuss later in this chapter and in the following chapters, rests on the postulate that ontology and epistemology of reality share a common framework. This necessarily eliminates the classical problem of epistemological access. The second feature of this model that is germane to the present discussion is that it links complexity, in the form of metric, to the dimensionality of the phenomenon under investigation. Moreover, if any phenomenon manifests itself in multiple dimensionalities simultaneously, which is certainly the case with the genome, then it is possible to define the relationship between different types of complexities within that phenomenon by assigning different dimensionalities to each of these complexities. This relationship is formalized using a generic category of operators of involution. Thus, it should be possible to explain the emergence of complexity during biological evolution and natural selection if one can define a set of operators of involution to represent natural selection. Secondly, since the proposed model offers a way to link ontology and epistemology, by analogy, it is possible to define the relationship between biological evolution and subsequent natural selection.

It is important to remember that the Darwinian paradigm, biological evolution as a priori. Darwin's theory was focused on the concept of descent with modifications and not on the scenario of how Life could have originated. Even during the subsequent paradigm shifts, the explanation for the origin of biological evolution remains nebulous. One of the most accepted hypotheses about the origin of Life employs RNA as a source of Life (Yarus 2010, see Chapter 2). This RNA world scenario has its own inherent merits. However, it fails to explain how the functionalities of Life could originate from chemical functionalities of RNA. The proposed model offers a way to link these chemical functionalities with the biological functionalities. It is tempting to think that the original objective of unifying ontology and epistemology finds a new interpretation in biological evolution. All that we need to do is to visualize the relationship between genotype and phenotype as the relationship between ontology and epistemology. If the ontological perspective in the domain of cosmology offers an explanation for the emergence of complexity of the manifest universe, then the proposed model links that ontological perspective to the epistemological perspective by a common framework. Similarly, we can think of the genome as a biological singularity and different genes as natural phenomena arising from that singularity. Thus, just like the evolution of Life, individual gene expressions must be viewed as an ontological paradigm. Correspondingly, we can think of the relationship between genotype and phenotype as an instance of epistemological access. Just as during epistemological processing, the information content becomes available, the process of gene expression enables the implicit information of genotype to become available in the form of phenotype. In other words, our formal concepts are phenotypes that arise from the nature of reality which is a genotype of the universe. Thus, it is possible to transcend the conventional definition of epistemology as a process by which knowledge becomes available and think of it as a process whereby the latent information finds its expression.

Thus, the cognitive conception of epistemology can be enlarged to include all the processes which enable the implicit information content to become explicit information content. Thus, we can aspire to have three different types of epistemology, cognitive epistemology, cosmic epistemology and genomic epistemology. Epistemology in that sense becomes a science of information transfers and transformations. This is a necessity if we wish to develop a truly naturalistic universal scientific theory. While this scenario has a certain romantic allure, it needs to be subjected to scientific rigor. Therefore, in the next section, we will look at the semantics of complexity in the context of the Darwinian paradigm. As mentioned above, it is indeed difficult to accommodate the emergence of complexity during what is an essentially random phenomenon of natural selection. As discussed here, it is intuitively clear that one can think of gene expressions as processes that transform the implicit information content of genotype into the explicit information content of phenotype. This perspective allows us to visualize that phenotype, in the form of functionalities, can have its own template and how it can be derived from the structural template of genotype. However, the problem is whether this can be formalized within the random nature of natural selection. This is essentially a problem of semantics, or rather a problem of the relationship between semantics and syntax. Just as a given semantics can find its expression in different syntactic templates, it is possible to visualize that a similar process connects different types of complexities to the semantics of a genome. We will explore this issue in the following section.

2.11 Semantics of Complexity in Biological Evolution

Normally, we prefer to deal with a scientific theory rather than with its semantics. This is because a scientific theory provides more than one verifiable (falsifiable according to the Popperian theory (Popper 1963)) consequences. This possibility of verifiability infuses confidence in one's mind. However, apart from the fact that this book is dealing with semantics of the Darwinian paradigm, Darwin's theory itself demands that one should evaluate it on the basis of its semantic consistency. This is because Darwin's theory is unlike any other scientific theories. To revert back to the Popperian paradigm, Darwin's theory doesn't predict anything, not even one verifiable prediction. In fact, Popper himself had raised this point in his writings. This inference is not intended to demean the Darwinian paradigm, rather it is meant to point out the exceptional nature of Darwin's theory. It is possible to argue that Darwin's theory doesn't predict anything, is not correct because genetics and population genetics have provided ample numbers of predictions that have been verified. However, such an argument is fallacious. Darwin's theory is not based on the mechanisms by which natural selection is produced (which is actually the focus of genetics and population genetics). Darwin's theory is about the semantics of natural selection. In other words, Darwin's theory is a semantic theory which deals with the biological underpinning of Life. It is this semantic nature of Darwin's theory that necessitates that we look at the semantics of complexity in biological evolution and natural selection.

Therefore, it is time to evaluate the semantic propositions of complexity in the context of the Darwinian paradigm. Once again, we will employ an information theoretical perspective of both: complexity and the Darwinian paradigm. However, we will sidestep the underlying mathematical or linguistic foundation of semantics. Instead, we will look at the abstract understanding of semantics based on the information content implicit in the notion of complexity and the Darwinian paradigm. The notion of complexity can be equated with the number of discrete units and their arrangements. We perceive a system to be more complex on the basis of the number of constituents of that system. A system having a lesser number of units will be perceived to be less complex than the system having a greater number of units. This perception of complexity is qualitative and acts as a thumb rule. However, from the epistemological perspective, we use the number of arrangements as a quantitative measure of complexity of any given system. This is the basic scenario of how we perceive the notion of complexity. Of course, we can refine these intuitive understanding of complexity by introducing mathematical constructs defining various types of arrangements and grading them. However, for the present discussion, this primitive understanding of the notion of complexity will suffice.

Once we define the notion of complexity in terms of number of units and their arrangements, it is intuitively clear why it is incongruent with the randomness implicit in the Darwinian paradigm. According to the conventional perspective of the Darwinian paradigm, phenotype emerges from genotype. Since natural selection requires that genotype keeps changing and thereby giving rise to newer phenotypes, it is inevitable that these changes in genotype and subsequently those of phenotype must be random. In the absence of such randomness, the Darwinian paradigm would become a teleological paradigm. Thus, the random nature of changes in genotype and phenotype is a semantic prerequisite for the Darwinian paradigm. This randomness is best formalized in genetics as mutations and in polygenic/pleiotropic influences on gene expressions (Dragini 1998). Since mutations are, by definition, random due to the inherent random nature of the environment and polygeny/pleiotropy would enhance the effects of these mutations, the randomness becomes a cornerstone of Darwinian semantics.

It is in this context that the notion of complexity needs to be evaluated. Prima facie, any random process should lead to more and more disorder. Therefore, the above mentioned scenario too should lead to more and more diverse structuralism of living organisms. Therefore, it is difficult to visualize the emergence of structural complexities during natural selection. However, there are two factors that influence the randomness as having arisen from the above mentioned scenario. Firstly, Darwin's own writings and the subsequent formalization of natural selection in genetics, endorse a common origin of Life. There is no either explicit or implicit suggestion that Life evolved several times on Earth. This singular origin of Life was implicit in Darwin's theory in the form of modifications with descent and has now been explicitly conceptualized as the Last Universal Common Ancestor (LUCA) (Bard 2016, see Chapter 9). This common ontology necessarily reduces the degree of randomness than one would have witnessed in case Life had originated several times on Earth. The most intuitive way to understand it is to think of a machine generating random numbers. Apparently, one cannot predict what the next number that machine would generate. This is because not only are there infinite numbers from which the machine has to choose from, but also there is no programming logic in selecting the next number by the machine. Now just think of a machine which is asked to choose a random number but only on the basis of some programming logic which uses the last generated number as a starting point. Even this modified machine can generate apparently random numbers. However, these numbers will not be really random but would appear to be random. This is because there exists some logic behind the generation of the next number.

Similar situation presents itself when we concede that all the living organisms arose from a single ancestor. The randomness of morphological features only appears to be random, but there exists a certain fixed pattern behind the diversity of forms. In fact, it is because of these visible similarities between a variety of morphological features that enabled Darwin to formulate his theory. Thus, if we were to concede that these patterns of variation of morphological features are by itself, a type of complexity, it is intuitively clear that complexity was always an integral semantic proposition of natural selection. It is more than likely Darwin's own aversion to the teleological arguments that forced him to emphasize on the random nature of natural selection.

Apart from the fact that the common ontology would diminish the randomness in natural selection, there is another feature of natural selection that also undermines the notion of randomness. This refers to the degree of discreteness of the units of inheritance and selection. If genomes were to be thought of as a string of independent genes, one would be justified in claiming that natural selection is random. This is because each of these genes would undergo independent mutations and an independent process of gene expression leading to a living organism with loosely connected organs and phenotypic features. However, the fact remains that the genes of any genome are not independent, either structurally or functionally. There exists a certain degree of interdependence among them. It is this modular architecture that compromises the random nature of gene expressions and that of the resulting phenotypes. This modularity is not confined to genotype but it extends to phenotype as well. Thus, every morphological feature of living organisms is modular in nature. It is this functional and structural modularity that limits the degree of randomness that one would have expected otherwise. Therefore, if we were to concede that the notion of modularity is itself a type of complexity, it is intuitively clear that the notion of complexity is an integral semantic proposition of the Darwinian paradigm. It is only because Darwin was keen to avoid design principles propounded by theologians that Darwin emphasized on the essential random nature of natural selection. Even when the theory of genetics was formulated, under the Mendelian influence (Darden 1991), the discreteness of the units of inheritance was the cornerstone of genetics. It was only after the advent of molecular biology and genomics that we have accepted that the discreteness of the units of inheritance and selection is at best, an approximation. Therefore, it is no longer fashionable to advocate randomness as the foundation of the Darwinian paradigm. Admittedly, it plays an important part in natural selection, but it is not the sole deciding factor.

Having looked at the historical perspective of randomness, let us look at the issue of the emergence of complexity during biological evolution and natural selection from the perspective of our post genomic era. One of the major changes in the semantics of natural selection in the post genomic era is the acceptance of the fact that the boundary between genotype and phenotype is not sharp. This loss of discreteness has also resulted in the articulation of a more nuanced semantics of natural selection. However, the problem of the incompatibility between the inherent randomness of natural selection and the emergence of complexity remains unresolved. As discussed above, if we were to define the notion of complexity in terms of the number of discrete constituents and their arrangements, then in the post genomic era, this definition of complexity gets blurred. This is because if the boundary between discrete units gets eliminated, it is difficult to quantify the degree of complexity.

However, as our knowledge of genomics increases, we realize that natural selection must be viewed not as the emergence of complexity, but rather as changes in the types of complexities. It is true that we no longer think of the genome as a string of independent genes. Rather, the genome appears to be a hierarchy of genes and different regulatory elements. However, this change in our conception of the genome has increased the number of constituents of the genome. More importantly, these constituents do not operate linearly. Therefore, we are forced to concede that the very notion of genotype is far more complex than thought earlier. Therefore, it is necessary to accept that natural selection actually alters the types of complexities along with the increase in the degree of complexity. Therefore, natural selection must be viewed as a selection process of certain types of complexities over the other. Axiomatically, natural selection cannot be an outcome of any passive role of the environment.

In view of our belief that biological evolution and natural selection are purely naturalistic processes, devoid of any design principles or teleology, there are two options for us to accommodate this nonpassive nature of natural selection. Firstly, we can continue to hold that the emergence of complexity during biological evolution and natural selection cannot be explained. This stoicism finds support from the conventional perspective of the Darwinian paradigm itself. According to the conventional perspective of the Darwinian paradigm, there is no biological explanation of the evolution of the functionalities of cognition. At least in the case of sensorimotor functionalities (Barth et al. 2012, see Parts II and IV), there exists a valid Darwinian argument for their selection (Of course, the evolution of the sensorimotor neurology cannot be explained adequately, but that is the different matter). Once such a functionality evolved, its selection would increase the survival of phenotype. Therefore, it is to justify natural selection of sensorimotor functionalities, once it evolved. However, there is no such argument available for our epistemological functionalities. Our ability to theorize can never aid our survival. Therefore, according to the conventional perspective of the Darwinian paradigm, our epistemological processes are, at best, an accidental payoff. Natural selection merely worked toward increasing our cognitive processing capabilities, only to enhance our sensorimotor functionalities. However, once evolved, the increased neocortex accidentally offered us with our epistemological functionalities. If this reasoning is valid, then we can see why we can't explain some of the most fundamental semantic ambiguities that we come across. The emergence of complexity during biological evolution and natural selection is one such ambiguity. We cannot explain these semantic ambiguities because Nature didn't "Plan" for this functionality, it just developed out of our increased neocortex accidentally.

The second option that we have is to revisit our conventional perspective of natural selection and define in terms of structural preference for certain types of complexities over other types of complexities. However, this eliminates any possibility of the process of natural selection being passive in nature. The process of natural selection is active in the sense that it operates on certain structural principles. It is possible to argue that once we concede that natural selection operates on certain structural principles, we are just one step away from reintroducing teleology. All that one needs to do is to demonstrate that this structuralism of the process of natural selection is actually a design principle embedded into the process of natural selection by some supernatural or transcendental powers.

However, this fear is unfounded. As implicit in the proposed model, the structural principle of natural selection could be due to natural causes, say, the fine structure of spacetime. It must be kept in mind that historically, the Darwinian paradigm acquired antipathy toward any design principle because it considered any form of teleology as an anathema. However, the design need not arise out of any teleological principles. It can arise naturally from the inherent structuralism of the manifest universe. For instance, theoretical physics is never accused of being a teleological discipline just because it accepts the standard model of the fundamental particles (Cottingham and Greenwood 2013). The structuralism of the standard model is never considered as a transcendental design. It is just taken as an instantiation of a mathematical template. Similarly, if we decide to cast the process of natural selection in some structural template, we are not conceding any teleological arguments. Admittedly, the relationship between mathematics and natural phenomena has its own set of ambiguities. To quote Wigner's famous phrase (Wigner 1960), "the unreasonable effectiveness of mathematics" in explaining natural phenomena is troubling. However, as discussed in the preceding monographs, this can be resolved by accepting a common framework for ontology as well as epistemology. Once we accept that evolution of Life and particularly the evolution of our cognitive faculty is a typical natural phenomenon absorbing and internalizing the structuralism of the manifest universe, the Darwinian paradigm can embrace any mathematical design principles without worrying about teleology or deism. We can go one step further and postulate that the structuralism of Life, like other natural phenomena, derives its structural template from the structural template induced by the symmetry breaking processes operating on the cosmic singularity.

Therefore, it is imperative that we must accept that the process of natural selection essentially employs some kind of structural template and try to formalize it without being worried about having to abandon naturalism which Darwin steadfastly adhered to. In fact, genomics, particularly proteomics, points toward such a structural template of natural selection. During the phylogenetic studies (Bromham 2008), we have discovered that it is the domain structures of proteins and not necessarily the underlying DNA sequence that is conserved during natural selection. Therefore, it is axiomatic that natural selection operates on structural templates of proteins which are essentially phenotypes in the language of the classical interpretation of Darwin's theory. Once we accept this, it is intuitively clear that the principles of natural selection too must be structural in nature. Therefore, these principles must be amenable to formalization using proper mathematical constructs. Therefore, in the following sections and in the following chapters, we will explore this issue in some detail.

2.12 Why Do We Need to Reinterpret Darwinian Theory?

Even if we accept this structural imperative of natural selection, we might feel reluctant to reinterpret the Darwinian paradigm. There is something undefinable puristic about the Darwinian semantics that stops us from exploring any possible reinterpretation. It is possible to argue that it is possible to continue to believe in the conventional perspective of the Darwinian paradigm even when we attempt to define some structural template of natural selection. After all, from the information theoretical perspective, natural selection, whether as defined in the conventional manner or whether defined using a structural template, is essentially about information transformation from genotype to phenotype and about information transfer from the environment to phenotype. Therefore, we can still adhere to the conventional perspective of Darwinian semantics even after developing a new template of natural selection. Admittedly, this is a valid argument. However, the question is whether the Darwinian paradigm is essentially a biological paradigm or it is a universal template of competitive survival?

As we have found out that the Darwinian paradigm can be applied to diversely different domains, it is possible to view the biological underpinnings of the Darwinian paradigm as some of the instantiations of some more fundamental conception of natural selection. It must be kept in mind that Darwin himself was inspired by the Malthusian model of economics of survival (Flew 2017, see Part III). Therefore, there is no need to hold on to the belief that biological evolution and natural selection are exceptions to other natural phenomena. Therefore, if we search for a domain-independent interpretation of Darwin's theory, it would help us to understand a more fundamental form of natural selection. Admittedly, there is a limit to which we can alter what the Darwinian paradigm means. Hypothetically, can we redefine Lamarckian theory as a variation of Darwin's theory? The answer is obviously not. However, we have found, much to our discomfort, that epigenetic phenomena are more like Lamarckian phenomena than Darwinian phenomena (Robert 2004). The best way to resolve this dilemma is to accept that it was the teleological imperative of the Lamarckian interpretation that was rejected by Darwin and his contemporaries. The mechanisms by which the information is transferred from phenotype to genotype can still be Darwinian in nature. Therefore, if we can redefine the process of natural selection in the language of information transfers and transformations, we should be able to transcend the biological perspective of natural selection. In such a scenario, the biological version of natural selection could still be different from other versions of natural selection. However, this is because biological evolution, as a natural phenomenon, is unique among the other natural phenomena. In order to explain this domain-independent version of the Darwinian paradigm, we will look at a topological model of natural selection in the next section.

2.13 Emergence of Complexity in the Involuted Model

In a normal course of discussion, we would prefer to outline a new model and then discuss its various nuances. However, we would reverse the sequence here and discuss one of the nuances of the new model outlined in the following chapters before giving the details. This is pardonable because in this chapter, we are looking at the semantics of the relationship between genotype and phenotype. Therefore, we will restrict ourselves to the particular perspective of complexity outlined above and how it is represented by the proposed model. Our objective, at this stage, is to see whether the relationship between genotype and phenotype can be defined using the notion of complexity as represented in the proposed model.

Therefore, without going into the formal description of the proposed model, let us reiterate two features of the proposed model. Firstly, it is possible to represent different types of complexities by assigning different dimensionalities to them. Secondly, it is possible to transform a given type of complexity into another type of complexity by defining an operator to bring about the changes in the dimensionalities. In other words, whenever the measure or type of complexity changes, it is accompanied by the corresponding change in the dimensionality of the system. Even with this minimalist description of the changes in complexity, it is intuitively clear that this structuralism can be used to describe the relationship between genotype and phenotype because, as mentioned above, this relationship is essentially that of information transformation. Genotype contains explicit structuralism and implicit functionalities. However, phenotype contains explicit structuralism as well as explicit functionalities. Therefore, the relationship between genotype and phenotype can be viewed as two parallel processes. Firstly, there is an information transformation of the explicit structuralism of genotype to the explicit structuralism of phenotype. Secondly, there is an information transfer from the implicit functionality of genotype to the explicit functionalities of phenotype.

What is significant is that this conception of the relationship between genotype and phenotype is generic and doesn't depend on any specific type of structuralism or functionalities. Therefore, this conception can be used to define the relationship between genotype and phenotype in an abstract sense. Now, let us see how this relationship gets represented in the proposed model. Let us begin with the information transformation from the explicit structuralism of genotype to the explicit structuralism of phenotype. Even without going into the details of the process of translation, it is intuitively clear that the structuralism of phenotype must be more complex than the structuralism of the corresponding genotype. This is because a typical phenotype will have a protein template the domains of which are assembled during the process of translation. Purely from the information theoretical perspective, this emergence of phenotype from genotype can be likened to the process of infolding. The information implicit in the timing of the corresponding gene expressions and the presence of proper initiators and facilitators gets incorporated into the explicit structuralism of phenotype. Therefore, the information content of phenotype is, by definition, more complex than that of genotype. Secondly, this additional complexity is not only in the structuralism of phenotype but also in the functionalities of phenotype. What remains unclear is whether the information present in the functionalities of phenotype consists entirely of the temporal and catalytic information or something else. We will look at the nature of information content of the functionalities in the following chapters.

When this process is represented in the proposed model, it is intuitively clear that phenotype will occupy a different dimensionality than that of genotype. This is because in the proposed model, each type of complexity is assigned different dimensionalities. Of course, there is one counterintuitive feature of the proposed model. It postulates an inverse relationship between dimensionality and complexity. Therefore, greater the complexity, lower the dimensionality. Admittedly, this is a counterintuitive feature. However, its justifications are discussed in the preceding monograph (Chhaya 2022a, see Chapters 2 and 3). At present, it will suffice to remember that phenotype, as a rule, will occupy lower dimensionality vis a vis its corresponding genotype. As a corollary, the complexity of the functionalities of a phenotype may not be visible from the dimensionality of a genotype. Thus, it is because the template of functionalities manifest in different dimensions than that of the corresponding structuralism, we have not been able to perceive the template of functionalities. However, as discussed in the first chapter, mathematics comes to our rescue. Though, we cannot perceive the template of functionalities, we can perceive mathematical constructs corresponding to the dimensionality of the functionalities. We can perceive different mathematical constructs belonging to different dimensionalities because the ontology of mathematical objects is congruent with their epistemology.

This generalization is also true for the functionalities of phenotype because unlike genotype, phenotype also possesses functionalities which have their own template, thereby adding to its complexity. Now, let us see how to resolve the enigma of the emergence of complexity during natural selection which is essentially a random process. As discussed above, the conventional perspective of the Darwinian paradigm has no explanation of this, except the phase space argument. However, the emergence of complexity during biological evolution and natural selection finds different representations in the proposed model. Instead of the phase space argument which relies on the inevitability of the earliest living organisms being simplest structures by default, the proposed model suggests that the very process of gene expressions leads to more complex outcomes because it entails an inward folding of the information content during gene expressions. The process of gene expression, according to this model, invokes not just the process of translation of DNA sequence to amino acid sequences, but it also entails the incorporation of temporal and catalytic information into phenotype. However, because this additional information is encoded as the information in the form of the functionalities of phenotype, we fail to account for it.

It is our historical bias of equating the information content with the structural template that misleads us into believing that biological evolution and natural selection are random phenomena. In reality, these are processes which transfer some structural information into functional information and therefore, they naturally give rise to complexity. There are two indirect supporting arguments available to support this reasoning. Firstly, the operator of involution which can be used to define the process of conversion of genotype into phenotype belongs to a mathematical category of group operators. This implies that these operators can be applied successively, but only in certain combinations as defined by symmetry principles. Thus, this restriction on the number of combinations of different operators, results in manifestation of only a certain type of complexities. Therefore, the process of gene expression is conceptualized as the process bringing about the changes in complexity and that too only a certain type of complexities. This restriction on combining arises from group theory. According to group theory, there are only certain ways in which these operators can be combined. Therefore, by definition, these operators will allow only certain kinds of complexity to be favored. This is precisely what we find in our phylogenetic studies (Bromham 2008). Out of infinite forms of morphological variations, only some variations are observed. One can extend this argument to the genetic code itself (Tamura 2018). It is intuitively clear that out of 60 odd possible combinations of the triad of genetic codes, only a handful of combinations of genetic code are observed in Nature. This selection must be based on this group theoretical consideration.

The second argument in favor of this reasoning is available from the thermodynamics of the open systems (Prigorgine 1968). The conventional rationalization for the open systems is that if we were to include both mass and information transfer in the computation of the entropy of an open system, there is always a net increase in entropy, particularly from the perspective of the open system and the environment taken together. While the organization of matter in an open system like living organisms decreases the entropy, the overall information transfer, including heat dissipation leads to increase in entropy of the system. Similarly, we can argue that the information transformation from temporal and catalytic information content provided by enhancers and initiators during the course of gene expressions is transformed into the information content of phenotype in the form of its functionalities, merely making the information content invisible. Therefore, in reality, what we perceive to be an increase in complexity is actually converting the visible form of complexity into an invisible form of complexity.

Moreover, the temporal and catalytic information content gets localized into the functionalities of phenotype due to the inward folding of information. Therefore, it leads to net increase in complexity as well. However, the resulting heat dissipation ensures that overall there is a net increase in entropy of the system. There is another aspect of the thermodynamics of the open systems that has not been appreciated. This refers to the need to include both the environment and the open systems while computing the increase in the overall entropy. While this fact is well known, its Darwinian context is often overlooked. In the Darwinian paradigm, the environment plays an important role. This is actually a thermodynamic imperative, and therefore, biological evolution and natural selection must be studied in conjunction with the environment and not separately. While the conventional perspective of the Darwinian paradigm, the role of the environment as a priori, it is the thermodynamic imperative of the open systems that explains why any formal description of natural selection must include the environment. More importantly, according to the rationale given above, the role of the environment must be active in nature and not passive as implicit in the conventional perspective of the Darwinian paradigm. This is possible only if we accept that the environment participates in natural selection by transferring some of its information content to either gene expressions or to phenotypes. This scenario is consistent with the entropic perspective as well.

In summary, one can say that the perception that complexity emerges from the random nature of natural selection arises because we are unable to perceive the process of transformation of the temporal and catalytic information content provided by enhancers and initiators during the course of gene expressions. This information is actually converted into the information content of phenotype in the form of its functionalities. Thus, during biological evolution and natural selection, there is a change in the types of complexities from genomic architecture to the functionalities of phenotype. These changes in the types of complexities are governed by the symmetry principles and its underlying mathematics. Since it is possible to define the changes in the types of complexities using mathematics, it is necessary that we define genotype and phenotype as ensembles of different types of information content having different types of complexities. Moreover, it seems reasonable to think that the environment consists of not just the elements of the ecosystem, but also spacetime itself. Admittedly, some of these aspects were mentioned above in passing. Therefore, in the next section, we will outline a well-defined definition of genotype and phenotype.

2.14 Redefining Genotype and Phenotype

In the preceding sections, we discussed several reasons why our present conceptions of biological evolution and natural selection are inadequate. It was pointed out that this inadequacy arises from the implicit semantics of the Darwinian paradigm and its structuralism. Admittedly, with every paradigm shift, the Darwinian paradigm has undergone rejuvenation and even new semantic nuances. However, with the advent of genomics, we have realized that the very conception of genotype and phenotype, particularly the boundary between them, need to be rearticulated. Admittedly, it is possible to argue that the core of the semantics of the Darwinian paradigm, which has survived unchanged during these successive paradigm shifts, is linked to the definition of genotype and phenotype. Therefore, any changes in their definitions would amount to alterations in the core semantic propositions.

This reluctance to revisit the core semantics is understandable and perhaps desirable because what has survived these paradigm shifts cannot be trivial and therefore shouldn't be trifled with. This reluctance to revisit the semantic core of any theory owes its origin to our lack of understanding of the relationship between structuralism and semantics of scientific theories (This is true in the case of natural languages as well.). This lack of understanding the relationship between semantics and structuralism arises from the mechanisms by which our cognitive faculty comprehends the reality. As discussed in the preceding monograph, we arrive at the semantic perspective by employing our noetic processes. However, the cognitive processes by which we formalize a theory (and which define the structuralism of the underlying natural phenomena) are distinct from our noetic processes. As a result, there is a fundamental dichotomy between the semantics and structuralism of every scientific theory. This dichotomy arises from the fact that our cognitive faculty operates from multiple types of cognitive processes simultaneously. Moreover, our cognitive faculty is characterized by a certain procedural opacity. This opacity ensures that we can comprehend the outcomes of our cognitive processing, but the principles on which our cognitive processing is based remains beyond our cognitive capabilities. Therefore, it is possible that these different cognitive processes which give rise to different templates of structuralism, are themselves based on certain semantic principles. However, these semantic principles remain unknown or rather unknowable.

Returning to the core semantics of the Darwinian paradigm, let us accept that the conceptions of genotype and phenotype, in some sense, encapsulate some propositions of this core semantics of the Darwinian paradigm. Therefore, in this section, we will try to retain some features of the conception of genotype and phenotype even while trying to infuse new features to their conceptions. This selective approach will ensure certain semantic continuity between the old and new conceptions of genotype and phenotype. Admittedly, at the first sight, this approach might appear to be ad hoc. However, since this selective approach is also driven by the previously cited unknowable semantic propositions, it might reveal, upon explication, hitherto unknown semantic content. Therefore, beneath this seemingly ad hoc approach, there exists certain implicit semantic continuity waiting to be made explicit.

Let us begin with Darwin's own interpretation of natural selection and accept that genotype stands for heritable attributes and phenotype stands for selectable attributes. This primitive description can hardly be faulted. Therefore, we will begin with it. Darwin's theory (Hodge and Radick 2009) and even the later paradigms accept that there exist certain mechanisms by which these heritable attributes can be transformed into selectable attributes. With this reasoning in place, let us redefine them in the language of information theory. Thus, now we can define genotype as an ensemble of information content which has three characteristic properties. Firstly, this information can be replicated with a certain degree of fidelity. Secondly, this information can alter its configuration and finally, it can be transformed into a different type of information. Apparently, this description is merely a reaffirmation of the classical interpretation of genotype. However, there are certain subtle nuances to this information theoretical definition. Firstly, it talks about replication with a certain degree of fidelity. This includes not just transcription but also the retention of genomic architecture that influences gene expression. Secondly, it talks about reconfiguration of information content. Thus, it treats mutations and large scale genome changes as equivalent. In other words, it focuses not only on the chemical changes brought about during mutations, but also on the influences that every reconfiguration has on gene expressions. Finally, it concedes that gene expressions do not necessarily represent the chemical structural perspective but also on the manifestation of functional features of phenotype. It must be kept in mind that functionalities of a phenotype are not represented by its chemical structure, but by the domains of the tertiary structures of proteins. As we have learned from phylogenetic studies, it is structuralism of various domains of proteins that are concerned during natural selection (Bryson and Vogel 1965, see Parts III, IV, and V). Therefore, we need to define genotype and phenotype not by the sequences of nucleotides and amino acids respectively, but by what is replicated, selected and conserved. These changes appear to be minor, but its significance would be apparent as the monograph progresses.

Now let us look at the information theoretical perspective of phenotype. Apparently, as mentioned above, phenotype should not be defined by the sequence of amino acids alone. In addition to its chemical template, phenotype consists of functionalities. Conventionally, these functionalities are presumed to be in the form domains that arise from the tertiary structures of proteins. However, we also need to include phenotypic potentials that become manifest when it tries to interact and survive in the environment. Since these features are potential and not manifest, we tend to ignore them. Just as it took our experience in phylogenetic studies to realize that it is the three-dimensional perspectives of protein domains that are selected and conserved (Bryson and Vogel 1965), similarly, we will discover in the decades to come that these potential features are also selected and conserved. It is not as fanciful as it sounds. Just think of how our immune response reacts to infections. Our immune system eventually finds a way to contain and eliminate infections. It is often forgotten that our immune system has potential functionalities which materialize only in the presence of infections (Kresina 1998, see Part I). Historically, we have overlooked the possibility that these potential features of phenotypes are also selected and conserved during natural selection. Therefore, just as we are getting reconciled with the fact that the three-dimensional structures of protein domains that are selected, we will discover that the potential features of phenotype too are selected.

It is possible to argue that there is no way to define potential features beforehand, and therefore, they can never be selected by any mechanisms. This argument rests on our historical bias toward the molecular perspective of genotype and phenotype. Just as we now concede that natural selection must have some mechanisms to select the three-dimensional features of protein domains, it is possible that natural selection has some hitherto unknown mechanisms to select potential features. All that we require to do is to redefine the role of the environment in natural selection. It is apparent that if the three-dimensional protein domains are selected, then it is intuitively clear that the process of natural selection would employ the inherent structuralism of spacetime while selecting protein domains. Similarly, if the potential features of phenotype are selected, then there must be some inherent features of spacetime (see Chhaya 2022b) that must be playing a role in natural selection of these potential features of phenotype.

There are two possible shortcomings of such an utopian perspective. Firstly, there is no way to formalize these features of spacetime playing a role of an arbiter in natural selection. Even in the case of selection of protein domains, we have rather reluctantly conceded that the three-dimensional structuralism that is selected by natural selection. However, we are totally clueless about the mechanisms by which spacetime can bring about the selection of domains of proteins. In the case of potential features of phenotype, the situation is worse. We have no means to define these potential features, leave alone the mechanisms by which these features get selected. This is precisely what this monograph has set out to achieve. If we can assign an active role to the environment in natural selection, then we can define a spacetime model that operates on three-dimensional mechanisms, thereby participating in natural selection of three-dimensional protein domains.

Similarly, it is possible to borrow a technique from quantum mechanics to define potential features of phenotype. It must be kept in mind that in quantum chemistry (Gupta 2016, see Chapter 4), we assign all the potential and manifest properties of a molecule to its wave function. Similarly, we can define potential features of phenotype using a suitable wave function. However, both these tentative suggestions have a serious semantic conflict with the Darwinian paradigm. Both these approaches induce a certain degree of determinism into the nature of the relationship between genotype, phenotype and the environment. This, it might appear at the first sight, to be against the essentially random nature of the Darwinian paradigm. However, as discussed in the following chapters, this apprehension is misplaced. Therefore, let us look at the template of the conversion of genotype into phenotype, the role of the environment in this conversion and the nature of natural selection using the perspective developed above. To simplify the description, we will employ a schematic diagram to understand these aspects.

Schemas 2.1, 2.2, 2.3, 2.4, and 2.5 try to summarize the intricacies of gene expressions and how each step in this process contributes to the generation of phenotype. The details of these schemas can be summed up as follows

1. Every step of gene expression adds complexity to phenotype.
2. There are two types of complexities created during the process of gene expressions.
3. The structural complexity of genotype is converted into a different type of structural complexities of phenotype. However, due to inherent increased stereochemical complexity of proteins, there is an increase in complexity during this stage.
4. In addition, the process of gene expressions gives rise to functionalities of phenotype which are created de novo. This adds to the structural complexities of phenotype.
5. Long-range influences of the genome, the precision of the transcription machinery and the plasticity of the open reading frame of RNA processing add complexity to phenotype.
6. Thus, we must view the process of gene expression as a process of increasing and augmenting complexity. The random nature of long-range influences and open frame reading of RNA merely contribute to the increase in complexity.

Having analyzed the reasons why gene expressions lead to more complex outcomes, let us look at the new explanation of the emergence of complexity during an essentially random phenomenon of natural selection. This is necessary because

TYPES OF INFORMATION CONTENTS	GENOTYPE	PHENOTYPE	ENVIRONMENT
STRUCTURAL TEMPLATE	NUCLEOTIDE SEQUENCE	AMINO ACID SEQUENCE	DISTRIBUTION OF NATURAL ELEMENTS, FOOD
FUNCTIONALITIES	TRANSLATION	CATALYSIS, SUBSTRATE PERMEABILITY	RANDOM FLUCTUATIONS OF PARAMETERS
EXPLICIT INFORMATION	CIS AND TRANS INFLUENCES	HYDROPHILIC , HYDROPHOBIC FOLDS	CHANGES IN TEMPERATURE, AVAILABILITY OF FOOD ETC.
IMPLICIT INFORMATION	TEMPORAL EXPRESSIONS OF TRANSCRIPTION MACHINERY	PLASTICITY OF RESPONSES TO ENVIRONMENTAL STRESSES	STRESS TO MEASURE FITNESS OF PHENOTYPES
TOPOLOGICAL DISTRIBUTION OF INFORMATION	FOUR DIMENSIONAL DISTRIBUTION OF GENE EXPRESSIONS	FOUR DIMENSIONAL TRANSPORT OF MOLECULES	FOUR DIMENSIONAL STRESSORS

Schema 2.1 Types of information contents

GENOTYPIC INPUTS	ENVIRONMENTAL INPUTS	PHENOTYPIC OUTPUTS	CHANGES IN COMPLEXITY
NUCLEOTIDE SEQUENCE	EXTERNAL STIMULUS	AMINO ACID SEQUENCE	INCREASE IN STRUCTURALISM AND FUNCTIONALITIES
LONG RANGE INFLUENCES	PRIOR EXPRESSIONS OF INITIATORS	TEMPORAL DISTRIBUTION OF PROTEINS	INCREASE IN PLASTICITY OF RESPONSES
TRANSCRIPTION MACHINERY	PRIOR EXPRESSIONS OF m-RNA	m-RNA SPLICING	INCREASE IN DIVERSITY OF PROTEINS
TRANSLATION MACHINERY	PRIOR ASSEMBLY OF RIBOSOMAL COMPLEXES	PROTEIN FOLDS	INCREASE IN SURVIVAL PROFILE

Schema 2.2 Conversion of genotype to phenotype

we can argue that the process of gene expression is not the only thing in natural selection. Natural selection operates on far more deeper principles. Therefore, in the next section, we will discuss a new interpretation of natural selection.

SOURCE OF INFORMATION	TYPES OF INFORMATION	CONTRIBUTION TO PHENOTYPE	INCREASE IN COMPLEXITY OF PHENOTYPE
DNA SEQUENCE	AMINO ACID SEQUENCE	PROTEIN FOLDS AND DOMAINS	STEREOCHEMICAL COMPLEXITY
LONG RANGE INFLUENCES	INITIATORS AND FACILITATORS	TEMPORAL INFORMATION	PLASTICITY OF RESPONSES
TRANSCRIPTION MACHINERY	PROTEIN SYNTHESIS	HYDROPHOBIC AND HYDROPHILIC SURFACES	ENZYMATIC CATALYSIS
RNA PROCESSING	OPEN FRAME TRANSLATION	VARIABLE PROTEIN SYNTHESIS	TEMPORAL PLASTICITY
ENVIRONMENT	ENVIRONMENTAL STIMULI	SURVIVAL FITNESS	MORPHOLOGICAL VARIATIONS

Schema 2.3 Information profile of gene expressions

SOURCE OF GENE EXPRESSIONS	TYPES OF ENVIRONMENTAL CONTRIBUTION	NATURE OF ENVIRONMENTAL CONTRIBUTIONS	TYPES OF INCREASE IN COMPLEXITY
DNA SEQUENCE	MUTATIONS	NEW PROTEINS	PHENOTYPIC PLURALITY
LONG RANGE INFLUENCES	INDELS AND DUPLICATIONS	LOSS AND GAIN OF FUNCTIONS	MORPHOLOGICAL VARIATIONS
TRANSCRIPTION MACHINERY	SIGNAL TRANSDUCTION	TEMPORAL CHANGES IN PROTEIN SYNTHESIS	VARIABLE SURVIVAL FITNESS
RNA PROCESSING	SIGNAL TRANSDUCTION	ALTERNATIVE GENE EXPRESSIONS	MORPHOLOGICAL VARIATIONS

Schema 2.4 Information profile of natural selection

2.15 New Explanation of the Emergence of Complexity

In the previous section, we looked at the emergence of complexity during the process of gene expressions. While the proposed explanation justifies the emergence of complexity during the course of gene expressions, it cannot be extrapolated to the process of natural selection. This is because the process of gene expression, in spite of its plasticity, is still a causal process. It is true that it gives rise to multiple outcomes depending on the cytoplasmic signals available at the time of gene expressions, but it is still governed by causality. The randomness of outcomes is

SOURCE OF COMPLEXITY	TYPES OF COMPLEXITY PASSED ON TO PHENOTYPE	TYPES OF COMPLEXITY CREATED IN PHENOTYPE	PHENOTYPIC FEATURES CREATED DURING GENE EXPRESSIONS	NET INCREASE IN COMPLEXITY
DNA SEQUENCE	AMINO ACID SEQUENCE	STEREOCHEMI CAL ORIENTATIONS	PROTEIN PROTEIN INTERACTIONS	MORE COMPLEX STRUCTURALI SM AND NEW FUNCTIONAL TEMPLATE
LONG RANGE INFLUENCES	TEMPORAL SEQUENCE	SEQUENCE OF PROTEIN SYNTHESIS	ORGANIZATION OF TISSUES AND ORGANS	DIFFERENT PROTEIN PROTEIN INTERACTIONS
TRANSCRIPTION MACHINERY	FOLDS AND DOMAINS	ENZYMATIC CATALYSIS	CORRECT FOLDING OF PROTEINS	NEW FUNCTIONAL TEMPLATE
RNA PROCESSING	ALTERNATIVE PROTEINS	MULTIPLE ENZYMATIC CATALYSIS	INCREASE IN SURVIVAL PROFILE	DIFFERENT PROTEIN PROTEIN INTERACTIONS

Schema 2.5 Relationship between structuralism and functionalities during gene expressions

essentially a Bayesian type (Press and Clyde 2003). If certain cytoplasmic clues are present, the process will give a certain outcome, if not, there will be different outcomes. Thus, there is a conditionality attached to every outcome of gene expressions. This scenario fits very well into the conventional Darwinian theory.

However, this explanation, by itself, is inadequate to explain the emergence of complexity during biological evolution and natural selection. The reasons for this inadequacy are simple. If natural selection is Life's response to an ever-changing environment, then natural selection will reflect the randomness of the environmental changes. Therefore, since the environmental changes are random, biological evolution and natural selection too should lead to random structuralism. However, this is not the case. We know that there is a distinct sense of continuity in structural complexity during biological evolution and natural selection. Admittedly, there are examples of natural selection that might have created simpler organisms (cf. Loss of function (Alberts 2004)), but these are exceptions to the rule. In general, biological evolution and natural selection have given rise to more and more complex organisms. Therefore, there must be an explanation for this emergence of complexity during an essentially random phenomenon of natural selection.

The conventional wisdom suggests (incidentally, it is expressed by Darwin himself) that the reason for this anomaly lies in the fact that living organisms pass on the details of structuralism to the next generation. The concept of descent with modifications encapsulates this continuity of complexity during natural selection. Thus, the postulate of the common ontology of all the living organisms, partly explains why the successive generations of living organisms are more and more complex. This is because the process of natural selection merely improves upon the existing complexity of the previous generation of living organisms. Therefore, axiomatically, the successive generations of living organisms will be more complex than their predecessors.

Before we delve into the reasons why this explanation is inadequate, we must realize that this argument, even if it were completely true, is not applicable to the evolution of Life. In retrospect, it is intuitively clear that the Darwinian paradigm (Grene 1986) assumes biological evolution as a priori. Its main focus is on the subsequent natural selection and the competitive survival. This is possibly due to the fact nothing much about the origin of Life was known, except perhaps the theological narrative of the origin of Life. It is also important to remember that even the arrival of Mendelian genetics didn't alter the situation because it too doesn't deal with biological evolution, rather it deals with how different traits, having arisen somehow, compete from their expressions. It was only after the advent of molecular biology that one could legitimately frame the question of the origin of Life Perhaps the best solution to the question of the origin of Life is the one provided by the RNA world scenario (Yarus 2010). With the help of modern techniques of molecular biology, it is possible to postulate and verify models of RNA could have evolved from the primordial soup of the cocktail of simple chemicals present on Earth at that time. Leaving aside the plethora of such models, let us see why the RNA world hypothesis cannot explain the emergence of complexity during biological evolution. More importantly, why the conventional wisdom described above also doesn't work in this scenario.

The reason why scientists veered toward the hypothesis that Life must have begun with RNA and not with DNA, is that while a DNA molecule is very good in storing and passing on the information content to the next generation, it cannot act on its own. It needs catalysts in the form of proteins to carry out these tasks. However, with the discovery of ribozymes, it was demonstrated that RNA is capable of performing these tasks of storing and passing on the information on its own. Therefore, it is axiomatic that RNA must be a predecessor of DNA as a unit of inheritance. This is plausible because RNA can give rise to several enzymes, including reverse transcriptase to transfer information to DNA itself. Admittedly, this germ of an idea has now become a full-fledged scientific research with deep insights into the origin of biological evolution.

However, this argument, though insightful, suffers from one drawback. It doesn't explain how RNA is capable of performing both these tasks of storing and relaying information by itself. Apparently, DNA possesses similar structuralism, and yet, it can perform only one function of storing information on its own. (To perform the other task of passing on information to the next generation, it needs the external help from proteins.) Admittedly, there is something unique about RNA that enables it to perform these two tasks on its own. However, we have not been able to formalize it. It is possible to argue that this is true not just for RNA, but also true for the entire range of catalysts. In fact, this is true for all the chemical functionalities. Therefore, we must accept these dual functionalities of RNA, viz., information storage and information transfer as a priori and try to deconstruct the origin of Life. This tacit acceptance has spurred the research in the hypothesis of the RNA world. However, the fact remains that we cannot formalize any structural basis for this dual functionality of RNA.

The analogy with other chemical functionalities like catalysis is not valid when it is applied to the dual functionalities of RNA. What RNA passed on to the next generation of phenotypes (the task which now DNA performs with the help of proteins) was not just the information content, but also its own functionalities. While from the chemical perspective, it is axiomatic that when a similar structure of a molecule is generated by chemical synthesis, the resulting molecules will have similar functionalities. There is a direct correspondence between chemical structural templates and its functionalities. Therefore, we have overlooked this anomaly in the functionalities of RNA. However, from the biological perspective and even from the information theoretical perspective, this is not a normal phenomenon. RNA didn't just pass on the information content but it passed on its own functionalities to the next generation. This is a classic case of self-reference. It was as if RNA "knew" what it was doing. It simply didn't pass on structural information to the next generation of phenotype, it passed on its ability to pass on the structural information as well.

In order to understand the distinction between the passage of structuralism and the passage of the ability to pass on this structuralism, let us step back and see what happens in chemical reactions. Suppose, we are using a catalyst to bring about a reaction. Our objective is not to produce a next generation of catalysts. However, if at the end of the reaction, by a stroke of luck, we discover that the product of this reaction also turns out to be a catalyst. While we feel happy with this good fortune of having accidentally discovered a new catalyst, we know from our experience that the new catalyst has certain catalytic functionality because it has a certain structural template. We never, not even for a moment, think that the catalytic functionality of the new catalyst is dependent on the original catalyst we had employed in carrying out the reaction. Thus, both these catalysts have their own functionality of catalysis, but in each case, it is strictly dependent on their own structuralism. We never formalize any theory that explains the origin of the catalytic functionality of the new catalyst on the structural template of the first catalyst that we had employed. The irony is that what we think is preposterous in chemistry is accepted as a priori in biology. We somehow concede that RNA can pass on the structural information as well as its own functionalities to the next generation.

As discussed in the first chapter, this situation arises because the template of functionalities is beyond our cognitive capabilities. While the complexity of structuralism is available to us, the corresponding complexity of functionalities is not available to us. However, our cognitive faculty is capable of deciphering underlying mathematics in both these cases. As discussed in the preceding monograph (Chhaya 2022d), our cognitive faculty operates from multiple dimensionalities simultaneously with each cognitive process operating from a unique dimensionality. Since each dimensionality has its own metric, our cognitive faculty can perceive several types of mathematical constructs, each belonging to different dimensionalities. Therefore, there are several natural phenomena whose dimensionality matches that of our cognitive processes and therefore, we are able to formalize them. The chemical complexity is one such example. At the same time, there are natural phenomena whose dimensionality is beyond the range of dimensionalities of our cognitive faculty. Therefore, those phenomena remain beyond our capabilities of formalization. Quantum phenomenon is an example of this category. However, since our cognitive faculty operates from multiple dimensionalities simultaneously, it has its mechanism of switching from one dimensionality to another. This mechanism is being sought to be formalized in the preceding monographs. It is the contention of this monograph that biological functionalities (and by implication, molecular functionalities) may not be amenable to conception, they can certainly be inferred using this formalization developed in defining changes in the metric details during the changes in the dimensionalities of our cognitive faculty.

When viewed from this perspective, it is intuitively clear that the complexities of phenotype will occupy different dimensionalities from those of corresponding genotype. Therefore, there exists a possibility of defining the complexity of phenotypes in the language of the complexity of genotype. In the context of the present discussion, it is possible to view natural selection not as a process whereby complexity emerges out of nowhere, but as a process whereby one type of complexity gives rise to complexity of a different type. This explanation also provides a semantic justification for employing the phase space argument. The conception of phase space rests on the belief that all types of complexities are interconnected and the variations arise due to "random walk" across the phase space. Now, according to this model, different types of complexities, or each point of the phase space, are connected to one another by these changes in the dimensionalities of respective complexity. Therefore, the "random walk" actually consists of walking across different dimensionalities. More importantly, there is a separate phase space representing functionalities which is connected to the phase space representing structuralism. In the context of the present discussion, one can understand the emergence of complexity during biological evolution and natural selection, by conceptualizing two phase spaces, one for genotype and phenotype each.

Interestingly, this model helps us to differentiate between Darwinian (Hodge and Radick 2009) and Lamarckian (Steele et al. 1998) interpretations of natural selection. The Darwinian interpretation suggests that it is the changes in the phase space of genotype that influences the changes in the phase space representing phenotype. On the other hand, the Lamarckian interpretation can be visualized as the phase space representing phenotype influences the phase space representing genotype. Admittedly, there is no evidence, either experimental or theoretical, to support the Lamarckian interpretation, we have yet to rationalize the epigenetic phenomena within the confines of the Darwinian paradigm. The proposed model offers a way to link epigenetic effects with the conventional Darwinian paradigm. We can postulate that the functional complexities of phenotype can influence the structural complexities of genotype through the influence between the two corresponding dimensionalities.

Admittedly, this interpretation implicitly opens the door for the design principle (and therefore to a teleological paradigm). However, there is a vital semantic distinction between the proposed model and any kind of teleology whatsoever. According to this model, there are any number of different types of functionalities in a given phase space. However, there are no preferred types of complexities (thereby eliminating design principles). It is just that the phase space representing functionalities occupies a higher dimensional space vis a vis its corresponding dimensionality of structuralism. Therefore, there exist certain mechanisms by which different functionalities are connected to the corresponding structuralism. However, just as different types of structural template are connected to one another, different types of functionalities are also connected to one another. In other words, there must be two types of "random walk," one in the phase space representing structuralism and another one in the phase space representing functionalities. Moreover, since structuralism and functionalities are connected to one another, the degree of randomness in each of these two "random walks" is limited. This is the reason why we observe only a certain kind of complexities in both these domains. It must be kept in mind that according to this model, the restriction to the randomness of the "random walk" would be more severe in the case of structuralism than in the case of functionalities. This is because of the different "granularity" of both these phase spaces. Since according to the proposed model, the phase space representing functionalities occupies a higher dimensionality, its granularity will be coarser than the granularity of the corresponding phase space representing structuralism. Therefore, functionalities will appear to some kind of continua, the corresponding structuralism will appear to be discrete jumps. This is best exemplified by our immune response. While the genotypic changes in the form of recombinations of different hypervariable segments of MHC complex appear to be random algebraic combinations of different DNA sequences, the corresponding changes in our immune response is a continuous spectrum (Kresina 1998). Thus, it is possible to rationalize the emergence of complexity during biological evolution and natural selection using the proposed model. In the next section, we will describe how this translates into the relationship between genotype and phenotype.

2.16 Role of Genotype and Phenotype in the Emergence of Complexity

In view of the reasoning given above about the emergence of complexity during biological evolution and natural selection, it becomes necessary to redefine the relationship between genotype and phenotype. This is all the more necessary because we wish to adhere to the Darwinian semantics of randomness. In fact, one of the significant fallout of the above reasoning is that it also accommodates the Lamarckian interpretation into its fold. However, as argued above, it doesn't lead to either any design principle or to any teleological paradigm. However, we trust that the justifications for the randomness implicit in Darwin's theory are available in the relationship between genotype and phenotype. Therefore, in this section, we will try to redefine this relationship, and in the next section, we will try to deconstruct the semantic imperative behind all the dualities present in the Darwinian paradigm, viz., the dualities of genotype/phenotype, DNA/RNA, and structuralism/functionalities.

In this section, we will focus on two aspects of the relationship between genotype and phenotype. These aspects are the nature of information transfers between genotype and phenotype and the contribution of genotype and phenotype to the increase in the complexity during biological evolution and natural selection. Let us begin with the nature of information transfers between genotype and phenotype. This is important because according to the Darwinian paradigm (Hodge and Radick 2009), this information transfer is one directional, whereas in the case of the Lamarckian interpretation (and therefore in the epigenetic phenomena as well), this information transfer is bidirectional (Steele et al. 1998). Therefore, it is vital to find out what the proposed model offers to resolve this dilemma.

As discussed in the preceding sections, the conventional perspective of the Darwinian paradigm is categorical about the one way influence of genotype on phenotypes. Admittedly, this view is further refined by molecular biology to redefine the boundary between genotype, phenotype and the environment. However, what changes in molecular biology is the definition of what constitutes a genotype and phenotype, but there is no dispute about the fact that the influence is always the genotype that influences phenotype and not vice versa. The proposed model, as discussed above, modifies this scenario. It suggests that genotype and phenotype each have their own templates. Therefore, although it is always genotype that influences the nature of phenotype, this influence is partial. While genotype gives rise to the complexity of phenotypes, the type of complexity of phenotypes is decided by the inherent template of phenotypes. Thus, even if genotype were to give rise to a large number of phenotypic variations, the choice between these phenotypic variations is decided not by genotype, but by the inherent template of phenotype. In developmental biology, this is usually referred to as canalization (Cross 2003).

However, in the conventional perspective, this process of canalization is governed by temporal genomic signals. However, according to this model, this choice is decided by the inherent template of phenotype. It is as if different options available for canalization have different free energies and depending on the nature of gene expressions, a particular phenotypic variation manifests itself. Upon a little reflection, it is intuitively clear that this is primarily what the environment is supposed to do in the classical Darwinian theory. In the classical interpretation of the Darwinian paradigm, the environment is nothing more than the combination of the temporal controls of genomic influences and the options from the phase space representing different free energies. Thus, the proposed model merely differentiates these two features of temporal influences (in the form of the sequence of gene expressions decided by the long-range influences of the genome) and spatial influences (in form of the free energies embedded in the phase space defining phenotypic variations). The proposed model instead creates a conception of dimensionalities to represent these temporal and spatial influences.

The proposed model formalizes different long-range influences of the genome into different dimensionalities. Similarly, it formalizes different points of the phase space representing phenotypic variations as different dimensionalities. More importantly, since these different dimensionalities are also interconnected to one another, we now have a topological equivalent of the conventional phase space. Moreover, we can change these dimensionalities by a uniform mechanism defined as the modified operators of involution. Therefore, let us see how genotype and phenotype contribute to the resulting complexity. In the conventional perspective, it is only genotype that contributes to the complexity by giving rise to complex phenotype. However, according to this model, genotype gives rise to structuralism of phenotype and therefore, adds to complexity. However, according to this model, even phenotype contributes to the complexity. Since every phenotype has its own functional template which is distinct from its structural template (and yet it derives itself from the structural template), phenotype increases the complexity. Thus, overall, there are two types of increases in the complexity at the end of the gene expressions. Firstly, phenotype is created de novo, thereby increasing the complexity of the organism. Secondly, phenotype also contributes to the complexity of the organism in the form of new functionalities.

The reason why these functional complexities have escaped our attention is that it manifests itself in a different dimensionality, the one that is not perceived by our cognitive faculty. However, as mentioned above, we are able to comprehend the nature of functionalities because the underlying mathematics of the complexity of genotype and phenotype is similar. Therefore, we can perceive the functional complexities, but cannot formalize it directly. This scenario also helps us to understand the nature of the role of the environment. While the role of the environment in the form of geophysical parameters does influence the outcomes of different gene expressions, the environment, in the form of spacetime (as embodied in its fine structure (see Friedman 1983)) influences the template of functionalities. This influence is indirectly manifest in the form of limited types of complexities that emerge during the course of evolution. This is implicit in the phase space argument. In the phase space, every isoform is linked to one another and the changes are seen as random walks. This linkage is because every kind of complexity is linked to one another at the higher dimensionality. Thus, any causal process which carries its own legacy (which is essentially what biological evolution does) will manifest only a limited range of complexity. Admittedly, this scenario appears to be intangible, if not superfluous. However, it is capable of explaining and clarifying semantic ambiguities of the Darwinian paradigm. One of the most fundamental semantic ambiguities of the Darwinian paradigm is why there should be different types of dualities. Therefore, in the next section, we will try to find out whether the proposed model offers any cogent explanation for these dualities. Having done that, we will conclude this chapter with some suggestions on how to verify the proposed model.

2.17 Semantics of Dualities

As discussed in the first chapter, the Darwinian paradigm has undergone several paradigm shifts. What is remarkable about the Darwinian paradigm is that it has not only survived these paradigm shifts, but it has found new semantic nuances to enrich its core semantics. Therefore, it seems reasonable to think that the Darwinian semantics captures some of the fundamental aspects of Nature. Therefore, it is interesting to find out why one of the semantic ambiguities of the Darwinian paradigm has remained unchanged during these successive paradigm shifts. More importantly, this semantic ambiguity has also found different expressions within each of these paradigm shifts. It is as if this semantic ambiguity, in the form of dualities, represents a hidden secret of Life. To borrow Darwin's own terminology, it is tempting to think that the semantic of dualities is a genotype itself and the various forms of dualities in different paradigms are the phenotypes of that semantic proposition.

These three forms of duality, viz., genotype/phenotype, DNA/RNA, and structuralism/functionalities, seem to represent the basic prerequisite for biological evolution and natural selection. However, till date, we have not been able to decode this prerequisite status of these dualities. Shorn of linguistic flourish, we can ask three fundamental questions and answers to these questions will tell why and how Life originated. These questions are: Why can't natural selection operate directly on genotype? Why can't either DNA or RNA alone perform the tasks of information storage and transfer? Why can't structuralism be synonymous with functionality? It must be kept in mind that we have several post facto rationalizations for these rhetorical questions, but these are really not answers to these questions. They are simply alibis for the real answers. For instance, it is conventionally argued that DNA/RNA duality is necessary because DNA has no measurable catalytic functionality which is necessary for information transfers. Similarly, it can be argued that RNA may be good for catalytic functionality, but its capacity to store information is not as good as that of DNA. However, these arguments miss the key point about biological evolution. Biological evolution can latch on to any feature and hone it to perfection during the course of natural selection.

Just think about our acquired immune response (Kresina 1998). It latches on to the correct epitope of the infection and hones its own specificity in a matter of days. Therefore, if biological evolution could have latched onto a perfect template of polymer that was good at both these activities of information storage and information transfer. However, the fact remains that biological evolution has chosen to maintain the duality of DNA/RNA. Therefore, there is something inherently fundamental structural prerequisite about biological evolution and natural selection that has escaped our attention. Therefore, our ignorance manifests itself as these three forms of the semantic ambiguity of dualities. As discussed in the first chapter, we have reconciled with these ambiguities in the mistaken belief that any such semantic resolution will compromise on the basic tenet of randomness implicit in Darwin's theory. However, this is a category mistake. Darwin's theory and for that matter, all the subsequent paradigms are essentially causal explanations. Therefore, it is inevitable that any causal explanation will have an underlying structural template. It is important to remember that Darwinian randomness doesn't arise from the non-causal nature of biological evolution and natural selection (which is the case with quantum phenomena). Rather, Darwinian randomness arises from the Bayesian probabilities of descent with modifications. Therefore, it is imperative that we must articulate the semantics of duality of the Darwinian paradigm. Admittedly, this is too profound a topic to be articulated in a small section such as this one. Therefore, we will briefly outline the core arguments here. We will return to this topic in the concluding chapters of this monograph.

Prima facie, in the absence of any knowledge of the semantics of dualities, we can think of two prerequisite conditions under which these dualities would axiomatically manifest themselves. Firstly, the structural requirements for natural selection are such that it cannot be fulfilled by either genotype or phenotype on their own. This is self-evidently true because, otherwise, natural selection would have operated directly on genotype or on phenotypes (in which case we would have labeled phenotype as genotype!). Therefore, we have to concede that the need for these dualities arises from the very structural template of natural selection. Secondly, there is something inherently inadequate in the templates of either form of these dualities that makes them dependent on their isoforms for their survival. In other words, these dualities are actually conjoined twins. Of these two inferences, the first one is unexceptional in the sense that it cannot be falsified unless it is articulated. The second inference is rather controversial because we know, at least in the case of the duality of DNA/RNA, that RNA can exist on its own. However, we will discuss this anomaly in the following chapters. Presently, we will accept these two inferences as serious scientific hypotheses.

If we were to accept these two inferences as provisionally valid, the next logical inference presents itself almost axiomatically. Since natural selection requires both isoforms of these dualities as prerequisites for its operations, it is intuitively clear that the process of natural selection operates at more than one level simultaneously. To understand this, let us employ the Socratic method of *ad absurdum reductio*. Let us assume that the process of natural selection operates separately on genotype and phenotype. In that case, descent with modifications will operate on genotype and phenotype in parallel. However, since we know that genotype and phenotype are causally connected to one another (albeit unidirectionally according to the Darwinian paradigm), it would result in a mismatch between genotype and phenotype. This is not the case. Although we are not able to explain every morphological or functional features of phenotype on the basis of genotype, we are convinced that this inability to explain is due to our lack of knowledge and not due to mismatch between genotype and phenotype. In fact, if phenotype were to modify independently of its genotype, the entire edifice of the Darwinian paradigm would collapse. At the heart of the Darwinian paradigm is the causal processes linking genotype, phenotype, and the environment. Therefore, to the extent we concede that genotype and phenotype are causally connected, there is only one inescapable inference available, i.e., the process of natural selection operates at different levels simultaneously. As a logical corollary, it is intuitively clear that genotype and phenotype must exist in different levels.

Thus, the semantic imperative of these dualities lies in simultaneity of the process of natural selection at more than one level. Because natural selection operates from multiple levels, we need the duality of genotype/phenotype, DNA/RNA and structuralism/functionalities. This scenario changes the semantics of the relationship between genotype and phenotype, between DNA and RNA and between structuralism and functionalities. More importantly, it changes the semantics of biological evolution and natural selection. We will expound on these aspects in the following chapters. Presently, let us summarize our discussion on the nature of the relationship between genotype and phenotype in the next section.

2.18 Conclusion

In the preceding sections, we discussed various aspects of the relationship between genotype and phenotype. Therefore, in this concluding section, we summarize the discussion given above. For the sake of simplicity, the summary is presented in a point-wise manner.

1. The central theme of Darwin's theory, descent with modifications necessitates the conception of genotype and phenotype.
2. The definitions of genotype and phenotype have evolved from a rigid bifurcation to contextual definitions.
3. With every paradigm shift, while the boundary between genotype and phenotype has shifted, the need for having separate units of inheritance and selection has remained unchanged.
4. This enduring need to have duality of the units of inheritance and selection, points toward a deeper semantic compulsions of the Darwinian paradigm.
5. The paradigm shifts from Darwin's own interpretation to genetic interpretation (including population genetics) and from genetics to molecular biology and from molecular biology to genomics have their own parallel dualities to replace the duality of genotype and phenotype.
6. However, all these dualities, viz., genotype/phenotype, DNA/RNA, and structuralism/functionalities, have retained a certain degree of semantic ambiguities about the relationship between themselves.
7. These semantic ambiguities are exemplified by the causes of the emergence of complexity during biological evolution and natural selection.
8. Irrespective of the paradigm that we employ, there is no satisfactory way to explain the emergence of complexity during what is an essentially random phenomenon of natural selection.
9. This lack of explanation for the emergence of complexity is further complicated by the way the role of the environment is formalized in each of these paradigms.
10. The proposed model offers a way to link the notion of complexity with the dimensionality.

11. It enables one to define genotype and phenotype in the form of their complexity of their information content.
12. Using this approach, it is possible to explain the emergence of complexity during biological evolution and natural selection, as the changes in the types of complexities rather than an increase in complexity.
13. Since different degrees of complexity are defined as different dimensionalities, it is possible to formalize natural selection as a process whereby the dimensionality changes.
14. Since the changes in dimensionality which defines the relationship between genotype and phenotype (and by implication between structuralism and functionalities), it is possible to define the role of the environment differently.
15. This chapter builds up a basic framework for redefining natural selection. These ideas will be elaborated and formalized in the following chapters.

References

Alberts B (ed) (2004) Molecular biology of cell. Garland Science, New York

Bard J (2016) Principles of evolution: systems, species and the history of life. Garland Science, New York

Barth FG, Giampieri-Deutsch P, Klein H (eds) (2012) Sensory perception: mind and matter. Springer, Dordrecht

Bonner JT (2013) Randomness in evolution. Princeton University Press, Princeton

Bromham L (2008) An introduction to molecular evolution and phylogenetics. Oxford University Press, Oxford

Bryson V, Vogel HJ (1965) Evolving genes and proteins. Academic Press, New York

Chhaya P (2022a) The origin and nature of mathematics. (Submitted)

Chhaya P (2022b) The origin and nature of spacetime. (Submitted)

Chhaya P (2022c) The origin of the counterintuitive nature of quantum phenomena. (Submitted)

Chhaya P (2022d) The origin and nature of cognition. (Submitted)

Chhaya P (2022e) On the origin and nature of languages. (To be Submitted)

Cottingham WN, Greenwood DA (2013) An introduction to standard model of particle physics. Cambridge University Press, Cambridge

Cross KTM (2003) The origins of the theory of genetic canalization and how the concept evolved including its impact on developmental and evolutionary biology overtime. Harvard University Press, Cambridge

Darden L (1991) Theory change in science: strategies from Mendelian genetics. Oxford University Press, Oxford

Delisle RG (ed) (2021) Natural selection: revisiting explanatory role in evolutionary biology. Springer, Dordrecht

Dennet DC (1995) Darwin's dangerous idea: evolution and meaning of life. Simon and Schuster, New York

DeVos J, Pluth E (eds) (2015) Neuroscience and critique: exploring the limits of neurological turn. Routledge, London

Dragini TA (ed) (1998) Human polygenic diseases. Harwood Academic Press, Amsterdam

Flew A (2017) Darwinian evolution. Routledge, London

Fodor J, Piatelli-Palmarini M (2011) What Darwin got wrong? Profile Books, London

Friedman M (1983) Foundation of spacetime theories: relativistic physics and philosophy of science. Princeton University Press, Princeton
Gould SJ (2002) The structure of evolutionary theory. Harvard University Press, Cambridge
Grene M (1986) Dimensions of Darwinism: themes and counter themes in the twentieth century evolutionary theory. Cambridge University Press, Cambridge
Gupta VP (2016) Principles and applications of quantum chemistry. Academic Press, London
Hodge J, Radick G (eds) (2009) Cambridge companion to Darwin. Cambridge University Press, Cambridge
Jackendoff R (2002) Foundations of language: brain, meaning, grammar and evolution. Oxford University Press, Oxford
Jeanteur P (ed) (2003) Regulation of alternative splicing. Springer, Dordrecht
Kimura M (1983) The neutral theory of molecular evolution. Cambridge University Press, Cambridge
Kresina TF (ed) (1998) Immune modulating agents. CRC Press, Boca Raton
Kurzfield NC (2009) Developmental gene expression regulation. Nova Biomedical Books, New York
Nei M (2013) Mutation driven evolution. Oxford University Press, Oxford
Okasha S (2010) Evolution and the levels of selection. Oxford University Press, Oxford
Pevsner J (2015) Bioinformatics and functional genomics. Wiley, Hoboken
Popper K (1963) Conjectures and refutations: the growth of scientific knowledge. Routledge, London
Press SJ, Clyde CM (2003) Subjective and objective Bayesian statistics: principles, models and applications. Wiley, Hoboken
Prigorgine I (1968) Introduction thermodynamics of irreversible processes. Wiley, Hoboken
Provine WB (2001) The origins of theoretical population genetics. University of Chicago Press, Chicago
Robert JS (2004) Embryology, epigenesis and evolution: taking development seriously. Cambridge University Press, Cambridge
Smith E, Morowitz HJ (2016) The origin and nature of life on earth. Cambridge University Press, Cambridge
Smullyan RM (1992) Godel's incompleteness theorems. Oxford University Press, Oxford
Sober E (2015) Ockham's razor. Cambridge University Press, Cambridge
Steele EJ, Lindley RA, Blanden RV (1998) Lamarck's signature: how retrogenes are changing Darwin's natural selection paradigm. Allen and Unwin, New South Wales
Tamura K (ed) (2018) The origin and evolution of the genetic code: 100th anniversary of the birth of Francis Crick. MDPI, Basel
Weinzierl ROJ (1999) Mechanism of gene expression: structure, function and evolution of the basal transcriptional machinery. Imperial College Press, London
Wigner E (1960) The unreasonable effectiveness of mathematics in natural science. Comm Pure Appl Math 13:1–14
Yarus M (2010) Life from an RNA world: the ancestor within. Harvard University Press, Cambridge

3 Nature of Genomic Architecture: A Topological Model of Genome

Abstract

Our present understanding of genomic architecture is based on incremental addition of the manner in which genes are expressed. At present, there is no theoretical model of genomic architecture. This is largely because genomics is focused on how the genome operates and not on how the genome has evolved. In the absence of any such ontological perspective, the genomic architecture cannot be formalized. This chapter seeks to fill in this lacuna. A topological model of the genome based on the formalism of the involuted manifold is outlined in this chapter. It seeks to formalize long-range influences involved in gene expressions by the class of operators called involutions. This chapter describes a possible course of evolution of genomic architecture using the operators of involutions as representing the forces of natural selection.

3.1 Introduction

The domain of genomics (Pevsner 2015, see p. 635) is characterized by the absence of any theoretical model of the genome. Our present knowledge of genomics is based on incremental additions to experimental observations. While this has resulted in a vast collection of data, there is no structural template which can accommodate this data. This lack of structuralism in genomics can be attributed to three historical reasons. Firstly, the Darwinian paradigm itself lacks necessary structuralism (Grene 1986). Secondly, molecular biology has always preoccupied itself with mechanical details of translation and transcription of DNA rather than on the long-range control elements (Weinzierl 1999, see Chapter 2). Thirdly, even in genomics, there is an absence of focus on the way the genomic architecture itself could have evolved.

The Darwinian theory (Hodge and Redick 2009, see p. 147), for instance, is unique as a scientific theory because it is devoid of any predictive methodologies.

© The Author(s), under exclusive license to Springer Nature Singapore Pte Ltd. 2023

P. Chhaya, *The Topological Model of Genome and Evolution*,
https://doi.org/10.1007/978-981-99-4318-0_3

This is because the Darwinian paradigm has steadfastly avoided structuralism or design in its semantics. The notion of randomness in the Darwinian paradigm (Bonner 2013) has been considered antithetical to any definitive structural template of natural selection. Thus, we can never predict the future course of evolution by using the Darwinian paradigm. Of course, using population genetics (Provine 2001, see Chapter 5), we can predict gene frequencies in any given population; however, we can never predict the course of evolution in that population. In some sense, Darwinian theory looks like a post facto rationalization rather than a proper scientific model. There are several reasons for this peculiar nature of the Darwinian paradigm and there exists a long standing debate on which of these reasons are primarily responsible for the peculiar nature of the Darwinian paradigm. However, there is one undisputed fact about the Darwinian paradigm that epitomizes this peculiarity. Darwinian theory has always been focused on the process of natural selection but never on evolution of life. In the edifice of the Darwinian theory, the biological evolution of life is taken as a priori. It is this absence of any ontological perspective that is the Darwinian paradigm (Grene 1986) that has been passed on to the subsequent domains of genetics, molecular biology and genomics.

Even after the advent of molecular biology, this nonontological perspective has remained intact. While this is understandable because molecular biology, by definition, is centered around the interactions at the molecular level. However, we cannot fail to notice that even at the molecular level, the long-range influences, in the form of control elements present in a given chromosome have not been formalized (Shmulevich and Dougherty 2014). Of course, one of the reasons for this lack of large-scale perspective is that at the time when molecular biology was being established, our computational capabilities were limited. Therefore, the necessary computational models were not available. However, it must be admitted that it was actually the inherent ontological bias of molecular biology that resulted in lacuna.

The domain of genomics however cannot offer any such alibis. It is honor bound to seek out a large scale model of genomic architecture. It has necessary computational wherewithal and an enormous amount of data available for such model building. If such a model of genome isn't forthcoming, then we ought to begin ab initio starting with the first principles. The problem with such an approach is that, as mentioned above, the Darwinian paradigm is silent on biological evolution itself. This is surprising, considering the fact that we concede that natural selection can occur at more than one level (Okasha 2010, see Chapter 2). Therefore, there is nothing in the Darwinian theory to prevent assuming that the complex molecules themselves could be subject to the process of natural selection. Admittedly, as discussed in the previous chapter, we have sought to develop a molecular perspective of biological evolution by postulating that RNA must be a good candidate for this purpose (Flew 2017, see Part III). However, it must be kept in mind that this approach is based on the ability of RNA molecules to bring about autocatalysis and not on Darwinian principles. It seems reasonable to think that the Darwinian paradigm aptly describes the process of natural selection and not necessarily biological evolution. Moreover, if natural selection operates at different levels simultaneously, it is axiomatic that a genome too could be a unit of selection. It must be kept in mind that the discipline of phylogenetics (Bromham 2008, see Chapter 5) is limited to study the changes in the DNA at the level of individual bases. However, there is no corresponding discipline to analyze large scale genomic changes. It is true that we routinely undertake genome wide association studies (Appasani 2016), but their focus is on the individual base pairs and rarely on higher structural elements.

In view of this situation, in this chapter, we will look at one such model of genomic architecture. In order to represent these unspecified large-scale structural elements of the genome and the long-range influences involved in control of gene expressions, a higher dimensional topological model has been outlined here. Moreover, in order to represent these long-range features and their influences on the individual genes, an operator of involution has been formalized. The resulting involuted manifold model seems capable of formalizing two aspects of genomics. It is suggested that this model formalizes both the functional and evolutionary perspectives of genomic architecture.

In this chapter, we will try to formalize genomic architecture based on three guiding postulates. These postulates are in consonance with our current understanding of genomics. However, there is no basis, either theoretical or experimental, to support these postulates. Since these postulates appear to be in agreement with our current understanding of genomics, they ought to be taken as serious scientific hypotheses, albeit subject to subsequent verification. As discussed in later sections, since these postulates represent a different semantic perspective, their eventual validation may lead to a new paradigm. These three postulates pertain to the nature of natural selection, nature of genomic architecture and nature of long-range influences involved in gene expressions. These three postulates, formally stated in the following sections, are quite mainstream by themselves. However, collectively, they give rise to a radically different scenario of the evolution of the genome.

This chapter has been further divided into 20 sections. Section 3.2: Conventional Perspective of Natural Selection, Sect. 3.3: Proposed Model of Natural Selection, Sect. 3.4: Postulate of Natural Selection, Sect. 3.5: Explicit Genomic Architecture, Sect. 3.6: Implicit Genomic Architecture, Sect. 3.7: Postulate of Genomic Architecture, Sect. 3.8: Functional Genome Versus Structural Genome, Sect. 3.9: Nature of Long-Range Influences in the Genome, Sect. 3.10: Postulate of Long-Range Influences, Sect. 3.11: Units of Selection Versus Units of Inheritance, Sect. 3.12: Topological Model of Genomic Unit Genotope, Sect. 3.13: Semantics of Genotope, Sect. 3.14: The Relationship Between Genotype and Phenotype Using Genotope, Sect. 3.15: Natural Selection and Genotope, Sect. 3.16: Genotopic Architecture of Genome, Sect. 3.17: Evolution of Genotopic Genome, Sect. 3.18: Phylogenetic Evidence for Genotopic Genome, Sect. 3.19: Semantic Congruence with the Conventional Perspective, Sect. 3.20: Semantic Incongruence with the Conventional Perspective, Sect. 3.21: Conclusion.

3.2 Conventional Perspective of Natural Selection

As discussed in the first two chapters of this monograph, the Darwinian paradigm is different from other scientific theories in two respects. Firstly, unlike other scientific theories, the Darwinian paradigm doesn't predict anything specific about the course of evolution. Therefore, it is, to borrow Popperian terminology, not falsifiable (Popper 1963). Secondly, in spite of its lack of falsifiability, the Darwinian paradigm has a semantic core that has enabled it to survive and thrive during the subsequent paradigm shifts. This points toward a possibility that the Darwinian paradigm encapsulates a fundamental law of Nature which is yet to be articulated in its totality. The fact that the rationale behind the Darwinian paradigm can be applied to a diverse range of systems, natural or man-made (Dennett 1995), points toward the fundamental aspect of reality being captured by the Darwinian paradigm.

It is tempting to think that this imbalance between the predictive and explanatory powers of the Darwinian paradigm arises because of the inherent epistemological compulsions of our cognitive faculty. It is possible that there exists a definitive structural template of biological evolution (with its inherent predictive powers) which has escaped our cognitive processing. Because we have not been able to perceive this putative structuralism of biological evolution, we have not been able to formalize Darwin's theory. Our intuitive understanding of the underlying semantics, however, helps us to understand the nature of biological evolution and natural selection, but stops short of formalizing its structuralism. Admittedly, even this possibility of a potential theory with its inherent predictive powers, is against the Darwinian semantics. The Darwinian paradigm rests on the inherent randomness of natural selection (Bonner 2013). Therefore, to postulate that there exists a yet unknown theory which has a certain structuralism capable of predictivity is antithetical to the Darwinian paradigm.

However, as discussed in the preceding chapters, it is a category mistake to equate the absence of any predictive theory with the inherent randomness implicit in the Darwinian paradigm. However, before we venture into a search for such a theory, it is necessary to understand the nature of natural selection as implicit in the Darwinian paradigm. Therefore, in this section, we will try to deconstruct the nature of natural selection and its relationship with randomness. It must be kept in mind that this search for a newer template of natural selection is not intended to introduce (or rather reintroduce) any design principle or teleology. The search is intended to reinterpret the inherent randomness of natural selection and make it compatible with the emergence of complexity during biological evolution and natural selection. Complexity arising from a random process is always problematic. In fact, it is unique because it is found only in the evolution of Life and subsequent natural selection. In fact, this unique feature has given impetus to the critics of the Darwinian paradigm to postulate some transcendental explanations for the evolution of Life. Therefore, our search for the mechanisms by which complexity arises during a random process of natural selection, is an attempt to reinforce the naturalistic foundation of the Darwinian paradigm.

Let us begin with the conventional perspective of natural selection of the Darwinian paradigm. As the most evocative phrase coined by Darwin himself suggests, the gist of natural selection is summed up as "descent with modifications." In the language of genetics, this can be expressed as genetic continuity with mutations. These two features are actually mutually incompatible because if the genetic continuity were to be total, there is no possibility of any mutations being passed on to the next generation. Similarly, if mutations were to dominate, there would be no genetic continuity during natural selection. Therefore, by definition, natural selection comprises balancing these two mutually incompatible features of genetic continuity and mutations. The randomness therefore must arise from the way natural selection balances these two features. If genetic continuity were to be 100%, there would be nothing to select from and natural selection would not manifest itself. Similarly, if mutations were to change the entire genetic contact of a given generation, there would never be selection, natural or otherwise. Thus, natural selection operates by balancing these two mutually conflicting processes.

This perspective of natural selection is important because it hides within itself, a solution to the problem of complexity. Unless genetic content was complex, it would not give rise to genetic continuity. Genomes must have structural and functional resilience to withstand the disruption caused by mutations and still retain its ability to pass on its complexity to the next generation. Similarly, a mutation has to be strong enough to alter some structural or functional features of the genome to make changes in what is passed on to the next generation. Thus, from either perspective, natural selection must be viewed as managing the types of complexities.

However, the conventional perspective (and this refers to not just Darwin's theory, but also to genetics and molecular biology paradigms as well) is categorically explicit that phenotype cannot influence genotype. This unidirectional influence is somewhat diluted in the genomic paradigm. The acknowledgment and systematic investigation of epigenetic phenomena (Robert 2004) is an admission of the more complex nature of influences operating during natural selection. This insistence on the unidirectional influence of genotype on phenotypes is further reinforced by the conception of a passive role of the environment in natural selection in the conventional perspective (Grene 1986). If the environment were to perform an active role in natural selection, we could have accommodated the epigenetic phenomena into semantics of the conventional perspective. All that we require is to define phenotype as an element of the environment. In such a scenario, phenotype too, being an integral part of the environment, could be thought of having influenced genotype. However, the conventional perspective of natural selection (the one that is implicit in Darwin's theory (Hodge and Redick 2009)) doesn't allow such a contextual definition of phenotype and the environment because then, it would be necessary to ascribe an active role of the environment in natural selection.

Thus, the semantic compulsions of rejecting any teleological arguments forced the conventional perspective to deny any active participation of the environment in natural selection. This was a reasonable postulate till the experimental evidence of epigenesis was found. In the present context, when epigenesis is a proven phenomenon, it is necessary to redefine the relationship between genotype, phenotype and the environment. We can understand that given the naturalistic foundation of science and given the random nature of natural selection, it is difficult to conceptualize any such design principles wherein phenotype starts influencing genotype and thereby giving rise to better and better design of itself. However, we can escape from this unpalatable scenario by accepting that the environment plays an active role in natural selection. By granting an active participation of the environment in natural selection, we can solve two problems that have remained unaddressed by the conventional perspective of the Darwinian paradigm. These problems are the emergence of complexity during an essentially random phenomenon of natural selection and observed epigenetic phenomena. In the next section, we will look at one such model of natural selection. The key point is that the proposed model is naturalistic and still resolves these two problems.

3.3 Proposed Model of Natural Selection

It must be admitted at the outset that this model was originally developed to formalize a naturalistic approach to epistemology (Chhaya 2022a, see Chapters 2 and 3). Therefore, we could legitimately question the validity of the grounds on which the same model can be extended to genomics and natural selection. However, upon a little reflection, it is intuitively clear that epistemology is essentially concerned about information transfer. The conventional definition of epistemology (Audi 1998) is a study of the ways in which knowledge becomes available to our cognitive faculty. Therefore, even in the classical interpretation of epistemology, its main objective was to formalize how transcendental knowledge is transferred to our cognitive faculty. In that perspective, knowledge already exists in the transcendental realm and our consciousness somehow finds access to that transcendental realm. Thus, right from the beginning, the objective of epistemology was to formalize information transfers.

Once we accept this theoretical definition of epistemology, it is intuitively clear that similar mechanisms can be postulated for any generic information transfers from one part of a system to another part of the system. In the case of genomics (which is the main focus of this monograph), it is intuitively clear that gene expressions involve information transfers in the form of long-range influences and in the form of temporal sequence. Therefore, it seems reasonable to apply the proposed model to genomics. While it is too early to label this approach as genomic epistemology, it aptly describes and defines the main objective of this monograph. Similarly, once we accept that the environment plays an active role in biological evolution and natural selection, it is intuitively clear that we can think of natural selection as a generic mode of information transfers. Therefore, it is axiomatic that we can legitimately extend the proposed model to deconstruct natural selection.

As discussed in the preceding chapters, we should accept that natural selection actually alters the types of complexities rather than allowing complexity to manifest de novo. Once we accept this proposition, it is intuitively clear that natural selection deals with information transfers. More importantly, it deals with the degree of congruence between the information content of genotype, phenotype and the environment. Survival can be seen as a certain degree of congruence between the contents of phenotype and the environment. Similarly, the process of gene expressions must be viewed as a process of information transformation from one type of complexity to that of another. In such a scenario, all that we require is a way to define information transfers and transformations. This is precisely what the proposed model offers. The details of this model are presented in the preceding monograph (Chhaya 2022a, see Chapter 3). Therefore, in this section, we will try to rephrase the key features of the model in the context of natural selection. For the sake of simplicity, the model is described in a point-wise manner.

1. The ecosystem consisting of the environment, genotype and phenotype is defined as a manifold existing in a nested hierarchy.
2. Keeping aside the notion of ecosystem in the present discussion, we can think of the environment as a parent manifold in which genotype and phenotype exist as submanifolds.
3. In view of the above discussion, we will assume that phenotype and genotype exist as separate submanifolds with a certain degree of overlap.
4. This provision of overlap between genotype and phenotype allows one to formalize gene expressions and epigenesis in the region of overlap between these two submanifolds.
5. Provisionally, one can think of biological evolution as having arisen from the ecosystem, as a whole devolving on to genomic complexity. However, we will defer the discussion on this aspect to later chapters. Presently, we will focus on natural selection after the first living organisms appeared in the ecosystem.
6. According to this model, any natural phenomenon is assigned a certain dimensionality depending on its complexity.
7. Any natural phenomenon can manifest itself in more than one dimensionality simultaneously if it possesses different types of complexities.
8. According to the proposed model, the relationship between dimensionality and complexity is an inverse relationship. The simplest form of complexities occupies the highest dimensionality, and the most complex phenomenon occupies the lowest dimensionality.
9. Admittedly, this is a counterintuitive feature of the proposed model. However, it is inevitable if we wish to develop a truly naturalistic model of biological evolution and natural selection. By assuming that the ecosystem itself (and no other transcendental agency) gives rise to biological evolution and subsequent natural selection, then it is imperative that the cause of biological evolution must arise from the ecosystem. Since living organisms are an integral part of the ecosystem, they must be placed inside the manifold representing the ecosystem. Therefore, the influence of the ecosystem (whatever its exact nature maybe) needs to be formalized as an inward influence. Therefore, it can be formalized using the operator of involution. Once we accept this framework, it is intuitively clear that the relationship between dimensionality and complexity is an inverse relationship.

10. Now, we can define gene expressions as an information transformation in the form of involution from the submanifold of genotype to submanifold of phenotype. However, since we know that the environment in the form of long-range influences of the genome also influences gene expression, we can define genotype as having multiple dimensionalities simultaneously with the genome occupying the highest dimensionality and individual genes occupying lower dimensionality depending on their relationships with the neighboring genes. According to the proposed model, the DNA sequence occupies the lowest dimensionality, below the dimensionality of the respective genes.
11. In this scenario, the long-range influences of the genome (in the form of sequence of gene expressions, initiators, enhancers and repressors) can be formalized as involutions.
12. The process of gene expressions, by itself, can be formalized as the changes in the types of complexities. At one level, it refers to the process of translation of DNA sequence into amino acid sequences. At another level, it refers to giving rise to complexity in the form functionalities of phenotype.
13. This process can also be formalized using the operator of involution because it represents a change from one type of complexity into another type of complexity. Since different types of complexities can be related to one another by symmetry operators, the proposed operators of involution are best suited to formalize the relationship between genotype and phenotype because these operators of involution too are connected to one another by symmetry principles.
14. There are three important consequences of this description of gene expressions. Firstly, since genotype and phenotype are embedded in the ecosystem, the operator of involution mentioned in point 13 can be equated with two operators, one operator linking genotype to the ecosystem in the form of inverse of an operator of involution. The second operator linking ecosystem to phenotype as a new operator of involution. Thus, the operator proposed in point 13 consists of two operators, one inverse operator and second operator connecting ecosystem to phenotype. This protocol allows one to accommodate active participation of the environment without worrying about teleology or design principles.
15. The second important consequence of point 13 is that when one defines an operator of involution connecting genotype to phenotype, as per rules of involution, the complexity of genotype gets converted into a different type of complexity. This is precisely what happens. The complexity of genotype gets transformed into a different type of complexity in phenotype.
16. Thirdly, according to the operator defined in point 13, the complexity of a phenotype itself is derived from the complexity of genotype. Moreover, different classes of complexity of phenotype itself must be related to one another. This paves the way for linking the template of functionalities of phenotype with its structural template. This is something that is sorely missing in the conventional perspective. Phenotypes have their functionalities but there is no correspondence between functionalities and structuralism in the formal sense in the conventional perspective. The proposed model fills this gap. More importantly,

it provides a way to formalize functional templates as distinct from the corresponding structural template.

The 16 points given above are merely indicative and need to be formalized. However, they serve the purpose of explaining the semantic nuances of natural selection according to the proposed model. In order to simplify and articulate key concepts of natural selection, in the following sections, we will articulate three principles which according to the proposed model define natural selection. These three principles are (1) Postulate of natural selection, (2) Postulate of genomic architecture and (3) Postulate of long-range influences. Together, these three postulates define natural selection according to this model.

3.4 Postulate of Natural Selection

As discussed in the first chapter and in the preceding sections, the conventional perspective of natural selection rests on three phenomena. Firstly, genotype gives rise to phenotype by certain mechanisms. Secondly, the environment provides parameters to define survival fitness of phenotype. Thirdly, the surviving phenotype passes on the genotype to the next generation. Since each of these processes can give rise to multiple outcomes, the final version of natural selection is governed by the Bayesian paradigm of conditional probabilities (Press and Clyde 2003). This conception of natural selection in the conventional perspective is valid irrespective of whether the environment actively participates or not, or whether there are any epigenetic influences or not, or whether there is any design principle embedded in the resulting complexity or not. Therefore, we will adhere to this conception of natural selection. The proposed model takes one step further and includes these three influences into the conception of natural selection and demonstrates that the inclusion of these three factors does not introduce any form of teleology or deism. Natural selection retains its nondeterministic and probabilistic nature. The proposed model merely restricts the types of complexities that emerge during the course of evolution. Therefore, the following postulate is proposed.

Postulate of Natural Selection
Natural selection manifests itself in any system that possesses multiple forms of complexity. Natural selection is a selection of congruence between different types of complexities and is independent of the materials employed. Natural selection is a structural law.

By itself, this postulate appears to be unobjectionable. However, as discussed in the following sections and in the following chapters, it gives rise to different semantics. More importantly, this revised semantics offers a way to inject a certain degree of predictivity in natural selection without compromising its essential nondeterminism. As mentioned above, this is the first of the three postulates of the proposed model. The second and third postulates arise naturally from the first postulate. Therefore, let us look at the second postulate and how it arises. In order to understand the background of the second postulate, in the next section, we will look at the genomic architecture which forms the basis of the second postulate.

3.5 Explicit Genomic Architecture

The conception of genomic architecture has been constantly changing. This is mainly because we don't have any theoretical perspective of what constitutes a genomic architecture. Our conception of it is based on the incremental increase in the information derived about individual genes and their interactions. The difficulty in arriving at the conception of genomic architecture is also due to the fact that the genome is not a static entity. Its configuration changes even during a cell cycle. Our present understanding of genomic architecture can be broadly summed up by two propositions. Firstly, we are convinced that genes do not act independent of one another. There are several sequences of gene expressions which are observed across the biological spectrum. This leads to belief that there are some hierarchies among the genes present in a given genome (Davidson 2006, see Chapter 1). Thus, not only genes but even their hierarchies are also conserved. This brings us to the second proposition. The genome can be visualized as existing at two levels. At the surface level, there exists an expressive genome and at a deeper level, there exists a regulatory genome. Since there are several excellent reviews on the regulatory framework of genome, we will not delve deeper into this topic. Of course, there are different conceptions of genomic architecture (Lynch 2007), each having its own semantic nuances. However, the propositions mentioned above are implicit in all these different conceptions of genomic architecture. Therefore, we restrict ourselves to these two propositions in this chapter because they are adequate for the present discussion. The proposed model offers a more detailed description of both levels of genome, but that will be discussed in the following chapters. Presently, these two propositions would suffice.

Even without going into the mechanisms by which gene expressions are timed, it is intuitively clear that if the regulatory framework of the genome is also conserved during natural selection, the mechanisms by which natural selection occurs must possess structuralism of its own. This is because when a regulatory framework is naturally selected, what is being selected is not the structural template, but a functionality. Therefore, the basis of natural selection must be a structural template and not a physical template. This is actually not new because we already know from our experience in proteomics that what is conserved during natural selection is not a sequence of amino acids but protein domains (Bryson and Vogel 1965, see Parts III, IV, and V). Therefore, we need to redefine mechanisms of natural selection to include some abstract conception of structuralism and not confine it to physical templates of molecules. Admittedly, such a redefinition would have its own set of problems. However, it is imperative that we face these problems and find a way to conceptualize natural selection from a purely structural perspective.

It is in this context that we should evaluate genomic architecture. Fortunately, our current understanding of genomic architecture is halfway through that goal. Thanks to functional genomics (Pevsner 2015), we have discerned a functional template of gene expressions. Therefore, we have reached the halfway mark in defining genomic architecture. We already know that genomic architecture doesn't limit itself to the distribution of genes on the genomic map. It now also includes a functional mapping of gene expressions. As discussed in the following sections, we don't know much about the exact mechanisms of long-range influences of the genome. However, it is intuitively clear that whatever the exact mechanisms may turn out to be, the genomic architecture must possess temporal and spatial patterns which underlie the present distribution of genes. While we have some idea about the nature of temporal patterns in the form of the sequence of gene expressions, our understanding of spatial patterns of the genomic architecture has not transcended the conventional molecular placement of genes. Therefore, at present, what we know should be labeled as explicit genomic architecture in which the molecular distribution of genes is known to us with a reasonable degree of certainty.

In addition to this, we must add the long-range influences in the form of the sequence of gene expressions into the explicit genomic architecture. This is because it too pertains to molecular interactions involved in gene expressions. The reason for naming these characteristics of the genome as explicit genomic architecture is based on the fact these features are natural consequences of our molecular biological paradigm. By implication, this nomenclature suggests that there must be corresponding implicit genomic architecture. At present, we can include the regulatory framework of the genome into the category of implicit genomic architecture. However, there must be something more to the conception of implicit genomic architecture. This "something" must be in the form of abstract structuralism which manifests itself in the regulatory framework of the genome. It is quite natural to be skeptical about any such abstractionist perspective of the genome. However, the history of genomics has taught us to be more open minded about these seemingly fanciful ideas. A couple of decades ago, the majority of scientists believed that noncoding DNA sequences were "junk." Today, we know that it is far from being junk. The discovery of several types of small RNAs like i-RNA indicates that the intergenic regions of the genome play a crucial role in regulation. Similarly, as large scale structures of chromatin have become discernible thanks to technological advances in genomics, we are now willing to concede that these large scale structuralism also contributes to genomic architecture (Rippe 2012), though we do not know how. For the present discussion, we will name the molecular distribution of genes and the regulatory framework as the explicit genomic architecture. Therefore, in the next section, we will try to outline what the implicit genomic architecture could be or rather what it should be.

3.6 Implicit Genomic Architecture

While it is intuitively clear that a genome must have a certain higher level of organization, it is not easy to discern the exact details of it. As mentioned in the previous section, we will rather arbitrarily divide this organization into two categories of explicit and implicit architecture. There are two features that must be taken as integral parts of the explicit architecture of genome, viz., the distribution of genes in the genome and the long-range influences that decide the sequence of gene expressions. Of these two, the distribution of genes on a genome is perhaps the most obvious feature. However, it must be admitted that we don't know much about the nature of this distribution of genes, but as a concept, it is very intuitive to think that this distribution of genes must have rationale. Therefore, it appears to be an integral to the conception of the architecture of the genome. The second feature of the explicit architecture, viz., long-range influences that decide the sequence of gene expressions, is a necessary structural element of any architectural design of genomes. However, even here we don't know much about these long-range influences that really influence gene expressions. Of course, there are several proposals postulating spatial distribution of chromosomes in the nucleus (cf. Chromosome territory (Fritz 2014)) which suggests that these long-range influences are in fact, short range influences due to particular orientations of different chromosomes. However, even if the exact mechanisms by which these long-range influences operate are not available, we find it intuitive as an element of architectural design. Therefore, we will label them as parts of explicit genome architecture.

However, there are two features of the genomic architecture that are yet to be articulated. Moreover, there is no template available on which we can define these features. Therefore, we will label them as implicit genomic architecture. These features are temporal aspects of gene expressions and the evolutionary perspective of genomic structuralism. Let us try to understand why they should be labeled as implicit genomic architecture. Firstly, purely from the semantic perspective, we agree that any genomic architecture should have these two features. They are in some sense, semantic imperatives. In other words, they are inevitable from the Darwinian perspective. The irony is that in spite of their semantic imperative, these features are overlooked. Therefore, it is legitimate to label these two features as being integral part of the implicit genomic architecture. Let us try to understand why these two features are overlooked in the conventional perspective of the Darwinian paradigm. This is important because once we deconstruct the neglect of these two features in the conventional perspective, we will be in a position to modify the structuralism of natural selection. Even at the risk of being called simplistic, it is postulated in the proposed model that our conception of genomic architecture is under an overwhelming influence of the molecular paradigm of molecular biology. Let us understand why our preoccupation with molecular biology has created this neglect of these two features of genomic architecture.

Purely from the semantic perspective, it makes sense to think of the regulatory framework of the genome as a separate entity by itself. However, we have selectively chosen to focus only on the molecular agents which regulate the sequence of gene expressions. Therefore, we have succeeded in defining these initiators, facilitators and even repressors of gene expressions (Weinzierl 1999). No doubt, these are critical to define long-range influences of the genome. Even from the semantic perspective, these discoveries are seminal works that have revolutionized genomics. However, the fact remains that this molecular paradigm focuses only on the molecular perspective. This is evident from the way in which these two features, viz., the temporal perspective and evolutionary perspectives or regulatory genome, are investigated from the molecules involved in these features. Thus, for the temporal control of gene expressions, we have focused on the molecules of initiators, facilitators and repressors. However, we tend to ignore that these molecules too are products of prior gene expressions. Therefore, now we know the agents that bring about temporal control. However, we do not know the mechanisms by which the gene expressions of these initiators and facilitators are placed prior to the gene expressions that they control. Apparently, there must be some mechanism which decides that the genes containing the information about the synthesis of these initiators and facilitators must be expressed before other genes. However, that insight is missing from this molecular paradigm. More importantly, it cannot be generated from this paradigm.

Let us look at the second feature of the evolutionary perspective of the genome. Even here, there is an enormous amount of literature on the phylogenetic studies which essentially tries to throw light on the evolution of the genome (Bromham 2008). However, the focus is on the molecules involved. Phylogenetic studies are focused on the changes in the DNA sequence during the course of evolution. Admittedly, this has been a very fruitful and illuminating field. However, as we know from our experience in proteomics, what is selected is not the amino acid sequences (and by implication, the DNA sequences). What is selected is the three-dimensional shapes (what we call protein domains) (Bryson and Vogel 1965). Therefore, it is axiomatic that natural selection must operate on structural terms and not on molecular terms. However, this doesn't get reflected in phylogenetic studies. It is important to remember that this has nothing to do with any deficiency in the molecular biological paradigm. It has turned out to be the most successful discipline to study natural selection. The deficiency, if any, lies in our inability to see beyond the molecular paradigm.

We must accept that genomic architecture contains features that are beyond molecular biology and we need to develop a different framework for studying them. The problem however arises from the fact that we don't know where to begin in building a new framework for this purpose. Therefore, now we will introduce a second postulate of genomic architecture. Without any historical burden, we will try to deconstruct the implicit genomic architecture using the Darwinian semantics and create a structural template. It is intuitively clear from the above discussion that both these features of the implicit genomic architecture, viz., the temporal perspective and structural perspective, transcend the molecular perspective. Therefore, we will try to frame the template of implicit genomic architecture in purely spatiotemporal terms. We will assume, without committing ourselves to any particular mechanisms, that natural selection operates using pure and simple spacetime features. Moreover, we will also assume that our present molecular paradigm is just one example of this spatiotemporal model of natural selection.

As a matter of precaution, we will try to eliminate any teleological or design explanation from this framework. Like physics, we will assume that spacetime has a certain inherent structuralism (which we call the fine structure of spacetime). It is this fine structure of spacetime that is a component of the environment as conceptualized in the Darwinian paradigm. Thus, spacetime itself behaves like an element of the ecosystem which operates as an arbiter in natural selection and decides survival fitness. With these caveats in place, let us formalize the postulate of genomic architecture.

3.7 Postulate of Genomic Architecture

Being preoccupied with the molecular perspective of genomics, we have overlooked the higher levels of genomic architecture. Therefore, while we concede the existence of different types of long-range influences, we do not incorporate them as structural elements of the genomic architecture itself. Instead, we focus on the consequences of these structural elements in the form of initiators and facilitators in our conception of genomic architecture. In order to redress this imbalance, we will postulate a new topological unit of the genome. This unit consists of spatiotemporal units of the genome. The conventional description of genomic architecture which describes distribution of genes as a unit, is just one facet of this new topological unit.. These spatiotemporal units are structural units and must be named as "Genotopes". A Genotope consists of assembly of molecules as well as the spacetime underlying this assembly.

Postulate of Genomic Architecture
A genome consists of spatiotemporal units surrounding the complex molecular assembly. These units named hereby as genotopes consist of topological spaces consisting of the units of spacetime and the molecules present within them. Functionally and structurally, a genotope is an independent entity.

The inclusion of spacetime in the definition of the unit of genome serves two purposes. Firstly, it suggests that spacetime plays an active role in terms of long-range influences and defining complexity of genotype. Secondly, it is to be taken as a part of the environment. The conventional perspective of the role of the environment was restricted to its element and their influences on the survival of phenotype. However, by including spacetime itself, the postulate resolves the debate on the active role of the environment. While the elements of the environment remain passive during natural selection, spacetime becomes a part of genotype and thereby contributes actively to the process of natural selection. Admittedly, at this stage, the concept of genotope is simplistic. However, in the following sections and in the following chapters, we will try to articulate various semantic propositions of this concept. Before we articulate the third and final postulate of this model, we will look at the relationship between functional and structural templates of the genome. Earlier we talked about explicit and implicit genomic architecture. However, these distinctions are not clearly defined. Therefore, we will revert to the conventional distinction between structuralism and functionalities of the genome.

3.8 Functional Genome Versus Structural Genome

As discussed in the first chapter, one of the unresolved ambiguities of the Darwinian paradigm is the relationship between structuralism and functionalities. Of course, in that chapter, the emphasis was on the epistemological difficulties in separating the template of functionalities from the corresponding structural template. Irrespective of our epistemological limitations, the relationship between structuralism and functionalities seems to encapsulate the relationship between reality and phenomenology (of course this includes the cognitive structuralism and functionalities; however, we will overlook this aspect from our present discussion). However, even if this relationship between structuralism and functionalities is manifest in every natural phenomenon, it seems to be unique in the case of living organisms. Therefore, it is necessary to understand why this relationship acquires a unique character in living organisms. Since there exists an enormous volume of research on genomics, particularly functional genomics (Pevsner 2015), it seems reasonable to deconstruct the relationship between structuralism and functionalities as it manifests in the case of genomics.

In the case of genomics, there exists another level of complexity in the relationship between structuralism and functionalities. Functional genomes seem to influence the structural genome. Normally, the relationship between structuralism and functionalities is unidirectional. Structuralism gives rise to functionalities and not the other way around. However, in the case of the genome, the functional genome seems to influence the structural genome and more importantly, the functional genome ends up by modifying itself, albeit indirectly through the structural genome. This is a classic case of self-reference. Therefore, it is possible that what separates Life from other natural phenomena is the self-referential functionalities of Life. Therefore, it is imperative that one must deconstruct this relationship between structuralism and functionalities in biology.

In the case of the genome, there exists an additional level of complexity. Normally, in any given natural phenomenon, the boundary between structuralism and functionalities is well-defined, and there is no ambiguity about the identities of structuralism and functionalities. However, in the case of the genome, it is very difficult to draw a line between structural genome and functional genome. As discussed in the first chapter, this ambiguity about differentiating structuralism and functionalities arises because the boundary separating the two is flexible. It is contextual. From one frame of reference, what appears to be a functionality turns out to be a structural element from another frame of reference. The long-range influences of the genome, both cis and trans influences, are actually parts of the structural template of the genome, and yet, they are functionalities. In fact, this is not an isolated example. It is quite a generic feature of the genome. Therefore, it is necessary to articulate these nuances of the relationship between structural and functional templates of the genome. Admittedly, we will discuss this aspect in the following chapters in greater detail; presently, we will outline general comments on the relationship between structural genome and functional genome. In this section, we will articulate three features that characterize the relationship between structuralism and functionalities of the genome. These are degrees of discreteness in both structuralism and functionalities of the genome; the need for chemical and physical channels of communication between structural and functional genomes and the nature of self-reference in gene expressions. Admittedly, these topics are very profound and greatly debated in literature. However, we will not analyze these topics. Rather, we will outline the information theoretical perspective of these features and lay the foundation of the third postulate.

Let us begin with the first topic of the different degrees of discreteness in structuralism or in functionalities of a genome. From the inception of genetics, when Mendel's pioneering work on genetics was published (Darden 1991), our conception of a gene has been that of a discrete unit of inheritance. It was only after the advent of molecular biology and genomics that we have realized that there exist several "coarser" modes of inheritance than the one implicit in our classical conception of a gene. As our laboratory techniques improved, we have realized that there are several levels at which information content is passed to the next generation and each level has its own degree of discreteness. Thus, our earlier conception of a genome being a string of independent genes has been replaced by a more nuanced hierarchy of genes and their expressions. It is intuitively clear that a similar question arises in the case of functionalities of a genome. There are genomic functionalities like epigenetic imprinting wherein the entire chromosomes behave as one unit (Robert 2004). On the other hand, there are genomic functionalities like cis and trans long-range influences wherein only a small portion of a chromosome influences the gene expressions. In both these types of functionalities, there exists a different degree of discreteness. However, there is no explanation available of the origin of these different degrees of discreteness.

However, in the context of the present discussion, the key issues are (1) whether these different degrees of discreteness are inevitable outcomes of biological evolution and natural selection? (2) whether these different degrees of discreteness are internally congruent? and iii) do these different degrees of discreteness represent different types of information content? These three issues have important bearings on the way we interpret the Darwinian paradigm. For instance, if we were to assume that the earliest living organisms (say, LUCA (Bard 2016, see Chapter 9)), had only one mode of information transfers and these different types of information transfers, each having its own degree of discreteness must have arisen during natural selection, then a different kind of a model of natural selection emerges. Alternatively, if we were to assume that these different modes of information transfers evolved separately and they converged due symbiosis like a typical eukaryotic cell with a mitochondria or a chloroplast as symbionts (Margulis 1970), then a different model of natural selection emerges. As mentioned above, this is too deep a topic to be discussed parenthetically. However, it is necessary to keep in mind that this information theoretical perspective of natural selection has something important to contribute to our understanding of natural selection. We will return to this topic in the following chapters.

Let us look at the second topic of different modes by which the information is transferred. Let us think of long-range influences of the genome on the individual gene expressions (Shmulevich and Dougherty 2014). Apparently, we have realized that very often, the prior expression of the genes representing initiators and facilitators is a general strategy which the genome employs to maintain a certain sequence of gene expressions. At the same time, we also know from our studies in the area of chromosome territories (Fritz 2014), that sometimes, the genome employs a particular topological alignment to ensure that a particular sequence of gene expressions is maintained. The conventional wisdom accepts this as parallel mechanisms. However, there is no explanation why both these mechanisms should be present in all the metazoan species. However, what is germane to the present discussion, viz., from the information theoretical perspective, is that these two different modes of information transfer represent two different types of information? This is important because one mode of information transfer refers to discrete units, in the form of molecules like initiators and facilitators (or even repressors) and the other mode of information transfer refers to nondiscrete modes of information. This problem of digital versus analog is quite generic (cf. Signal transduction across a synaptic junction). It will be interesting to deconstruct this dichotomy between digital and analog information transfers and their role in natural selection. We will return to this topic in the following chapters. Presently, the key point is why should the genome manifest these two forms of information transfers? Is it because of evolutionary legacy? Or is it because both these forms of information transfers are inherently different and therefore not interchangeable?

There is another aspect of this distinction between digital and analog information transfers in the long-range influences of the genome. This refers to their nature of influences. The information transfer arising from prior presence of initiators and facilitators of gene expressions is essentially a temporal influence. On the other hand, the nondiscrete influences arising from conformational proximity of otherwise distant DNA sequences are essentially spatial influences. This could not be a coincidence. Perhaps, this duality of the modes of information transfers represents the distinction between spatial and temporal changes in the genomic architecture. Therefore, it is tempting to think that we should define genomic architecture not in terms of molecules involved in it, but in terms of spatiotemporal orientations of the genome. We will explore this idea further in the following sections and chapters.

Presently, let us return to the third topic of the functionality of self-reference. There are two types of self-reference that are manifest in the gene expressions and the role of genomic architecture in these gene expressions. Firstly, by arranging to express genes responsible for initiating, facilitating or even suppressing other gene expressions, the genome is indulging in self-reference. The genome, as an entity, is trying to modify itself. Therefore, this sequence of gene expressions constitutes a self-reference. Secondly, during the epigenetic phenomena like X chromosome silencing (Sado 2018) or during homologous pairing during meiosis (Zhang 2012), chromosomes seem to behave as a single unit and try to change their own structuralism and functionalities. Generally, the changes in genes occur at an individual level. They involve the interactions with different types of proteins. However, in the instances cited above, the changes occur at the level of chromosomes and not at the level of individual genes. It is as if chromosomes are acting as autonomous entities. Admittedly, as we know more and more about homologous recombination during meiosis, we will find out more about the mechanisms by which these phenomena manifest themselves. However, from the information theoretical perspective, it is inevitable that we must concede that these information transfers occur from the frame of reference of the chromosome and not at the level of individual genes or proteins. This phenomenon also points toward an alternative design of genomic architecture. Before we articulate such an alternative design, let us look at the details of these long-range influences of the genome.

3.9 Nature of Long-Range Influences in Genome

It must be admitted at the outset that there exists a humongous amount of literature on the topic of long-range influences. It is not possible to summarize it here. Instead, we will restrict ourselves to the concepts behind these long-range influences. This selectivity is justified because we are trying to understand the nature of information content and its modes of transfers. As mentioned in the previous section, long-range influences manifest in two forms, viz., in the form of prior expressions of gene initiators and facilitators and in the form of conformational proximity of otherwise distantly placed DNA sequences. However, instead of worrying about the semantics behind these two types of influences mentioned above, we will try to deconstruct the nature of these two influences. Admittedly, our objective is to understand the underlying architectural design behind these two features, but in this section, we will focus on the structural templates behind these two types of long-range influences.

Let us begin with the first feature of prior expressions of initiators and facilitators of gene expressions. Firstly, these initiators and facilitators (and even repressors for that matter) are themselves products of gene expressions. Therefore, it is difficult to conceptualize any such design which necessarily results in the prior expressions of these agents. Of course, as we know from developmental biology, some of these agents could be in the form of signals already present in the ovum, either in the form of maternal RNAs which can translate into active molecules or in the form of active molecules themselves (Marlow 2010, see p. 131). However, this is not a universal pattern. Even in the case of maternal inheritance of these signals, the original question about how Nature decided this particular sequence of gene expressions remains unanswered. The key question is whether there is anything inherent in the genomic architecture itself that decides a particular sequence of gene expressions? The conventional perspective of this topic is that this particular sequence of gene expressions is not designed, but it arose from natural selection. In other words, in the earliest living organisms, there wasn't any particular sequence of gene expressions (possibly because they were unicellular and less complex organisms). Therefore, the present sequence of gene expressions has been selected by natural selection.

However, upon a little reflection, it is intuitively clear that this explanation is logically inconsistent. Even in the simplest organisms (say Archaea), there exists a preferred sequence of gene expressions. In fact, even these organisms manifest genomic architecture which is sufficiently complex to be explained without any prior assumption about their complexity. In fact, because all the three types of living organisms, viz., Archaea, Prokaryotic, and Eukaryotic, possess enough complexity of their genomes that we are still not able to define a tree of Life and how Life originated (Garrett and Klenk 2007, see Chapter 3). Therefore, sooner or later, we will have to concede that there exists a definitive mechanism by which the sequence of gene expressions is decided. What is possibly selected is which type of sequence of gene expressions is good enough for survival. The prior existence of a sequence of gene expressions is a logical necessity. It must be kept in mind that this admission of having a prior sequence of gene expressions, doesn't involve any design principle. It is inherent to any genome that contains several genes waiting for their turn of expressions. Therefore, it is possible that a particular sequence of gene expressions gets selected among many such sequences, thereby eliminating any design principle.

When we think of a genome having multiple genes and how it could have evolved, it is intuitively clear that there is no rational explanation available. Conventionally, it is possible to think that this could have happened when Life decided to switch from RNA as information carriers to DNA as information carriers. It is possible to imagine that these different genes existing in their RNA versions might have found it favorable to transfer their information content to a single DNA sequence, thereby creating a prototype of a genome which under the pressure of natural selection must have evolved into the present day genomes. However, there is no empirical or theoretical basis for such a speculation. In addition to the fact that RNA and its precursors are unstable under aqueous conditions, the problem of having very long RNA sequences is also problematic. Our current understanding of the origin of Life is ambivalent about it (Yarus 2010, see Chapter 2). We can similarly think of proteins as candidates for this role because of their inherent catalytic properties and information carrying properties. However, there is no method of synthesizing proteins without any catalytic agents (which is what RNA molecules do).

In view of these ambiguities, there is no clarity on the course of evolution by which the genome could have evolved. Therefore, it is difficult to deconstruct the reasons why there are two types of long-range influences involved in determining the sequence of gene expressions. Of course, as discussed in the first chapter, we have chosen to ascribe these ontological ambiguities to the inherent random nature of natural selection. However, this is a category mistake. Each transition from prebiotic chemicals to the earliest living organisms and later on, to more complex living organisms, was shaped by causal processes. The resulting randomness arises because at each transition, there were multiple possible outcomes, and therefore, there must be some causal explanations for the origin of these two types of long-range influences, although the knowability of such explanations is debatable.

Even if we were to be ignorant of the ontological scenario of these long-range influences, it should be possible to infer their ontological aspects from the nature of these influences. This is where the conventional perspective fails to provide any insights. Our current understanding of these long-range influences can be summarized in three propositions. Firstly, there are two types of influences, discrete and nondiscrete. Secondly, they represent temporal and spatial influences respectively. Finally, both these long-range influences operate through the interactions between the outermost electrons of different molecules operating either individually or collectively as a polymer. Fortunately, thanks to our understanding of quantum chemistry (Szabo and Ostlund 1989, see Chapters 2 and 3), it is possible to describe in detail the nature of these interactions. It is these complex and sophisticated computations of stereochemical orientations that has helped functional genomics to make impressive progress. At the same time, it is important to remember that this perspective, by definition, is incapable of defining the ontological perspective of these long-range influences and their role in natural selection.

In view of this situation, we will try to formalize an information theoretical perspective of long-range influences and try to develop an ontological perspective of these long-range influences. We will begin with accepting the three propositions mentioned above. This will ensure that there is a semantic congruence between the conventional perspective and the proposed model. In addition, by adhering to the conventional perspective, we should be able to avoid any additional hoc design principle which is subject to any teleological justifications. It was suggested above that one way to resolve the semantic ambiguities of the Darwinian paradigm is to think of spacetime as an integral part of the environment. In continuation with that belief, we will try to formalize these long-range influences in the language of spacetime. This is also congruent with our current understanding of quantum chemistry wherein the orbital interactions are defined in the language of spacetime in the form of the wave function. Therefore, in the next section, we will articulate the third postulate of long-range influences.

3.10 Postulate of Long-Range Influences

As discussed in the section dealing with the postulate of genomic architecture, we have conventionally kept the molecular perspective of genomes at the foundation of genomic architecture. Therefore, we have not been able to incorporate long-range influences of the genome into our conception of genomic architecture. As a result, we define these long-range influences in the language of molecular interactions between the molecules involved in these long-range influences. Admittedly, these long-range influences eventually manifest as molecular interactions, such a perspective doesn't tell us anything about the architectural design of genomes. Therefore, as it was suggested in Sect. 3.7, it is necessary to formalize these long-range influences in the spatiotemporal context and not in the molecular context as conventionally understood. It is possible that natural selection must have preserved these spatiotemporal templates of these long-range influences and what we observe in the form of molecular interactions is some kind of phenotypic expression. There is an exact parallel to this hypothesis in proteomics. While we try to think of amino acid sequences as being conserved during natural selection, in reality, it is the sequence of protein domains that are conserved during natural selection (Bryson and Vogel 1965). Similarly, postulates of long-range influences suggest that these long-range influences are essentially spatiotemporal in nature and the molecular interactions are their phenotypic expressions.

Postulate of Long-Range Influences
Genotopes interact with one another through topological surfaces of the spacetime. The long-range influences in the form of molecular interactions are four-dimensional projections of these topological influences.

This postulate completes the trilogy of postulates. Therefore, let us summarize the combined implications of these three postulates. Firstly, they emphasize that genomic architecture must be defined using larger structural and functional concepts rather than on the molecules participating in making a genome. Admittedly, the molecular perspective is important, but it is not the fundamental perspective. The fundamental perspective of a genome must be topological ensembles of which the molecular perspective is an offshoot. Secondly, the molecular perspective is completely derived from this topological perspective. In other words, the proposed topological perspective is not adjunct to the molecular perspective. On the contrary, the molecular perspective of a genome is completely defined by the underlying topological perspective. This might sound preposterous, but it is not. Just think of medicinal chemistry (DeVillers and Balaban 1999). We intuitively accept that the medicinal properties of molecules are totally dependent on the underlying wave function which is essentially a topological perspective to molecules. Just as the medicinal properties of molecules are phenomenology arising from the wave function of these molecules, the molecular interactions (which we now consider as the bedrock of genomic functionalities) of various constituent molecules of the genome are phenomenology arising from these topological units of genomes.

Thirdly, by incorporating spacetime itself in the conception of the structural unit of the genomic architecture, the model incorporates two semantic propositions. It suggests that if spacetime is included in our conception of the environment, then the environment plays an active role in natural selection. Therefore, if spacetime has a certain structuralism having a certain degree of complexity, the complexity arises naturally during natural selection. This is because according to this proposition, natural selection merely alters the types of complexities. It doesn't introduce complexity de novo. The second semantic proposition of this model is that it implicitly places biological evolution and natural selection in the purely naturalistic domain. Just as the fine structure of spacetime gives rise to quantum mechanical, physical and chemical phenomenology, it also gives rise to biological (and by extension to cognitive) phenomenology. Thus, the proposed model serves three purposes. Firstly, it integrates Life with other natural phenomena. Secondly, it gives a naturalistic explanation of biological evolution and natural selection. Finally, it provides a template for explaining the origin of complexity during biological evolution and natural selection.

Admittedly, these are tall claims and we will try to find out how justified these claims are. We will try to deconstruct various features of biological evolution and natural selection, particularly those features which have remained ambiguous in the conventional perspective of the Darwinian paradigm. In the next section, we will begin with the deconstruction of the semantic necessity of having two separate units of inheritance and selection. Apparently, as discussed in the first chapter, this ambiguity remains unresolved in the conventional perspective of the Darwinian paradigm.

3.11 Units of Selection Versus Units of Inheritance

The question of why the units of inheritance and selection are necessary for natural selection remains unresolved. Earlier, during the time when Darwin's theory (Hodge and Redick 2009) was published, one could understand that since the science of genetics wasn't available, this question could not be answered. However, the fact remains that even after the advent of genetics and even molecular biology, this question still remains unanswered. There is no explanation why natural selection cannot directly operate on genotypes. It is possible to argue that in the conventional sense of genotype and phenotype, it is a phenotype which is exposed to the environment and therefore, natural selection operates only on phenotypes. According to this perspective, genotype is firmly ensconced within the nucleus, and therefore, it didn't have to face the changes in the environment. Therefore, natural selection could not operate directly on genotype.

However, as discussed in the preceding chapters, our current understanding of the distinction between genotype and phenotype is not rigid. We now know that it is the intracellular context that defines which is genotype, which is phenotype, and what constitutes the environment. However, even under such a contextual definition of genotype and phenotype, the question of the need for separate units of inheritance and selection remains unresolved. There is something inherently unique about natural selection which has escaped our attention. Without committing ourselves to any particular mechanisms, it is intuitively clear that the answer to this question must lie in the structural template of natural selection. There must be some structural mechanism of natural selection which can distinguish between genotype and phenotype. More importantly, this mechanism must operate selectively on phenotypes. If such a structural mechanism was indeed operational, then it must carry within itself, a context defining capabilities which decides the intracellular context as well.

The proposed model starts with this rationale and articulates the information theoretical perspective of the definition of genotype, phenotype and the environment. Once we define these three entities in the form of different types of information content, it will be possible to formalize natural selection purely as a process of information transfers and transformations. Moreover, if natural selection is defined as an information transfer and transformation process, it can be selective in its application to only a particular type of information content. This will ensure that natural selection operates on phenotypes. More importantly, this selective ability to distinguish between genotype and phenotype will always be contextual because the basis of natural selection is the type of information content and not where it is located.

The key point of the proposed model is that it replaces the structural perspective of defining genotype, phenotype and the environment by a semantic perspective. Moreover, it defines the semantic context using the information theoretical perspective. Therefore, structuralism axiomatically enters into the conception of natural selection. This is because from an information theoretical perspective, it is possible to link structuralism with the semantic criteria. Information content per se provides structuralism and the relationship among the information content provides semantic criteria. This approach to unify structuralism and semantics is not available in the conventional perspective. This information theoretical unified perspective has its own importance in defining the genomic architecture, particularly the one proposed here, viz., the genotopic architecture. Before we define this architecture, let us define the topological unit of Genotope in the next section. Having done that, we will elaborate the exact relationship between genotype and phenotype according to this information theoretical perspective.

3.12 Topological Model of Genomic Unit Genotope

The conception of Genotope is based on a topological model of modified involuted manifold model. The details of this model are presented in the preceding monograph (Chhaya 2022a). Therefore, we will begin this section with a brief point-wise description of the proposed model. Having done that, we will look at the formal description of the unit of Genotope. The proposed modified involuted manifold model can be described as follows.

1. A topological manifold having the highest dimensionality is defined as a parent manifold. It is singular in nature in the sense that it has no metric and its dimensionality is some positive integer value. The exact value of the dimensionality of the parent manifold can be adjusted according to the object sought to be formalized. The numerical value is not important because it is the changes in this numerical value of dimensionality that defines the consequences of any natural phenomenon including biological evolution.
2. A new operator of involution is introduced in the model. This modified operator of involution is a special case of the conventional operator of involution. In the conventional definition, the operator of involution connects the parent manifold to any of its submanifolds. However, the modified operator of involution is a special case because it applies to the parent manifold itself and not to any other of

its submanifolds. Since every manifold is its own submanifold, the modified operator of involution is still an operator of involution.
3. This definition of the operator of involution brings about two important consequences. Firstly, every operation of this modified operator results in the decrease in the dimensionality of the parent manifold by one. Secondly, every operation of this modified operator of involution leads to an increase in the complexity of the metric of the resulting involuted manifold. Since the operators of involution, collectively, constitute a mathematical group, every operation of this modified operator brings about a certain type of changes in the metric details of the resulting involuted manifold.

The details of this formalism and its ontological and epistemological perspectives are described in the preceding monograph. Therefore, we will take this description of the proposed model at a face value and see how we can define a Genotope. As discussed above, we wish to include temporal and spatial long-range influences into the conception of genomic architecture. Therefore, we will define Genotope as a unit holding molecules (including DNA sequence, chromatin, and other proteins like adhesins) and the spacetime surrounding it. At this stage, we will not define the relationship between these molecules and spacetime, but assume that they act as a structural and functional unit. Now using the proposed model, we will try to describe a Genotope.

1. As mentioned above, we will assign the spacetime component of Genotope the highest dimensionality.
2. The molecules which constitute a Genotope would be assigned lower dimensionalities. We will assign different dimensionalities to different molecules later on when we have developed the genomic architecture in more detail. Presently, it will suffice to say that these molecules will occupy lower dimensionalities than that of spacetime.
3. In accordance with the relativistic model of spaçetime, we will assume that spacetime in this case also is a blend of time-like and space-like features. Thus, spacetime will occupy the highest dimensionality in a Genotope, but each of its dimensions will have different degrees of blending of time-like and space-like features. This ensures that the passage of time arises from the changes in the blendings of these dimensions. Thus, during the passage of time, the dimensionality of spacetime is unchanged, only its blend of time-like and space-like features changes.
4. This feature ensures that genomes too would evolve during these changes in the blendings of time-like and space-like features because these changes eventually pass on to the lower dimensionalities through involutions. Thus, phylogenetic models too can be formalized using the proposed model.
5. Both types of long-range influences, viz., temporal and spatial influences, can be formalized as an involution. Thus, the temporal long-range influences can be formalized using the operator of involution on a dimension of spacetime which has predominant time-like features. Similarly, the spatial long-range influences

can be thought of as an involution of one dimension of spacetime which possesses space-like features predominantly. Thus, the temporal and spatial long-range influences can be formalized using a single operator of involution. More importantly, these two types of long-range influences remain distinctly different because they arise from different dimensionalities.

6. It must be kept in mind that since each dimension of spacetime is a blend of time-like and space-like features, every temporal long-range influence will be accompanied by a minor spatial influence and vice versa. This is interesting because it provides a way to verify the proposed model.

At this stage, this description of a Genotope is adequate for the present discussion. We will add more details as we develop more arguments. A more detailed description of genotopic architecture will be discussed in the final chapter. At this stage, let us look at what are the semantic implications of this conception of the topological unit of genome.

3.13 Semantics of Genotope

As discussed in the preceding monograph (Chhaya 2020) where the proposed model was first articulated, the structuralism of scientific theories is derived from mathematics. However, its semantics are derived from our intuitive understanding of reality. Therefore, there is an inherent mismatch between the structuralism and semantics of scientific theories. Quantum theory is perhaps the best example of this dichotomy between our formal and intuitive understanding of natural phenomena (Chhaya 2022c). Of course, the proposed model explains the origin of this dichotomy between structuralism and semantics of scientific theories. However, from the perspective of the present discussion, it is intuitively clear that every change in the structural template of a scientific theory must be accompanied by the corresponding change in its semantics. Therefore, in this section, we will discuss the semantics of Genotope. Therefore, let us begin with the structural changes that the conception of Genotope brings about in the conventional perspective of genomic structuralism. Having done that, we will deconstruct the semantic implications of the proposed model of genomic architecture. For this purpose, we will select three structural changes that the proposed model imposes on the genomic architecture. These changes are inclusion of spacetime into the genomic architecture, parity between temporal and spatial long-range influences and the postulate of higher dimensional architecture. Here again, we will adhere to simpler consequences of these changes. We will add more and more nuances to this primitive semantics.

Let us begin with the first structural change proposed by this model, viz., the inclusion of spacetime in genomic architecture. There are two important semantic implications of this change. Firstly, by incorporating spacetime into the genomic architecture itself, the proposed model firmly places biological evolution in the naturalistic domain. Admittedly, the earlier teleological arguments or some kind of transcendental entities like the "Vital Force" are no longer taken seriously. However, the tendency to invoke some kind of design principles is still actively pursued (Fodor and Piattelli-Palmarini 2011). Earlier mysticism has been replaced by a romanticism of Nature. The point is not whether there are some design principles manifest in biological evolution and natural selection. The point is that what is the source of any such design principles? Naturalism, which is the foundation of modern science, doesn't deny such design principles. In fact, it is possible to think *a la Spinoza,* (Morgan 2002, see p. 108) that all the fundamental laws of Nature deal with the symmetry and its breaking. Therefore, the conception of design emerges naturally in modern science. However, biological evolution (and to a lesser extent natural selection) have remained outside such a naturalistic interpretation. Therefore, the inclusion of spacetime in the conception of genomic architecture allows such a naturalistic interpretation of biological evolution and natural selection. It addresses semantic ambiguity about the origin of complexity.

However, this inclusion of spacetime in the conception of genomic architecture is not dictated by the individual ideological predilection. It is based on the semantic imperative of the conventional perspective of the Darwinian paradigm. The proposed model merely formalizes what was always implicit in the conventional perspective of the Darwinian paradigm. Conventionally, the long-range influences are thought of having arisen from stereochemical proximity of otherwise distant DNA sequences (cf. Chromosome territories (Fritz 2014)). Similarly, as mentioned above, prior expressions of initiators and facilitators are actually a temporal influence. However, we try to define it in terms of spatial patterns of gene expressions. Therefore, the conventional perspective implicitly equated both these forms of influences. The proposed model takes one step further and formalizes the parity between temporal and spatial influences respectively. Once we accept this parity, it is intuitively clear that this parity is best formalized using spacetime itself. This is because in the theories of spacetime, spatial and temporal features of spacetime are equivalent (cf Minkowski space (Naber 2003)). Moreover, as mentioned above, this inclusion of spacetime also helps to explain the origin of complexity during biological evolution. This is because the fine structure of spacetime has its own degree of complexity which can be passed on to the complexity of genotype. This rationale is also evident from the second structural change proposed by this model, viz., the parity between temporal and spatial long-range influences during gene expressions.

While we may or may not agree with the proposition of spacetime being an active participant in natural selection, the best way to confirm this proposition is to formalize the relationship between spacetime and biological evolution and natural selection. More importantly, it can be verified by the method by which the parity between temporal and spatial influences is formalized. Even if there is uncertainty about the semantic justifications for such a proposition, it is best resolved by putting in place an exact template of this parity. Therefore, let us look at the semantics and structural basis of the parity between temporal and spatial influences. With reference to the semantic justifications for this parity, it is intuitively clear that if the theories of spacetime find it necessary to treat spatial and temporal dimensions as undistinguishable (Schutz 2009, see Chapters 7 and 8), there is no reason for a biological model to overturn it, provided there is a suitable mechanism in place. Therefore, let us see what kind of structural template the proposed model offers for this parity between spatial and temporal influences on the gene expressions.

According to the conception of a Genotope, it is an integrated unit of genomic architecture. Therefore, it doesn't distinguish between spacetime and the molecules present in the Genotope. Similarly, it doesn't distinguish between different molecules that constitute a Genotope. This might sound unconventional, but it arises naturally from the conception of Genotope. In order to understand this, it is important to remember that according to this model, as discussed in the preceding monograph (Chhaya 2022b, see Chapters 2 and 3), every fundamental particle is an isomorph of spacetime. Therefore, by definition, even atoms and molecules are isomorphs of spacetime. Thus, the proposed conception of Genotope is consistent with the model proposed earlier for spacetime. Furthermore, this model doesn't say that these molecules are identical. Rather, it suggests that what separates two different molecules present in Genotope, say, a DNA sequence and chromatin or even a protein like adhesin, is their information content. Conventionally, we think of these molecules in terms of electron density in a given stereochemical orientation. We don't think that these molecules are different unless they have different distributions of electron density in a given three-dimensional space. Thus, from the perspective of chemical reactivity, molecules are recognized to be different only when the distribution of their electron density is different (cf. Quantum chemistry (Szabo and Ostlund 1989)). Similarly, from the genomic perspective, these molecules like DNA sequences, chromatins or different proteins are identical unless their information content is different.

Essentially, this is the rationale behind the emergence principle (Smith and Morowitz 2016). At every level of organization, there emerges a different type of structural units having their own structuralism and semantics. Let us consider the case of different chemical elements. When we think of, say, a carbon atom, we don't think in terms of protons and neutrons present in the carbon atom. We don't do it because we have trained ourselves to think about the carbon atom from the atomic perspective. We could extend this logic in both the directions, toward the constituents of neutrons and protons or toward amino acids and nucleotides. It is the frame of reference of the organization that defines its structural units. This situation is further reinforced by our cognitive processing which is grounded in the four-dimensional spacetime (Chhaya 2022d, see Chapter 4). Just as we cannot perceive time in the same manner as we perceive space, we cannot perceive higher dimensional perspectives of reality. If we had a cognitive functionality of perception of reality from a higher dimensionality, maybe our perception of time could have been different. Similarly, if we had such a higher dimensional perspective, we would have formalized a structural unit of the genome in the form of Genotope (or some other such unit). However, mathematics, particularly topology provides a way to overcome this limitation. Thanks to our understanding of topology, it is possible to conceptualize any such higher dimensional perspective. This is precisely what this proposed model offers. Therefore, we must accept that genomic architecture has its own unit which we have chosen to call a Genotope.

Having said that, let us see how spatial and temporal influences can be formalized using a single framework. There are two conflicting requirements for this formalization. Firstly, it must establish a formal parity between temporal and spatial influences. Secondly, at the same time, it must manifest these two types of long-range influences in different types of phenomenology. The proposed model offers a way to fulfill these two conflicting requirements. It defines a higher dimensional representation of long-range influences which is singular in nature. Then, using an operator of involution, it devolves this singular functionality to lower dimensionalities in two different types of influences mentioned above. Thus, at a higher dimensionality, there is only one mode of long-range influences. This is transferred to lower dimensionalities by an operator of involution. As discussed in the preceding monograph (Chhaya 2022b), the representation of spacetime according to this model is such that the time-like dimension is of a higher dimensionality than that of space-like dimensions. (It is a different matter that during the epistemology of spacetime, due to inherent structuralism of our cognitive faculty, we can't perceive this difference in the dimensionalities of time-like and space-like dimensions. Albeit, we perceive the passage of time, but only through our psychological memory, but never directly. Now, according to this model, the operator required for the involution from the singular long-range influences to the time-like dimension is different from the operator of involution needed to devolve from the singular long-range influences to the space-like dimensions. Therefore, it is possible to have a unified representation of long-range influences of a Genotope which can be devolved into separate temporal and spatial influences by two different operators of involution. This is necessary because in the conventional perspective, long-range influences manifest in two different ways. Moreover, in the proposed model, this duality arises naturally because according to the proposed model, space-like and time-like dimensions occupy different dimensionalities, thereby requiring two different operators of involution. Thus, the proposed model fulfills the conflicting requirements for unitary representation of binary form of long-range influences.

It is possible to argue that while the scenario described above may be mathematically feasible, it still needs to be translated into the details of biochemical description of these two types of long-range influences. To understand this, let us revert back to the conventional manifestation of these two types of long-range influences. As discussed above, the spatial long-range influences occur in the form of the placement of the facilitators and initiators of any particular gene expression at a reasonable distance from the concerned gene (For the present discussion, we will try to overlook the distinction between cis and trans influences). On the other hand, the temporal long-range influences manifest in the form of prior expressions of genes encoding these facilitators and initiators. Admittedly, this is a simplistic scenario, but it would suffice for the present discussion.

The question is how these two different phenomena can be shown to have arisen from a single source, viz., a higher dimensional structural unit like Genotope? For the spatial long-range influences, it is somewhat easier to deconstruct how they arise from the higher dimensional topological units. We will pick up an example from developmental biology (Gilbert and Barresi 2020). Once the three germ layers of an embryo are established, the embryo undergoes a process of invagination. During this process, an outer germ layer pierces the inner germ layer and starts spreading inside the cavity (blastocoel). At the end of the invagination, the adjacent cells present in the inner germ layer (in the blastocoel) are now separated by the intrusion of the cells from the outer germ layer. Thus, the points that were contiguous earlier now appear to be separated from one another. Similarly, if the proposed higher dimensional unit of genomic architecture like Genotope were to exist as a compact unit, it would appear to be stretched and separated in the lower dimensionality, viz., the four-dimensional spacetime from which we observe genomic architecture. The only difference between these two scenarios is that in the former, a germ layer is the invader, while in the latter, the spatiotemporal dimension is the invader. In fact, this scenario can be verified by mapping different long-range spatial influences and backtracking them to higher dimensional models where they will appear to be contiguous. When we find all the known long-range spatial influences contiguous in any particular dimensionality, that is the dimensionality in which a topological unit like a Genotope must be formalized.

Now let us look at the temporal long-range influences. These influences are slightly difficult to conceptualize. However, we will try to visualize it in the simplest possible form. The difficulty in conceptualizing the long-range temporal influences is that our cognitive faculty can't directly perceive time as the fourth dimension. Therefore, we will try to visualize this in an indirect manner. For this purpose, we will employ the model of spaçetime that is employed by physicists. In the general theory of relativity (Naber 2003), it is generally conceded that there are no separate time and space dimensions. Every dimension of spacetime is a blend of time-like and space-like features. Admittedly, this conception of spacetime gives rise to a host of counterintuitive features which have been investigated extensively. However, we will simply use this description of spacetime for our discussion without going into its mathematical formalism or its metaphysical nuances. Let us assume that such a spacetime influences the gene expressions by some unknown mechanism. Naturally, since every dimension of spacetime is a blend of time-like and space-like dimensions, the temporal influences will result in two types of effects. These effects will be expressed in terms of different timings of gene expressions and different positions in the genome. More importantly, since these influences are correlated with one another, they will have discernible patterns. This is because at a higher dimensionality wherein a Genotope exists, both these effects will originate from a single point. It is only when the effects arising from a single point devolves into lower dimensionality, say, four-dimensional spacetime, that they would appear to be two separate types of effects. Thus, it is possible, at least in principle, to correlate the sequence of gene expressions and their spatial distribution on the genome by constructing a higher dimensional topological model where these two spatial and temporal maps converge. Therefore, it is possible to verify this model even in the absence of its formal description by simply superposing temporal and spatial distances between different gene expressions and finding out a dimensionality wherein these two maps coalesce.

In retrospect, it is intuitively clear why the conventional perspective didn't factor the contribution of spacetime in biological evolution and natural selection. This omission occurred because the conventional perspective is grounded in the sensibilities of the four-dimensional world wherein time-like features are not directly perceptible. More importantly, in the four-dimensional world, the distinction between time-like and space-like dimensions is manifest, unlike the higher dimensional model of spacetime wherein they are unified. Admittedly, what has been discussed in this section appears to be vague, without any experimental evidence. In the following sections and chapters, we will try to provide more information that seems to be congruent with this rationale. Before doing that, it is time to deconstruct several conventional concepts using the proposed model, viz., the relationship between genotype and phenotype, natural selection, and genomic architecture. We will begin with the relationship between genotype and phenotype in the next section.

3.14 The Relationship Between Genotype and Phenotype Using Genotope

In the preceding chapters, we outlined various shortcomings and ambiguities of the relationship between genotype and phenotype in the conventional perspective of the Darwinian paradigm. In the preceding sections, we outlined a simplistic description of a new topological structural template of genomic architecture. The proposed unit of that model of genomic architecture, viz., Genotope, must also be capable of defining the relationship between genotype and phenotype. Therefore, in this section, we will try to deconstruct this relationship using Genotope as a starting point. For this purpose, we will also include spacetime in deconstructing the nature of relationships between genotype and phenotype.

To begin with, we will define genotype, phenotype and spacetime in terms of their information content. Moreover, we will adhere to the conventional perspective of respective definitions of genotype and phenotype. Since the conventional perspective doesn't permit any active participation of the environment, we will employ the notion of spacetime, as defined above, instead of the environment in this deconstruction. When viewed from the information theoretical perspective, it is apparent that the types of information content of genotype and phenotype (and their degrees of complexity) are different. Therefore, according to this model, they must be represented in different dimensionalities. For instance, the information content of genotype consists of the information content in the form of DNA sequence. Admittedly, as discussed above, genomic architecture too should be taken into account while defining genotype. However, since we are focusing on the conventional perspective of genotype, we will adhere to this conception of genotype. Similarly, the information content of phenotype must be conceptualized in the form of amino acid sequences. Therefore, according to the conventional perspective, both genotype and phenotype must possess a comparable complexity of their information content. This is in congruence with the molecular perspective wherein the nucleotide sequences and amino acid sequences of DNA and protein form complexes. Such molecular complexes, which are the backbone of cellular physiology, would not have been naturally selected unless they had structural and conformational compatibility. This is essentially the conventional perspective as seen from the information theoretical perspective.

Now, we will try to modify it using the proposed model. According to the proposed model, the complexities of the information content of genotype and phenotype are of different types. For instance, the informational complexity of phenotype consists of not only its amino acid sequences, but also its functionalities, say, catalytic properties. Therefore, a correct representation of phenotype according to this model must include the functionalities of phenotype as well. Therefore, according to this model, phenotype must occupy more than one dimensionality, one for its molecular complexity and another one for its functional complexity. It is possible to argue that similar rationale must be applied to genotype as well. This is where we need to introduce genomic architecture. While in the classical perspective, genotype was conceptualized as a discrete entity of a gene (with its DNA sequence), in the genomic perspective, genotype must include a functional unit of gene plus the DNA sequences of long-range influences of cis and trans varieties. In such a scenario, the functionalities, in the form of long-range influences, must occupy different dimensionalities than the one occupied by the gene itself.

Therefore, according to this model, genotype and phenotype occupy multiple dimensionalities simultaneously and their relationship must be represented by the changes in their dimensionalities. Now, let us introduce spacetime as a participant. In the conventional perspective, spacetime (or rather environment) has no active role to play. Therefore, from the information theoretical perspective, the complexity of spacetime has only a marginal, if any, role to play. However, according to this model, spacetime has its own degree of informational complexity and therefore, it will play an active role in defining the relationship between genotype and phenotype. Moreover, according to the arguments presented above, spacetime contributes to genotype and phenotype. Therefore, it is intuitively clear that it must occupy a dimension which is higher than those of genotype and phenotype. Therefore, in this model, one can represent genotype and phenotype as submanifolds of spacetime. This is necessary because as discussed in the preceding monographs, every form of matter must be considered as an involuted state of spacetime. Therefore, DNA or its nucleotide sequences, too, must be seen as involuted forms of spacetime. Thus, genotype derives its structural template from that of spacetime through a series of involutions. Admittedly, from the computational perspective, the computation of structuralism of DNA can't be derived from the fine structure of spacetime because it is beyond our current computational capabilities. However, in principle, it is possible to define the structural complexity of genotype (and its information content) as having arisen from the fine structure of spacetime.

Similar arguments can be made for phenotype as well. However, there is one subtle difference in the case of phenotype. Phenotype also derives its functional template from the fine structure of spacetime as well. Thus, we will need to formalize natural selection using a hierarchy of manifolds. There will be a parent manifold representing spacetime. Then, we can represent genotype and phenotype as submanifolds of the manifold of spacetime. Next, we will have to define genotypes and phenotypes having multiple dimensions simultaneously. Moreover, there will be certain dimensionalities which are common to both genotype and phenotype (representing their molecular description), and there will be separate characteristic dimensionalities of genotype and phenotype each, representing their functionalities. The conventional molecular biological transformations will occur in the dimensionalities which are common to both genotype and phenotype. However, the relationship between molecular templates of genotype and phenotype will have separate relationships with their respective functional features which occupy separate dimensionalities, other than those dimensionalities which are common to both. Moreover, since genotype and phenotype occupy different submanifolds, it is axiomatic that the relationship between genotypic functionalities and phenotypic functionalities cannot occur by any direct interactions between these noncommon dimensionalities. This relationship can only manifest via higher dimensionality representing the environment. Thus, this model ensures that only the one-way influence of genotype on phenotypes is manifest. The corresponding reverse influences from phenotype to genotype cannot occur directly. It can occur indirectly through the common dimensionalities of genotype and phenotype through epigenesis. Such a hierarchy is adequate for deconstructing the process of natural selection. It must be kept in mind that this scenario is congruent with the conventional perspective of natural selection except for the inclusion of spacetime. Therefore, in the next section, we will try to understand how complexity arises during natural selection. We will try to distinguish between the conventional perspective of the emergence of complexity during biological evolution and natural selection and the proposed model.

3.15 Natural Selection and Genotope

In the previous section, we outlined a simplistic template of the relationship among genotype, phenotype, and spacetime. Using this model, we will try to deconstruct natural selection in this section. We will begin with the conventional perspective and try to understand why it is inadequate to provide an explanation for the emergence of complexity during biological evolution and natural selection. Having done that, we will describe natural selection from the perspective of the proposed model and demonstrate how it offers a natural explanation for the emergence of complexity. In the conventional perspective (Hodge and Redick 2009), the environment only acts as a passive role in natural selection. Its role in natural selection is indirect. It plays a role in changing genotypes by inducing mutations. Similarly, the environment provides parameters to define survival fitness of phenotype but doesn't directly choose phenotype. Of course, as discussed in the preceding chapters, there were historical and valid reasons for this insistence on the passive role of the environment in natural selection. However, we will sidestep this issue in the present discussion. According to the conventional perspective, particularly in the post genomic era, it was conceded that the environment plays an active role in the process of conversion of genotype into phenotype. In fact, it was this realization that enabled us to understand the notions of genotype and phenotype are overlapping and the boundary between them is flexible. However, even after this realization, there was no explanation why the successive generation of species turned out to be more and more complex.

With the advent of genomics, we realized that the genome as a whole also adds to the complexity during gene expressions. However, we still don't know the level of complexity present in the genome and how it gives rise to complexity during natural selection. This is evident from the fact that we don't have any formal description of genomic architecture. Leave alone the genomic architecture itself, we do not know what is the relationship between the number of genes and the complexity of organisms. Our understanding of the emergence of complexity during biological evolution and natural selection is confined to the conception of gene expressions and the resulting functionalities of the protein molecules.

Therefore, when we try to represent this understanding in the structural template outlined above, we can see that there is a structural complexity of genotype which through the process of gene expressions, gives rise to a different kind of structural complexity of proteins. Therefore, we have two manifolds occupying multiple dimensionalities simultaneously. Some of these dimensionalities are common to genotype and phenotype (the dimensionality representing molecular perspective). Moreover, phenotype has several dimensionalities which it doesn't share with its genotype. These dimensionalities refer to the functionalities of phenotype. Admittedly, there are some dimensionalities in which genotype too manifests its functionalities. Interestingly, some of these dimensionalities representing functionalities could be common to both genotype and phenotype. However, these are relatively few in numbers. These dimensionalities can be assigned to RNA or ribozymes that are presumed to be original replicators in the RNA world hypothesis (Yarus 2010). When visualized in this manner, it is intuitively clear that the conventional perspective cannot offer any explanation for the emergence of complexity during biological evolution and natural selection. This is because it cannot formalize biological evolution itself. It is important to remember that the Darwinian paradigm is more concerned with natural selection than with biological evolution. In fact, it takes biological evolution as a priori. Therefore, the conventional perspective cannot explain the origins of the complexity of genotype itself. Therefore, it also cannot explain why different kinds of complexity arise during gene expressions and what is the origin of functionalities of phenotype.

Admittedly, this shortcoming of failing to explain functionalities is not confined to the Darwinian paradigm, and it manifests itself in genetics and molecular biology as well. In fact, it also manifests itself in biochemistry too. What is the origin of chemical functionalities like enzymatic catalysis can never be explained by biochemical theories. The only explanation that is available is in the form of quantum chemistry wherein one postulates that the wave function of these molecules also contains the details of its chemical functionalities. It is important to remember that quantum chemistry, particularly the molecular wave function, is merely an affirmation of our belief that molecules have inherent functionalities. Quantum chemistry

(Szabo and Ostlund 1989) is never meant to be an ontological theory. Therefore, the failure of the conventional perspective of natural selection to explain the emergence of complexity is far deep rooted in the very semantic foundation of modern science.

There are two reasons why the conventional perspective of natural selection cannot explain the emergence of complexity. Firstly, in common with other disciplines, it assumes that structural template is also a functional template. Secondly, as an extension of the belief, the conventional perspective assumes that the molecular template is the only structural template of genotype. Although with the advent of genomics, there is a gradual realization that there could be higher levels of structural elements which supervene the molecular framework. In order to overcome this limitation, the proposed model seeks to define everything from the information theoretical perspective. Therefore, not only the conventional molecular paradigm, but also higher level structural elements are defined in a single framework. More importantly, the proposed model also seeks to define a template for functionalities using the same framework.

Once we define a unified framework for structuralism and functionalities in a single framework, it is self-evident that complexity doesn't emerge de novo, but it undergoes transformations from one type of complexity to another. Therefore, it is tempting to think that natural selection can be formalized using the information theoretical perspective. In other words, natural selection must be thought of as being a structural process operating in any system having a requisite level of complexity. Moreover, if natural selection is a structural process, it can be formalized as a mathematical operator capable of changing one type of complexity into another. It must be kept in mind that even if we were to attempt such a formulation of natural selection, the real problem is that we don't have a common framework in which we can define genotype and phenotype. In the conventional perspective, we can attempt to define genotype and phenotype using, say, molecular perspective. This is precisely what present day genomics is all about. However, because the molecular perspective cannot assign a separate framework for molecular functionalities, we have not been able to formalize molecular functionalities (cf. Enzymatic catalysis (Hammes 1982, see Chapter 5)). Therefore, the molecular perspective is inadequate to formalize natural selection as a structural process. It must be kept in mind that the molecular perspective has been very successful in defining the molecular changes that happen when genotype is converted into phenotype. However, because it cannot formalize functionalities of phenotype (as a separate entity from its molecular structuralism), the molecular perspective cannot formalize natural selection as a structural process.

In contrast, the proposed model defines all structural elements, including those of functionalities of phenotype, in a single framework. Therefore, the proposed model is inherently capable of defining natural selection as a purely structural process independent of the domain. This is where the significance of incorporating spacetime into natural selection lies. Once we include spacetime as a participant in natural selection, it is intuitively clear that it must be integral to genotype and phenotype. Thus, when we define the structural template of genotype and phenotype (and by implication, structuralism and functionalities) in the single framework of the structural template of spacetime, all that we require is a way to formally represent genotype and phenotype in the language of structuralism of spacetime. Once we have this formal description in place, one can define natural selection as mathematical operators which connect different types of metrics of genotype and phenotype as well as those of structuralism and functionalities.

One of the important consequences of this approach is that it provides a template for predictive methodologies into natural selection without compromising its essential nondeterminism (Bonner 2013). Since genotypes will have only certain types of complexities, in the form molecular structuralism and correspondingly, phenotypes will have certain types of complexities in the form of their functionalities, both these types of complexities are connected to one another by only a special class of operators of involution. This enables one to formalize a methodology for predictions. However, since the operators of involution defined for this purpose do not apply to any particular molecular structures (but to a type of complexity), this model is not deterministic. It merely introduces a limit to the range of changes that natural selection can cause.

Moreover, since the operators of involution, by definition, are Abelian, they also provide for loss of complexity during natural selection. This is important because such conception of natural selection explains the phenomena of gain of function as well as the loss of function (Alberts 2004). Moreover, since the model employs spacetime instead of the environment, it is neutral toward the changes in the environment and its influence on natural selection. At the same time, since the structural template of spacetime is constant (at least in the present context), it provides a continuity of forms. In other words, the proposed model allows phylogenetic legacy to continue, thereby reinforcing Darwin's original propositions of descent with modifications.

Admittedly, it is possible to argue that while such an approach appears logical, there is no justification to abandon the conventional perspective for something which is purely speculative. After all, the conventional perspective has withstood several paradigm shifts and there is no need to replace it. This reasoning needs to be refuted on two grounds. Firstly, as discussed above, it fills in several semantic lacunae of the conventional perspective. Secondly, if we can demonstrate that genomic architecture is indeed based on the proposed unit of Genotope then, it would be necessary to abandon the conventional perspective. Therefore, in the next few sections, we will explore the possibility of the genome being indeed configured on the basis of this conception of Genotope.

3.16 Genotopic Architecture of Genome

At the outset, it must be admitted that till date, we do not have any formal description of genomic architecture. Therefore, what we will discuss in this section is purely conjectural. Therefore, we will begin with the implicit genomic architecture obtained by inference from the available evidence. Moreover, we will simply look at the broad outline of the conventional perspective. Having done that, we will outline the proposed genomic architecture. In the following section, we will compare both these architectures from the evolutionary perspective. Let us begin with a list of propositions about the genomic architecture implicit in the conventional perspective. These propositions are not explicitly articulated but are always implicit in literature.

1. A genome is a collection of discrete units of genes.
2. The distribution of genes on a genome is not governed by any logic or rationale.
3. At the same time, some clusters of genes appear together in genomes indicating either functional or phylogenetic relationship between them.
4. However, as a rule, the placement of different genes within the genome is not indicative of functional linkages.
5. Between these two extremes of a group of genes found in proximity of one another and the random placement of genes, there is a more common scenario wherein gene initiators and repressors are present in the vicinity of the concerned genes. These DNA sequences are usually found upstream (cf. cis influences).
6. Similarly, it is possible to have initiator and repressor DNA sequences not in the proximity of the concerned gene, but in different chromosomes (cf. trans influences).
7. Therefore, there exists some kind of higher level organization of the genome wherein the notion of proximity is defined not in the geometric sense but in the topological sense.
8. Since the processes of insertion and deletion (indel) in genomes occur regularly during natural selection, it is not possible to formalize the notion of proximity or mechanisms of long-range influences.
9. However, since genomes display different levels of organization during a cell cycle, the genomic architecture cannot be based on the parameters like the length of DNA sequences. There must be some higher level of organization.
10. Since the processes of insertion and deletion are random, it is difficult to think of these processes as instruments of genomic architecture. Rather, they cause variations in the genomic architecture itself.
11. These different architectural designs of genomes enable natural selection to operate at the level of the genome. In other words, the genome is a unit of natural selection just like its constituent genes.

These propositions give us a broad outline of the conventional perspective of genomic architecture. Admittedly, there are several semantic ambiguities in these propositions, but they merely represent the ambivalence of our conventional perspective. For instance, the conventional perspective is ambivalent about the distribution of genes on a genome. On the one hand, the conventional perspective accepts that a certain group of genes like Homeobox are always present together on genomes (Duboule 1994). However, on the other hand, the conventional perspective is not willing to concede that this logic should be valid for all the genes. Therefore, the conventional perspective cites the processes like insertions and deletions to justify this ambivalence. The logic being that only the functional groupings of genes like

Homeobox have survived these *indels* because the resulting segregated Homeobox genes would never survive. Therefore, *indels* ensure that only the functionally related genes retain their organization while the rest of the genes keep on shifting their positions in the genome (provided they remain functional even after these *indels*).

However, this scenario is logically inconsistent for two reasons. Firstly, if this reasoning is correct, the earliest genomes must have been more compactly configured and the *indels* must have resulted in these less organized or loosely structured genomes. This is not true. If anything, genomes have become more organized and more complex during the course of natural selection. Secondly, if the *indels* were really random, why would natural selection favor trans long-range influences? The fact that the ideas like chromosome territory (Fritz 2014) make sense, suggests that the genomic architecture is not limited to the linear proximity to genes in the genome. Genomic architecture must be based on a different notion of proximity. However, the conventional perspective has continued with the notion of linear proximity and has retained this ambivalence toward the distribution of genes on a genome.

Having looked at the implicit genomic architecture according to the conventional perspective, let us look at the proposed model of genotopic architecture. This architecture is based on two semantic propositions. Firstly, natural selection merely alters the types of complexities and doesn't give rise to complexity de novo. Secondly, the notion of linear proximity must be replaced by a more fundamental notion of proximity of which linear proximity is just one example. As discussed above, in order to accommodate these two semantic propositions, the proposed model employs spacetime as a participant in natural selection. Using these semantic propositions now, we can provide a simple outline of genotopic architecture. Admittedly, this outline is primitive and even simplistic at this stage. However, in the following chapters, we refine this model. For the sake of simplicity, we will outline this model in a point-wise manner.

1. A Genome exists in multiple dimensionalities simultaneously. Therefore, its formalization must be defined in the language of topology.
2. Since the proposed model postulates that spacetime actively participates in biological evolution and natural selection, the genomic architecture must be defined in terms of spatiotemporal dimensions.
3. However, in accordance with the current spacetime theories, the proposed model assumes that spacetime exists in multiple dimensions simultaneously and that the four-dimensional spacetime is just one projection of this higher dimensional spacetime.
4. According to the proposed model, time-like and space-like features are present in every dimension of spacetime. The blend of these two types of features changes from one dimensionality to another. Similarly, the long-range influences in the form of spatial and temporal influences are one and the same but they appear to be different in different dimensionalities.

5. Similarly, a structural template of a genome, say its molecular perspective and its functionalities occupy different dimensionalities. Moreover, according to this model, every dimensionality has its own metric. Therefore, every genome has two separate templates, one each for structuralism and functionalities. However, just like long-range influences, there is a dimensionality wherein both these templates are unified.
6. It is at this dimensionality than one can define a unit of genomic architecture called "Genotope ".
7. Genotope must be viewed as a topological object and individual features of genes must be viewed as different dimensions within a Genotope in the dimensionality in which it is defined. .
8. Genotope is characterized by the linkages with different dimensionalities by a single type of a group of operators of involution.
9. Since spacetime is a constituent of Genotope, spatial and temporal changes are also defined by this set of operators.
10. Genomes must be formalized not as a string of Genotopes but as a nested hierarchy of Genotopes, with one Genotope encapsulating another. For analogy, we can think of an onion which has different layers, one encapsulating the other. However, This conception of Genotopic architecture of the genome differs from this analogue by the fact that unlike onion, this architecture is characterized by active interactions between different layers. Each layer, a Genotope in fact, is communicating with the neighboring higher and lower layers or Genotopes.

We will refine this conception of genomic architecture and the conception of Genotope in the following chapters. Presently, let us see how this conception agrees with the conventional perspective of genomic architecture and particularly with the phylogenetic evidence.

3.17 Evolution of Genome

Our current understanding of natural selection and its Darwinian interpretation, suggests that there must be more than one level at which natural selection occurs. This parallelism opens up a possibility of the genome itself being subject to natural selection. The idea of a genome being a unit of selection has been generally conceded. However, in the absence of any genomic architecture, it has not been possible to verify this concept. This is because the signals from natural selection at the level of individual genes contaminate the signals available from natural selection of the genome itself. Therefore, though our phylogenetic studies (Bromham 2008) provide enough hints about genomic evolution, as distinct from the parallel evolution of individual genes, it is not possible to segregate the changes from these parallel evolutions. Therefore, in this section, we will try to define a method of separating genomic and genetic evolution. For this purpose, we will employ the Genotopic architecture proposed above. In the following section, we will employ this method to the phylogenetic evidence available in literature.

For this purpose, we will overlook the molecular perspective that is commonly used in phylogenetic studies. Instead, we will try to employ a topological perspective. The reason for this choice is the above mentioned contamination of different types of signals. Admittedly, irrespective of the model we choose, there is bound to be some mixing of signals emanating from genomic evolution and genetic evolution. However, the problem with the molecular perspective is that it is biased toward genetic evolution because our conception of genes is molecular. Therefore, the proposed topological model seems to be a better tool to distinguish between genomic evolution and genetic evolution.

The idea that a higher level of organization could be a unit of selection is within the mainstream research on evolutionary biology (Okasha 2010). Protein domains are found to be more conserved during natural selection than their amino acid sequences (Bryson and Vogel 1965). Therefore, natural selection must operate on structural terms and our conventional molecular perspective is just one aspect of that structural template of natural selection. Therefore, in principle, any topological unit could also be a unit of selection. In this section, we will look at two aspects of any topological unit being a unit of selection. These aspects are: what kind of topological elements could be thought of as units of selection? and what kind of selection rules can be formulated for natural selection. Having done that, in the following section, we will try to deconstruct what kind of evidence one should look for in this topological model of natural selection.

Let us begin with the question of what kind of topological units can act as units of selection. Even in the absence of any template of natural selection, it is intuitively clear that the candidate topological elements must be decided on the basis of their symmetry or rather, lack of symmetry. There is a corresponding principle in the conventional perspective. For instance, during transcription and translation of DNA sequences, the machinery for these processes, in the form of molecular ensembles of ribozymes and proteins (Weinzierl 1999), use molecular shape as a yardstick. Therefore, it is legitimate to think that symmetry principles must be inherently involved in natural selection. The only difference in the proposed model is that it seeks to define a topological principle rather than a stereochemical principle. The key difference between these two processes is that the notion of proximity is defined differently in the topological sense than in the stereochemical sense. The stereochemical connotation of proximity rests on the linear metric of spacetime. On the other hand, the topological connotation of proximity rests on adjacency of different dimensionalities. Stereochemical proximity is an instance of more general topological proximity. More importantly, there could be instances wherein the topological proximity would not manifest itself in the corresponding stereochemical proximity. Therefore, it is possible that due to our possible preoccupation with molecular perspective, we have not sought out any of these nonstereochemical, but purely topological, instances of proximity. (Interestingly, this distinction also provides a method of distinguishing genomic evolution from genetic evolution.)

Returning to the present discussion, if genome is really organized in the form of onion like overlapping Genotopes, what kind of topological elements could be thought of as units of natural selection. At this stage, it is important to remember that these overlapping units consist of not just molecules but also slivers of spacetime. Therefore, we must include spacetime features into the conception of the units of selection. In the conventional perspective of natural selection, the unit of selection is a phenotype. However, in the conventional perspective, phenotype must be considered as an outcome of gene expressions under the influence of the environment. However, in the proposed model, genotype and phenotype are not separate categories, but they are separate types of complexities. Therefore, the unit of selection in the proposed model is the types of complexities. Thus, the correct way to visualize a Genotope is to think of it as a type of complexity. The part of a Genotope which alters due to its inherent propensities must be seen as an equivalent of the conventional idea of genotype. Similarly, the part of Genotope which undergoes changes in its complexity due to outside influences must be taken as an equivalent of the conventional idea of phenotype. Finally, the part of a Genotope which does not undergo any changes either under its inherent influences or due to any external influences, must be taken as an equivalent of the conventional idea of the environment.

This categorization gives us a more intuitive understanding of natural selection. Let us take an example of genetic mutations. In the conventional perspective, they are considered to be outside the purview of natural selection. However, according to this model, mutations occurring due to the shortcomings of the chemical processes of replication are considered to be part of genotype. On the other hand, mutations occurring due to the environment would be treated as a part of natural selection and therefore, the resulting DNA sequence will be treated as a phenotype subject to natural selection. Thus, the proposed model offers a way to define genotype and phenotype in the contextual manner. This is because it employs different types of complexities in defining various changes. The changes happening on their own cannot be included in natural selection. Similarly, changes happening due to external influences, irrespective of where they are happening, must be included in natural selection. This essentially removes a fixed boundary between the conception of genotype and phenotype and replaces it with a contextual semantics.

Returning to the unit of selection, it is intuitively clear that the unit of selection too must be defined in a similar contextual semantic sense. As mentioned above, the plurality of the units of selection is a well-established concept. Therefore, the proposed model offers a universal definition of the unit of selection by replacing the earlier categories with the types of complexities. With this preamble, the time has come to formalize the definition of the unit of selection according to the proposed model. For this purpose, once again, we will define the unit of selection in a point-wise manner.

1. Increase and decrease in complexity of a system are governed by the principle of symmetry and therefore, it cannot be governed by natural selection. However, in the case of plurality of possible outcomes, an increase or decrease in the complexity of a system will be governed by the Bayesian paradigm of conditional probabilities.

2. In other words, natural selection operates separately from the Bayesian paradigm. However, it is intuitively clear that both can operate in parallel or in an antiparallel sense. Therefore, the origin of the semantic ambiguities of the Darwinian paradigm must be seen as having arisen from our mistake of equating both these processes.
3. Natural selection is distinctly different from the Bayesian paradigm of conditional probabilities. However, since natural selection operates in parallel with the Bayesian paradigm, we have borrowed the Bayesian semantics to explain natural selection.
4. The distinction between natural selection and any model of conditional probabilities lies in the fact that natural selection arises from the interactions between two Bayesian processes.
5. When any two Bayesian processes interact with each other in a causal sense, the resulting outcomes are governed by structural laws defined by the principles of symmetry.
6. Since the principles of symmetry dictate discontinuity (in the form of asymmetries), outcomes of natural selection acquire a directionality. Therefore, due to the conjunction with the Bayesian probabilities, natural selection gives rise branching during the process of evolution. What has been classically defined as a "punctuated model of evolution (Gould 2007)" is a testimony of this inherent dichotomy between natural selection and the Bayesian paradigm. The same phenomenon is also responsible for several such discontinuities like speciation, separation of genes in a given genome, separation of a given genome into chromosomes etc.
7. A unit of natural selection can now be defined as a smallest possible component of the genome wherein two different Bayesian processes interact with each other.
8. To employ the conventional classification, genotype can be a unit of selection if its Bayesian process of mutations interacts with its genomic environment. This is because mutations merely change the complexity of genotype, but not its type of complexity. Mutations, per se, cannot be governed by the law of natural selection. However, if genotype were to change due to its interactions with genome or other products from gene expressions, say, gene silencing via methylation or imprinting (Robert 2004), then such changes would result in a change in the complexity as well as in the change in the types of complexities. Therefore, it is governed by natural selection. It must be kept in mind that this definition of genotypic changes is consistent with the conventional perspective and it is generally conceded. The key point of this way of looking at natural selection is that it defines genotype and phenotype in a contextual way. It suggests that the mutation, per se, must be viewed purely from the Bayesian perspective. However, the changes in genotype can be thought of as a part of natural selection if it brings about changes in the types of complexities as in the case of epigenetic processes.
9. Similarly, phenotypic changes must be differentiated on the basis of the changes brought about. For instance, phenotypic changes leading to either gain of

function or a loss of function would naturally be governed by natural selection. However, the phenotypic changes like synonymous changes in protein domains would be governed by the Bayesian paradigm. It is intuitively clear that the line separating natural selection and the Bayesian changes is often blurred. Therefore, Darwinian semantics has conventionally conflated both these influences.
10. In such a scenario, it would be legitimate to question the need to segregate both these influences. This is where the conception of complexity comes into the picture. The emergence of different types of complexities and the rules of their interconversion are governed by symmetry principles and therefore deterministic.
11. More importantly, the rules of interconversions of different types of complexities are defined at a systemic level, which in this case is at the genomic level. Secondly, since the environment (as incorporated here in the form of spacetime) plays an active role in the execution of symmetry breaking processes, natural selection needs to be defined at a systemic level. Therefore, the proposed model seeks to define a unit of selection which contains a genomic unit including the molecular components as well as the underlying spacetime itself. It is this unit which is sought to be named as the Genotope.
12. Once we accept this topological conception of the unit of selection, it is intuitively clear that its interactions with other Genotopes will be essentially spatiotemporal in nature. These interactions can now be defined in the language of topology which will axiomatically obey the symmetry principles.
13. In addition, the conventional molecular interactions which are the subject matter of genomics will be a subset of these spatiotemporal interactions. Therefore, if we could expand the genomics to include the topological perspective of the underlying spacetime, it is possible to recast the theory of natural selection and even the Darwinian paradigm into a scientific theory capable of some degree of predictivity.

There are two possible ways to respond to this model. Firstly, we can be dismissive and think about it as a superfluous trivialization of the Darwinian paradigm. Alternatively, we can think of refuting/ validating it on the basis of evidence. The trouble with the second option is that evolutionary biology is not amenable to such experimental verification. However, fortunately, we have at our disposal an enormous amount of literature on the phylogenetic studies (Bromham 2008). Therefore, in the next section, we will try to deconstruct the available evidence using the proposed model.

3.18 Phylogenetic Evidence of Genotopic Genome

Phylogenetics occupies the center stage of genomics. There are two reasons for this. Phylogenetics provides an ontological perspective of genomes. Secondly, it provides a structural template for genomic architecture by outlining the course of evolution of genomes. While the first aspect of providing an ontological perspective of genomes has been extensively used in genomics, the second aspect of using phylogenetics as a tool for mapping genomic architecture is less explored. One of the reasons for this imbalance lies in the fact that we don't have any theoretical model of genomic architecture.

This monograph offers one such model of genomic architecture. Therefore, it is possible to seek its affirmation from the phylogenetic evidence. The key question is what kind of evidence phylogenetic studies can provide for genomic architecture? In order to understand this question, let us look at the conventional approach to develop an ontological perspective of the genome using phylogenetic studies. A typical phylogenetic study involves comparison of DNA sequences of several species which we suspect to have arisen from a common ancestor. Such a comparative analysis suggests different degrees of separation between the species under investigation. These degrees of separation consist of the number of mutations required to convert the DNA sequence of one species to another. By using this parameter, it is possible, at least in principle, to place different species in different slots in the tree of evolution. Admittedly, in reality it is difficult to ascertain which of these species under investigation happens to be the oldest or most ancestral species. In the absence of any other evidence, it is not easy to decide which species must be considered as a common ancestor. This is usually referred to as the problem of rooting, i.e., the problem of where to place the roots of the proposed tree of evolution (Akalin 2021). Of course, conventionally, it is possible to employ different sources of information about the ancestry of the given set of species. In addition, we can employ the principle of parsimony to decide between different trees that one can generate from a given phylogenetic data.

On the other hand, the use of phylogenetics in decoding genomic architecture is indirect. This is because comparative study of different DNA sequences cannot offer any insights into the structural template of how different DNA sequences can be organized within a genome, at least not directly. However, using modern informatics, it is possible to discern higher level correlations among different genes and their interactions (Fertin et al. 2009). With this limitation in mind, let us look whether the proposed model can be investigated using informatics. Since this model is being proposed for the first time, there are no actual studies available in literature. Therefore, in this section, we will merely point out a couple of strategies for the verification of the proposed model. For this purpose, we will look at two features of the proposed model which seem to be amenable to the techniques of informatics. As discussed in the preceding sections, there are two key features of the proposed model which distinguishes it from the conventional perspective. Firstly, according to this model, different genes are connected to one another by topological proximity and not linearly (as implicit in the conventional perspective). Therefore, it should be possible to distinguish between topological proximity and geometric proximity using phylogenetic models. Secondly, the proposed model differs from the conventional perspective by its insistence on equivalence between long-range spatial and temporal influences. While the conventional perspective merely accepts the distinction between cis and trans influences, there is no higher order hierarchy of these long-range influences. Therefore, it seems reasonable to think that phylogenetic techniques can help us to distinguish between the proposed model and the conventional perspective.

Let us begin with the task of distinguishing between topological and geometry proximities. Conventionally, it is possible to establish linkages between different genes studying *indels* in the intergenic regions of any two neighboring genes. By looking at the degree of conservation in such intergenic regions (Fertin et al. 2009), it is possible to establish linkages between neighboring genes. In fact, this approach has laid the foundation of the first level of genomic structuralism. Therefore, let us see whether any such conservation of topological proximity can be investigated. In order to observe such proximities, all that we need to do is to define larger and larger intergenic regions. Admittedly, we don't have, as yet, any particular topological model of a genome. However, irrespective of the topological model chosen, it would reflect in some lengths of intergenic regions. This is possible because the higher dimensional unit of genomic architecture, say Genotope, will be contiguous at the higher dimensionality. However, it would devolve into nonlocal distribution of topologically contiguous genes. Since it is possible to define the relationship between topological continuity and its corresponding geometric dispersion, topologically contiguous genes would disperse linearly by a fixed distance. In the absence of any prior knowledge of the topology involved in the genomic architecture, it is not possible to predict the extent by which these topologically contiguous genes would be separated from one another in the geometric sense. However, if we keep increasing the length of intergenic regions, at some stage, we will find such correlations. In fact, should we find such a length of intergenic regions, it is possible to configure the exact dimensionality of the genomic architecture. Moreover, according to this model, there must be a nested hierarchy among different genotopes. Therefore, there ought to be more than one such correlation between different lengths of intergenic distances. Thus, the proposed model can be verified by empirical evidence available from the current archives for genomes by mapping the hierarchy of the lengths of intergenic distances. .

Now, let us look at the second verifiable aspect of the proposed model, viz., the equivalence between temporal and spatial long-range influences. As discussed above, according to this model, spacetime itself actively participates in biological evolution and natural selection. Therefore, the structural template of spacetime must also influence the process of natural selection. At the simplest level, these influences must reflect the key feature of spacetime, viz., the equivalence between time-like and space-like features. Therefore, if spacetime were to participate in the process of natural selection, this parity must be reflected in the form of spatial and temporal long-range influences of the genome. However, the problem is how would it reflect in the genomic architecture and in phylogenesis? Apparently, this parity must manifest in the form of modularity. However, our current understanding of genomic modularity (Peter and Davidson 2015) is based on functional modularity. Therefore, the question is whether a functional modularity necessarily represents the equivalence between temporal and spatial long-range influences? If not, is there any other type of modularity which characterizes the equivalence between temporal and spatial long-range influences?

In order to understand this dilemma, let us understand what kind of functional modularity is evident in genomes. A prototype of functional modularity is best exemplified by the operon (Miller and Reznikoff 1980). An operon consists of a couple of genes whose expressions are interlinked. More importantly, this link between two gene expressions consists of at least one product of either of these genes and an environmental signal. The *lac operon* is the first and the most typical example of an operon. It is intuitively clear that the sequence of gene expressions or which of the genes must be expressed is subject to some kind of feedback available from a signal from the environment. The environmental signal could be in the form of the messenger molecule or a metabolite (as a lactose molecule in the case of *lac operon*). The functionality in this case is to decide which of these genes must be expressed. Now, let us deconstruct this simple description of an operon using the framework of temporal and spatial influences.

The temporality of signal consists of the availability of the environmental signal in the form of a messenger molecule. Depending on the concentration of that molecule, a particular gene would be expressed. In case of excess of the messenger molecule, one gene would be expressed; otherwise, the second gene would be expressed. The key point is that temporal influences are not decided by genes themselves (or by the products of their expressions). It is decided by the environment. While this indicates an active participation of the environment, it merely translates the temporal influence into the spatial influence. The messenger molecule is actually a temporal signal because of its varying concentration. However, in the conventional perspective, it is translated into the spatial signal due to its characteristic rate of diffusion. Thus, even in the conventional perspective, there is some sort of equivalence between temporal and spatial influences. What is actually a temporal signal manifests itself in the form of spatial signal.

However, it must be admitted that the conventional perspective doesn't acknowledge the temporal perspective of modularity in its formal conception of genomic architecture. The temporal aspect merely manifests itself as a byproduct. Of course, it is possible to argue that the rate of diffusion of the messenger molecule itself is derived from the temporal influences. This is because the underlying kinematics integrates temporal and spatial influences. However, the proposed model places the equivalence between temporal and spatial influences at the core of genomic architecture. Therefore, we need to look out for different facets of modularity which characterizes this equivalence.

There is one example of modularity in the conventional perspective which is a stronger indicator of this equivalence between temporal and spatial influences. In the case of operon, there is a tacit acknowledgment of this equivalence. However, trans effects provide much stronger evidence of the equivalence between temporal and spatial influences. This refers to the notion of chromosome territories (Fritz 2014). Chromosome territory refers to the three-dimensional arrangements of a genome in the nucleus. Since during the process of cell division, chromosomes do not remain in their condensed form characterized by nucleosomes, they appear to be in the form of long intertwined threads. However, these chromosomes seem to occupy a particular spatial orientation that gives us the impression that each chromosome occupies a well-marked territory in the nucleus. In other words, entanglement is not random, but follows certain patterns. The key point is that this arrangement brings together different, otherwise distant sequences present on different chromosomes into close proximity. It is this short lived proximity that allows distant DNA sequences to influence gene expressions. Evidently, this is a case of temporal influence being translated into a spatial influence. A temporal influence of when to begin a gene expression is brought about by spatial proximity.

However, the conventional perspective doesn't acknowledge this as evidence of the equivalence between temporal and spatial influences. Instead, it explains this feature as having arisen by random chance, followed by natural selection. However, the proposed model suggests that this phenomenon is caused not by random chance, but by the active participation of spacetime. According to the proposed model, because space-like and time-like features are integrated in each of the dimensions of spacetime, they manifest their influences in the equivalent manner. Therefore, we need to define a different kind of modularity which incorporates this equivalence between temporal and spatial influences.

The proposed model offers a different conception of modularity. Let us look at this conception of modularity and its possible consequences which can be verified by phylogenetic studies. Once again, we will look at this conception in a point-wise manner.

1. Genotope must be thought of as a module by itself. It spans from the higher dimensional spacetime to the four-dimensional spacetime. It is this higher dimensional spacetime that gets projected onto the four-dimensional spacetime giving rise to time-like and space-like features of spacetime. It is the same projection that gives rise to temporal and spatial long-range influences. Since it arises from the higher dimensionalities, both these influences would operate in parallel giving rise to a module which we have labeled as Genotope.
2. The modular structure of Genotope is different from the modular structure of spacetime itself. While spacetime has been postulated to possess a foliated structural template, the modular structure of Genotope is slightly different.
3. Different sheaves in the case of Genotope are also interconnected. This is unlike the foliations of spacetime wherein each sheaf is connected to its neighbors at the seams (with the seams being defined as the inertial frames of reference). In the case of Genotope, the linkages of each sheaf with its neighbors occur at multiple points and not just at the seams.
4. This is because according to the proposed model, atoms and molecules are to be treated as isomorphs of spacetime. Since this isomorphism arises through the inward folding, the number of molecules present in a given unit of spacetime decides the overall topology of the module. Therefore, since Genotope consists of assembly of molecules, with each molecule being an involuted state of spacetime, the resulting topology of Genotope results in multiple connections between the neighboring sheaves of a given Genotope.
5. Since there is a gap between the dimensionalities of Genotope and the corresponding DNA sequences, the intervening dimensionalities also have

phenomenology of their own in the form of the templates of functionalities. This is where one can place our conventional perspective of genomic architecture.
6. According to the conventional perspective, the only structural template of genomic architecture is based on structural elements; it cannot accommodate a different template for its functionalities. As discussed in the first chapter, there is an inherent need to redefine the functional template of the genome as an independent entity distinct from its corresponding structural template. Therefore, this model provides the necessary wherewithal. One can define a functional architecture in any of the intermediate dimensionalities.
7. In this conception of Genotope, natural selection operates between different dimensionalities. On the other hand, the changes within each of these dimensionalities are governed by the Bayesian paradigm.
8. Thus, one can define genotype as any template that is capable of changing within the single dimensionality. Similarly, phenotype can be defined as any template that arises from the changes in the dimensionalities.
9. Accordingly, natural selection must be defined as the process responsible for the changes in the dimensionalities.

This minimalist scenario is adequate for the present discussion. We will refine it as we discuss more and more topics in the following chapters. In the next couple of sections, we will try to compare the semantics of the proposed model and the conventional perspective. We will begin with the discussion on congruence between these two approaches. We will discuss how the proposed model is consistent with the implicit semantics and the explicit structuralism of the conventional perspective.

3.19 Semantic Congruence with the Conventional Perspective

As discussed in the first chapter, the Darwinian paradigm, like any other scientific theories, has an explicit structuralism and an implicit semantics. More importantly, there is an inherent mismatch between its structuralism and semantics. Therefore, if we were to think of replacing the Darwinian paradigm, the replacement framework must be compared to the Darwinian paradigm from the structural as well as semantic perspectives. In the preceding sections, we discussed some of the structural features of the proposed model and compared them with the corresponding features of the Darwinian paradigm. It is tempting to think that these structural changes proposed in the new template are superficial or even superfluous. Of course, as discussed above, the model offers some consequences which are amenable to empirical verification, particularly from the phylogenetic perspective. However, historically, the most characteristic feature of the Darwinian paradigm has been its semantics. As discussed in the first chapter, the Darwinian paradigm has faced a series of paradigm shifts. On every occasion, it has not only rejuvenated itself, but has also found new semantic nuances (Grene 1986). Thus, semantics plays a primary role in the conception of the Darwinian paradigm. Therefore, it is necessary to evaluate the semantics of the proposed model with respect to the semantics of the conventional perspective.

Therefore, in this section, we will try to deconstruct the congruence between the semantics of the proposed model with the semantics of the conventional perspective. In the following section, we will outline the semantic incongruence between these two approaches.

Admittedly, the Darwinian semantics has innumerable facets and it is not possible to cover them even in a separate monograph, let alone a separate section. Therefore, for the present discussion, we will pick up two semantic pillars of the edifice of Darwinian semantics. They are the absence of design and randomness. It is possible to argue that the absence of design and the emergence of design are two sides of the same coin and therefore, there is no need to discuss them separately. However, as discussed below, there is a subtle but definitive difference between these two features.

Let us begin with the semantic proposition that natural selection doesn't obey any design principle. Moreover, if any design elements arise from natural selection, they were not planned for. As discussed in the first chapter, this proposition arose mainly because of Darwin's insistence on keeping any teleological arguments out of the theory of natural selection. This aversion to teleological arguments could be construed as a blend of two connotations. Firstly, it refers to aversion to any kind of theological interpretation of natural selection. Secondly, it can also be construed as an aversion to the Lamarckian type of interpretation of utility as a design principle (Steele et al. 1998). It was central to Darwin's view that natural selection must be conceptualized as devoid of any theistic belief as well as of any progressive belief of natural selection leading to perfection. In Darwin's view, there is no sense of direction in natural selection, either from the past (in the form of divine purpose) or from future (in the form of goal-directed changes).

Let us see how the proposed model compares with this semantic proposition. As discussed in the preceding sections, the proposed model does not introduce any form of design. However, it does suggest that the underlying spacetime has its own structuralism which may influence the course of natural selection. Since according to the proposed model, the structural template of the underlying spacetime itself consists of mathematical constructs, the question of any goal-directed, utility-based improvements doesn't arise. Moreover, if mathematics itself were to be construed as a design principle, then even the conventional perspective of natural selection obeys the Bayesian statistics. Therefore, as far as the Lamarckian type of explanations is concerned, the proposed model shares the relevant semantic propositions with the conventional perspective.

It is possible to argue that by suggesting that mathematics is woven into the fabric of spacetime amounts to some kind of design principles from the past that influences the course of natural selection. However, this objection applies to most of the scientific theories. This is because once we accept that the spatiotemporal universe emerged from the cosmic singularity (Eden et al. 2013), it can be argued that the cosmic singularity acts as a source of all natural phenomena, and therefore, the cosmic singularity acts as a design principle for the entire spatiotemporal universe. Modern science has voluntarily refrained from asking questions about reality prior to the cosmic singularity. Therefore, in some sense, the notion of the cosmic singularity is a placeholder for our ignorance. However, it doesn't stand for any design principle. Moreover, as mentioned above, if the Darwinian paradigm obeys some statistical model, this congruence is hard to explain. This is true for all scientific theories. As discussed in the preceding monographs, there is no explanation why mathematics is capable of formalizing natural phenomena. It is all the more perplexing because modern science can't explain the origins of mathematics. Thus, as far as the proposition of design principle is concerned, the proposed model is congruent with the conventional perspective of natural selection. Admittedly, there are semantic ambiguities in both these approaches, but the proposed model offers a more naturalistic foundation than the conventional perspective because it doesn't need to explain the origin of mathematics which is taken as a priori in the conventional perspective.

Let us look at the second semantic proposition of randomness. It has to be admitted at the outset that randomness is at heart of the Darwinian paradigm. Therefore, any reinterpretation compromising this feature of randomness must be treated with skepticism. Surprisingly, the proposed model does seem to weaken this notion of randomness by limiting the possible outcomes of natural selection. However, this is not the case. Let us understand why. To assert, as the proposed model does, that mathematical structuralism influences biological evolution and natural selection is stating the obvious. Let us begin with the simplest example of mathematical structuralism shaping biological evolution. As of now, we know that there are 120 odd elements in the universe. However, biological evolution has happened (at least to the best of our knowledge) only when the element of carbon is involved. Obviously, this peculiarity can be traced back to the chemical properties of carbon compounds. However, the key point is that the chemical properties of carbon compounds (or any chemical for that matter) are defined by their atomic and molecular structuralism. This structuralism, in turn, is governed by the underlying mathematics. Thus, not just biological evolution and natural selection, but every natural phenomenon is governed by the underlying mathematics. Therefore, when the proposed model postulates an active role of spacetime (and by implication that of mathematics), it is not reducing the randomness implicit in the conventional perspective. To say that the proposed model reduces the randomness of biological evolution and natural selection is equivalent to suggest that Darwin's theory doesn't manifest randomness just because it doesn't deal with biological evolution involving other elements of the periodic table.

The principal semantic assertion of the conventional perspective of natural selection is that the changes in genotype and phenotype are essentially random in the sense that they can't be predicted. The same is true for the proposed model. It does not claim to predict the course of evolution. All that the proposed model offers is an explanation for why only certain types of complexities are produced during natural selection. In any case, the conventional perspective of natural selection finds it difficult to explain discontinuities like speciation (Coyne and Orr 2004), different rates of evolution in different species (Stallmann 2008) or loss of function during the evolution (Alberts 2004). The proposed model offers a unifying semantic foundation to accommodate these diverse forms of evolutionary discontinuities. Some of these aspects will be discussed in the following chapters. Presently, it seems reasonable to assert that the proposed model does not compromise on the semantic proposition of randomness. The proposed model simply postulates that the randomness during natural selection arises from the Bayesian conditional probabilities which overlie the structurally causal processes defined by the structural template of spacetime. Thus, according to this model, Darwinian randomness arises from Bayesian logic and it is not the foundation of natural selection.

Thus, it is evident from this discussion that the proposed model is in consonance with the conventional perspective of natural selection. More importantly, it provides a cogent explanation of why mathematics plays an integral part in natural selection and explains why the discontinuities like speciation, separation of genes and different pace of evolution takes place. However, there is one aspect of the proposed model that prima facie appears to be incongruent with the conventional perspective, viz., the emergence of complexity. Therefore, in the next section, we will look at the emergence of complexity and find out a way to reconcile both these approaches.

3.20 Semantic Incongruence with the Conventional Perspective

Admittedly, this issue of the emergence of complexity during biological evolution and natural selection has defied any resolution. Before we articulate how the proposed model resolves this key lacuna of the conventional perspective, it is necessary to understand the nuances of the conventional perspective that has resulted in this lacuna. Prima facie, the conventional perspective of natural selection, doesn't have any explanation for the emergence of complexity. However, it is reasonably argued that since the earliest organisms must be the simplest possible forms, any mechanism of natural selection would lead to more and more complex forms. This is essentially a phase space argument (Bonner 1988). Since the region of the phase space of natural selection is already occupied by the earliest living organisms, all the later organisms will have to perforce occupy the phase space representing complex life forms. The problem with this argument is that Life, as a natural phenomenon, is the only phenomenon that obeys this rationale. No other natural phenomenon follows this rationale. In fact, historically, it used to be a point of contention. Every natural phenomenon, following the second law of thermodynamics, would unfold from more complex structuralism to less complex structuralism (Haynie 2008). Life, on the other hand, progressed from less complex forms to more complex forms. Therefore, Life, as a natural phenomenon, was an exception to the second law of thermodynamics. Admittedly, it seemed to justify our ancient belief in the transcendental origin of Life.

However, with the development of thermodynamics of open systems, it was easy to demonstrate that Life is a natural phenomenon like any other natural phenomena. However, the problem of the emergence of complexity remains unresolved. This problem is also linked to another semantic problem of the type of complexities. Natural selection not only results in the increase in complexity, but it results in only certain types of complexities. It is also important to keep in mind that this emergence of complexity is linked with the problem of the design principle (Fodor and Piattelli-Palmarini 2011). As discussed in the previous section, Darwin's conviction to keep out any teleological arguments out of the theory of natural selection also resulted in the lack of articulation of any explanation for the emergence of complexity. The topic of complexity was wrongly conflated with the design principle and by extension to teleology.

However, as the Darwinian paradigm underwent a series of paradigm shifts, it became self-evident that natural selection obeys certain mathematical templates and certain causal mechanisms. Therefore, one would have expected that some kind of explanation for the emergence of complexity would have been articulated. There are two factors which prevented us from doing it. Firstly, most of us intuitively feel that Darwin's instinct on nondeterminism was spot on. Therefore, we have treated this topic as a semantic ambiguity rather than a problem. Secondly, the statistical framework of population genetics (Provine 2001) and its spectacular success deters us from revisiting this semantic ambiguity. Of course, as discussed in the previous section, our ambivalence toward the origin of mathematics also contributed to our reluctance to deconstruct this feature of the emergence of complexity from a causal perspective.

As discussed in the preceding monograph (Chhaya 2022a), the key point is that if any mathematical formalism explains a nature of any natural phenomenon, it must point toward some ontological explanation for this congruence between mathematics and Nature. However, under the Cartesian influence (Cottingham 2008), we have put the problem of the origin of mathematics in an academic limbo. We have chosen to treat mathematics as a priori. While this absolves scientists from explaining why they use mathematics in explaining Nature, the question of the "unreasonable effectiveness of mathematics" in explaining Nature remains unanswered. However, a naturalistic model of the origin of mathematics was outlined in the preceding monograph. Therefore, it is necessary to understand why mathematical templates operate in natural selection. It is important to remember that even statistical templates are mathematical templates and originate like any other mathematical constructs from spacetime itself. Therefore, if there exists a naturalistic interpretation of the origin of mathematics which places mathematics into the fabric of spacetime, then it is imperative that we must seek out causal processes even behind statistical templates.

Thus, the emergence of complexity may be governed by the Bayesian logic and the resulting statistical template, but it will possess a causal explanation. It may not be possible to predict the complexities that emerge during natural selection, but it is possible to define what kinds of complexities are permitted in natural selection. Thus, by postulating an active role of spacetime in biological evolution and natural selection, the proposed model offers a naturalistic explanation of the emergence of complexity during biological evolution and the prevalence of only certain kinds of complexities during natural selection. More importantly, it does so without compromising the inherent randomness of natural selection implicit in the Darwinian paradigm. In addition, it doesn't invoke any implicit or explicit design principles. The proposed model delinks the design principles from the ontology of complexity.

Thus, what appears to be incongruence between the Darwinian paradigm and the proposed model turns out to be a dialectical difference between these two approaches. Our cognitive predisposition toward pattern recognition and linking sequential events to causality has led us to conflate the notions of complexity and design. Having looked at the semantic congruence and incongruence between the proposed model and the Darwinian paradigm, it is necessary to outline a brief sketch of a new paradigm of genomic architecture. Admittedly, we will expand this description in the following chapters as we explore different aspects of genomic architecture.

3.21 Conclusion

In the preceding sections, we discussed several aspects of genome and genomic architecture. We began with the conventional perspective of natural selection and demonstrated that it leads to certain explicit and implicit elements of genomic architecture. Then, we tried to deconstruct the mismatch between these implicit and explicit genomic architectures in the light of natural selection. In order to reconcile these two templates, a new model of genomic architecture was introduced. The key postulate behind this architecture is that spacetime plays an active role in biological evolution and natural selection. However, as our understanding of spacetime suggests, there is no distinction between time-like and space-like dimensions. Accordingly, the long-range influences of both types, viz., temporal and spatial influences, must be redefined in formalizing genomic architecture. Therefore, a topological model of the genome was put forth. A new topological unit of Genotope was introduced in this architectural plan. A Genotope can be thought of as an ensemble of proteins, DNA and the underlying spacetime. Instead of any linear arrangements of molecules, a Genotope can be visualized as a series of topological surfaces in a nested hierarchy, much like an onion. Each topological surface occupies different dimensionalities. However, these surfaces interact with each other through the processes which can be formalized as the changes in the dimensionalities. A brief mathematical description of the operators representing the changes in the dimensionalities and their outcomes were tentatively outlined. According to this scenario, structuralism and functionalities occupy different dimensionalities. However, due to our preoccupation with molecular perspective, we have mistakenly assumed that structuralism and functionalities share a common framework. However, according to this model, they possess different metrics and both these metrics are connected by the operators representing the changes in dimensionality.

It is proposed that the changes in dimensionality and its attendant changes in metrics are directly influenced by the structural template of spacetime. However, the changes in the metric within a given dimensionality are governed by thermodynamic and Bayesian processes. This distinction between two types of changes (the ones involving changes in the dimensionalities and the ones happening within a single dimensionality) helps us to resolve various semantic ambiguities of the Darwinian paradigm. In the following chapters, we will explore this conception of genomic architecture from different perspectives.

References

Akalin A (2021) Computational genomics in R. CRC Press, Boca Raton
Alberts B (2004) Molecular biology of cell. Garland Science, New York
Appasani K (ed) (2016) Genome wide association studies: from polymorphism to personalized medicine. Cambridge University Press, Cambridge
Audi R (1998) Epistemology: a contemporary introduction to the theory of knowledge. Routledge, London
Bard J (2016) Principles of evolution: systems, species and the history of life. Garland Science, New York
Bonner JT (1988) The evolution of complexity by means of natural selection. Princeton University Press, Princeton
Bonner JT (2013) Randomness in evolution. Princeton University Press, Princeton
Bromham L (2008) An introduction to molecular evolution and phylogenetics. Oxford University Press, Oxford
Bryson V, Vogel HJ (1965) Evolving genes and proteins. Academic Press, New York
Chhaya P (2020) The universe within: and a view from that within. Universal Press, Irvine. (Provisionally accepted)
Chhaya P (2022a) The origin and nature of mathematics. (Submitted)
Chhaya P (2022b) The origin and nature of spacetime. (Submitted)
Chhaya P (2022c) The origin of the counterintuitive nature of quantum phenomena. (Submitted)
Chhaya P (2022d) The origin and nature of cognition. (Submitted)
Cottingham J (2008) Cartesian reflections. Oxford University Press, Oxford
Coyne JA, Orr HA (2004) Speciation. Sinauer Publishing, Sunderland
Darden L (1991) Theory change in science: strategies from Mendelian genetics. Oxford University Press, Oxford
Davidson EH (2006) The regulatory genome: gene regulatory networks in development and evolution. Elsevier, Amsterdam
Dennett DC (1995) Darwin's dangerous idea: evolution and meaning of life. Simon and Schuster, New York
DeVillers J, Balaban AT (eds) (1999) Topological indices and related descriptors in QSAR and QSPR. CRC Press, Boca Raton
Duboule D (1994) Guidebook to homeobox genes. Oxford University Press, Oxford
Eden H, Moor JH, Soaraken JH, Steinhart E (2013) Singularity hypothesis: a scientific and philosophical assessment. Springer, Dordrecht
Fertin G, Labarre A, Rusu I, Tannier E, Vialette S (2009) Combinatorics of genome rearrangements. MIT Press, Cambridge
Flew A (2017) Darwinian evolution. Routledge, London
Fodor J, Piattelli-Palmarini M (2011) What Darwin got wrong. Profile Books, London
Fritz A (2014) Elucidating chromosome territories organization during the cell cycle and in breast cancer cells. Proquest Publishing, Ann Arbor
Garrett RA, Klenk HP (eds) (2007) Archaea: evolution, physiology and molecular biology. Blackwell Science, Oxford
Gilbert SF, Barresi MJF (2020) Developmental biology. Oxford University Press, Oxford
Gould SJ (2007) Punctuated evolution. Harvard University Press, Cambridge
Grene M (1986) Dimensions of Darwinism: themes and counter themes in the twentieth century evolutionary theory. Cambridge University Press, Cambridge
Hammes GG (1982) Enzyme catalysis and regulation. Academic Press, New York

Haynie DT (2008) Biological thermodynamics. Cambridge University Press, Cambridge
Hodge J, Redick G (eds) (2009) Cambridge companion to Darwin. Cambridge University Press, Cambridge
Lynch M (2007) The origins of genome architecture. Oxford University Press, Oxford
Margulis L (1970) Origin of eukaryotic cells. Yale University Press, New Haven
Marlow FL (2010) Maternal control of development in vertebrates: my mother made me do it. Morgan and Claypool Life Science, San Rafael
Miller JH, Reznikoff WS (1980) The operon. Cold Spring Harbour Laboratory Press, New York
Morgan ML (ed) (2002) Spinoza: complete works. Hackett Publishing, Indianapolis
Naber GL (2003) The geometry of Minkowski spacetime: an introduction to the mathematics of the special theory of relativity. Dover Publications, Mineola
Okasha S (2010) Evolution and the levels of selection. Oxford University Press, Oxford
Peter IS, Davidson EH (2015) Genome control processes: development and evolution. Academic Press, London
Pevsner J (2015) Bioinformatics and functional genomics. Wiley, Hoboken
Popper K (1963) Conjectures and refutations: the growth of scientific knowledge. Routledge, London
Press SJ, Clyde CM (2003) Subjective and objective Bayesian statistics: principles, models and applications. Wiley, Hoboken
Provine WB (2001) The origins of theoretical population genetics. University of Chicago Press, Chicago
Rippe K (ed) (2012) Genome organization and functions in cell nucleus. Wiley VCH, Weinheim
Robert JS (2004) Embryology, epigenesis and evolution: taking development seriously. Cambridge University Press, Cambridge
Sado T (ed) (2018) X chromosome inactivation: methods and protocols. Springer, Dordrecht
Schutz BF (2009) A first course in general relativity. Cambridge University Press, Cambridge
Shmulevich I, Dougherty ER (2014) Genomic signal processing. Princeton University Press, Princeton
Smith E, Morowitz HJ (2016) The origin and nature of life on earth: the emergence of the fourth geosphere. Cambridge University Press, Cambridge
Stallmann RR (2008) The pace and processes of early divergence and stasis: morphological evolution in isolated populations. University of California Press, Davis
Steele EJ, Lindley RA, Blanden RV (1998) Lamarck's signature: how retrogenes are changing Darwin's natural selection paradigm. Allen and Unwin, New South Wales
Szabo A, Ostlund NS (1989) Modern quantum chemistry: an introduction to advanced electronic structure theory. Dover Publications, Mineola
Weinzierl ROJ (1999) Mechanism of gene expression: structure, function and evolution of the basal transcriptional machinery. Imperial College Press, London
Yarus M (2010) Life from an RNA world: the ancestor within. Harvard University Press, Cambridge
Zhang W (2012) Regulation and coordination of homologous pairing and synapsis during Caenorhabditis elegans meiosis. Stanford University Press, Redwood

Biological Algorithm of Involution: Ontology of Gene Expressions

4

Abstract

In the previous chapter, a topological model of genomic architecture using the formalism of the involuted manifold was discussed. It was suggested that all the long-range influences involved in the gene expression can be formalized as involutions. However, this model needs to be corroborated by the evolutionary perspective. In this chapter, we will demonstrate that the protocol of gene expressions is unitary. Its subsequent temporal and spatial separations in the genomes are defined by the Darwinian paradigm. In accordance with the phylogenetic perspective, it is possible to map out these separations onto a genomic singularity. The model of genomic architecture presented here seems capable of explaining the functional as well as evolutionary perspectives of the genome.

4.1 Introduction

As discussed in the preceding chapter, the domain of genomics lacks an evolutionary perspective at a genome level. This has resulted in a lack of structural and functional models of genomes. One such model of genomic architecture based on the formalism of the involuted manifold was articulated in that chapter. It was argued that any model of genomic architecture must not only be congruent with the functionalities of the genome, but it must also be congruent with the evolutionary perspective. While this evolutionary perspective is available for individual genes in the form of phylogenetics, there is no corresponding theory of comparative study of evolution of genomes. The problem in developing such comparative analysis of genomes is not lack of data (Appasani 2016), but lack of a structural template of genomic evolution. This lacuna is more acute in the case of long-range influences (Shmulevich and Dougherty 2014) involved in the gene expressions. For instance, it is possible to compare different genomes for differences in base pair sequences. However, it is not

© The Author(s), under exclusive license to Springer Nature Singapore Pte Ltd. 2023

P. Chhaya, *The Topological Model of Genome and Evolution*, https://doi.org/10.1007/978-981-99-4318-0_4

easy to analyze the distribution of genes in different genomes from the evolutionary perspective. In fact, it is also not possible to formalize distribution of genes in any given genome. Admittedly, genes often tend to relocate themselves during cell cycle and during speciation. However, it ought to be feasible to map the distribution of different genes in any given genome and compare this with other genomic distributions of genes.

However, such an approach requires a universal structural template of genomic architecture. Such an architecture must be based on functionality and not on the DNA sequence. In such a model, it should be possible to define the distribution of genes on purely topological terms. Moreover, such a template would also enable us to develop an evolutionary course of genomes per se. The topological model of genomic architecture proposed in the preceding chapter was essentially a static perspective. In order to understand the perspective mentioned above, we would have to define a dynamic model of involuted manifolds representing the changing genomic architecture.

Prima facie, there are two dynamic processes that are relevant to the present discussion. Firstly, there are long-range influences involved in the gene expressions. While the phenomenon of initiators and promoters of gene expressions being some distance away from the gene sequence is fairly well-documented (Weinzierl 1999, see Chapter 2), there is no general theory of explaining the extent and the nature of separation between the genes themselves and these promoters and initiators. However, if we were to accept that genomes could also be units of selection then it is axiomatically true that genomic architecture too must reflect natural selection. In such a scenario, the location of transcription factors and their topochemistry with initiators and promoters would be determined by the genomic architecture. Therefore, it seems intuitive that these long-range influences too must have some evolutionary explanation behind their extent of separation from the target genes. Thus, the assumption that genomic architecture is a product of natural selection leads to inference that all such long-range influences present in a genome too must have a common structural template. In other words, the algorithm of gene expressions must be amenable to formalization in the language of topology. We would explore this idea in this chapter. Of course, a more detailed scenario of the regulatory framework of a genome would be discussed in the following chapter.

The second dynamic process that must influence genomic architecture is the evolution and differentiation of genomic functionalities. As mentioned in the preceding monograph (Chhaya 2020, see Chapter 9), the belief that there exists a common ancestor to all the living organisms can be formalized as a genomic singularity. Such a singularity could embody the formal representation of living systems. In such a scenario, it is intuitively clear that the evolution of different functionalities of life and by analogy, different genes could be traced back to this genomic singularity. Therefore, it is tempting to think that this breakdown of such a genomic singularity into different functionalities and different genes could be formalized using some formal principles. In the preceding monograph, it was proposed that life must be treated as any other natural phenomena, and therefore, it is possible that it shares a common ontology with the other natural phenomena starting from the cosmic singularity. In another monograph, a topological model of the involuted manifold was shown to be applicable to the breaking of the cosmic singularity into different fundamental particles (Chhaya 2022b, see Chapter 5).

Therefore, in this chapter, one would look at the possibility of whether the postulated genomic singularity could be broken down into different functionalities and their corresponding genes. In the model proposed in that monograph (Chhaya 2022a, see Chapter 3), the operator of involution is capable of fragmenting the singularity into different types of fragments. This model is also congruent with the notion of modularity in evolutionary biology. Therefore, we would explore the possibility of different types of functionalities forming different types of modules and thereby generating different classes of living organisms. Incidentally, this evolutionary perspective of segregation of functionalities is congruent with the separation of genes from their promoters and initiators. Therefore, we would try to define a general schema of segregation of both these dynamic processes. Perhaps, the algorithm that separates different functionalities also separates the genes from their initiators. Therefore, we will formalize a topological model of gene expressions which is congruent with the topological model of genomic architecture. This ought to establish a scale free protocol of biological evolution which acts like an algorithm.

For the sake of simplicity, these deeply entwined issues have been discussed in a linear fashion. Accordingly, this chapter has been further divided into 20 sections. Section 4.2: Genomic Singularity, Sect. 4.3: Origin of Modularity from Singularity, Sect. 4.4: Separation of Structuralism from Functionalities, Sect. 4.5: Separation of Regulatory and Expressive Features of Genomic Architecture, Sect. 4.6: Structural Modularity, Sect. 4.7: Functional Modularity, Sect. 4.8: Regulatory Modularity, Sect. 4.9: Expressive Modularity, Sect. 4.10: Topological Model of Modularity, Sect. 4.11: Involutive Formalism of Evolution of Modularity, Sect. 4.12: Topological Model of the Relationship Between Structuralism and Functionalities, Sect. 4.13: Topological Model of the Relationship Between Regulatory and Expressive Genome, Sect. 4.14: Involutive Formalism of Genomic Expression, Sect. 4.15: Involutive Formalism of Gene Expression, Sect. 4.16: Unitary Biological Algorithm, Sect. 4.17: Semantics of Biological Algorithms, Sect. 4.18: Conventional Semantics of Darwinian Paradigm, Sect. 4.19: Revised Semantics of Darwinian Paradigm, Sect. 4.20: Biological Algorithm as a Special Case of General Involuted Algorithms, Sect. 4.21: Conclusion.

4.2 Genomic Singularity

There are several semantic arguments that have prompted this conception of genomic singularity. Admittedly, these arguments arise from different disciplines and therefore, carry different semantic nuances. As a result, there could be several ways to define genomic singularity. In this section, we will try to understand different nuances of this conception. There are three different semantic arguments that are sought to be integrated into the conception of genomic singularity. Firstly, the notion of singularity is adopted from cosmology. The conception of the cosmic singularity

(Montani et al. 2011, see Chapter 1) is a potent amalgam of several explicit and implicit propositions. Some of these propositions are sought to be incorporated into the conception of genomic singularity. There are three propositions of the cosmic singularity that are applicable to the proposed genomic singularity. They are, ontological primacy, potential for giving rise to different structural and functional templates and absence of any native structuralism. From the epistemological perspective, these attributes of the cosmic singularity reflect as much on the nature of the cosmic singularity as much on our epistemological limitations. By analogy, when we ascribe ontological primacy to genomic singularity, we are simply asserting that no entity prior to genomic singularity needs to be conceptualized. Similarly, when we ascribe a potential for engendering different structural templates to genomic singularity, we are simply asserting that all the structural and functional templates found in biological organisms must have originated from the genomic singularity. Finally, when we assert that the genomic singularity is essentially a nonstructural entity, we are simply asserting that it has no phenomenology of its own. In other words, we are simply asserting that the genomic singularity has no observable features.

Let us now look at the second source of the conception of the genomic singularity, viz., the notion of common ancestor employed in several biological models. The notions of mitochondrial eve (Hamilton 1989) and LUCA (last universal common ancestor) (Bard 2016, see Chapter 9) are typical examples of this tendency to objectify a phenomenon. Similarly, we will assert that the conception of the genomic singularity need not represent any particular genome that existed in the distant past. The entity of genomic singularity must be taken as an abstraction of a unitary representation of the original form of the genome.

Finally, there is an epistemological proposition that is incorporated into the conception of the genomic singularity. This refers to the anthropic bias that permeates our scientific theories. Firstly, as formalized in the axiom of choice (Herrlich 2006, see Chapter 2), there could be structural elements of genomic architecture that are beyond our cognitive capabilities. These would be excluded from the conception of the genomic singularity. Similarly, it is possible that our quest for the universal theory of science could itself be a cognitive artifact arising from the limitations of our cognitive faculty. Therefore, it is possible that this conception of the genomic singularity too could be an artifact arising from our cognitive bias toward finding a common framework in everything that we examine. Thus, it is possible that this conception of the genomic singularity might represent our incipient cognitive singularity. However, we will acknowledge this bias beforehand and still pursue this model with an open mind.

Having put in place these caveats, it is time to outline what the genomic singularity stands for. To understand this, let us look at the conception of modularity. Purely from the structural perspective, modularity occupies the middle position. On the one end, we have individual genes and on the other end, we have genomic singularity. While we have conceded that the conceptions of a gene and a module represent the structural template of the genome, the conception of the genome as a unit has remained nebulous. Just as we accept that an individual gene or an individual module can be a unit of natural selection, we cannot accept the genome as a unit of selection, at least not with the same conviction. It can be argued that the reason for this ambiguity lies in the fact that we don't know how a genome could have evolved without any prior evolution of units of modules or genes. However, this is not true. In the conception of a module, similar ambiguity prevails. Yet, we are convinced about our conception of a module.

For instance, ab initio, we can think of two scenarios for the emergence of modules during the course of biological evolution. In the first scenario, we can postulate that a genomic module evolved from the synchronous functionalities of its constituent genes. In such a scenario, this process of integration would lead to different genomic architectures depending on different paths of integration of modules. However, this is not the case. The fact that as our phylogenetic studies demonstrate, we can conceptualize a common ancestor of all the living organisms, suggests that this scenario is not supported by facts. The second scenario is that different modules evolved and acquired their identities from a single common ancestor genomic framework. Just like speciation occurred, different modules evolved out of evolutionary contingencies. From the semantic perspective, speciation and the emergence of different modules are discontinuities and are the results of varying environmental influences.

Upon a little reflection, it is intuitively clear that the second scenario implicitly endorses a primordial entity which we have chosen to call genomic singularity. The real reason why this conception of genomic singularity, both as a primordial structural template of genome and as a unit of natural selection, hasn't been articulated is our cognitive bias toward a particular type of structural patterns. Because our cognitive faculty is ingrained with the structural template of the four-dimensional perspective of reality, we are not able to conceptualize any higher dimensional perspective. As discussed in the previous chapter, our perception of genomic architecture is predominated by the molecular perspective (Lynch 2007), even though we concede that there exists long-range influences which defy this molecular paradigm. Similarly, we are comfortable with the conception of individual genes as discrete entities (just as we are comfortable with the discrete entities like atoms and molecules), but we find it difficult to conceptualize the genome as a discrete entity. Incidentally, from the semantic perspective, we are ambivalent about the conception of modules. This is because we are able to comprehend the functional unity of the constituent genes, but we can't comprehend the structural template of modules.

Therefore, in this chapter, we will assign several features to the entity we have chosen to name as a genomic singularity. These semantic propositions are novel, but unexceptional in the sense that they are integral to our understanding of genomics. Having done that, we will demonstrate that each of these semantic propositions carry inherent structuralism, and therefore, they collectively lead to a structural template of genomic singularity. For simplicity, we will define these semantic propositions in a point-wise manner.

1. There exists a common ancestry of every functional feature of the genome.
2. There exists a common ancestry of every structural element of the genome.

3. There exists a common ancestry of all the relationships between structuralism and functionalities of the genome.
4. There exists a formal relationship between each structural element and its corresponding functionality.
5. Every change in the functional features of the genome can be formalized as a measurable unit.
6. Every change in the structural elements of the genome can be formalized as a measurable unit.
7. There are no restrictions on the changes in either functional features or structural elements during the course of evolution and natural selection.
8. However, the relationship between structural elements and their corresponding functional features is constrained by some principles of symmetry.
9. As a result of unlimited range of changes in structuralism and functionalities being restricted by fixed nature of relationships between structuralism and functionalities, the formal description of genomic architecture will necessarily be in topological language and not in geometric format.
10. The common ancestor which follows these propositions is named as the genomic singularity.
11. Genomic singularity undergoes transformations and differentiations to give rise to topological units defined earlier as Genotope.
12. Therefore, genomic singularity too must possess a comparable topology to that of Genotope.

These features of genomic architecture are in consonance with the conventional perspective. However, they have not been articulated in this manner, not at least collectively. Before we look at the structural implications of this description of genomic singularity, we must look at some of the unorthodox consequences of this description. There are two such consequences that must be deconstructed, viz., the assertion that the relationship between every structural element and its corresponding functionality is constrained by some fundamental principles and that this restriction necessarily gives rise to topological conception of genomic singularity (and by implication that of genome).

Since the proposed model doesn't limit the changes in either structuralism or in functionalities per se, this imposition of restriction of only a certain kind of relationships between structuralism and functionalities appears arbitrary, if not unwarranted. Therefore, let us look at its semantic necessity. Admittedly, the absence of any restrictions on the changes in either structuralism or in functionalities is intuitive and congruent with the nondeterminism implicit in the Darwinian paradigm (Bonner 1988). Therefore, it is reasonable to doubt the validity of this restriction on the nature of relationship between structuralism and functionalities. Surprisingly, this restriction arises not from some peculiar nature of genomic architecture. It arises from the laws of Nature. Beginning with the cosmic singularity, every natural phenomenon is governed by this limitation. Every fundamental particle has some fundamental properties which are manifestations of its internal structure. It is intuitively clear that these properties are nothing but functionalities and that they arise from a certain internal structure of a given particle. While there could be multiple types of particles, each representing a certain structuralism (Of course in this case the plurality of particles is not random but decided by some symmetry principles, once a particular structuralism manifests, its properties are defined in certain fixed types of relationships (Cottingham and Greenwood 2013)). Similarly, we can think of medicinal properties of molecules (Roy 2015, see Chapter 4). We can think of an infinite number of molecules each having its own medicinal properties. Similarly, we can think of a large number of medicinal properties that are relevant to our therapeutic priorities. However, once we select a particular class of molecules or a particular group of medicinal properties, the relationship between structuralism and functionalities in each of these choices is definitive. Once again, this relationship is governed by the symmetry principles. Thus, it is possible to imagine or postulate infinite variations in either structural elements or in functionalities. There are no restrictions on the possible variations. However, once we select either a particular structural element or a specific functional feature, its possible counterparts are limited. Moreover, they are limited by the underlying symmetry principles. Thus, this restriction that the relationship between structuralism and functionalities follows only symmetry driven variations is universal and is being included in the conception of genomic singularity as an ontological imperative.

Now, let us look at the reason why this limitation leads to a topological perspective in contrast to any geometric perspective of genomic architecture. The most intuitive way to understand the distinction between topological model and geometric model is to think about molecules and their properties. Molecular structures can be defined by three-dimensional arrangements of the nuclei present in a given molecule. This is a typical geometric model. However, if we were to employ this model to deconstruct or predict properties of molecules, it would lead to very crude approximations. This is precisely what we do while defining medicinal properties of molecules using molecular mechanical or quantum chemical models (Roy 2015; Shi 1989). However, it is possible to define molecular properties of any given molecule by creating a model of the electrons circling around the nuclei of the given molecule. This model of electrons is a typical example of a topological model. The question that we need to ask is why we need different types of models for molecules and their properties? The reason is self-evident. In a molecular model one assigns a point representation to the nuclei of a given molecule. Therefore, the relationship between different nuclei can be defined by assigning different coordinates to the constituent nuclei. Thus, intramolecular interactions can be adequately formalized using a set of algebraic equations. One can extend this approach to intermolecular interactions as well. This constitutes the heart of quantum chemistry (Szabo and Ostlund 1989, see Chapters 2 and 3). However, it is important to keep in mind that the quantum chemical paradigm lacks a higher dimensional perspective. Therefore, it can merely postulate that a wave function is a repository of all the possible types of properties of a given molecule. However, because of the lack of higher dimensional perspective, it cannot formalize this postulate.

However, if we want to determine the molecular properties, this model is found to be inadequate. It is important to remember that we can use the geometric perspective while formalizing molecular properties, but it is inadequate. This is precisely what is happening in the case of genomics. Our geometric perspective of genomic architecture is not wrong, but it is inadequate. Now, let us look at the reason why the topological model of molecular properties works better. (Admittedly, even this approach has its own limitations. However, they mainly arise from our limited computational capabilities. The inadequacies of topological approach arise to a lesser degree from the ontological ambiguities of the relationship between geometry and topology (Brendon 1993, see Chapter 2). Incidentally, this is precisely what the proposed model seeks to formalize). In the case of molecular properties, there are two inherent advantages. Firstly, topological models are more easily amenable to formalize the movements of electrons around the nuclei. Secondly, topological models allow us to define the movements of electrons in different dimensionalities simultaneously. For instance, when we define a wave function, the wave function incorporates the dynamics and distribution of electron density simultaneously. Thus, a topological model offers a way to connect different dimensionalities in a time variable manner. This is something that doesn't arise naturally in the case of geometric models. This is because we need to impose a framework of dynamics on the geometric model in an ad hoc manner. Albeit, this ad hocism is more problematic in case of electrons because they move around the nuclei with the speed comparable to that of light.

Thus, in the geometric approach, we have two frameworks, spatial arrangements of nuclei and the dynamic picture of electron density. However, these two frameworks cannot be satisfactorily blended into a single framework. Something similar is manifest in genomic architecture. We have two frameworks, the three-dimensional arrangements of nucleotides and the higher dimensional functional template. While nucleotide sequences can change, there are parallel higher dimensional changes in genomic architecture. However, these higher dimensional changes are essentially functional in nature. Therefore, unless we postulate that the relationship between structuralism and functionalities are governed by some symmetry principles, it is not possible to define genomic architecture. This is possible only when we opt for a topological model which can accommodate the structural and functional templates in a single framework. This inevitability of having a topological perspective justifies our earlier postulate that spacetime plays an active role in natural selection. It does so because it inherently defines the relationship between time-like and space-like features. Thus, the dynamic nature of genomic functionalities can only be integrated with the static structural template of the genome if we employ spacetime as the underlying template. Having looked at the structural and semantic nuances of genomic singularity, it is necessary to understand how this singularity gives rise to different heterogeneous discontinuities. Therefore, in the next section, we will look at the next level of discontinuity, viz., modularity.

4.3 Origin of Modularity from Singularity

As discussed in the preceding sections, the notion of modularity has been generally conceded both in biology and genomics (Pevsner 2015, see p. 635). However, its evolutionary justifications and the particular evolutionary path of its emergence remain ambiguous. As mentioned above, purely from the systemic perspective, modularity could arise in two possible ways. Firstly, it could arise by integration of contingent structural or functional elements during the course of natural selection. Alternatively, it could arise by segregation of a singular system based on the environmental contingencies. However, both these scenarios point toward multiple modular templates. While the phylogenetic evidence points toward a common ontology, it doesn't support any such singularity, at least not directly. This ambiguity arises perhaps because of the underlying ambiguity about the relationship between structuralism and functionalities of Life. As discussed in the first chapter, the evolutionary need to have separate units of inheritance and selection points toward hitherto undefined semantics of natural selection. Thus, the problem with the notion of modularity is not about its manifestation. Rather, the problem is about the evolutionary significance of modularity.

In the preceding chapters and in the preceding sections, we looked at an evolutionary model based on genomic singularity and its formalization. Therefore, it is important to find out how the proposed model conceptualizes the notion of modularity. Therefore, we will try to understand how a genomic singularity can give rise to modularity and if so, which kind of modularity. While outlining these aspects, we will sidestep the current models of biological modularity. Instead, we will try to deconstruct the emergence of modularity from two perspectives, viz., modularity arising from the peculiar nature of the relationship between structuralism and functionalities and modularity arising from the temporal and spatial configuration of the genome. Admittedly, these two perspectives could be causally and functionally connected to one another (after all, this is implicit in the conception of genomic singularity). However, for the sake of simplicity, we will treat them as separate features.

Let us begin with the first topic of the influence of the relationship between structuralism and functionalities on the nature of modularity. There are two implicit assumptions in this issue. Firstly, we will assume that the relationship between structuralism and functionalities is somewhat peculiar in nature and that this peculiarity is responsible for the kind of modularity that has evolved. Secondly, we will assume that modularity has arisen during natural selection because it increases the survival quotient of the organisms possessing it. While the second assumption must be taken as self-evident (at least in the Darwinian sense), the first assumption is ad hoc. It is hoped that the arguments presented below would justify this ad hocism.

Let us begin with the assumption that the relationship between structuralism and functionalities is peculiar in nature. By peculiar, we mean unique and yet universal at least in the biological context. The relationship between structuralism and functionalities can be viewed as unique because the functionalities produced in living organisms are unique. No other natural phenomenon gives rise to features of replicative memory, goal-directed behavior, and, in some cases, sentience. At the same time, the structural template of genomes is not unique. Genomic architecture can be taken as a highly organized arrangement of atoms and molecules. However, such precise arrangements are found in several natural phenomena, including minerals and crystals. These natural objects do not manifest any such unique functionalities. Thus, the structural template of living organisms is not unique (exceptionally organized, but not unique), and yet, it gives rise to unique functionalities. Therefore, it is the unique relationship between structuralism and functionalities that must be taken as a source of these unique functionalities.

Now let us speculate what features of this relationship make it unique. Let us take an example of another example of relationships between structuralism and functionalities which is not unique and see how it is different from the relationship between biological structuralism and functionalities. For this, let us look at the example cited above, viz., minerals (Theng 2019, see Chapter 8). Let us look at the mineral like clay which has catalytic properties. This choice of clay is important because there is a school of thought that subscribes to the view that Life originated due to catalytic properties of some types of clay synthesizing proteins and other complex molecules from the primordial soup available on Earth (Yarus 2010, see Chapter 2). As mentioned above, clay can be thought of as an organized arrangement of its constituents, viz., metal salts and silicates. Apparently, it is the particular arrangement of these chemicals that gives rise to its catalytic properties. Thus, we have an example of a particular structuralism giving rise to a functionality of catalysis. More importantly, the functionality in this case is not unique. The functionality of catalysis can never regenerate itself. It can't remember its past catalytic reactions. In short, it doesn't possess any replicative memory. Therefore, this constitutes a good example for comparison.

Let us focus on the relationship between the structuralism of clay and its catalytic properties. The catalytic properties arise solely due to the stereochemical orientations of these chemicals among themselves. Of the several such orientations, only some manifest catalytic properties while the rest of the orientations do not manifest any catalytic properties. Thus, there is a clear correlation between structuralism and functionalities. It is possible to quantify this relationship. It is possible to define electron densities or hydrophobic regions of any variety of clay. Based on such estimates, it is possible to predict the kind of catalysis that a candidate clay might possess. Thus, it is possible to manipulate functionalities like catalytic properties simply by manipulating the structural template of clays. (Admittedly, in practice chemists do not resort to this methodology because of our limited computational capabilities, but it can be done, at least in principle.)

However, such manipulations are not possible in the case of biological functionalities. This is because the relationship between biological structuralism (say genomic architecture) and its functionalities is not amenable to factorial analysis. In other words, it is not possible to segregate the structural elements of biological structuralism and combine them to obtain desired functionalities. As we have realized when we progressed from genetics to genomics, a genome is not a series of independent genes strung together. There exists a hierarchy of genes in which individual genes derive their power from where they are placed in the hierarchy of genes. It is this hierarchical nature of biological structuralism that gives unique characteristics to the relationship between biological structuralism and biological functionalities.

If this multilayered organization of structuralism of biological systems is responsible for the peculiar nature of biological functionalities, then it is intuitively clear that it is nature of hierarchy or the relationship among different layers of structuralism of biological systems that decides the kind of bundling of different functionalities manifests itself during the course of natural selection. The key question is whether it is possible to define this hierarchical structuralism in a manner that we can predict the resulting biological functionalities or not. Our conventional wisdom says that one cannot predict the biological functionalities by looking at the genomic architecture. We will return to this topic in the following chapters.

This brings us to the next topic of the nature of modularity and whether it arises from any particular genomic architecture. Let us assume temporarily that there exists a definitive genomic architecture that is a product of natural selection. In other words, while individual genes and their expressions might vary from one organism to another, the overall architecture is retained across all the organisms. Admittedly, this is a sweeping generalization, it is not contrary to our understanding of genomics as available from phylogenetic studies (Bromham 2008, see Chapter 5). Once we accept this universal template of genomic architecture, the question is whether and how it can be reflected in the type of modularity which characterizes living organisms. Even in the absence of any prior knowledge of genomic architecture, it is intuitively clear that any type of modularity will have two guiding principles. A module will be separated from the rest of the genome in spatial and temporal sense. As we know in the case of Homeobox genes (Mazza 2007), while there are changes in the individual member genes of Homeobox in different species, the genes present in Homeobox retain their contiguity with one another. Similarly, there is a definitive order of gene expressions within the genes of Homeobox. This exemplifies the conception of a module as being a spatially and temporally contiguous unit which is functionally segregated from the rest of the genome.

As discussed in the preceding sections, since we cannot perceive a temporal template of a genome, we have not been able to formalize genomic architecture beyond the conception of the modules. Even in the case of individual modules like Homeobox, we could conceptualize the module because of our knowledge of a particular sequence of gene expressions of the member genes of Homeobox (see Chap. 5). Thus, our conception of temporal patterns arises from the functionality of gene expressions. Therefore, conventionally, we have replaced the notions of temporal and spatial templates with those of structuralism and functionalities. Upon a little reflection, it is intuitively clear that this analogy is not exact. There exists a definitive distinction between a temporal template of a genome and its functionalities. However, from the epistemological perspective, our cognitive bias toward the molecular perspective has created this cognitive dissonance. A temporal template of a genome need not be synonymous with a functional genome, but due to the above mentioned cognitive bias, we have conflated both these nuances.

Therefore, if we were to deconstruct the modularity of a genome in the conventional perspective, we would perforce deconstruct it through the language of structuralism and functionalities. However, as discussed in the first chapter, this where the conventional perspective has failed to distinguish between temporal and spatial elements of genomic architecture. It has failed because it is preoccupied with the molecular perspective. However, the proposed model suggests that the temporal and spatial elements of genomic architecture are equivalent and operate as one. Albeit, this unitary approach is made possible by the fundamental postulate of the proposed model that spacetime actively participates in biological evolution and natural selection. Therefore, since spacetime consists of four dimensions each having its own blend of time-like and space-like features, this would be reflected in the temporal and spatial elements of genomic architecture as a unitary template. Thus, according to this model, genomic architecture is unitary in the sense that the temporal and spatial elements are identical. Rather, genomic architecture is essentially topological in nature. More importantly, it exists in multiple dimensionalities simultaneously. It is only when the effects of these higher dimensional structural elements are manifest in the four-dimensional perspective that we observe a separate manifestation of temporal and spatial elements of genomic architecture.

It is legitimate to question such an intangible scenario by pointing out that after all Life manifests itself in the four-dimensional spacetime, and therefore, what is germane to the conception of genomic architecture is how the temporal and spatial influences manifest. The conjecture of a higher dimensional model of genomic architecture is redundant and even unnecessary. However, the fact remains that our attempts to formalize genomic architecture using the four-dimensional manifestations of the temporal and spatial influences has not provided us not even a semblance of any architectural design of genomes (Lynch 2007). Therefore, it is necessary to conceptualize any such higher dimensional perspective, if only to derive a decent plan of genomic architecture.

Returning to the present discussion, let us make a second assumption that there exists a higher dimensional topological model of genomic architecture wherein the temporal and spatial structural elements are identical. In such a scenario, the question is, how would it be reflected in the type of modularity of genomic architecture? Prima facie, there is one inevitable consequence of any higher dimensional model having unified temporal and spatial patterns. Irrespective of the mechanisms by which these unitary elements devolve into the four-dimensional spacetime, there will always be some parallelism between now separated temporal and spatial patterns. Thus, if the proposed model is correct, the genomic modularity will be symmetric in its patterns of gene expression and the distribution of genes within the genome. It is important to remember that this does not imply that gene expressions and the gene distribution are parallel. Rather it implies that the pattern of gene expressions of different modules and the patterns of gene distribution in different modules will be parallel. Therefore, in the following sections, we will try to deconstruct our present knowledge of genomics to find any such evidence to support or to refute this possibility. However, since our literature is coded in the language of structuralism and functionalities, we will adhere to this conception of genomic architecture.

However, it must be kept in mind that we are using this terminology only for the sake of convenience. We will revert back to the topological perspective of higher dimensionality whenever it is necessary. Therefore, we will try to deconstruct the modularity of genomic architecture from two conventional connotations, viz., structuralism vs. functionalities and regulatory vs. expressive genome. In the following sections, we will try to deconstruct these two dualities from two perspectives. Firstly, we will try to deconstruct how a genomic singularity can give rise to these dualities. Secondly, we will try to deconstruct our present understanding of modularity in the context of these dualities. Whenever required, we will describe a topological perspective of these topics.

4.4 Separation of Structuralism from Functionalities

In the preceding chapters, we discussed the centrality of the separation of structuralism and functionalities in biological evolution and natural selection. It was suggested that because of our cognitive compulsions, we have conflated both these features of Life. Moreover, due to inherent structuralism of the four-dimensional spacetime which is woven into our cognitive faculty, we can't perceive any higher dimensional perspective of any natural phenomenon. Therefore, we have chosen to focus only on the molecular perspective which finds its natural expressions in the four-dimensional spacetime. As a result, we have overlooked two important possibilities of defining Life. Firstly, we have overlooked the possibility of the genomic functionalities having its own template which is distinctly different from the structural template of the genome as defined by the molecular perspective. Secondly, we have overlooked the possibility of the genome having a higher dimensional architecture. This monograph outlines one such model of genomic architecture. In the preceding chapters, we have discussed various aspects of the relationship between structuralism and functionalities of Life. These aspects referred to the difficulties in formalizing Life and to their role in biological evolution and natural selection.

However, in this section, we will take a purely structural view of the nature of separation between structuralism and functionalities as implicit in the conventional perspective of genomic architecture. Having done that, we will try to formalize it using the proposed model. For this purpose, we will focus on three aspects of the separation between structuralism and functionalities. Admittedly, there are a myriad of nuances of this separation, but we will focus only on these three aspects. The reasons behind this choice would become apparent as the arguments unfold. These three aspects are: the separation of structuralism and functionalities as an evolutionary imperative, the separation as a structural imperative and the separation as a semantic imperative. Upon a little reflection, it is intuitively clear that this choice implies certain inevitability of this separation. Prima facie, any suggestion that the separation of structuralism from functionalities is inevitable, is antithetical to the Darwinian paradigm which is founded on the tenet of nondeterminism. However, as discussed all along in this monograph, this dichotomy between randomness and complexity is at the heart of the semantics of the Darwinian paradigm. Therefore, if the present discussion refers to imperatives, it is merely acknowledging the fundamental paradox of the Darwinian paradigm and that of Life. In the later chapters, we will try to reconcile the paradox. Presently, we will simply accept it as a *fait accompli.* Returning to the present discussion, let us begin with the first aspect, viz., the separation of structuralism from functionalities as an evolutionary imperative.

Firstly, let us understand what is meant by an evolutionary imperative. The semantic proposition of this evolutionary imperative can be formulated as: the separation of structuralism and functionalities arises because of the inherent structuralism of the process of natural selection. Admittedly, it is a radical proposition on several counts. Firstly, it implies that natural selection has a certain fixed structuralism that inevitably leads to this separation. Secondly, if as suggested in the first chapter, biological evolution and natural selection are like any other natural phenomenon, then it is axiomatic that something parallel must be happening in other natural phenomena. In other words, outcomes of all the natural processes must manifest comparable separations. Finally, this proposition implies that every natural phenomenon is governed by the conflicting requirements of chaos and order. We will not be discussing these broader issues here, except for its evolutionary perspective.

If the separation of structuralism from functionalities arises because of a certain fixed template of the process of natural selection, it must leave behind three types of evidence in the genomic architecture. Firstly, the relationship between structuralism and functionalities must be unitary. In other words, this relationship must be amenable to formalization. Secondly, the types of complexities of either the structuralism or of functionalities must be limited in numbers. Finally, it must introduce a certain degree of predictivity in natural selection. Let us briefly look at each of these consequences from the phylogenetic perspective. Admittedly, the conventional perspective is silent on the topic of any formal relationship between structuralism and functionalities. However, as discussed in the first chapter, that is because the conventional perspective doesn't acknowledge the possibility that functionalities can have their own template which is distinctly different from the corresponding structural template. However, as discussed in this monograph, there exists a definitive relationship between structuralism and functionalities in not just biological systems, but in every natural phenomenon. This relationship has escaped our attention only because of the fact that this relationship requires a conceptualization from a higher dimensionality, something that is not available to our cognitive faculty. However, with the recent advances in topology, particularly in the higher dimensional topology, it should be possible to define a higher dimensional perspective of the relationship between structuralism and functionalities (Laudal 2021). Although we do not have any conception of this relationship, there are enough phylogenetic indications of its existence.

This brings us to the second topic of the limitations in the types of complexities of structuralism and functionalities that can arise in the process of natural selection. As discussed in the first chapter and above, there are several discontinuities like speciation, separation of genes which cannot be explained by the conventional perspective. We normally invoke contingencies to justify these discontinuities. However, the proposed model offers a unitary explanation of all such structural and functional discontinuities. It postulates a certain mechanism by which natural selection occurs. This along with a formal description of the relationship between structuralism and functionalities, will give rise to discontinuities. For instance, we could postulate an entity which is named here as a genomic singularity. This entity, among other things, represents a functional and structural continuum. However, this continuum exists in a higher dimensionality. Therefore, when this continuum devolves into a four-dimensional spacetime, it breaks in discrete fragments. In the case of structuralism of the genome, this results in the manifestation of discrete genes. More importantly, this model suggests that the intergenic regions, in the form of nonfunctional DNA sequences, are inserted by the process of involution. Still more importantly, this process of involution (of the insertion of intergenic regions) itself is a mechanism of natural selection. We will return to this topic later on, but presently, it is important to remember that the origin of all such discontinuities which are not easily explained by the conventional perspective, can be explained by a generic mechanism of involution. This should be taken as indirect evidence of the proposition that natural selection operates using a certain fixed structural template.

This brings us to the third topic of predictivity. It is legitimate to argue that if natural selection operates using a certain mechanism, it should introduce a certain degree of predictivity in natural selection. In fact, historically, it has been argued that because it is not possible to predict the outcomes of natural selection, it is a nondeterministic process. After all, randomness is the foundation of the Darwinian paradigm (Bonner 1988). However, as discussed in the preceding chapters, there are two levels at which natural selection operates. To quote Darwin's phrase, natural selection operates using descent with modifications. While the descent is a causal and therefore predictable process, the modifications are random events. Therefore, even in the conventional perspective, causality of descent is accommodated. It is axiomatic that any causal processes must be amenable to predictive methodologies. However, as discussed in the first chapter, it was Darwin's own aversion to any kind of theological or teleological arguments that has resulted in the neglect of the predictivity implicit in the causal nature of descent. However, we now know from our experience in population genetics (Provine 2001, see Chapter 5), a degree of predictivity can coexist with nondeterministic semantics of the Darwinian paradigm. The question is what kind of predictivity is permissible within the Darwinian paradigm.

The proposed model suggests that the Bayesian perspective of conditional probabilities (Press and Clyde 2003) which is successfully applied to natural selection arises from the parallel evolution of structuralism and functionalities happening in different dimensionalities. When these parallel modifications converge in a single dimensionality through the process of involution, we observe the Bayesian patterns. Thus, while there is direct evidence of natural selection operating through a single mechanism, there are enough indirect indications even in the conventional perspective, to suggest that natural selection must operate using a single framework.

This brings us to the second semantic proposition of the structural imperative. This refers to the assertion that biological structuralism and functionalities have certain definitive templates that inevitably result in their separation. As mentioned above, the conventional perspective accepts the structural template as the only way to formalize genomic architecture. Even within this structural perspective, only the molecular perspective is taken as a basis of genomic architecture. It has to be accepted that this approach is by no means incorrect. After all, it has led to enormous advances in our understanding of genomes. The problem with this exclusive use of the molecular paradigm is that it is inadequate and more importantly, it has reached a dead end. Therefore, if we wish to expand the formal description of genomic architecture, we will have to attempt a paradigm shift. As discussed in the first chapter, even within the conventional perspective, we can think of several lacunae. The most glaring lacuna is the lack of understanding of the origin and nature of functionalities. In the conventional perspective, functionalities are assumed to be originating from the structuralism of genes as defined by their molecular description and therefore, functionalities owe their features solely to the molecular structuralism of individual genes.

Admittedly, the first part of this assumption is self-evidently true. If functionalities were to originate from any other sources, we will have to abandon the naturalistic foundation of biology and regress back to some transcendental explanations like vital force theory. Therefore, it is axiomatic that functionalities owe their origins to the structuralism of genes. However, the second part of the assumption that functionalities owe their features also to the structuralism of genes, is something that is debatable. Even if we decide to adhere to this conventional wisdom, it is intuitively clear that the individual features of functionalities do not have any one to one relationship with the features of the structural elements of genes. In other words, it is not possible to predict what kind of functionalities a given DNA sequence can give rise to, at least not completely. Admittedly, as discussed in the first chapter, this mismatch between the features of functionalities and structuralism is not confined to biology, but manifests itself in all types of phenomenology. What is unique about this mismatch between the features of functionalities and structuralism in biological systems is the kinds of functionalities produced. We will return to this topic while discussing the semantic imperative.

Presently, the key point is that functionalities, once developed, acquire a template of their own. As discussed in the first chapter, medicinal properties of molecules are classic examples of this (Roy 2015). As we have learnt in medicinal chemistry, it is possible to estimate medicinal properties of fragments of molecules separately (Howard and Abell 2015). However, even if we know the properties of individual fragments of a given drug molecule, it is not possible to predict the medicinal properties of that molecule. This is true in the case of individual genes. The functionality of a gene doesn't add up with those of its neighboring genes to give rise to predictable functionalities. The key point is that structural details cannot help us to know the nature of the resulting functionalities beforehand. In fact, this is acknowledged in the conventional perspective in the form of the emergence principle (Smith and Morowitz 2016, see Chapter 4). This principle says that there exists characteristic functionality at every level of organization.

This is precisely what is meant by structural imperative. In every natural phenomenon, there exists a mismatch between functionalities and structuralism. However, in the case of biological systems, this mismatch is such that it creates a separation of structuralism from its functionalities. The core argument of this monograph is that we need to redefine templates of structuralism and functionalities separately and define their relationships before formalizing natural selection. Once we accept this separation of structuralism from functionalities, it is logical to ask what is so unique about the separation of structuralism from functionalities in biology that makes natural selection unique among other natural phenomena? This is where the semantic imperative comes into the picture.

The semantic imperative can be stated thus: The features of biological functionalities represent semantic propositions that cannot be derived from the semantic propositions implicit in the structural template of the respective biological systems. This formal statement of semantic imperative raises a few questions. Firstly, what are those unique features of biological functionalities that force this separation between structuralism and functionalities? Secondly, why invoke semantic nuances in a scientific theory when the concerned theory, viz., the Darwinian theory has worked well without any reference to such anthropic conception of semantics. After all, the Darwinian paradigm has set out to steer clear of other anthropic conceptions like deism or teleology or design principles. Finally, do semantic propositions possess their own structuralism that forces this separation of structuralism from functionalities? Therefore, in the following paragraphs, we will try to answer some of these questions.

Let us begin with the first question of what are those features of biological functionalities that necessitate this separation? Prima facie, we can have no objections to the assertion that features of functionalities are different from the features of the underlying structuralism. After all, because these two sets of features are different, we can distinguish between functionalities and structuralism. The problem with the above assertion is that it suggests that the features of biological functionalities are per se different from the features of other functionalities observed in Nature. Even this suggestion is acceptable because as discussed in the first chapter, it is very difficult to formalize the definition of Life because it seems to be unlike any other natural phenomenon. The real problem is we can't define what is so unique about the features of biological functionalities that make Life unlike any other natural phenomenon. Therefore, we will pick up some of the arguments presented in the first chapter to highlight this aspect.

There are two fundamental features of Life which are essentially the functionalities arising from the genomic perspective. These are the ability to replicate and the feature of self reference. Admittedly, both these functionalities are connected to one another, but we will treat them as separate features. As discussed in the first chapter, no other natural phenomena have this functionality of replication. Therefore, the question is what kind of semantic proposition this functionality represents that it cannot be congruent with the underlying structuralism? In other words, there is something about this functionality of replication that is not present in the structural template, say DNA sequence of a given organism. The obvious answer is the proteins responsible for the transcription. Though these proteins too are encoded in the DNA sequence of the organism (with the exception of maternal RNAs), the timing of their syntheses is not. Thus, it is the genomic architecture that decides the sequence of gene expressions which enables the prior syntheses of these proteins, thereby actualizing the functionality of replication. Thus, if we accept that the DNA sequence of an organism is its structuralism and its genomic architecture is its functionality, it is intuitively clear that genomic architecture cannot be present in the DNA sequence, though it arises from it. In fact, as discussed in this chapter, genomic architecture supervenes the DNA sequence. This is precisely what is implicit in the above assertion that biological functionalities are separated from structuralism due to semantic differences. Incidentally, this scenario is congruent with the notion of variable definitions of genotype and phenotype discussed in the preceding chapter.

Now, let us look at the second feature of self reference. Upon a little reflection, it is intuitively clear that the above mentioned feature of replication couldn't have arisen unless the genome acts upon itself. Thus, replication is a kind of self reference. However, we are looking for a broader class of self reference here. We are looking at self reference that pertains to the perception of self as distinct from the environment. It is self-evident that some of the species of animal Kingdom possess this perception of self to a varying degree. However, even beyond that narrow definition of self, there is a wider and perhaps more incipient form self-perception is manifest in every living organism. It is this subtle form of self-perception that is at the heart of the Darwinian paradigm. The competition for survival would not take place unless every participant can perceive itself as separate from others. No other natural outcome possesses this functionality, and therefore, it is unique biological functionality. This functionality is not encoded in either the genomic or morphological structuralism of an organism and yet it manifests in every organism. Once again, this functional feature exists separate from the underlying structuralism and yet, it arises from the underlying structuralism.

Therefore, it is reasonable to infer that biological functionalities have inherent features that cannot be placed within the corresponding structural template. Thus, there is an inherent semantic imperative of biological functionalities that they must exist separately from their corresponding structuralism. This brings us to the second question of why invoke semantic nuances when the Darwinian paradigm has consciously avoided anthropic perspective since its inception? This question is based on the assumption that semantics, per se, is an anthropic construct. In a conventional sense, this assumption is true. Semantics is concerned with meaning and to the extent meaning refers to our understanding of reality, semantics is an anthropic perspective of reality (Stamenov 1992, see Part I). However, there exists another way to conceptualize semantics (Portner 2005). This pertains to the information theoretical perspective. If we can define information content of any theory, then we can define semantics as a sum of all the relationships among the elements of information content of the theory. Of course, it is a different matter that we can never define all the information content of any scientific theory. However, in an abstract sense, semantics can be defined using this information theoretical definition.

Now, we can answer the question posed above from two perspectives. From the information theoretical perspective of semantics, it is intuitively clear that we can assign semantic propositions to biological functionalities and structuralism. Therefore, if we can demonstrate that these semantic propositions are mutually incompatible, then we can justify the separation of structuralism from functionalities. The second way to answer this question belongs to the conventional connotation of semantics as a science of meaning, as available to our cognitive faculty. The traditional objection to any ascription of anthropic values to the Darwinian paradigm rests on two grounds. Firstly, the founders of this paradigm, including Darwin himself, consciously kept away any religious or theistic arguments, if only to maintain the naturalistic foundation of the Darwinian paradigm (Grene 1986). Even Darwin's own rejection of Lamarckian types of explanations was partly based on the suspicion that such a teleological argument would eventually lead to a nascent form of theism. Secondly, as our understanding of natural selection (and by implication, that of the Darwinian paradigm) increased, this initial aversion to any anthropic perspective was refined by the statistical foundation of natural selection. Nondeterminism and the Bayesian conditional probabilities now sum up our reasons for rejecting any anthropic perspective.

However, our own preoccupation with the nonanthropic perspective has obscured the fact that our own conception of semantics is based as much on our anthropic values as it is on the information theoretical perspective. It is a category mistake to think that anthropic semantics (Stamenov 1992) and information theoretical semantics (Portner 2005) are mutually exclusive. There exists certain commonality between these two perspectives. There are instances, for instance the interpretations like mutual altruism, where both these perspectives are conflated into a single framework (Sussman and Cloninger 2011). This is true for concepts like group selection (Borrello 2010) (and social biology in general (Alcock 2001, see Chapter 3)) as well. The key point of this digression is that semantics, even if conceptualized in a nonanthropic perspective, carries structuralism that is parallel to the anthropic perspective of semantics. From a broader perspective, this is a central dilemma of epistemology. Our knowledge of reality might be clothed in anthropic perspective, but it enjoys a considerable degree of congruence with nonanthropic knowledge, say mathematics. The epistemological dilemma is not that our knowledge is incomplete, but rather that it is congruent with the nonanthropic knowledge, in spite of being incomplete.

Thus, infusion of semantic propositions, even if they are anthropic propositions, in any scientific theory is a legitimate exercise if it leads to insights into our understanding of that theory. What should bother us is whether semantic incongruence between any two propositions should force us to alter the theoretical framework or not. For instance, let us assume, even if temporarily, that the semantic propositions of biological functionalities and structuralism are mutually incompatible. Should that be a reason to alter the theory of natural selection? This brings us to the third question: do semantic propositions possess their own structuralism on the basis of which we can decide whether they are mutually incompatible or not?

As discussed in the preceding monograph (Chhaya 2020, see Chapter 12), the problem with semantics of natural languages is that our conception of semantics is based on two different modes of cognition. Our understanding of semantics is primarily intuitive. This mode of cognition is noetic, in the sense that we don't know how we arrive at it. However, our formal conception of semantics is derived from our formal theory of syntax. Historically, there is linear progression from natural languages to syntactic grammar (Arnove 2008), to compositional semantics (Jackendoff 2002, see Chapter 8), to denotational semantics (Allison 1986), to artificial languages. However, the exact relationship between syntax and semantics remains unarticulated. Therefore, at present whatever we know about semantics is based on our understanding of the underlying syntax, a domain which is amenable to formalization. As mentioned above, there is another way to define semantics and it is based on information theory and the underlying mathematics. Leaving aside the relative merits of these two approaches, it is intuitively clear that both these approaches tacitly ascribe some structural template to semantics, even though by its congruence with the corresponding syntax.

We can argue that if there exists a definitive relationship between syntax and semantics, it implies a certain structuralism of semantics itself (as distinct from the syntactic structuralism). This is because congruence between any two domains must be defined using some structural template. While in the case of linguistic connotation of semantics, this structural template of semantics must resemble the corresponding syntactic template, in the case of information theoretical connotation of semantics, the semantic template must resemble the corresponding mathematical template of information theory. Once we accept this scenario, the answer to this question suggests itself. Semantics, in spite of its apparent abstract nature, must possess a structuralism of its own. As argued in the preceding monographs, if semantics appears to be an abstract or a nonstructural entity, it is because of the inherent limitations of our cognitive faculty. Therefore, to the extent the functionalities of Life are unique their semantic template must be different from its structural template. This justifies the earlier assertion that structuralism and functionalities must be separated from one another because the underlying semantics demand it.

Thus, the arguments presented in this section provide us with three imperatives why biological functionalities are separate from the underlying structuralism of living organisms. Before we explore the nature of this separation of structuralism from functionalities, let's look at another type of separation manifest in genomic architecture. This refers to the separation of the regulatory framework of the genome from the ensemble of gene expressions. Therefore, in the next section, we will look at the separation of the regulatory genome from the expressive genome.

4.5 Separation of Regulatory and Expressive Features of Genomic Architecture

In the previous section, we looked at the separation of structuralism from functionalities of genomic architecture. However, it would be wrong to think that this separation defines genomic architecture. There exists another and perhaps more fundamental, separation in the genomic architecture, viz., the separation of the regulatory framework of the genome from the expressive framework. Therefore, before we look at the nature of modularity, it is necessary to deconstruct this separation as well. Therefore, in this section, we will look at the nature of the separation of the regulatory genome from the expressive genome. We will also try to deconstruct the semantic and evolutionary perspectives of this separation in this section. This separation of regulation from expressions shares one feature with the separation of structuralism from functionalities. Just like genomic functionalities, the regulatory framework of the genome is not represented in the DNA sequence of the genome. It is present in higher levels of genomic organization, just like genomic functionalities. In fact, it is tempting to think that the regulation of gene expressions must be classified as a genomic functionality. In some sense, this is true. However, as discussed below, there are reasons why the feature of regulation should not be treated as a genomic functionality. While we will discuss this aspect a little later, prima facie, we can think of regulation of gene expressions as a systemic functionality and therefore, it needs to be treated as a separate category by itself.

Having said that, let us see the nature of separation between the regulatory and expressive genome. In order to understand this separation, we will focus on its four specific aspects in this section, viz., the separation as a structural imperative, the separation as an evolutionary imperative, the separation as a semantic imperative, and the separation as a special case of a general separation of structuralism from functionalities. The core argument of this section is that this separation is in a certain sense, an inevitable outcome of natural selection. More importantly, these four aspects that we will discuss here are merely different perspectives of a single framework of biological evolution and natural selection. These four aspects merely point toward a theory of biological evolution and natural selection that is founded on semantic propositions. Thus, while the Darwinian paradigm is characterized by uncertainty and nondeterminism, its foundation lies in the set of semantic propositions. Moreover, these different semantic propositions are connected to one another through the semantic singularity. This coexistence of singularity and its fragmentation into different semantic propositions, is reflected in the genomic architecture as well. There exists a genomic singularity representing the semantic singularity. By analogy, different types of separations and different types of modularities, cohabit this genomic singularity. Thus, nondeterminism and uncertainty that characterize the Darwinian paradigm arise, not from the nature of Life, but they arise from the coexistence of different semantic propositions which are individually deterministic. The Bayesian paradigm (Press and Clyde 2003) defines the conditional probabilities of these different semantic propositions. This sums up the basic dichotomy of Life, an emergence of order amidst chaos. The genomic architecture, or rather these different modularities, embody this dichotomy.

Returning to the present discussion, let us look at the structural imperative of the separation of regulatory and expressive genomes. The structural imperative of the separation can be defined as a mutual incompatibility between the structural prerequisites of regulation and gene expression. It postulates that the structural template of gene expressions, per se, cannot perform a regulatory function and vice versa. In order to understand this incompatibility, let us look at the conventional perspective of gene expressions and their regulation (Weinzierl 1999, see Chapter 2). Admittedly, there are subtle nuances of gene expressions and their regulation. However, for the present discussion, we will look at a simplest scenario and that too, in an abstract manner. One can rest assured that this simplistic scenario doesn't compromise the validity of the argument presented below. We will omit the details only because they are not integral parts of the logic presented here.

A gene expression begins when a signal in the form of a molecule attaches itself to the DNA sequence upstream of the DNA sequence of the gene being expressed. Once the initiator attaches itself, the subsequent assembly of the transcription machinery begins to be followed by the actual transcription. As mentioned above, each of these three steps, viz., the attachment of the initiator, recruitment of the constituent molecules of the transcription machinery, and the transcription itself, are highly nuanced and are tailored differently for different types of genes and cellular contexts. However, purely from the perspective of logic, these three steps are well-defined and have their own structural prerequisites. Therefore, the question is whether each of these three steps can act as an instrument to regulate gene expressions? The answer is both, yes and no. The answer is yes in the sense that in their absence the gene under investigation will not be expressed. However, if we take regulation as a capability to initiate, regulate and terminate a gene expression, then the answer is no. None of these molecules, either singly or collectively, can fine tune the extent of gene expressions. They can initiate or terminate gene expressions by their presence or absence, but they cannot control the extent of gene expressions. They operate on a binary logic. Their presence switches on the gene expressions and their absence switches off the gene expressions. Albeit, because there are several molecules involved in these processes and therefore, these binary switches acquire a more nuanced process of the gene expressions. Instead of on and off positions, the gene expression becomes a Bayesian process. However, the bottom line is that there exists a certain finality about gene expressions. If these molecules are present, the gene expressions would inevitably follow and if any of these molecules are absent, the gene expressions will not take place.

It would be legitimate to wonder that in this simplistic scenario, how can regulation of gene expressions operate? The answer is by controlling the presence and the absence of these molecules. To continue the analogy of binary logic, one can distinguish between gene expressions and their regulation as a relationship between the digital and analog signals. If a gene expression is controlled by discrete signals in the form of the presence or absence of these molecules, then the process of gene regulation is controlled by analog signals. These analog signals consist of where and when these molecules must be present. Thus, if molecules are discrete signals then the analog signals are space and time coordinates of the gene expressions. Once we realize this distinction between the structural prerequisites of gene expressions and their regulations, it is intuitively clear that they are not compatible. Thus, the distinction between discrete and analog signals constitutes the structural imperative of the separation of the regulatory and expressive genomes.

We would be justified in questioning this scenario on the grounds that with the advent of iRNA (Cretoiu et al. 2020), this distinction is not valid because this RNA is a discrete signal and yet it regulates the gene expressions. However, as discussed in the following chapters, RNA interference arises from another feature of genomic functionalities. It arises from the functionality of self reference. Because the genome has a topologically complex structuralism, the functionality of self reference arises automatically. The manifestation of RNA interference points toward another semantic proposition. Genomic architecture is characterized by modularity and not by compositionality. A genome is not constructed of several subunits. It is composed of several modules which share certain common functionalities. Thus, the very conception of modularity contains the presence of singularity at higher level. Therefore, the regulatory and expressive genomes are not separate genomes welded together. Rather, they have segregated themselves out of a genomic singularity during the course of evolution and natural selection. Therefore, they not only share a common ontology, but they can communicate with one another through their deeper structural connectivity. The phenomenon of RNA interference is a manifestation of a deeper level of communication between these two frameworks. It is important to keep in mind that RNA interference manifests itself only in the genomes that have a certain degree of structural complexity, an indication of incipient genomic singularity (Howard 2013). However, during the course of evolution, as genomic architecture evolved, different mechanisms like RNA splicing evolved, thereby removing the need for self reference. Thus, it is intuitively clear why structurally the separation of regulatory and expressive genomes was inevitable. It is inevitable because the regulatory framework operates in an analogue fashion, whereas the expressive genome operates in a digital fashion.

Now let us look at the second imperative, viz., the evolutionary imperative. We will look at three different perspectives of this issue. Firstly, let us look at the phase space perspective. If we were to think of this separation of regulation from expressions of genes as a measure of complexity, the evolutionary context of this separation becomes clear. The conventional perspective on the emergence of complexity during biological evolution and natural selection uses a phase space type of arguments. If Life indeed evolved by random chance, it is axiomatic that the first living organisms must have possessed a minimum degree of complexity because the chances of evolving a complex living organism out of a primordial soup are lesser than those of simple organisms. Therefore, once Life evolved and the process of natural selection became operational, the successive generations will turn out to be more complex than their ancestors. Thus, the emergence of more and more complex organisms during natural selection happens by default. In the language of phase space, if the ancestors occupy the phase space representing less complexity, then the successive generations will have to perforce occupy the domain of the phase space that represents more complexity. Thus, once we accept that the separation of regulatory and expressive genomes is a type of complexity, the phase space argument explains why they should separate out.

However, let us sidestep this argument and try to understand why evolution should lead to this separation of regulatory and expressive genomes from the perspective of implicit structuralism of the conventional interpretation of natural selection. This is necessary because as mentioned in the preceding chapters and in the preceding sections, the proposed model asserts that the process of natural selection itself has its own structuralism. Therefore, without committing ourselves to any particular structuralism of natural selection, let us try to deconstruct what kind of structural template could necessitate this separation. Prima facie, we can safely assume that the evolutionary compulsions of this separation must be in the form of a better survival quotient of this separation. In other words, the separation of regulatory and expressive genomes increases the fitness of the organisms possessing these separated functionalities. Admittedly, the conventional perspective doesn't offer any clear evidence for this proposition. However, we can see that it was always implicit. Prima facie, the emergence of complexity and the subsequent emergence of modularity point toward overall benefits in the form of better survivability. Moreover, the separation of structuralism from functionalities can afford larger scope for variations in the structural template, thereby providing an opportunity for natural selection to operate effectively. Incidentally, if we deny that natural selection has a certain structuralism of its own, then it is not possible to explain why the regulatory framework should separate from the expressive genome. Thus, it is less problematic to think that natural selection possesses a native structuralism of its own.

Let us sidestep this implicit argument as well and consider it from the third perspective of the proposed model. Since the proposed model asserts that natural selection has its own structuralism, let us now look for any specific features of the mechanisms of natural selection that necessitate this separation. While we don't know much about the mechanisms by which natural selection operates (except that there is a random process of mutations the outcomes of which are selected for their fitness quotients), it will be reasonable to think that such mechanisms require separate units of inheritance and selection. Therefore, if we consider the separation of regulatory and expressive genomes as an analogous requirement for natural selection to operate, then obviously, this separation must have survival benefits. Moreover, since simpler forms of living organisms do not manifest this separation of regulation from expressions (Howard 2013), it is reasonable to think that this separation is necessary for more complex forms of Life. When viewed from this perspective, an interesting possibility suggests itself. The separation of regulatory and expressive genomes must have occurred in parallel with the separation of structuralism from functionalities and the separation of genotype and phenotype. It is tempting to think that at the beginning, there was a unitary template which we have chosen to call genomic singularity. It is this genomic singularity that the process of natural selection must have acted upon. Therefore, due to the unitary structural template of natural selection, different types of separations must have evolved out of this genomic singularity. Moreover, it is possible that due to the structural imperative of natural selection (which we discussed above), different types of separations must have evolved at different stages of biological evolution.

Let us now look at the third imperative behind the separation of regulatory and expressive genomes, viz., the semantic imperative. It is possible to argue that these two imperatives, structural and evolutionary, are essentially post facto rationalization rather than valid interpretations of natural selection. After all, natural selection operates by pure chance and necessity. There is no reason why it should be amenable to such anthropic constructs. However, it would be a different thing if we can demonstrate that there exists a semantic imperative that necessitates this separation. Semantic imperative, if established, can negate the nondeterministic justifications for the absence of any theoretical perspective of natural selection. This semantic primacy rests on two grounds. Firstly, evolution of Life and subsequent natural selection are essentially information transfers and transformations. (In the post molecular biology era, it is impossible to refute this assertion.) Therefore, to the extent information content necessarily carries semantic propositions, any system dealing with the information transfers and transformations must have semantic propositions. Therefore, if we can establish a semantic imperative behind the separation of regulatory and expressive genomes, it must be accepted that our ascription of nondeterminism to natural selection must be an epistemological artifact.

Secondly, the reason why semantic imperative enjoys primacy lies in the very conception of biological evolution and subsequent natural selection. Both these processes are essentially causal processes. Upon a little reflection, it is intuitively clear that there can't be any causality without any inherent semantics. The relationship between cause and effect is essentially a semantic proposition. Therefore, to the extent biological evolution and natural selection are causal processes, they have the semantic imperative. Admittedly, both these grounds of semantic primacy fly in the face of our universal acceptance of nondeterministic semantics of the Darwinian paradigm (Bonner 1988). It must be kept in mind that, as discussed in the preceding chapters, one of the reasons why the Darwinian paradigm rejected such anthropic constructs was its own aversion to any kind of theological or teleological arguments. Therefore, to an extent it is natural that any imputation of any such anthropic construct like semantics, too would be rejected. However, now that we have realized an information theoretical perspective of semantics (Winter 2016), there is no valid ground to reject semantic imperative.

This resistance to semantic imperative arises from the fear that it would undermine the characteristic randomness of the Darwinian paradigm (Bonner 1988). However, this need not be the case. Randomness and its attendant nondeterminism could very well turn out to be epistemological artifacts. Quantum mechanics offers an exactly parallel example (see Chhaya 2022c, for further discussion). Presently, we are convinced that the foundation of quantum mechanics rests on noncausality. EPR experiments and the subsequent debate on the possibility of hidden variables have convinced us that there are no such hidden variables. However, as we know from Bell's theorem that a possibility of nonlocal solutions always exists. Therefore, just as the possible existence of nonlocal solutions does not undermine the statistical nature of quantum mechanics, the causality and its implicit semantics do not undermine the Darwinian randomness. As far as our ability to comprehend is concerned, Darwinian randomness and its lack of predictability are permanent features.

Returning to the present discussion, let us see why it is semantically imperative that regulatory and expressive genomes must be separated from one another. As mentioned above, from the information theoretical perspective, the types of information required for regulation and expression are analogue and digital respectively. However, Nature is capable of employing this duality of discrete and nondiscrete information content in an interchangeable manner. This scenario is best exemplified by quantum phenomena. Therefore, purely from the information theoretical perspective, biological evolution and natural selection per se, do not have to manifest this duality. Nature could, at least in principle, could have used the same information content to execute both these functionalities of regulation and expression. Therefore, there has to be something else that necessitated the separation of these functionalities. This is where the semantic imperative comes into the picture. There must be something inherently different between the semantics of regulation and the semantics of gene expressions.

Upon a little reflection, it is intuitively clear that this semantic difference consists of presence or absence of the functionality of self reference. Let us deconstruct the functionalities of regulation and expression in the context of the functionality of self reference. It is intuitively clear that gene expression, per se, doesn't require any self reference. This is because the trigger for gene expression lies outside the gene. Similarly, the rate and end of the gene expression lies outside it, in the form of the kinetic and thermodynamic profile of these molecular triggers. Thus, a gene expression requires promoters, enhancers in the form of molecules which are external to the gene's own molecular template. Similarly, the rate of the gene expressions depends on the conformational changes and the proximity of the catalytic centers. This also is governed by different molecules from outside. Similar logic applies to the end of the gene expressions. Thus, any given gene will express itself every time these molecular agents are in the vicinity. In other words, gene expression is governed by external signals.

The gene regulation, on the other hand, requires a self-reference. Admittedly, we are still unaware of the mechanisms by which long-range influences involved in the gene expressions are executed (Shmulevich and Dougherty 2014). However, prima facie, it seems reasonable to think that these influences must arise from larger frameworks involving larger templates of the genome, if not the entire genome. Admittedly, the conventional wisdom behind this that it is the manner in which chromosomes are entwined with one another that brings about these long-range influences into force (cf. Chromosome territory (Fritz 2014)). Therefore, the conventional perspective also implicitly suggests that this placement of different chromosome territories must have been naturally selected. This is a reasonable assumption because the number of conformations that different chromosomes of a given genome may exhibit are limited in number and therefore, natural selection would select a particular distribution of chromosome territories that maximizes the long-range influences involved in the gene expressions.

However, our primary concern here is not how long-range influences must have evolved. Rather, our primary concern is what kind of information content is necessary to execute these long-range influences to regulate gene expressions and how this information content is different from the information content necessary to execute gene expression per se. It is in this context that we must think of the semantics of regulation and whether it requires self reference. The basic fallacy in the conventional perspective of the long-range influences (the one leading to a right combination of chromosomes territories) is that it overlooks this semantic perspective. It is important to note that natural selection merely singles out one particular (or perhaps a few favorable) conformation of different chromosomes. Natural selection never invests any semantic content to any particular way of entwining different chromosomes. The long-range influences which manifest from different conformations of a given genome are intrinsic to the genomic architecture. Natural selection merely optimizes this feature. It is because of our misplaced conflation of Darwinian randomness with the lack of knowledge of genomic architecture that we continue to perpetuate this category mistake.

For those who are trying to figure out genomic architecture, it is imperative that they formalize the notion of chromosome territories as an offshoot of genomic architecture. The key question is not whether different conformations are subject to natural selection (obviously, they must be). Rather, the key question is how genomic architecture ensures the propagation of these long-range influences. Even in the absence of any prior knowledge of genomic architecture, it is intuitively clear that whatever the mechanism by which these long-range influences are propagated, it must entail a higher dimensional entity. Assuming that this entity should naturally be a genome (or its module), it must be able to sense its different fragments. Therefore, in order to achieve this functionality of being able to recognize its own different fragments, the genome must possess a functionality of self reference.

This functionality of self reference is characterized by two features. Firstly, it must be semantic in nature. This is because it must be able to distinguish between self and nonself. Secondly, it must be something other than chemical reactivity. As mentioned above, chemical reactivity is a prerequisite for initiation, continuation and termination of gene expressions. However, all these activities occur in short range and at the molecular level. However, self reference cannot be in the form of the expression of any such chemical reactivity. Moreover, whatever the nature of the functionality of self reference may be, it has to be physicochemical in nature. Any other type of self reference would introduce some kind of supernatural or quasisupernatural forces like the earlier vital force. Therefore, the only way to reconcile the prerequisites of semantic nature of self reference and the essential physicochemical framework is to think of some kind of topological (or to be more precise, topochemical) framework.

The trouble with investing such a functionality of self reference at the molecular level is that it does not manifest anywhere, except for the genome. Therefore, we need to be more circumspect (or more conservative) while formalizing such a scenario. Therefore, it seems reasonable to think of a higher dimensional model for such a functionality of self reference. In such a scenario, one can think of the functionality of gene expressions which is essentially a molecular functionality, to be a feature of the four-dimensional spacetime and by implication, the functionality of gene regulation which requires long-range influences involving self reference, must be taken as a feature of a higher dimensionality. Once we accept that the regulatory framework entails self reference and that it is essentially an analogue information transfer, the most obvious candidate as an agent of this information transfer is spacetime itself. This is precisely what the proposed model postulates. Thus, the semantic imperative suggests that the expressive and regulatory genomes must coexist, but at different dimensionalities of a given genome.

As discussed in the preceding sections, genomic architecture is characterized by two different types of divergent influences. On the one hand, the genome must strike a balance between singularity and its modularities. On the other hand, the genome must separate its regulatory and expressive machineries. This dichotomy between structuralism and functionalities gets spread over these two divergent tendencies. Therefore, if we wish to develop the concept of different dimensionalities possessing different features, it is necessary to deconstruct both these divergences. Therefore, in the following sections, we will look at four different types of modularities, viz., structural, functional, regulatory, and expressive modularities. There are two aspects of these different types of modularities that we will discuss. Firstly, are these four types of modularities unified by a common ontology? This is important because a common ontology would point toward a notion of genomic singularity, a notion that is not openly supported by the conventional perspective. Secondly, is there any common structural template of these four types of modularities? This is important because it might help us to deconstruct the nature of natural selection. Moreover, it might help us to rationalize the evolutionary prerequisite for the dualities discussed in the preceding chapters. It must be kept in mind that it was suggested in the preceding chapters that these dualities, viz., the DNA/RNA, Phenotype/Genotype, and structuralism/functionalities, are the evolutionary compulsions for biological evolution. However, we do not know as yet the reasons for these compulsions. Therefore, it is tempting to think that these modularities also arise from the same compulsions that gave rise to the above mentioned dualities. Let us begin with structural modularity.

4.6 Structural Modularity

Ironically, the most obvious example of this structural modularity, viz., the segregation of genomes into different numbers of chromosomes, remains unexplained by the conventional perspective. There is no explanation why any given genome should fragment itself into a particular set of chromosomes. Admittedly, we know from comparative genomics that during biological evolution, the number of chromosomes in any given evolutionary lineage has changed. There have been instances of polyploidy during natural selection (Soltis and Soltis 2012). Similarly there have been instances where chromosome numbers have increased or decreased during biological evolution. Thus, there is no cogent explanation for linking genomic architecture with the number of chromosomes present in a given genome. More importantly, there are no correlations between the complexity of genomic functionalities and the number of chromosomes present. Therefore, prima facie, to suggest that there exists a well-defined principle of structural modularity of genomes is unwarranted. This apathy toward articulating the principle of structural modularity can be traced back to the original exposition of the Darwinian paradigm (Hodge and Redick 2009).

During the initial stages of the articulation of the Darwinian paradigm, the idea of any fixed structuralism was opposed for the fear of allowing some kind of design principles usurping the Darwinian paradigm. Later on, when the notion of Darwinian nondeterminism was very well-articulated (Bonner 1988), this lack of any correlations between the chromosome number and genomic architecture was taken as a proof of the nondeterministic foundation of natural selection. It is only after the advent of molecular biology and the subsequent emergence of genomics that we have come to realize that in spite of the random nature of natural selection, there exists a set of causal mechanisms by which biological evolution manifests. More importantly, these mechanisms employ the functional consequences of genomic architecture. Thus, we have been forced into the situation wherein we are convinced about the random nature of natural selection even when it is apparent that there exists a definitive structuralism of genomes.

The problem is not that we have to choose between these two mutually incompatible semantic propositions. Rather, the problem is how to reconcile them. This is necessary because both these propositions, viz., the random course of biological evolution and definitive structuralism leading to predictable outcomes, are true by themselves. There is no way we can abandon the Darwinian paradigm given its validity. At the same time, there is no way to deny the existence of architectural design of genomes. Therefore, we must find a semantic statement which accommodates both these mutually incompatible propositions. This dialectical approach is inevitable in some sense. It is in this context that we must deconstruct the structural modularity. However, as discussed above, we cannot think of the number of chromosomes present in different species as a basis for structural modularity. Things would have been simpler if each chromosome was a structural module by itself. However, this is not the case. In fact, our experience with chromosome territories (Fritz 2014) suggests that different regions of different chromosomes influence the sequence of gene expressions in a purely topological manner. Thus, different regions of chromosomes influence the gene expressions of different regions of other chromosomes. This is because during the process of gene expressions in nucleolus, different chromosomes are arranged in such a way that hitherto unrelated and structurally separated segments of chromosomes present in different chromosomes come into proximity. This proximity is good enough for one segment of chromosome to functionally influence the gene expression of another segment of another chromosome.

While conceding that chromosome territories are operating units of gene expressions, it is important to keep in mind two things. Firstly, there is no evidence that the generation of chromosome territories follows any predetermined pathways. This precludes any design principles being responsible for the exact arrangement of chromosome territories in any given species. Secondly, it is generally conceded that once a particular arrangement of chromosome territories is formed and if it is beneficial for the survival of a species, that arrangement will be naturally selected. What is germane to the present discussion is that this example embodies how the conflicting demands of Darwinian randomness and the compulsions of structuralism of modularity, are accommodated by natural selection.

The key point from the perspective of structural modularity is that while we cannot use organization of chromosomes as a basis for defining the structural modularity, the structural modularity is manifest nonetheless. There is another structural template that we can think of as a basis for defining the structural modularity. This refers to the notion of an operon. However, the trouble with the conception of an operon is that it is basically a semantic unit which is transposed onto the structuralism of the genome. Logically, its conception is elegant. However, when we impose this logic onto the genome, we are also ascribing anthropic perspective on the genomic architecture. The problem with this ascription of anthropic perspective to either Nature or genomic architecture is not that it is wrong, but rather, we are ascribing either Nature or the genome with semantic processing. This is something that no diehard follower of the Darwinian paradigm would want.

In addition to the objection of the conception of operon being an anthropic artifact, there is another and perhaps more fundamental problem with the conception of operon (Miller and Reznikoff 1980). Upon a little reflection, it is intuitively clear that even if the conception of operon is true, it can't exist in isolation. From the Darwinian perspective, there are two prerequisites. Firstly, natural selection must be able to favor the selection of this semantically loaded unit. For this to happen, the process of natural selection too must operate on semantic principles. This is something we are not comfortable with. Even if we were to overlook this evolutionary perspective of operon, there exists another structural problem. This refers to the placement of a given operon in the architecture of the genome. Apparently, an operon can't exist in isolation within the genome. Therefore, it is less problematic to think of an operon as a structural unit of the genomic architecture. In other words, our conception of the genomic architecture must be that of an ensemble of operons connected with one another by the same principle that governs an operon internally.

Once we accept that there exists a hierarchy of operons in the genomic architecture, it is intuitively clear that we can think of structural as well as functional modularity. As a logical corollary, we must think of the genomic architecture as the one consisting of several levels of operon-like architectures existing in a hierarchy. Thus, an operon could consist of several smaller operons placed in the hierarchical relationship among themselves. Once we accept this scenario, it is intuitively clear that this scenario also embodies another form of duality, viz., the structural and functional modularity. Purely from the information theoretical perspective, it is simpler to think of this duality (and also the other dualities mentioned above) as a Janus-faced construct. It operates as a structural unit at one level and as a functional unit at another level. The advantage of this conception of genomic architecture is that it reduces the ontological burden of having to explain two hierarchies of functionalism and structuralism. It also reduces the semantic burden of the process of natural selection. Natural selection can possess only one mechanism which works for both structuralism and functionalism. More importantly, it is also congruent with our belief in the shared ontology of all the living organisms.

Having looked at the basis of structural modularity, let us develop this notion further. We will incorporate, albeit speculatively, two structural elements to this conception of operon-like units. Firstly, we will postulate that structural changes that manifest after an operon is activated follow certain principles that limit the number of outcomes. Secondly, we will assume that these principles are scale independent. Let us elaborate these two postulates. Conventionally, when we think of an activation or a suppression of gene expressions, we normally conceptualize that process in terms of stereochemistry. We conceptualize how a suppressor or an activator fits on the site upstream of the DNA sequence encoding a protein. This is true in the case of preexisting molecules (either maternal RNA or proteins) or different parts of the genome itself (in the case of trans and cis long-range influences). The key point is that it is always the stereochemical congruence between two molecules that brings about the change in the order or rate of gene expressions. Therefore, it is intuitively clear that whatever the mechanism an operon may employ, it would always be through the three-dimensional arrangements of the atoms of molecules involved in the gene expressions. Axiomatically, this justifies our postulate that the outcomes of the operon control will be limited in numbers.

Now, let us look at the second postulate of scale independence of the mechanism by which operons operate. Till now, we have thought of an operon as a functional unit of the control of the gene expressions. Therefore, we have restricted our conception of an operon operating on a single gene expression. However, if we wish to think of the genomic architecture as consisting of a hierarchy of operons, we will be required to provide for a mechanism which is not limited to one or two genes. Incidentally, this notion of scale independence is implicit in the conception of chromosome territories. Of course the mechanism is still local (at proximal distance between two different DNA sequences of a given genome), but it is scale independent in terms of the separation of these two DNA sequences on the genome. Thus, we need to enlarge the notion of proximity to include all the possible interactions between different DNA sequences.

The key question is how we can define a structural modularity that yields these two features envisaged in these two postulates? Ab initio, we can think of some kind of nested hierarchy of the DNA sequences of the genome. However, the problem with this reasoning is that a genome is a dynamic entity and not a rigid framework. A genome undergoes several structural compressions and decompressions during the cell cycle. Therefore, whatever the nature of the structural modularity may be, it has to be something other than the one based on coiling and uncoiling of the spliceosomes. Our conventional perspective stops at this point. Since our conception of the genomic architecture is preoccupied with the notion of DNA sequences and their transcription, we cannot conceptualize anything that excludes this centrality of molecular perspective of gene expressions. Historically, there is another reason behind our reluctance to think beyond this molecular perspective. Since the days of "new synthesis (Delisle 2021, see Part II)" and population genetics (Provine 2001, see Chapter 5), we are convinced that the origin of Darwinian nondeterminism (which was originally conceived to keep out teleological explanations and any design principles) lies in the random nature of mutations of DNA sequences. The exposition of neutral or near neutral rate of mutations provided the necessary semantics of Darwinian nondeterminism (Kimura 1983). Therefore, we are always apprehensive that any elaborate conception of genomic architecture would undermine this cherished core of the Darwinian paradigm.

However, this need not be necessarily true. Even if genomic architecture were to possess a well-defined hierarchy of controls (say in the form a hierarchy of operons), it would still be nondeterministic, if only because of its inherent Bayesian logic. Therefore, here we will consider one such hierarchy which will still be nondeterministic. Our experience with scientific modeling (Laudal 2021) suggests that if we wish to achieve scale independence and still retain a unitary mechanism for defining any natural phenomenon, the best way to do so is to employ a topological perspective. In topology, it is possible to formalize a natural phenomenon without imposing any predefined metric. Therefore, such a model would always be scale independent. Secondly, it is possible in a topological manifold to connect its different submanifolds with a single operator of involution. Thus, a model of genomic architecture based on the formalism of the involuted manifold can be an ideal choice. We will explore this possibility in the following chapters. Presently, let us look at what kind of a hierarchy of operons can represent the structural modularity.

We have two requirements from such a hierarchy. Firstly, as mentioned above, we need a unitary mechanism which would give rise scale independence. Secondly, we need this unitary mechanism to afford only a finite number of outcomes. Besides these two prerequisites, there are a few implicit features of this mechanism that would make it compatible with the process of natural selection. Firstly, it should be congruent with the mechanism of natural selection (about which we don't know much). Secondly, this unitary mechanism must arise from the very beginning of Life. This is necessary because our phylogenetic studies (Bromham 2008, see Chapter 5) show that almost every aspect of genomes obey continuity. Therefore, if this mechanism were to evolve much later during biological evolution, there would always be some parts of the genome that would have different types of architecture as compared to the remaining genome. However, we have yet to come across any genomic feature that defies phylogenetic continuity. Therefore, this putative unitary mechanism must arise very early in biological evolution.

In fact, the implications that such a unitary mechanism should arise very early in biological evolution is consistent with the conception of structural modularity. In view of our discussion on the semantic compulsions of conceptualizing genomic singularity, it makes sense to think that this unitary mechanism must be applicable to the genomic singularity itself, thereby ensuring its universality across the genome.

The next implicit feature of this hierarchy of operons is that of congruence between this hierarchy of operons with the hierarchy of functionalities. As mentioned in the preceding sections, there exists a definitive relationship between biological structuralism and functionalities. Therefore, unless this hierarchy of operons finds its analogue in the hierarchy of functionalities, this model cannot be congruent with the process of natural selection because natural selection operates on functionalities and not on structuralism of the living organisms. Finally, the most important implicit feature of this unitary mechanism is that it must give rise to different operons (or operon-like units) that are complementary to one another. Admittedly, this is another way to justify the conception of genomic singularity, but it has always been implicit in phylogenetics. This feature is necessary not only for explaining the phylogenetic continuity, but it is also necessary to justify our belief that natural selection operates using a certain mechanism which is universal in nature.

The key question is that even if we are able to formalize such a hierarchy of operons, how is it manifested in Nature? Obviously, it must be instantiated in every living cell. Moreover, as mentioned above, we know that the genome does remain fixed during different stages of the cycle of cell division. Therefore, this hierarchy cannot be present in any of the coiled and uncooked states of the DNA Sequence of the genome. This is an important factor that has prevented us from postulating any model of genomic architecture. This monograph offers a way to resolve this paradox. We will return to this concept in the following chapters. However, it is intuitively clear that this hierarchy which is a prerequisite for defining genomic architecture (and which cannot be formulated with our DNA centric paradigm of genomics) has to be sourced from somewhere else. It is proposed that this source of hierarchy of operons must come from the structural template of spacetime itself. Once we accept that some higher dimensional model of spacetime can be used to formalize this hierarchy of operons, it is intuitively clear that chromosome number is not of any significance. In fact, events like the emergence of polyploidy could arise out of Bayesian processes, just like mutations. We will elaborate this radical idea in the following chapters. However, before doing so, let us look at the nature of functional modularity.

4.7 Functional Modularity

Understanding the nature of functionalities, per se, is problematic on several grounds. Therefore, defining functional modularity is further complicated. Therefore, in this section, we will briefly mention the problems associated with the nature of functionalities. Having done that, we will sidestep these problems and try to deconstruct the modularity implicit in the conventional perspective. The objective is to pinpoint the underlying semantic ambiguities of the conventional perspective. Let us begin with the problems with the conventional perspective of biological functionalities. As discussed in the preceding chapters, functionality, per se, is a nebulous concept. This is all the more true for biological functionalities. By definition, functionality is defined in the context of the surroundings and not on the entity possessing that functionality. Therefore, our scientific theories try to formalize a functionality in terms of the structural template of the entity possessing that functionality, even though it is the environment that induces the functionality. Therefore, the structural template of the entity possessing that functionality gets incorporated into the structuralism of the functionality. This is best exemplified in the study of the medicinal properties of molecules. Typically, we pick up a library of molecules that are known to possess the medicinal property under investigation. Then, we define fragments of these molecules and try to factorize their contribution to the desired medicinal property (Howard and Abell 2015). Admittedly, this protocol works reasonably well. However, it doesn't explain the nature of the medicinal property under investigation. It merely tells us which molecules are potential drugs.

The basic problem with this empirical approach is that it tries to formalize a continuum in the language of discrete entities. As we know from our studies in acquired immunity (Kresina 1998), our genome, at least in principle, is capable of responding to almost any type of infection. Therefore, in principle, it should be treated as a continuum. The most wonderful feature of the human genome is that it is able to translate this continuum of immune response into discrete entities of molecules. However, our expertise in medicinal chemistry is no match for genomic plasticity. This is because we have focused on the discrete units of genomic architecture and not on the higher dimensional features of the genome. The key point is this. There exists a definite relationship between a continuum and its constituent discrete elements. However, this relationship defies our cognitive capacity. This is true not just for medicinal properties of molecules, but it is also true for a wide variety of disciplines like Linguistics, Mathematics, Computation Theory and even Quantum Mechanics. These difficulties in different domains have been deconstructed in the accompanying monographs.

Returning to the present discussion, given the nebulous conception of a functionality, the problem with the notion of genomic functionalities becomes more intractable. There are two additional reasons in the case of genomic functionalities that compound these semantic ambiguities of formalizing a given functionality. Firstly, as we know from our studies in polygeny (Reavey 2013) and pleiotropy (Lozano 2017), the basic principle of gene expression is the existence of many to many relationships between genomic architecture and the individual genes. Therefore, it is not possible to employ any kind of algebraic model to define the relationship between functionalities and structuralism of the genome. Secondly, in the case of genomes there is a problem of self reference. To understand this problem of self reference, let us go back to the earlier example of medicinal properties. As discussed above, the functionality is defined in terms of the medicinally active molecules. However, the target of this medicinal activity is a different and perhaps unrelated molecule present in the infective agent. Therefore, there exists a clear separation between the functionality of a drug and the infective agents.

However, this is not the case with genomic functionalities. The agent that manifests a functionality and the agent that becomes a recipient of this functionality are one and the same. The genome acts upon itself. Therefore, in the case of genomic functionalities, it is all the more difficult to formalize it. This is best exemplified in our studies in autoimmune diseases (Cohen and Miller 1994). The human genome is well-equipped to counter almost any type of target infections. However, when it comes to its own constituents, it occasionally fails. This is a case of the functionality of self reference being imprecise. Thus, it seems reasonable to think that defining a functionality per se and particularly a genomic functionality, is difficult because of the semantic ambiguities inherent in the conception of what constitutes a functionality and where it is located. With these caveats in place, let us look at the notion of functional modularity of genomes. It is important to keep in mind that in spite of these semantic ambiguities, there is no dispute about the fact that natural selection has given rise to modularity of functionalities as well as that of structural elements of the genome.

Therefore, it seems reasonable to conceptualize the notion of modularity which starts with these two features of many to many relationships and self reference and then it can be demonstrated that different modularities can be derived from such a primordial entity. This Monograph tries to explicate this entity in the form of genomic singularity. There are two advantages to this approach. Firstly, we can rationalize how different modules evolved from such a singularity. This evolution of different modules can be conceptualized within the semantics of the Darwinian paradigm. We don't need to cope with the problem of the emergence of complexity and the implicit design principles behind the complexity. Admittedly, this approach necessitates that the putative genomic singularity must be considered as a source of all the manifest complexity and plurality. However, as discussed in the following chapters, this is not a serious problem. The second advantage of this approach is that it also explains why we have parallelism in the form of these dualities, viz., the functional and structural duality; the expressive and regulatory duality; and, on a broader level, the duality of genotype and phenotype. We will return to this topic in the following sections.

Presently, let us look at the conventional perspective of functional modularity. Prima facie, we can divide biological functionalities into three broad categories, viz., the replicative functionalities, the adaptive functionalities, and the defensive functionalities. Admittedly, these three categories are not rigidly delineated in the genomic architecture, but each category has its own semantics and mechanism. Therefore, it makes sense to segregate them into separate categories. Of course, if the proposed model of genomic singularity is correct, then these categories would have a lot more commonalities. In the conventional sense, it is possible to define different semantic content for these three types of functionalities. For instance, the replicative functionalities are meant to regulate growth, both during the developmental stages and regenerative stages of a given living organism. Therefore, these functionalities are concerned about the sequence and tempo of gene expressions. Their primary focus is on synchronization of the expressions of different genes. Therefore, the kind of modularity required in these cases is essentially a temporal modularity.

Similarly, in the case of adaptive functionalities, the key component of such a modularity is the ability to perceive different stressors from the environment and activate different types of genes. The primary focus in this functionality is the ability to quantify the magnitude of different stresses and respond accordingly. Therefore, the semantics of such a functionality consists of threshold values of different stressors. Obviously, this semantic content is different from the semantic content of replicative functionalities. Finally, in the case of defensive functionalities, it is intuitively clear that its semantic content consists of threat perception and activation of defense mechanisms, including apoptosis.

Upon a little reflection, it is intuitively clear that there exists overlaps between the semantic content of these three types of functionalities. Any defects in the replicative functionalities could cause stress or even a threat to the organism's survival, thereby activating the remaining functionalities. Similarly, any stressor when present in excess could trigger the defensive functionalities. It is also important to keep in mind that each of these three types of functionalities consist of several types of networks of genes. In other words, these three types of functionalities are modular in nature. Therefore, it is intuitively clear that these different functionalities and their different modules exist in a hierarchy of functionalities. The key point is that in order to define the genomic architecture, we need a unitary mechanism linking these different types of functionalities. More importantly, the reason why we have not been able to formalize genomic architecture is that we have failed to define this unitary mechanism of linking different functionalities to one another. It is the contention of this model that once we accept that these different functionalities have evolved from a common ontology, it ought to be possible to formalize this unitary mechanism. We will return to this topic in the following sections.

It is possible to question the wisdom behind such a classification of functionalities. However, to the extent these functionalities are characterized by a particular semantic content, it seems reasonable to classify these functionalities in such a manner. Of course, it will be evident as more and more details are outlined below that this classification is justified not only from the perspective of the proposed model, but it is also justified from the conventional perspective. What is germane to the present discussion is that these different functionalities have to depend on the same DNA sequence to achieve their objectives. Therefore, the above mentioned many to many relationships arise naturally. These criss-cross pathways of different functionalities trying to exploit the same DNA sequence makes it difficult for us to formalize the genomic architecture. In addition to this complex network of pathways, there exists another problem with the conventional perspective of functional modularity. This is actually the most fundamental problem of architecture and more so in the case of genomic architecture. These functionalities must have a definite mechanism to be manifest. However, for that to happen, every functionality needs a structural template. A genomic functionality cannot exist on its own. It needs a material entity to act as its vehicle of expression. To put it differently, genomic functionalities are not transcendental entities which influence the mundane world of molecules (genes in this case). Rather, these functionalities are the phenomenology of the DNA sequence of genomes.

It is because the same DNA sequence of the genome is required to possess structuralism and functionalities that the conventional perspective of the genomic architecture has not been formalized. Our mathematical techniques cannot accommodate the duality of structuralism and functionalities. This is partly because structuralism consists of discrete elements combined in some algebraic fashion and the functionalities are conventionally accepted as some kind of continua. The second reason for this failure of conventional mathematical techniques in formalizing this duality of structuralism and functionalities, is that structuralism and functionalities seem to possess different semantic propositions, something that Mathematics is incapable of handling. It is important to note this problem of the dual role of the DNA sequence acting as a functionality as well as a structural unit can be solved if we can develop Mathematics of self reference. This is precisely what the proposed model offers. Before we look at the details of Mathematics of self reference, let us look at another pair of modularity, viz., the regulative and the expressive modularities.

4.8 Regulatory Modularity

It is tempting to think that the regulatory framework of the genome is actually nothing but functional modularity. In a sense, it is true, but not entirely. Therefore, in this section, we will try to distinguish between the functional modularity and the regulatory modularity. In order to understand this distinction, we need to ask ourselves about the kind of functionalities that are not regulatory in nature. It is intuitively clear that any functionality that endows a genome with an ability to regulate its constituent genes must be treated as regulatory modularity.. Prima facie, it is intuitively clear that any such regulatory feature would consist of temporal and spatial control of gene expressions. Therefore, any functionality that entails temporal and spatial control would be a part of the regulatory framework of the genome. More importantly, if in multicellular organisms these temporal and spatial controls would be perforce modular in nature. Therefore, regulatory modularity must be an essential part of genomic architecture.

As discussed in the previous section, it is possible to view this regulatory modularity as a part of overall functional modularity. However, it would be wrong to equate these two modularities. There are two specific reasons why we need to treat regulatory modularity as a separate category of modularity and not as a part of overall functional modularity. Firstly, it is not difficult to realize that the regulatory functions must override other functionalities per se. There are any number of instances where lack of regulation of the expressions of any given functionality leads to pathology. Therefore, it is intuitively clear that among the different levels in a given hierarchy of the genomic architecture, the regulatory framework must occupy the highest level. Thus, from the semantic perspective, it is possible to categorize the regulatory framework as a part of overall functionalities. From the evolutionary perspective, it is the regulatory framework that enjoys centrality. This mismatch between the semantics of architecture and the semantics of biological evolution is crucial in understanding the origin of Life. It is tempting to think that Nature follows the semantic perspective and creates a particular type of hierarchy of functionalities, but Life subverts this hierarchy to place the regulatory framework at the heart of this hierarchy of functionalities. Thus, Nature allows natural phenomena to unfold regardless of their consequences, but Life tries to impose a regulatory framework on the natural phenomena to achieve the objective of survival. Of course, it is a different matter that in the end Nature wins over this subversion of semantics of hierarchy of functionalities.

The second reason why we should not treat the regulatory framework as a part of the functional modularity is that functionalities in general, either as singularly or as modules, do not integrate the spatiotemporal perspective of the gene expressions while unfolding. It is always the regulatory modularity that cognizance of the spatiotemporal features of the gene expressions. Prima facie, this may appear to be a tautological and even a trivial distinction. However, from the evolutionary perspective, this is a fundamental issue. If biological evolution is essentially an evolution of a functionality of responding to the features of the environment, then the nature of spacetime must be treated as the most fundamental feature of the environment. Therefore, from the evolutionary perspective, this conception of spatiotemporal features must have been incorporated into genomes much earlier than any other functionalities. Thus, from the evolutionary perspective, the first class of functionalities to evolve must be that of cellular housekeeping functions and the concomitant cellular architecture. (In principle, even these functionalities could be dictated by the temporal and spatial control elements. However, we will overlook this in the present context.) In order to achieve multicellular complexity, biological evolution must incorporate the temporal and spatial features of spacetime into genomes. The remaining functionalities must be thought of as having arisen subsequent to this regulatory framework.

Once we accept the evolutionary primacy of the regulatory modularity in the genomic architecture, it also gives us a solution to the hitherto difficult problem of the genomic architecture. Conventionally, modularity is accepted as the characteristic feature of all the features of biological complexity. However, there is no clarity on the evolution of modularity. Ab initio, there are two possible ways modularity could have evolved. Firstly, it could have arisen from the separate evolution of different types of functionalities and then, eventual bundling of these separate functionalities into different modules. Admittedly, this is a plausible scenario because it makes sense to think that subsequent bundling of different functionalities could be governed by the Darwinian processes and the most suitable plan of modularity would survive and through genetic drift, monopolize the population. The phenomenon of mitochondrial incorporation into a eukaryotic cell is a classic example of this reasoning (Margulis 1970).

However, once we accept the primacy of regulatory framework over all the other types of functionalities, it is inevitable that it could arise only if we concede a common ancestry of every living organism. Our phylogenetic studies support this conjecture of the primacy of the regulatory framework over all the functionalities. Incidentally, the proposed model goes one step further and suggests that this common ancestry of every living organism must be formalized as a genomic singularity. The conception of the genomic singularity not only provides an intuitive explanation for the evolution of different types of modularities, but it also provides a semantic explanation for the modularization. Before we look at the details of this semantics of modularization, let us look at the last of modularities, viz., the expressive modularity.

4.9 Expressive Modularity

This type of modularity is rather difficult to conceptualize. This is partly because it is defined by default and partly because it is passive. This feature of definition by default refers to the fact that there are no demarcations, either structural or functional, to define this modularity. It's existence is inferred only after gene expressions occur. Similarly, the feature of passive definition refers to the fact that this modularity does not have any characteristic features. Once again, its existence is inferred only after gene expressions occur. Therefore, it is legitimate to wonder about the need to define this modularity. This apprehension is further strengthened by the fact that a large portion of genomes remain untranslated. However, there are two reasons why we must accept the existence of this expressive modularity. Firstly, none of the other modularities can exist without the existence of this modularity. For instance, there cannot be any functional modularity unless it has a definitive template of its own. This template would not have arisen had it not been for any underlying expressive modularity. Unless the expressive genome had a definitive pattern, functional modularity could not have evolved. Similarly, structural modularity can only be defined in the context of the underlying expressive modularity.

The conception of expressive modularity is difficult to articulate because we are conditioned by the primacy of the molecular perspective of genomics. When we think of genomic architecture, we instinctively visualize the underlying DNA sequence. We often fail to distinguish between the fact that DNA sequence is a structural unit and not the structure of a genome. The structure of a genome is not its DNA sequence, but it is the way in which its DNA sequence is organized in its higher dimensional perspective. Once we accept that the genome is something more than its DNA sequence, it is intuitively clear that there must be some corresponding expressive modularity that is congruent with the other modularities mentioned above. The existence of expressive modularity is also evident from phenomena like RNA interference (Cretoiu et al. 2020). What was earlier thought to be a "junk" DNA, seems to be playing an important role in shaping the regulatory framework. Therefore, it is intuitively clear that this expressive modularity must exist in genomes.

It is possible to think that such a higher dimensional architecture might undermine the nondeterminism implicit in natural selection. For instance, it is possible to argue that it was earlier thought that the presence of "junk" DNA was necessary for maintaining a certain rate of mutations on which natural selection operates. Since these mutations are essentially random, the outcomes of natural selection too would be random, leading to Darwinian nondeterminism. Therefore, if this so-called "junk" DNA too has a functionality of its own, there is hardly any room for randomness in natural selection. However, as discussed in the following chapters, this fear is misplaced. In accordance with the Bayesian Logic, so far as there are multiple outcomes, there will always be some randomness in the process of natural selection.

In this section and in the preceding Sects. 4.6–4.8, we outlined several types of modularities. It is important to keep in mind that these modularities are integral to conventional wisdom. However, rather surprisingly, there is no formal description of these modularities available in the conventional perspective. As discussed in the preceding chapters, this type of lacunae arises from the mismatch between the implicit semantics and the explicit structuralism of the Darwinian paradigm. Without going into the semantic disambiguation, we will take a pragmatic view and try to find a way to accommodate these four types of modularities in a single framework. Therefore, in the next section, we will outline a topological perspective of accommodating these modularities in a topological model.

4.10 Topological Model of Modularity

It is apparent from the discussions presented above that these four types of modularities are operating in the functioning of genomes. However, there is no clarity either about their structuralism or about their relationships among themselves. It is true that in isolated cases the concept of operons can be demonstrated. However, there is no systemic template of operons available. As discussed in the preceding chapters, there are several legacy problems in our conception of genomic architecture. As a result, our present conception of genomic architecture doesn't seem to throw any light on this topic. More importantly, it doesn't seem capable of offering any formal description of these modularities and their relationships among themselves, at least not at the systemic level. In view of this, it makes sense to begin ab initio and conceptualize the underlying structuralism of these modularities. When viewed from this perspective, following propositions suggest themselves.

1. There are sufficient reasons for believing that these four types of modularities operate in genomes.
2. The nature of these modularities is such that they enjoy a certain degree of autonomy even when they are dependent on one another.
3. Therefore, it seems reasonable to think that they collectively constitute a hierarchy of modularity.
4. In spite of their autonomous and distinctive nature, each of these modularities seems to follow a unitary mechanism for their functioning.
5. This postulate of a unitary mechanism is necessary to explain their interdependence. In the absence of any such unitary mechanism, we would require an interface between different modularities to formalize their interdependence. This will necessarily increase the number of levels in the hierarchy of controls. Therefore, following the thumb rule of Ockham's Razor (Sober 2015, see Chapter 1), we will assume the existence of a unitary mechanism for all the operations of between and within modularities. This necessarily eliminates the

need for several such interfaces and simplifies the hierarchy (Bard 2016, see Chapter 9).
6. This postulate of a unitary mechanism is consistent with the notion of common ancestor employed in phylogenetics.

Let us try to visualize a possible hierarchy of levels of genomic architecture using these propositions. Once we accept that genomic architecture should be defined as a hierarchy of controls, it is intuitively clear that the best way to do so is to employ a topological perspective. Admittedly, this monograph is based on one such model, the suitability of a topological modeling of any hierarchy is intuitively clear. However, it must be admitted that the objective behind proposing this particular model is twofold. Both these reasons were essentially semantic imperatives. Firstly, it was thought necessary to disambiguate the semantic ambiguities of the Darwinian paradigm like the emergence of complexity during biological evolution without resorting to any design principles. Therefore, it was necessary to demonstrate that structural complexity can arise naturally if we can define the process of natural selection as a mathematical formalism. In such a scenario, complexity would arise as a result of this formalism and not as a result of any such design principles, deistic or otherwise. Secondly, even if we want to avoid any topological perspective of the genome, the classical perspective of the Darwinian paradigm requires that we include the role of the environment in the formal description of biological evolution. However, if we wish to include the environment and the competing species in a single framework, it is inevitable that we must employ some kind of a topological model. With these caveats in place, let us try to think of a topological model of these modularities.

Once we accept that we need to develop a topological model of the genomic architecture and particularly that of different modularities present in the genome, it is intuitively clear that each modularity must be assigned a different dimensionality. While conventionally, we assign dimensionalities we employ epistemological justifications. Thus, the numerical value of each dimensionality is decided by the number of parameters necessary to define the semantic perspective. However, in the case of the proposed model, the notion of dimensionality refers to the actual dimensionality of spacetime. Admittedly, this physical interpretation is consistent with the epistemological perspective dimensionalities conventionally used. In addition, none of these modularities should occupy the highest dimensionality. Let us understand why these propositions are necessary. Firstly, if each modularity were to influence other modularities, it is imperative that they must occupy different dimensionalities. This will also enable us to assign different characteristics to each dimensionality, thereby defining distinct features of each of these modularities. Secondly, it is possible to define an operator for formalizing the influence of one modularity on the other modularities. Such an operator can be defined as an operator bringing about the changes in the dimensionalities. This would ensure that whenever one modularity influences another modularity, it can only do so when it operates in the dimensionality of the second modularity. The most intuitive example of this phenomenon is the long-range influences (both, cis and trans). Admittedly, in these cases the changes in the dimensionalities takes a literal meaning of the changes in the conformations. This need not always be the case. Prior syntheses of inhibitors and enhancers also exemplify this phenomenon. It is just that, in these cases, the changes in the dimensionalities occur in the form of priority of gene expressions.

The second advantage of defining the influences of one modularity on another modularity as the changes in the dimensionalities, is that it ensures that these influences obey a unitary mechanism. This is because the proposed operator would only define a mechanism for the changes in the dimensionalities which is independent of the dimensionalities involved. Thus, it ensures that the underlying mechanism for changing the dimensionalities (and by implication, the nature of influence of one modularity on another modularity) remains unchanged. This also ensures that whatever the nature of the influence of one modularity on the other modularities may be, it has to be in the form of characteristic features of the recipient modularity. For instance, let us say that chromosome territories operate from a particular conformational arrangement (Fritz 2014). However, its influence must always be in the form of steric hindrance or facilitation of gene expressions. This is because the characteristic feature of gene expressions is stereochemical. Similarly, in the case of RNA interference (Howard 2013), the sequence of gene expressions of the noncoding regions of the genome could be decided by long-range influences, but the process of RNA interference always manifests itself at the level of transcription or translation of genes.

This brings us to the second proposition that none of these modularities should occupy the highest dimensionality. Prima facie, this appears to be an ad hoc prerequisite. However, this is not the case. There are two semantic imperatives behind this prerequisite. Firstly, if any of these modularities were to occupy the highest dimensionality, naturally, that modularity would be immune to the influences of the remaining modularities. Our current understanding of genomics doesn't support such one-way influences. Therefore, till we know more about the genomic architecture, it is necessary to impose this condition. Secondly, even if we choose to place one of these modularities in the highest dimensionality, it would be incongruent with the Darwinian paradigm. This is because it would entail some kind of determinism in the way the remaining modularities can evolve. The proposed model, instead, offers a way to resolve this dilemma. The proposed model suggests that the highest dimensionality of this representation of the genomic architecture must be called genomic singularity. Admittedly, this is a notional entity, at least for the present. Genomic singularity can represent either the universal common ancestor or it could represent hitherto unknown functionalities of genomes which acts as a central processing unit. Given this ambivalence, it is proposed that the highest dimensionality of this model ought to represent the notion of genomic "Self Reference." With this minimalist design of the topology of the genomic architecture, we will try to fill in the necessary details in the following sections. Therefore, in the next section, we will outline the involuted manifold model of a genome and its modularities. The emphasis will be on the evolutionary perspective.

4.11 Involutive Formalism of Evolution of Modularity

Having outlined a schematic outline of a topological model that could be useful to formalize the genomic architecture in the previous section, we will try to outline a broader picture of that architecture in this section. However, there is a slight change of perspective here. This refers to the fact that while we outlined the semantic imperative of accommodating different modularities in a single framework in the previous sections, we didn't commit to any particular ontology of the proposed model. However, in this section, we will begin with a particular ontology of genomic architecture. Since this shift in emphasis is a major semantic proposition, by itself, it is necessary to explain it.

There are two reasons for this shift in emphasis. Firstly, the proposed model, as mentioned in the previous section, provides for a hitherto unassigned dimensionality, the highest dimensionality, for an entity named here as a genomic singularity. Therefore, it is intuitively clear that this genomic singularity must be treated as a predecessor of all the modularities mentioned above. Therefore, it is necessary to include the highest dimensionality and its relationship with these modularities. However, once we include an entity named here as a genomic singularity in our discussion, it is inevitable that it defines a particular origin of these modularities. Thus, ontological perspective cannot be excluded from the description of these modularities in the proposed model. Secondly, even if we were not to subscribe to ontology implicit in the invocation of the genomic singularity, the fact remains that these modularities must have definitive relationships among themselves. These relationships could not have arisen without any ontological explanation. Therefore, whenever we choose to formalize the relationships among the modularities, we will be forced to postulate some kind of ontology of these modularities. Therefore, it makes sense to begin with a particular ontology of these modularities and test it against the available empirical evidence. Thus, this inclusion of an ontological perspective in formalizing these modularities is welcome because it provides us with a tool for the verification (or falsification (Popper 1963)) of the proposed model. With these caveats in place, let us look at the topological perspective of these modularities and their relationships among themselves. For the sake of brevity, we will present this description in a point-wise manner.

1. Genomes exist in multiple dimensionalities simultaneously. This architecture supervenes the DNA sequence and its various types of coiling.
2. Even though genomes exist in a higher dimensional topological space, the influence of these different dimensionalities are strictly represented by stereochemical forces. There are no hidden or mysterious forces, other than the stereochemical forces, that play any role in the evolution, selection and the expressions of genomes.
3. Whatever may be the nature of influence of the higher dimensionalities of the genome, its manifestation at the four-dimensional topological spaces will always be stereochemical in nature.

4. The highest dimensionality of a genome is named as the genomic singularity. This dimensionality represents all the possible genomic functionalities, expressed or unexpressed.
5. The genomic singularity undergoes a series of involutions leading to the manifestation of different modularities. These modularities, in turn, undergo further involutions to the individual or clusters of genes.
6. Each involution from the genomic singularity downward right up to the individual genes, gives rise to two important consequences.
7. Firstly, each involution entails a lowering of the dimensionality. Secondly, at the end of the involution, genomic functionalities of the lower dimensionality manifests. Thus, what is unexpressed in the genomic singularity, finds its expressions due to the process of involution. Since there could be any number of functionalities in the genomic singularity, there will always be many possible combinations of these functionalities that would arise during different evolutionary pathways. Different evolutionary pathways can be thought of as combinations of different types of involutions.
8. At the intermediate level, after a limited number of involutions, the genomic architecture would consist of several interconnected modules. These modularities are broadly classified here as structural modularity, functional modularity, regulatory modularity and expressive modularity.

   With this background let us look at the relationship between different modularities and their relationships according to this model. Once again, we will follow a point-wise presentation.
9. As discussed above, the highest dimensionality must represent the entity named here as a genomic singularity. Therefore, it is axiomatic that all the other modularities must have arisen from it. This emergence of different modularities during the course of evolution can be formalized as involutions in this model. However, the key question is that of precedence.
10. In view of the conventional wisdom that the mutations must be the source of variations and subsequent natural selection, it seems reasonable to think that structural modularity, in its primitive form, must have arisen first.
11. This must be followed by subsequent functional modularity arising from this preexisting structural modularity.
12. However, given the fact that it is the phenotype that is subjected to the process of natural selection, it is intuitively clear that it is the functional modularity that would start evolving at a faster rate than the corresponding evolution of the structural modularity.
13. Prima facie, this inference is contrary to the conventional wisdom. However, according to this model, functional modularity also arises from the genomic singularity. Therefore, it should not be treated as a derivative of the structural modularity. If at all, both these modularities must be treated as a conjoined twin. However, the dimensionality of functional modularity ought to be higher than that of structural modularity because its metric is coarser than that of structural modularity.

14. In other words, the relationship between structural modularity and the functional modularity must be defined by a pair of operators of involution.
15. In order to explain this contradiction between the conventional perspective and the proposed model, it is necessary to understand the significance of Epigenetics. Once we accept that the environment shapes the nature of phenotypic variations, it can only do so using the epigenetic imprinting of genotype. Incidentally, epigenetic processes can be viewed as the inverse operator of involution. Therefore, to the extent phenotype is synonymous with genomic functionalities, it is intuitively clear that functionalities would evolve faster than the evolution of the underlying structuralism.
16. Therefore, by analogy, functional modularity too would evolve faster than the corresponding structural modularity. It is possible to argue that this scenario undermines the Darwinian paradigm, particularly the absence of any design principles. However, since functional modularity cannot exist independent of the genomic singularity, it is axiomatic that its template must be derived from the template of spacetime which itself derives its template from the cosmic singularity. This aspect will be discussed in the following chapters.
17. In addition to this bidirectional influences between structural and functional modularities, there is another feature that separates them. By definition, the structural modularity would possess finer granularity in the form of molecular structuralism. In comparison, functional modularity would possess a coarser granularity. This is anyway implicit in the conventional perspective in the form of pleiotropy. (Incidentally, the manifestation of polygeny should be treated as an evidence of bidirectional relationship between these two modularities.) Therefore, according to this model, functional modularity must occupy higher dimensionality as compared to structural modularity.
18. Once we accept that functional modularity can evolve on its own, it is intuitively clear that this evolution should lead to the emergence of regulatory modularity. More importantly, this regulatory framework need not reflect on the structural modularity. This is because according to this model, the regulatory framework is the product of natural selection of functionalities and not that of structuralism. This is evident from the fact we can't define functionalities of a given genome either from the number of chromosomes or from the length of the DNA sequence.
19. In comparison to the functional modularity, the regulatory modularity must possess a finer granularity if only to represent its limited scale of influence. Thus, according to this model, the regulatory modularity must occupy the dimensionality intermediate level between the dimensionality of the functional modularity and the dimensionality of the structural modularity.
20. This placement of the regulatory modularity between the dimensionality of the functional and structural modularities suggests that the regulatory modularity is a subset of the genomic functionalities. However, its scope is more local than the scopes of other genomic functionalities. At the same time, it is slightly more nonlocal than the structural modularity.

21. The structural modularity needs to be further modularized into several modules of expressive genomes. It would be legitimate to question the need for expressive modularity in addition to structural modularity. However, upon a little reflection, it is possible to discern the need for an additional modularity. The structural modularity refers to the distribution of genes in any given genome. Expressive modularity, on the other hand, refers to the distribution of coding and noncoding regions of the DNA sequence of any given genome.
22. Admittedly, at present, we do not know whether different genes are distributed according to a certain pattern. Similarly, we do not know whether there is any operative principle behind the insertion of noncoding intergenic regions between any two genes. However, irrespective of whether these distributions are based on some principles, it is intuitively clear that both these distributions occur independent of one another. The proposed model merely provides a room for such a distinction.
23. More importantly, according to this model, the starting point is the genomic singularity. Therefore, if this postulate of a genomic singularity is valid, then it is axiomatic that there exists some principles which define the distribution of genes in any given genome and the interspersed intergenic noncoding sequences. In fact, this provides a way to verify the proposed model. If we can demonstrate that there exists a definitive pattern of the distribution of genes, then we will have to accept the proposed model as true.
24. While we don't know whether the above mentioned distributions are based on any mathematical compulsions or not, there is one interesting possibility to verify it even without working out the underlying Mathematics. This refers to the cases where two different genes share a partial common DNA sequence. In all such cases, it should be possible to develop higher dimensional models. At later stages, when we find the underlying mathematical template of the distribution of genes, it seems reasonable that the template will be reflected in the pattern of RNA interference.

With these preliminary comments on the topological arrangement of different modularities in a typical genomic architecture (Fig. 4.1), let us delve deeper into some features of this modular architecture in the following sections.

4.12 Topological Model of the Relationship Between Structuralism and Functionalities

In the previous section, we looked at the overall topology of the proposed genomic architecture and particularly the placement of different modularities. In this section, we will try to outline the generic relationship between structuralism and functionalities. It is generic in the sense that it is not confined to modularity proposed earlier. It is also generic in the sense that it refers to the properties of functionalities and structuralism and not to any particular instance of these two properties. For instance, in the case of regulatory framework, it is possible to think of regulatory framework as a structural template as well as a functional template. Thus, the

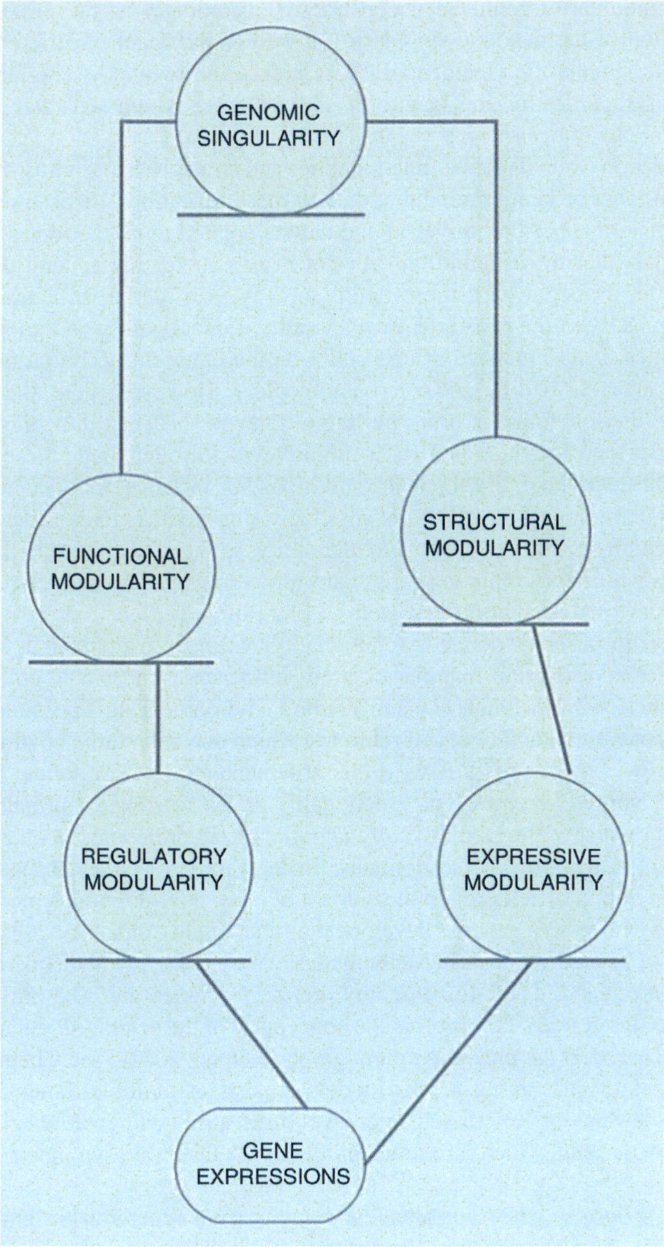

Fig. 4.1 Typical genomic architecture

regulatory modularity could be deconstructed accordingly as the product of the differentiation of the genomic singularity as well as the agent of differentiation of the genomic singularity. (Incidentally, this makes the regulatory modularity as a unique entity manifesting the feature of self reference which separates Life from other natural phenomena.)

In this section, we will define functionalities and structuralism slightly differently. However, it must be kept in mind that though these definitions are used only in this section, their semantics are universal and can be applied to all the topics discussed here. It is just that these definitions are not necessary for the general discussion. Therefore, let us begin with these definitions. We can define structuralism as a relationship between the constituents of an entity. Functionality, on the other hand, can be defined as an influence of one entity on the other entity. Prima facie, these definitions are self-evident and even tautological. However, from the semantic perspective, these definitions bring out the differences between the conceptions of functionalities and structuralism. It is intuitively clear that both these concepts require more than one entity for their definition. For instance, there cannot be any meaningful definition of the structuralism of any single entity. Even if we choose to impose a certain structuralism on a singular entity, we need to conceptualize several components within that entity to propose any structuralism. Therefore, structuralism is always construed as a unifying system of plurality.

Similarly, in order to define a functionality, we need a domain in which the system under investigation manifests certain influences. Any single entity cannot exhibit a functionality without any surrounding environment. It is possible to argue that there is one functionality of self reference which doesn't require anything else to manifest. However, there is a flaw in this argument. In order to define "Self," we need other "Nonself" entities. The functionality of self reference can manifest only when that influence is directed inwardly and that too in preference to other external entities. This is true for all the domains. In Psychology (Kline 1990), an infant develops the notion of self only in the context of other nonself entities present in the vicinity. In Genomics, a genome develops the functionality of self reference only in the context of its environment. In Mathematics, we can define lambda operators only when they are typed. Thus, functionality, per se, is contextual. The only possible exception to this conception of self is the conception of the cosmic singularity. Even here, as discussed in the preceding monograph (Chhaya 2022b, see Chapter 1), the cosmic singularity according to this model, consists of multiple dimensionalities. Therefore, the functionality of self reference in the cosmic singularity arises in the form of involutions which are conventionally defined as symmetry breaking processes.

Now, let us look at what separates functionality from structuralism. Though both require the presence of more than one entity for their definitions, there is one critical difference between them. A functionality is a process and structuralism is not. The functionality is dynamic and structuralism is static. In the language of formalization, we can define a functionality only in the context of the spacetime it occupies. However, we can define structuralism purely in the spatial description. The key point is that because we exist in spacetime itself, we often conflate the definitions of these two concepts. Moreover, while formalizing any structuralism, we employ cognitive processes. Therefore, this act of ostension gives us a crystallized conception of structuralism. However, we often overlook the fact that this conception of structuralism is a product of a process, viz., the cognitive process. This problem is compounded in the conception of functionalities. A functionality, as mentioned above, is a process. However, during its conceptualization, our cognitive faculty objectifies it. Thus, our conception of a functionality is that of an object and not that of a process. Thus, it is our epistemological compulsions of objectifying every entity while overlooking the underlying process of conceptualization that is responsible for the semantic ambiguities between what constitutes functionality and what constitutes structuralism. This is best exemplified by our attempts to define functionalities by ascribing them with certain structural templates. Thus, purely from the epistemological perspective, it is our cognitive compulsions to objectify everything that is responsible for our tendency to objectify and structuralize the processes.

It is important to keep in mind that this tendency to objectify everything has helped Science to understand the things which are otherwise not easily understood. However, the point is that this semantic overlap between our conception of structuralism and functionalities has made it difficult for us to formalize complex phenomena like Life. If only we could perceive time in the same way we perceive space. In such a scenario, maybe we could formalize processes differently. This rather long preamble to the distinction between functionalities and structuralism is necessary because these cognitive compulsions have prevented us from formalizing genomic architecture. In the context of the present discussion, if we could define structuralism as spatial arrangements and functionalities as processes connecting these spatial arrangements, we will obtain an intuitive layout of genomic architecture.

If our cognitive compulsions of objectifying processes could be formalized and if we could define processes in the dual forms of objects and processes, it would help us to deconstruct the genomic architecture as well. One such model of epistemology has been discussed in the preceding monograph (Chhaya 2022a, see Chapter 3). Therefore, we will employ it here to deconstruct the genomic architecture. The key point of this model is that it is possible to treat any object as a discrete entity and a continuum in a single framework. This is because this model connects the degree of discreteness with the dimensionality of the system. Thus, according to this model, our cognitive faculty perceives processes as objects from one dimensionality. At the same time, it perceives them as a continuum from another dimensionality. It is just that our ability to formalize using Mathematics (and Logic) also occurs from the dimensionality wherein we perceive processes as objects. Therefore, our formalisms try to formalize processes as discrete entities.

It is important to keep in mind that our cognitive faculty is capable of perceiving continua. Therefore, we can perceive not just processes, but also different types of continuities like quantum fields, rational numbers or even infinity. However, the dimensionality from which we perceive these continuities is incapable of formalizing these continuities. Therefore, whenever we wish to formalize these continuities, we revert back to the cognitive dimensionalities wherein we employ

Mathematics (and Logic) as discrete entities and their algebras. Therefore, it is intuitively clear why our formal conception of continuities is incomplete. This is evident from our theories of quantum fields, continuous functions and infinity.

In order to avoid this ambiguity about whether a given entity is a discrete element or a continuum, the proposed model offers a mathematical formalism that connects the two forms. Therefore, in this model, we can represent a process as a continuum and discrete entity simultaneously. More importantly, we can convert one form into another at will. Thus, this model is ideally suited to formalize genomic architecture because it can accommodate functionalities (which as we discussed above, are processes) and structuralism (which is an ensemble of discrete elements) in a single framework of a topological manifold. With this background, let us see how functionalities and structuralism can be formalized using this model. At present, we will eschew mathematical perspective and describe the model in the language of genomics.

The proposed model postulates that the degree of discreteness of any natural phenomenon is inversely related to the dimensionality from which we observe the phenomenon. Therefore, it is intuitively clear that when we conceptualize functionalities as objects, we are placing them in the lower dimensionalities (which also happen to be the dimensionalities we employ for formalizing these functionalities). Therefore, according to this model, these functionalities must exist as continua in higher dimensionalities. Once we accept this scenario, then it is axiomatic that the genomic architecture must postulate that functionalities occupy higher dimensionality as compared to the dimensionality of the structuralism of genomes. However, they need to be related to the underlying structuralism by some fixed mechanism, otherwise we will end up with functionalities having some mystical origins. This relationship between structuralism and functionalities can be defined by the operator of involution. It is important to keep in mind that since the proposed model postulates that the genomic singularity is the source of these different types of modularities, it is intuitively clear that the operator of involution connects a given functionality with its corresponding structuralism. Since this operator refers to an inward influence, it connects the functionality with the corresponding structural template. However, from the epistemological perspective, we cannot perceive this influence since our cognitive faculty operates from the dimensionalities wherein the structural template is formalized. Therefore, when we wish to define functionalities from a given structural template (say, while predicting new drug molecules), we need to define an inverse of an involution to formalize the functionalities.

Returning to the modularities mentioned above, this scenario offers an intuitive understanding of the genomic architecture. We can think of the genomic singularity as having evolved during biological evolution, into a more and more complex entity through the process which can be formalized as an involution. Thus, the separation would result in the distinction between functionalities and structuralism. The second stage would be that of separation of regulatory modularity from the rest of the functional modularity. In parallel, the structural modularity would have differentiated into expressive modularity and what is conventionally called noncoding regions of a genome.

It is possible to argue that this ontological primacy of functionalities over structural templates amounts to revival of some kind of design principles. However, there are two reasons why it is not the case. Firstly, the genomic singularity could have undergone different types of involutions (in the form of different environmental influences). Therefore, it was the environment that must have guided biological evolution toward its present epoch. Since the changes in the environment have been chaotic, there is no room for any design principles or even the teleological arguments in the proposed model. Secondly, this model proposes an inverse of an involution to represent the influence of the structural template on the functionalities. Therefore, as mutations kept changing the structural template, the corresponding changes in the respective functionalities would have arisen randomly. Thus, the proposed model does not allow any design principles or any teleological arguments. Admittedly, it postulates that functionalities enjoy ontological primacy and this has significant bearings on the semantics of biological evolution and even the Darwinian paradigm. We will return to this topic in later chapters. Incidentally, it must be mentioned that this provision of an inverse of an involution can also be used to formalize epigenetic influences. While the scenario presented here is plausible, it is still vague. In order to obtain clarity, we will discuss the case of regulatory modularity in the next section. This is because as mentioned above, it represents a feature of self reference, but also because it has a more easily comprehensible influence on the structuralism of the genome.

4.13 Topological Model of the Relationship Between Regulatory and Expressive Genome

As discussed in the previous section, the proposition that functionalities ought to occupy higher dimensionality as compared to the dimensionality of the corresponding structural template seems to be consistent with our intuitive understanding of the relationship between structuralism and functionalities. Admittedly, our intuitive understanding doesn't directly suggest any topological perspective of the relationship between structuralism and functionalities. However, we intuitively know that functionalities per se are more abstract that the structural template which is available to our sensory perceptions. Therefore, even in the absence of any formal proof, it is easy to accept that proposition. Given the present understanding of the genomic architecture, it is not possible to either prove or disprove the proposition that functionalities occupy higher dimensionality as compared to the dimensionality of the corresponding structural template. However, it is possible to establish that this proposition is logically consistent with the conventional perspective of genomics.

The best way to do so is to deconstruct the regulatory framework of the gene expressions using this proposition. Therefore, in this section, we will try to deconstruct the regulatory modularity vis a vis the relationship between structuralism and functionalities. This is necessary because as mentioned above, the regulatory modularity happens to be both a functionality and a structural template. It is a functionality because it influences the process of gene expressions. At the same time, it has its own structure in the form of a sequence of inhibitors and enhancers of gene expressions. Therefore, if we can demonstrate that this dual nature of the regulatory modularity is consistent with the proposition that the dimensionality of functionalities is higher than the dimensionality of the corresponding structural template, then we can legitimately claim that the proposed model is at least a serious scientific hypothesis.

The simplest way to begin deconstructing the dual nature of the regulatory modularity is to assume that it must occupy a dimensionality which is intermediate to the dimensionalities of the corresponding functionality and the structure template. To establish that this is correct, let us employ the Socratic method of *reductio ad absurdum*. Accordingly, let us assume that the dimensionality of the regulatory modularity is higher than the dimensionality of the functional modularity. Then, given the formalism of the proposed model, it should act as a regulatory framework of the functionalities. This amounts to suggesting that the regulatory modularity is a functionality of functionalities. While it is counterintuitive, there is already one such entity proposed in this model, viz., the genomic singularity. However, by definition, the genomic singularity cannot possess any structuralism of its own. Therefore, it is intuitively clear that the dimensionality of the regulatory modularity cannot be higher than the dimensionality of the functional modularity. Upon a little reflection, it is intuitively clear that using this method, it can be established that the regulatory modularity cannot occupy the dimensionality which is lower than the dimensionality occupied by the structural modularity.

This leaves one more option. Let us assume that the regulatory modularity occupies the same dimensionality of either the functional modularity or the structural modularity. In either case, it can be established using the method of *reductio ad absurdum* that it is logically inconsistent. This is because we know from our studies in genomics that while the structure template of genomes is manifest in the four-dimensional spacetime, the corresponding functionalities are required to be inferred from the subsequent phenomenology. Therefore, to the extent the regulatory modularity possesses the dual nature of being a functionality having its structuralism, it can't occupy either the dimensionality of structuralism or the dimensionality of the corresponding functionality.

This logical imperative that the regulatory modularity must be placed in the dimensionality which is at an intermediate level between the dimensionalities of functionalities and structuralism is actually a consequence of the semantics of the proposed model. The moment we postulate that the dimensionality decides the role of any entity present in that dimensionality, we are constructing a hierarchy of units of genomes. The placement of the individual elements of the genome follows axiomatically. Perhaps, it has something to do with the way our cognitive faculty fractionates reality into a hierarchy of elements. However, what characterizes the proposed model is that it asserts that the singularity and its fragments are connected with one another by a unitary mechanism. Because the proposed model employs a

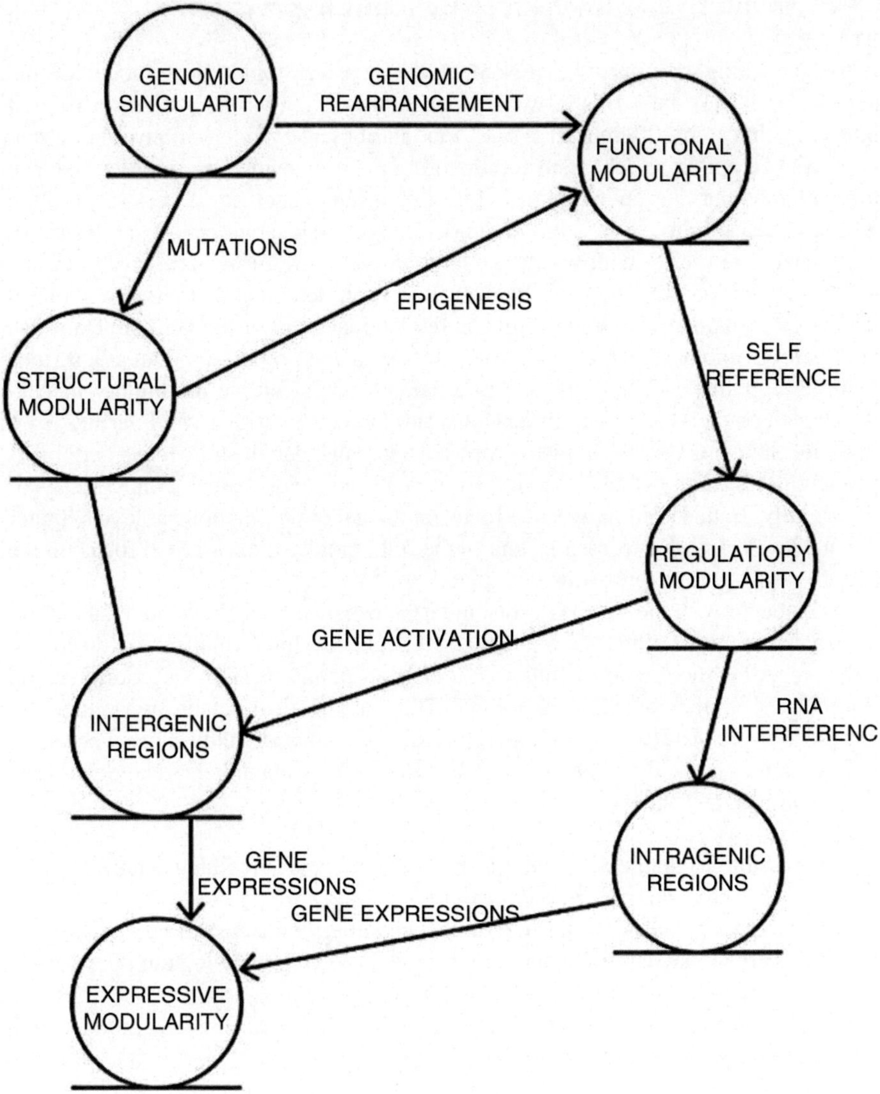

Fig. 4.2 Modular design of genomic architecture

topological perspective, it defines this unitary mechanism in the form of the changes in the dimensionalities.

Therefore, we now can map this relationship between different modularities in the language of dimensionalities. This is schematically shown in Fig. 4.2.

4.14 Involutive Formalism of Genomic Expression

In the preceding sections, we looked at the possible topological model of the genome. It was based on two propositions. Firstly, it assigns different dimensionalities to different types of modularities. Secondly, it postulates that there exists a unitary mechanism for changing the dimensionality in order to make different modularities operational. While this arrangement has a certain intuitive appeal, it doesn't make it a good model of genomic architecture, at least not necessarily. In order to endow this model with credibility, it is necessary to define the above mentioned unitary mechanism. If we can demonstrate that there exists a definitive mechanism that transforms the operational level of the genome from one modularity to another, then it is possible to consider the proposed model as a serious scientific hypothesis. Therefore, in this section we will outline the unitary mechanism which connects different dimensionalities (and their resident modularities) by a single mechanism. The mechanism proposed here must demonstrate that a genome is a functional system which operates from multiple levels either concurrently or sequentially. In this section, we will focus on the genomic functions at levels higher than the level of gene expressions and in the following section, we will focus on the level in which gene expressions take place.

Let us begin with the details of this unitary mechanism which we have labeled so far as an involution. While the mathematical details of this formalism are discussed in the preceding monograph (Chhaya 2022a, see Chapter 3), in this section, we will reformulate it in the context of genomics. The mathematical details are omitted not because they are different from the ones originally proposed, but because we wish to focus on the biological perspective of that formalism. The mathematical details can be summarized as follows.

1. It is possible to define any natural phenomenon as an involuted manifold. This manifold is characterized by two features.
2. The manifold consists of multiple dimensionalities connected to one another.
3. It is possible to switch from one dimensionality to another using the operators of involution.
4. Upon application, this operator brings about two changes. Firstly, it reduces the dimensionality of the system by one.
5. Secondly, it increases the complexity of the resulting dimensionality.
6. The features described in 4 and 5 are connected to one another by the mechanism by which the operator of involution brings about the change.
7. The operator of involution can be visualized as an inward folding of one of the dimensions of the manifold into the remaining dimensions of the manifold. Since one of the dimensions is folded inwardly, the dimensionality of the manifold is obviously reduced by one. Similarly, since every dimension of the manifold contains information, the information content of the dimension undergoing involution gets smeared over the information content of the recipient dimensions. This axiomatically leads to the increase in complexity of the resulting manifold.

With this brief description of the mathematical details of the formalism of the involuted manifold model, let us apply it to the hierarchy of modularities proposed in Fig. 4.1. The resulting schema is given in Fig. 4.2. Now let us understand what happens when a genome switches functioning from the genomic singularity to the functional modularity. To begin with, we will use the term genomic singularity in the functional sense. This is necessary because in the preceding and the following chapters, we have used the term genomic singularity to denote the evolutionary perspective as well. However, in this chapter, we will use the term genomic singularity to denote all the expressed and latent functionalities of the genome.

Let us assume that a genome has just come into existence and has begun functioning. (This is to be taken as an idealized scenario of a diploid cell produced during fertilization. It is idealized because in real life such a cell would be surrounded by the maternal RNAs and proteins.) In such a scenario, it is reasonable to think that the genome operates as a single entity. Therefore, all the functionalities of the genome, both latent and active, are governed by what is labeled here as a genomic singularity. The genomic singularity can be characterized by six features.

1. Genomic singularity is a repository of all latent functionalities of the genome.
2. Genomic singularity undergoes topological transformations in the form of involutions to give rise to functionalities.
3. Genomic singularity can undergo different topological transformations giving rise to different types of functionalities.
4. Every transformation of the genomic singularity can be defined as the changes in the dimensionalities.
5. Each transformation is formalized as an operator of involution which brings about the change in the dimensionality.
6. Each dimensionality represents different functionalities.

Admittedly, these features are generic in nature. This is because formalism of the involuted manifold model is applicable to a wide range of disciplines (including different types of genomes) and forms the foundation of yet to be articulated universal theory of science. Therefore, in this section, we will try to apply it to the modularities discussed above. To begin with, we can now think of the genomic singularity as a representation of a genome in its highest dimensionality. Apparently, according to this model, it doesn't have any manifest property other than its high dimensionality. For any functionality of the genome to manifest, the genomic singularity must reduce its dimensionality. This reduction in the dimensionality is essentially a topological process (formalized as involutions in this model). Each involution can be triggered by external factors (like maternal RNAs) or internal factors like topological constraints. Moreover, at every dimensionality (including the highest dimensionality representing the genomic singularity), there are more than one possible involutions. This ensures that the central requirement of the Darwinian paradigm, viz., nondeterminism is obeyed. In that sense, the proposed model doesn't alter the underlying Darwinian paradigm, but it merely provides a new template of diverse functionalities and their modular design.

As discussed above, the first involution results in the manifestation of the functional modularity. Here again, depending on the type of the operator invoked, we can activate different types of functional modules. From the conventional perspective, we can think of the first involution as a genomic rearrangement without any changes in the DNA Sequence of the genome. However, since the conventional perspective doesn't incorporate any higher dimensional perspective, we are forced to think of genomic rearrangement in terms of polyploidy or aneuploidy. However, this is not the case with the proposed model. It suggests that despite the appearance of random distribution of genes in different chromosomes (and even the number of chromosomes per se), there is a well-defined structural template of the distribution of genes. However, it ought to be perceptible from the higher dimensions. Therefore, the genomic transformations that the first involution can bring about may not manifest in the three-dimensional arrangements of DNA sequences. Moreover, these transformations from the genomic singularity to functional modularity need not always be triggered by external factors. They could arise from the topological considerations as well.

It is possible to question such a proposition. After all, all the known changes in the genomes are governed by thermodynamic processes. This is true not just for mutations occurring at the level of the individual nucleotides, but also for the duplications and inversions of DNA sequences. In each of these cases, there is always an agent that alters the changes in free energy and thereby enables these changes. Therefore, to suggest that genomic transformations can occur purely for topological reasons is speculative at best. However, there are two good reasons why topology should play a role in genomic transformations. Firstly, even though the higher dimensionalities may not possess time-like features (and therefore there is no thermodynamic impetus for the changes), topological compulsions can be as forceful. This is because according to this model, spacetime exists in multiple dimensionalities simultaneously. Therefore, its information content can determine the topological preferences. These higher dimensional tendencies of spacetime can be compared with the stereochemical forces that determine the conformational preferences. Moreover, as discussed in the preceding chapters, spacetime plays an active role in biological evolution and natural selection. Therefore, this suggestion is congruent with the earlier assertion about the active participation of spacetime in biological evolution. Admittedly, there is no evidence of either higher dimensional architecture of genomes. Nor is there any evidence of the active participation of spacetime in biological evolution. However, if these inferences provide cogent explanations for the otherwise unanalyzable features of the genomic architecture, these assertions must be taken seriously.

The second justification for suggesting that topological considerations influence the genomic transformations comes from a more fundamental aspect of biological evolution and natural selection. As discussed in the preceding chapters, the emergence of complexity during biological evolution and the subsequent natural selection has always been problematic in the Darwinian paradigm. This is largely because of the tenet of nondeterminism (Bonner 1988). Historically, Darwinism has resisted any design principles for the fear that it might resurrect teleological or Deistic explanations. Therefore, the emergence of complexity in biological evolution has remained unresolved. Therefore, this assertion that biological compulsions of the higher dimensional architecture of genomes provides an explanation for the emergence of complexity. More importantly, it does so without undermining the nondeterministic foundation of the Darwinian paradigm. This is because in topology, it is possible to have more than one equivalent configuration. Therefore, if the genomic singularity were to devolve into lower dimensional functional modularity, it could do so in more than one way, thereby eliminating any determinism whatsoever. At the same time, the physicality of spacetime (and that of information content present in spacetime) provides the necessary impetus for these changes.

We will end this section with a few remarks on this transition from the genomic singularity to the functional modularity and then to the regulatory modularity. As mentioned in the title of this chapter, the gene expressions, either singly or collectively, must be viewed as algorithms. This necessarily implies that individual steps are discrete operations. This is precisely what the proposed model offers. It offers a hierarchy of steps from the genomic singularity to the expressive modularity. Moreover, like a good algorithm, it offers a decision tree at each level. When the genomic singularity undergoes an involution, it can activate any of the functional modularities. This choice depends on the trigger that induces the genomic singularity to undergo transition to any of the functional modularities. Similar logic applies to the transition from the functional modularity to any of the regulatory modularities that is activated. The only difference in this routine is when any of the regulatory modularities acts on the expressive modularity. This difference arises because the activated regulatory modularity must act upon itself to synthesize gene regulators in the form of either activators or suppressors. Since this choice depends on the molecular signals available in the nucleus, this part of the algorithm resembles do-loops of the programming language.

This distinction is important because until a regulatory module is activated, the prior transitions are not governed by feedback mechanisms. From the genomic singularity to any of the functional modularities and from a functional module to any of its regulatory modularities, the mechanism is unidirectional. There are either chemical triggers or the topological constraints that bring about the changes in dimensionality and the corresponding activations of different levels of genomic architecture. It is only when any of the regulatory modules is activated that the feedback mechanism or the functionality of self reference is manifest. This is semantically significant and we will return to this topic in Sect. 4.17. Finally, it is intuitively clear that the proposed algorithmic approach is capable of parallel processing within and outside the activated modules. Considering the number of genes present in the genomes, it is self-evident that the resulting algorithm for the entire genome would be beyond manual construction. However, using modern programming techniques like machine learning, the approach outlined here is amenable to formalization. The only prerequisite for such a methodology is that it must provide a topological model for programming. This is precisely what the proposed model offers. Having laid out a general schema, it is now time to look at a typical example in some detail. Therefore, in the next section, we would look at a single gene expression scenario. Obviously, we will deal with this in an abstract manner. The specific examples of the individual genes and their expressions will be discussed in the future communications.

4.15 Involutive Formalism of Gene Expression

For the sake of simplicity, we will try to idealize the scenario of single gene expression by assuming the following features of the hypothetical gene.

1. This gene, called IG (idealized gene) is situated in a cluster of genes.
2. However, none of these genes share overlapping DNA sequences with IG. This also includes activator DNA sequences as well. In other words, IG doesn't have any functional or structural overlap with its neighbors.
3. The gene IG is activated by a single chemical messenger (AM) that attaches itself to the DNA sequence (US).upstream of the DNA sequence of IG.
4. Similarly, the gene IG is suppressed by a single chemical messenger (SM) which also operates by attaching itself to the upstream DNA sequence US.

Admittedly this is a highly idealized scenario and perhaps simplest of the operon models. However, it will suffice for the present discussion. We are not interested in the chemical details of any of the agents involved in this scenario. We are interested in the operating logic behind this algorithm. We are familiar with the conventional scenario of a typical gene expression (Weinzierl 1999). A signal attaches itself to the DNA sequence upstream of the DNA sequence of a gene assembles the translation complex and the gene expression begins. Admittedly, the corresponding suppression of the gene expression in the conventional scenario has multiple options. The suppressor could attach upstream as idealized above. However, it could act in a couple of other ways. It could interact with the messenger RNA and stop the protein synthesis. Alternatively, it could block the DNA sequence of the gene. However, for the sake of simplicity, we will adhere to the idealized scenario mentioned above. Once we have outlined the new model, it will be intuitively clear that these different pathways of gene suppression can be accommodated in the proposed model.

Let us now look at the proposed model of gene expressions as simplified for the above mentioned idealized scenario. According to this model, the regulatory module could operate in two ways, both involving the process of involution. Firstly, it could undergo further involution to become a structural template for gene expressions. This is analogous to the conventional method of synthesis of gene expression activators and suppressors (including the pathways involving RNA interference (Howard 2013)). However, this mode of gene regulation requires a self-reference. This may have its origin in the prebiotic world or what is popularly called the RNA world (Yarus 2010). The second mode of gene regulation in which a regulatory module can participate is via the stereochemical transformations. In this pathway, there is an involution in which the higher dimensional configuration undergoes involution leading to the corresponding stereochemical conformational changes. In such a scenario, the regulatory module, due to stereochemical proximity, influences the structural template of the genome and thereby controls gene expressions. This is perhaps illustrated by the conventional phenomenon of chromosome territories (Fritz 2014).

There are three takeaways from this scenario. Firstly, from the evolutionary perspective, the regulatory modules had evolved a functionality of self reference. This could be in the form of self replication and autocatalytic functionality. It is important to keep in mind that this emergence of the functionality of self reference is purely due to stereochemical forces. However, as the complexity of the genome increased, it also evolved higher dimensional configurations. These higher dimensional configurations enabled natural selection to partition different functionalities into different dimensionalities. This is the basic reason why the most fundamental dualities, viz., the separation of structuralism and functionalities, the separation of genotypes and phenotypes, and the separation of DNA and RNA came about. In other words, as the molecular complexity of RNA molecules increased, it acquired higher dimensional configurations. This scenario is consistent with the notion that spacetime plays an active role in natural selection. It is from these higher dimensional configurations that modularization and the emergence of dualities originated. This eliminates any teleological or design arguments.

The second takeaway from this scenario is that the process of involution is generic in nature and it happens in biological evolution just as it manifests itself in other natural phenomena. Thus, the mechanism of involution provides a way to unify living organisms with all the other natural phenomena. The third takeaway from this discussion is something that is of immediate interest in the present discussion. This refers to the unitary mechanism of involution that links various levels of genomic architecture into a single framework. In the following section, we will delve deeper into this unitary mechanism that links genomic singularity to the DNA sequence of the genome.

4.16 Unitary Biological Algorithm

One of the major obstacles in formalizing the template of genomic architecture has been that there is no semantic or structural theme which could hold diverse features into a single framework. It is possible to argue that as far as semantics is concerned, this is not a valid argument because we already have an in-depth deconstruction of the Darwinian paradigm (Grene 1986). Similarly, it can be argued that even from the structural perspective, we have enough depth of knowledge in the form of phylogenetics to think of a unifying structural template (Bromham 2008, see Chapter 5). However, these arguments represent misplaced optimism. Let us see why. As discussed in the preceding chapters, the Darwinian paradigm has historically failed to explain the emergence of complexity during biological evolution and natural selection. Therefore, to think that the Darwinian paradigm, by itself, can help us to formalize genomic architecture is unrealistic, at least in the sense we understand that the Darwinian paradigm. Admittedly, the Darwinian paradigm provides the necessary semantics of evolution and natural selection, but it lacks a structural template of its own.

Similarly, to think that phylogenetic models, by themselves, can help us to formalize genomic architecture is also unreasonable. This is because phylogenetics (Bromham 2008) (or at a fundamental level, Molecular biology) is a reductionist discipline. To assume that every biological functionality can be reduced its molecular structuralism is unwarranted. It is true that molecular structuralism should be the basis of every domain of biology. However, it will be a category mistake to think that every biological functionality can be reduced to its molecular perspective. The basic premise of phylogenetics is that the genome is essentially its DNA sequence (along with the accompanying proteins). In reality, phylogenetics should be concerned about the relationship between structuralism and functionalities and their evolution. However, because we have opted for molecular structuralism, we think of genomic architecture in the language of stereochemistry. Instead, if we decide to focus on the relationship between structuralism and functionalities and their evolution, we ought to look at topology and not stereochemistry. Thus, it is imperative that we must formalize genomic architecture on a framework which incorporates two principles, viz., the generic relationship between structuralism and functionalities and a unifying mechanism that connects different types of structural complexity with different types of functionalities. In other words, a good model of genomic architecture must expand the conventional molecular perspective into a wider structural perspective and expand the chemical perspective of functionalities into a wider perspective of phenomenology. The proposed model offers precisely such a framework.

However, in this section, we will look at the second imperative of formalizing a unitary mechanism that connects structuralism and functionalities in a broader sense mentioned above. The proposed model connects structuralism (and not just molecular structuralism) with phenomenology (and not just chemical functionalities). It postulates that functionalities and structuralism are conjoined twins. Their manifestation depends on the dimensionality. Thus, what happens to be a functionality can become structuralism at a lower dimensionality. This is best exemplified by the regulatory modularity as discussed in the preceding sections. Though this relationship between structuralism and functionalities is generic, it doesn't manifest in every natural phenomenon. This is because of the limitations of the range of dimensionalities that our cognitive faculty can process. For instance, a similar relationship between a quantum and a continuum exists in the case of quantum fields (see discussion in (Chhaya 2022c, see Chapter 1)). As discussed in the preceding monograph dealing with the nature of quantum reality, the dimensionalities in which a quantum and its corresponding continuum exist are amenable to cognition. However, it is only the dimensionality of a quantum that is amenable to formalization, but that of the corresponding continuum is not.

However, Life is perhaps the only natural phenomena that have the dimensionality of structuralism and the dimensionality of functionalities that are within the range of dimensionalities that our cognitive faculty operates from. More importantly, both these dimensionalities (of structuralism and functionalities) are capable of generating self reference. There are two prime examples of this, viz., the genome and our cognitive faculty. The key point in the present context is that this self reference, in the form of linking two dimensionalities of structuralism and functionalities with one another through the operation of involution is generic, at least in the case of genomic architecture. Thus, different dimensionalities in the genomic architecture can communicate with one another through the operator of involution which formalizes the changes in the dimensionalities. Thus, there is a two-way communication between these different dimensionalities of the genome. When a functionality present in the higher dimensionality communicates with the structuralism of a lower dimensionality, it does so by undergoing involution. The activation of gene expressions is a typical example of this. Similarly, when a structuralism at a lower dimensionality communicates with the higher dimensional functionality, it does so by undergoing an inverse of an involution. Epigenetic processes are typical examples of this.

It is important to realize that in order to formalize the generic nature of the mechanism, we have to abandon the geometric approach of stereochemistry and adopt a topological approach involving dimensionalities. Similarly, we have to abandon the chemical reactivity and adopt functionalities. Admittedly, it is difficult to accept this transition because all of us have been trained to think in the molecular perspective. However, if it can be demonstrated that the proposed unitary mechanism of changes in dimensionality leading the changes in the functionality is consistent with the core semantics of biological evolution and natural selection, it would enable us to accept this formulation of the genomic architecture. It is important to keep in mind that any such unitary mechanism can help us in formalizing the process of biological evolution and natural selection in the form of an algorithm. It is intuitively clear that like any typical algorithm, the process of natural selection works at many levels in parallel. Therefore, in the next section, we will discuss the semantics of biological evolution and natural selection in the context of the proposed unitary mechanism.

4.17 Semantics of Biological Algorithm

It must be admitted at the outset that it is impossible to articulate the semantics of the Darwinian paradigm. The voluminous literature on this topic is an indication of this impossibility (Grene 1986). Therefore, we will take a modest position and instead select a couple of semantic propositions that are universally accepted as being prerequisite for the Darwinian paradigm. Secondly, even in the case of these two semantic propositions, we will discuss them only in the context of their relevance to genomic architecture. For the present discussion, we will focus on two semantic propositions, viz., nondeterminism and complexity. We will discuss whether the proposed model of genomic architecture undermines Darwinian nondeterminism or not. Secondly, we will try to deconstruct the conception of biological complexity and how the proposed model explains the emergence of complexity. Admittedly, even these narrowed down topics are extensively articulated in literature. Therefore, in this section we will confine ourselves to a brief discussion. We will return to these topics in detail in the following chapters.

To begin with, nondeterminism is at the heart of the Darwinian paradigm. As discussed in the preceding chapters, historically, the emphasis on nondeterminism was to counter any teleological or deistic explanations for biological evolution and natural selection (Hodge and Redick 2009). Later on, with the advent of population genetics (Provine 2001), nondeterminism found a structural expression for itself. This formalism also added heft to nondeterminism implicit in the Darwinian paradigm. In the next paradigm shift to molecular biology, the random nature and causes of mutations provided further confirmation of the primacy of nondeterminism to the Darwinian paradigm. Therefore, if there is a new model, it must either reinterpret nondeterministic semantics of the Darwinian paradigm (Bonner 1988) or it must offer valid reasons for rejecting it. As mentioned above, one of the reasons why we have not been able to formalize genomic architecture is that we are apprehensive that such universal architecture might undermine Darwinian randomness. It is in this context that we must evaluate the proposed model on two grounds. Firstly, the notion of a universal architecture for all the genomes might compromise the nondeterminism implicit in the Darwinian paradigm. Secondly, the universal mechanism proposed above might be semantically incompatible with the conventional perspective of the Darwinian paradigm.

Let us begin with the postulate that genomes, per se, have a common architecture. Prima facie, this has always been implicit in our studies in Phylogenetics. The moment we accept the conception of LUCA (Last Universal Common Ancestor) (Bard 2016, see Chapter 9), we accept that the primordial structuralism of the genome of LUCA too must have been passed to the later organisms, albeit modified by natural selection. Thus, the question is not about whether there is a universal architecture of genomes or not? Rather, the question is whether any proposed genomic architecture conforms to the semantics of the Darwinian paradigm? Let us see whether the proposed model stands up to this yardstick. As mentioned in the preceding sections, there is one operator of involution which links different levels of genomic architecture. Therefore, it might introduce a certain degree of predictivity in the outcomes, thereby undermining the inherent randomness. However, this is not the case. Firstly, as discussed in the preceding monographs, mathematics underlying this model, provides for multiple involutions and even multiple types of involutions. Therefore, whenever there is an involution during genomic transformations, there will always be multiple outcomes. The same logic applies to different modularities. Each modularity contains several modules. Therefore, there are multiple ways in which even a single involution will lead to multiple outcomes. Thus, the proposed model does not undermine the nondeterministic semantics of the Darwinian paradigm.

In fact, it is possible to argue that if there are multiple outcomes at every level of genomic architecture, then how effective is this architecture? We might as well abandon this architecture. The answer to this question is that the universal architecture implicit in the proposed model is not optional. It manifests because of the nature of the complexity of genomes. In fact, as discussed in the preceding monograph

(Chhaya 2020), this structural template is applicable to all natural phenomena and not just biological evolution. The merit of the proposed model lies in its three features. Firstly, it allows us to formalize Life as any other natural phenomenon. Secondly, it offers a certain degree of predictivity to genomics. Finally, it explains the emergence of complexity during biological evolution and natural selection, something that is not available in the conventional perspective of the Darwinian paradigm. Similarly, it is possible to argue that the existence of any such putative unitary mechanism, by itself, cannot undermine Darwinian randomness. To use Darwin's own phrase "descent with modifications" is also reflective of the internal paradox of the Darwinian paradigm. According to this model, the unitary mechanism refers to the descent, but not necessarily to predetermined descent. Thus, even if natural selection operates using a certain unitary mechanism, the outcomes are not singular. These outcomes depend on the starting point and not on the mechanism.

This brings us to the second semantic propositions of the proposed model. It suggests that complexity arises naturally in any system obeying involuted architecture. The key point is that the conception of involution points toward inward flow of information. Therefore, according to this model, Life has evolved due to the ability of the earliest living organisms to absorb information about their surroundings and retain it. However, because of the Cartesian influence (Cottingham 2008), we have assumed that this information content of surroundings is an abstract entity. However, once we accept that information content is a physical entity. It is intuitively clear that biological evolution could not have taken place without the ability of the earliest living organisms to absorb the information content of their surroundings. Thus, the basic principle of biological evolution is the inward flow of information. Once we accept it, it is axiomatic that biological evolution will inevitably lead to more and more complex living organisms. The key point is that since there is a plurality of the information content of the surroundings, there will be multiple ways in which this information is absorbed by the living organisms. This ensures that outcomes are nondeterministic. However, since living organisms have only a limited number of absorbing information content of the surroundings, there will always be some common structuralism of these multiple outcomes.

This has always been implicit in the conventional perspective of the Darwinian paradigm. However, since we thought of information content as being abstract, we had difficulty in formalizing Life and its evolution. The proposed model postulates that information is a physical entity and it gets incorporated into living organisms. We might wonder about the nature of information content that can physically be absorbed by the living organisms. According to this model, this information comes from spacetime itself, in the form of its fine structure. Once the living organisms became complex, the inward flow of information content came from within the genomic architecture. Thus, once LUCA evolved, further biological evolution was simply reorganization of the genome by inward foldings, albeit augmented by natural selection and the changes in the environment. Thus, the proposed model explains the emergence of complexity during biological evolution and natural selection, in an intuitive manner. However, this scenario requires that we revisit the semantics of the conventional perspective of the Darwinian paradigm and reinterpret it. We will briefly do this in the following two sections.

4.18 Conventional Semantics of Darwinian Paradigm

Once again, we would follow the caveats mentioned above. The topic of the semantics of the Darwinian paradigm is vast and well-discussed in literature (Grene 1986). Therefore, we will not even try to summarize it here and instead, we will take it as having been read. For the sake of brevity, we will focus on two aspects of the Darwinian semantics, viz., the neutral or near-neutral rate of mutations (Kimura 1983) and its role in natural selection and the units of selection. Admittedly, even these two topics are well-researched and have been articulated in a nuanced manner. Therefore, we will restrict ourselves to the implicit semantics behind these two topics as conventionally understood. We will also point out certain semantic ambiguities of the conventional wisdom in this section. In the following section, we will try to deconstruct these semantic ambiguities using the proposed model.

Let us begin with the concept of neutral rate of mutations. As discussed in the preceding chapters, the Darwinian paradigm has survived several paradigm shifts and has found newer versions after each of these paradigm shifts. This history of paradigm shifts is well-documented. The topic of neural rate of mutations can be placed in the era when genetics was being used to deconstruct natural selection. However, this topic precedes the molecular biological paradigm and the subsequent genomics era. This topic of the neutral rate of mutations (and population genetics in general) is significant because it refined the semantics of natural selection in the absence of any molecular perspective. Moreover, by application of statistical methodology, it connected Darwinian randomness with the essential randomness of the universe at large. The success of population genetics convinced most of us about the centrality of nondeterminism in the Darwinian paradigm. This also made us realize the metaphysics of randomness. As an aside, it is important to keep in mind that when a natural phenomenon obeys a particular statistical model, does it do so out of a more fundamental aspect of reality or is it just a coincidence (albeit a statistical coincidence!!)? The metaphysics of statistics is troublesome. If Nature follows the structural template of the statistical models, does she have a structural template which coincides with the structural template of these models? The key point is that in spite of its remarkable success, the discipline of population genetics (including the theory of neutral rate of mutations) leaves the origin of Darwinian randomness unsolved.

Returning to the present discussion, the neutral rate of mutations tries to answer the question about the nature of mutations. Prior to the advent of this theory, the conventional wisdom about mutations was that harmful mutations will be weeded out during the course of natural selection. Similarly, the beneficial mutations will spread through the population. Moreover, as mentioned above, since the molecular perspective was not available, the emphasis was on the frequency distribution of phenotypes in the population. The inherent beauty of this concept of the neutral rate of mutations is that it translates the core of the Darwinian randomness into the most intuitive proposition that mutations are neither good nor bad. Incidentally, this is precisely what molecular perspective suggests. Mutations are oblivious to their consequences. It is the process of natural selection that assigns the attributes of being good or bad to these mutations. In other words, it is the outcome of natural selection that decides the merit of mutations. This nonjudgmental or value agnostic semantics of this theory captures the heart of the Darwinian paradigm.

Admittedly, this theory has certain problems with the quantitative aspects of genetic drift and its relationship with the size of the population. Therefore, a modification of this theory, in the form of a near neutral rate of mutations has been put forth (Hartle and Clark 1989). However, this theory has a slightly different interpretation of natural selection than the conventional perspective of the classical view of the Darwinian paradigm. It relegates the phenomena of positive and negative selections, which are central to Darwin's theory, to a secondary status. According to this theory, it is the rate of mutations that alone decides natural selection (albeit subject to the size of population). In a sense, it brings out the inner contradiction of the classical Darwinian perspective. The conventional view has been that the process of natural selection employs both positive and negative selections during biological evolution. At the same time, according to the classical view, the process of natural selection is random or nondeterministic. However, as this theory points out, for natural selection to be nondeterministic, it must be based on a neutral rate of mutations. In other words, mutations occur in a truly blind fashion. It doesn't depend on the survival quotient of the resulting outcomes. Thus, our conception of positive and negative selections are based on post facto judgments. Nature has nothing to do with it. Nature allows mutations of genes irrespective of their benefits or handicaps. Natural selection is merely a consequence of the prevailing environment. Thus, this theory reinforces the central tenet of nondeterminism in natural selection but in a slightly different way.

Now, let us look at the second feature of the units of selection. At the time when Darwin put forth his ideas on biological evolution and natural selection, he focused on the observable traits of different species. These traits are called phenotypes in our current jargon. This was because the science of genetics was not established. However, as our understanding of genetics expanded, scientists found a new context of the Darwinian paradigm (New Synthesis (Delisle 2021)). Since then, there is no unanimity about the units of selection. Apparently, the process of natural selection operates on phenotypes and not on the corresponding genotypes because the environment interacts with phenotypes and rarely, if ever, with genotypes. As discussed in the accompanying chapters, the need to have two frameworks of genotypes and phenotypes operating in parallel is enigmatic. There must be some semantic implications of this duality of units of inheritance and the units of selection. However, its complete import has eluded us so far.

However, as molecular biology replaced classical genetics, this problem was further aggravated. Firstly, genes are no longer thought to be discrete sequences of DNA. There is a considerable degree of overlap between genes. Secondly, we have now realized that DNA sequences are also subject to different types of environment in the form subcellular milieu. Therefore, they ought to be subject to natural selection. The key point is that if the classical Darwinian perspective is about competitive survival among different species, the same logic should apply to any group of competing entities and not just biological species. Therefore, it is not impossible for different genes to compete among themselves. At the other extreme, we have the phenomena of different groups of individuals competing among themselves. Therefore, there is no reason why we can't have natural selection for genes or natural selection of groups. This confusion arises from the semantic ambiguities of the classical Darwinian paradigm. If as originally conceived, natural selection would manifest whenever there are competing entities fighting for limited resources (Flew 2017, see Part III), why should the Darwinian paradigm be restricted to different biological species? Apparently, literature is replete with such theories of group selection, intergenic selection and even neuronal selection. Therefore, it is imperative that we must try to deconstruct the underlying semantics of the units of selection. It is important to keep in mind that both aspects of natural selection, viz., what drives natural selection and what gets selected, have remained enigmatic in the conventional perspective of the Darwinian paradigm. Therefore, it will be interesting to find out whether the proposed model has anything to offer, particularly because of the algorithmic model of the genome. This will be discussed in the next section.

4.19 Revised Semantics of Darwinian Paradigm

In the previous section, we looked at two semantic propositions of the conventional perspective of the Darwinian paradigm, viz., primacy of mutations and the units of selection. In the preceding sections and in the preceding chapters, we also discussed the emergence of complexity during biological evolution. Therefore, in this section, we will look at these three topics from the perspective of the proposed model. However, we will restrict ourselves to the deconstruction of these topics from the limited perspective of the genomic architecture proposed above. A more general discussion on these topics will be discussed in the following chapters.

Therefore, let us look at how the proposed model of the genomic architecture views the primacy of mutations vis a vis the primacy of positive and negative selections. Let us begin with the question of what is of primary importance in natural selection, mutations or the positive/negative selections? It is important to keep in mind that neither the conventional perspective (Grene 1986) nor the theory of neutral rate of mutations (Kimura 1983) deny that these three phenomena operate in natural selection. What they differ in is which of these three phenomena are the driving force behind natural selection. According to the model of genomic architecture proposed above, mutations (including the higher dimensional genomic transformations which are not postulated in the conventional perspective of the genomic architecture) are prior to natural selection. Therefore, the positive or negative selections must operate only after the structural changes via mutations have taken place. More importantly, according to this model, mutations (including the higher dimensional genomic transformations) are based on topological compulsions. Therefore, the nature of subsequent positive or negative selections will not influence the topological transformations. Thus, the proposed architecture is congruent with the theory of neutral rate of mutations. Admittedly, there are broader issues with the Darwinian paradigm, including the positive and negative selections in which the environment plays an important part (possibly via epigenetic processes). However, these issues will be discussed at a later date. As far as the genomic architecture is concerned, the proposed model sides with the theory of the neutral rate of mutations. However, there is a different nuance to this primacy of mutations. According to the conventional perspective, mutations are random and governed by thermodynamic principles. However, the proposed model goes one step further and suggests that the mutations are governed by the inherent template of spacetime and the thermodynamic connotation is one of the many influences of the structural template of spacetime.

Let us now look at the question of the units of selection. Since we are dealing with the genomic architecture here, we will exclude the topic of group selection from the discussion. Instead, we will try to understand what could be thought of as a unit of selection. Prima facie, the proposed model defines genomic modules as units existing in different dimensionalities. Even within a given module, we can have multiple dimensionalities each representing different structural units. Thus, the overall architecture must be visualized as a nested hierarchy of dimensionalities. There are two important consequences of this architecture that are relevant to the present discussion. Firstly, there are independent structural units in every dimensionality. Secondly, due to the involutive nature of interactions between two different structural units, a higher dimensional unit can influence several units present in the lower dimensionality. Thus, there will always be competition in each of these dimensionalities. Therefore, it is intuitively clear that there will be natural selection at each of these dimensionalities. Thus, the unit of selection is not a predefined unit. The identity of the unit of selection depends on the dimensionality in which natural selection operates. Therefore, according to this model, there are multiple definitions of the unit of selection. It is possible to argue that this rationale can also be applied to group selection as well. In fact, this rationale ought to be applicable to the entire domain of social biology (Alcock 2001, see Chapter 3). However, we will refrain from commenting on this generalization here.

Now, let us look at the third problem of the emergence of complexity during biological evolution. Admittedly, we have discussed some aspects of this topic in the preceding chapters and we will return to this topic in the following chapters as well. In this section, we will confine ourselves to the complexity genomes, both functional and structural. As discussed elsewhere in this monograph, the emergence of complexity during biological evolution has been problematic in the Darwinian paradigm. This problem can be traced back to two semantic propositions of the Darwinian paradigm. Firstly, the Darwinian paradigm has consciously avoided any teleological explanations or design principles being a guiding force behind natural selection. Secondly, the Darwinian paradigm has insisted upon the essential randomness behind natural selection. Therefore, to the extent we perceive complexity as a measure of design, it is difficult to justify its emergence during biological evolution, at least while believing in the conventional wisdom behind the Darwinian paradigm. Therefore, let us see how we can justify the emergence of complexity during biological evolution and natural selection using this model. Admittedly, we cannot discuss the evolutionary perspective of genomic architecture here due to lack of space. However, we will discuss here how complexity arises within a genome. The logic behind this choice is simple. If we can demonstrate that complexity arises naturally within a genome itself, then at least in principle, it should arise naturally during biological evolution of genomes.

According to this model, every genome contains the highest dimensionality that is assigned the least degree of complexity. However, due to the involutive nature of genomic transformations, a genome can express itself at several lower dimensionalities simultaneously. Now according to this model, every involution leads to an increase in complexity. Therefore, in principle, every genomic expression requires an increase in complexity. Thus, the genomic singularity which is the least complex entity becomes operational only when its complexity is increased via involution to give rise to functional modularity. The same logic applies when the functional modularity influences the regulatory modularity. Thus, the functioning of the genome is defined by the increase in its complexity. The key takeaway is this. The source of complexity is the information content of the genome itself. Therefore, complexity doesn't emerge de novo. Rather it manifests de novo every time a genomic functionality is manifest. In other words, complexity never emerged during biological evolution, only it became manifest during biological evolution. In the following chapters, we will discuss the evolutionary perspective of the emergence of complexity.

4.20 Biological Algorithm as a Special Case of General Involuted Algorithms

As discussed in the preceding chapters, science has failed to formalize Life. Therefore, in spite of our knowledge of biology, we cannot perceive Life as a natural phenomenon. Life seems to be an exception among natural phenomena. This failure has given room for all varieties of creationist or deistic explanations to flourish. Historically, it was Darwin's theory that tried to conceptualize Life and its evolution as a natural phenomenon. Therefore, it was inevitable that it resisted teleological and deistic explanations right from its conception. However, even today, it is difficult to justify that Life is like any other natural phenomena. Primarily, there are two reasons for it. Firstly, due to the influence of the Cartesian paradigm (Cottingham 2008), science has not been able to formalize human consciousness. True, we know a lot about neurology and psychology, but we have not been able to formalize consciousness. The second reason why we can't justify the naturalism of Life lies in our failure to formalize the origin and evolution of Life. Therefore, the Darwinian paradigm and its correct interpretation is crucial to demonstrate that Life is like any other natural phenomenon. Some aspects of this reinterpretation of the Darwinian paradigm was outlined in the preceding monograph (Chhaya 2020, see Chapter 8) and some more aspects will be discussed in a future communication. It was this objective of naturalizing Life that was one of the motivations behind writing this monograph.

It is intuitively clear that if we can formalize genomic architecture, it would constitute a major advancement in naturalizing Life and understanding biological evolution and natural selection. When we try to find out why we have not been able to demonstrate that Life is like any other natural phenomena, it is intuitively clear that this is because we have not been able to demonstrate that the processes occurring in a living organism are actually like any other natural processes. This might appear to be a false assertion because we have been able to formalize most of the processes to be ordinary chemical and electrochemical processes. However, upon a little reflection, it is intuitively clear that while at the ground level, every cellular process is indeed a chemical or electrochemical process, their control and coordination is something that still defies our analytical skills.

For instance, in the case of genomics (Pevsner 2015), we have a detailed knowledge of how a gene is transcribed or translated. We know the complexity of the transcription machinery and the proteins involved in it. Similarly, we also know some of the details of how polygenic or pleiotropic gene expressions are fine-tuned (Reavey 2013; Lozano 2017). However, as we move higher in this hierarchy of controls, we find that our analytical skills are inadequate to explain these higher levels of control. A similar argument can be put forth to explain our failure to connect the neurological paradigm to the cognitive paradigm (DeVos and Pluth 2015, see Part I). This certainly points toward the inadequacies of our conventional analytical methods. This monograph and the preceding monographs are aimed at redefining our analytical methodology. The model of genomic architecture proposed here must be evaluated in this context.

Returning to the present discussion, if we wish to demonstrate that Life is like any other natural phenomena, it is intuitively clear that the processes occurring in living organisms must be demonstrated to be like any other natural processes occurring in other natural phenomena. In order to achieve this unification, we ought to employ a universal abstraction of natural processes. In view of the fact that there is an increasing tendency to formalize natural processes as computations, it seems appropriate to employ an algorithmic template for the unification of living processes with other natural processes. When thinking of biochemical processes operating in a biological cell, this is not necessary because it is self-evidently true. Therefore, in order to extend this computational perspective to genomic processes, the above mentioned model of genomic architecture is proposed. Therefore, in this section, we will try to deconstruct how this model of genomic architecture is actually an algorithm and how a genome can be thought of as a computational device. Admittedly, these are sweeping assertions. However, instead of going into the polemics, we will focus on four specific topics, viz., genome as a computational device, structural modularity as hardware, functional modularity as software, and genetic disorders as halting problems. Admittedly, each of these four topics requires a separate chapter (if not a separate monograph) for itself. However, we will restrict ourselves to a few general observations on these topics.

Let us begin with the notion that a genome is a computational device. At the outset, it must be admitted that our current conception of computation, as formalized in the universal Turing machine, needs to be reinterpreted in view of quantum computation. However, this topic has been dealt with in the accompanying monograph dealing with the universal computation theory (Chhaya 2022e). However, in the present context, we will adhere to the conventional view as articulated in the universal Turing machine. All the same, it is interesting to note that the proposed new template of the universal computation is based on the same model that has been employed in formalizing genomic architecture in this monograph. The model was originally proposed to formalize naturalistic epistemology of mathematics (Chhaya 2022a). It is only in hindsight that it is obvious that be it an epistemological process or a genomic expression, both deal with information transfers and transformations.

When we think of a genome as a computational device, it is intuitively clear that it is not easy to visualize all the genomic expressions as simple operations. Even if we were to map out individual genes present in a genome and define their locations, we still can't formalize their collective behavior. This is because we have realized that some of the processes involved in gene expressions are nonlocal, nonlinear and parallel. Therefore, it is the sheer complexity of enumerable gene expressions that poses a real challenge. However, let us assume that we have access to very sophisticated computers (say, quantum computers) with infinitely large computational capabilities. The question is will we be able to formalize a genome as a computational device? The key point is that, what prevents us from thinking of a genome as a computational device is not really the lack of computational capabilities (Of course, this is a serious handicap, but it is only a short term problem.), but it is also a lack of semantic clarity. As mentioned above, somewhere deep within our conception of Life, there exists a holistic perspective. It is this cognitive bias that prevents us from thinking about Life as a natural phenomenon and genome as a computational device. This bias of treating holistic approach as something mystical, is at the core of our theories of science. As discussed in the preceding monograph (Chhaya 2022d), this arises because the cognitive processes responsible for our analytical skills are only a part of our cognitive processing. We can perceive reality through multiple modes of cognition and logical or mathematical mode is one of them. We arrive at pre analytical propositions through noetic processes. In fact, these noetic perceptions precede and influence our analytical mode of cognition. Therefore, the above mentioned bias toward a holistic paradigm of Life is an outcome of noesis and should be seen as a guiding force rather than a conflicting force in formalizing phenomena that appear to be too complex to our analytical skills.

Having said that, let us see how a genome can be thought of as a computational device. A computational device can be thought of as a system having a logical operator, data that needs to be processed using the given logical operator and a place for storing the output. Once we accept this simplistic definition, it is intuitively clear that all the three requirements are met within a genome. We can think of DNA sequences as data to be processed, functional modularity as a logical operator and the messenger RNA as a place to store the outputs. The only problem is the sheer complexity of gene expressions. However, the proposed model offers a way out of this impasse. By formalizing different modularities in different dimensionalities, the proposed model converts a genome into a parallel series of computational devices. More importantly, to account for the inherent complexity, the proposed model postulates that these different dimensionalities also interact with one another through a computational protocol. Thus, intradimensional and interdimensional operations are accounted for in this model. More importantly, each of these operations are defined by a unitary mechanism. Thus, prima facie, there should not be any problems with conceptualizing a genome as a computational device. It is tempting to think that just as a Turing machine paradigm operates on the Boolean algebras of discrete elements, a genome operates on the Boolean algebras of discrete dimensionalities.

However, there is one fundamental problem with this reasoning. In a conventional computer (or even with a quantum computer), there exists a definitive boundary between the logical operator and the data to be processed by that logical operator. In the case of a genome, this is not the case. The logical operator and the data to be processed are both one and the same. This is precisely what prevents us from accepting that a genome is a computational device. However, the proposed model offers a way to resolve this dilemma. But postulating that a genome exists in multiple dimensionalities simultaneously, the proposed model lessens the complexity arising from self reference. Thus, the conception of dimensionality serves two purposes. Firstly, it gives rise to a better model for computational complexity and secondly, it enables us to define self reference which has so far remained an intractable concept.

Another problem with the notion that a genome as a computational device is that a typical computer has a clearly defined boundary between software and hardware. This is obviously missing in the case of a genome. DNA sequence acts as a structural unit and as a functional unit at the same time. Therefore, let us see what the proposed model offers to resolve this dilemma. It is intuitively clear that both these problems of separation, viz., the separation of programming instructions and data and the separation of software and hardware, are the artifacts of our conception of computation (as formalized in the universal Turing machine (Herken 1995)). More importantly, these separations are emblematic of our cognitive mechanisms as well. It will be correct to say that the Turing machine paradigm merely transfers the dichotomy of our cognitive processing to the theory of computation. This topic needs a more detailed deconstruction and is outlined in the forthcoming monograph (Chhaya 2022f).

Presently, let us see how the proposed model resolves these issues and abolishes these two separations. According to this model, the genome exists in multiple dimensionalities simultaneously. Therefore, in a given genome, there will always be some dimensionalities wherein it acts either as a programming instruction or as a software. Similarly, there are dimensionalities wherein the genome will act either as a data or as a hardware. It is important to keep in mind the distinction between data and hardware because in the case of genomes both happen to be a DNA sequence. However, this need not be necessarily true in all the other types of computations. As discussed in the forthcoming monograph dealing with cognition (Chhaya 2022d), data and hardware are different entities, and therefore, according to this model, they occupy different dimensionalities. However, in the case of genomes, both are the same and therefore, they occupy the same dimensionality.

Once we accept this scenario, it becomes clear that a genome is a computational device but it operates as a topological computational device. This is in contrast to the conventional Turing type of computational devices (including quantum computers) which are essentially algebraic in their conception and therefore, they are geometric devices. Thus, the proposed model offers a new conception of a genome as a computational device, but it is a topological device. Therefore, it is possible to assert that Life is like any other natural phenomenon because it too is amenable to formalization using computational perspective.

Let us now look at two propositions, viz., functional modularity as a software and the structural modularity as a hardware. In the conventional perspective of computation (as formalized in the universal Turing machine (Herken 1995)), the boundary between software and hardware is well-defined, and this boundary is never breached. Surprisingly, this is true even in the case of quantum computers, at least in the way they are constructed presently. It is often mistaken that software consists of mathematical constructs, say, a programming language. The real software in a Turing type of computational device is not really the underlying mathematical formalism, but the phenomena of quantum spin present in a magnetic field. Mathematics or the programming language, is merely a representation of that behavior of the quantum spin. Thus, in a Turing machine, the boundary between hardware and software is unbreachable. In fact, we can say in the hindsight that this separation is a prerequisite for formalizing a computational process.

According to this model, there is another way to maintain this separation of software and hardware. This consists of placing them in different dimensionalities. This is precisely what Nature does in the cases of genomic expressions, quantum superposition state and in the noetic processing by our cognitive faculty. Thus, the genome, from the fact that it exists in multiple dimensions simultaneously, can achieve the separation of software and hardware in the form of functionalities and structuralism. It must be kept in mind that the evolution of modularities at each of these dimensionalities is a product of natural selection whereas the separation of structuralism from functionalities is a product of biological evolution. This distinction is important because most of us overlook the fact that the Darwinian paradigm is more about natural selection and less about biological evolution. Evolution of Life has remained beyond conceptualization. Our best model is that of the RNA world hypothesis which still remains to be formalized (Yarus 2010).

This brings us to the last point, viz., genetic disorders as halting problems of computation theory. Our present conception of computation theory has an inbuilt elegance in the form of the universal Turing machine. However, the computation theory based on the Turing machine paradigm is also plagued by the problems of undecidability. There is a class of computation problems which can never be solved by any Turing type of machine. This has nothing to do with the computational powers of a real computer, but these problems arise from the inherent structuralism of the Turing machine. In other words, these problems entailing undecidability can never be solved, not even in principle. This has something to do with what is known as halting. The very conception of a universal Turing machine requires that the machine must stop when it reaches the correct answer. However, there are several types of computations, particularly the numerical calculations, which leads to an infinite number of operations. In such cases, a Turing type of machine will not halt and there will be no answer. Therefore, if we can demonstrate that genetic disorders are the examples of this phenomenon of computational undecidability, then it is axiomatic that the genomes are computational devices.

Prima facie, the notion that genetic disorders are results of lack of controls of gene expressions is intuitive and even self-evident (Weinzierl 1999). This has been clinically proven in many of the cases, particularly those involving a single gene and its expression. Therefore, it is reasonable to think that there is no need to defend this proposition that genetic disorders are actually instances of programming errors. However, we will take a different approach to this topic. This is necessary because the origin and nature of pathologies arising from genetic disorders can be explained conventionally without resorting to any computational perspective of a genome. What we are looking for is a phenomenon, a genetic disorder that arises not because it arises from improper gene expression, but because of the long-range influences involving genomic participation. If such a case can be shown to have arisen from the lack of controls arising from long-range influences, then we can reasonably argue that this scenario is an example of halting problems that the Turing machine paradigm has formalized. Admittedly, we will not go into any specific examples of such disorders in this section. This will be discussed in a succeeding monograph dealing with therapeutic application of this model. However, in this section, we will present a general argument about how genetic disorders must be thought of as halting problems.

Prima facie, there are two classes of genetic disorders that could be shown to be instances of halting problem, viz., cancer and immune response. In both these types of disorders, it is intuitively clear that the control elements of the concerned genes and the genes themselves are very well separated from one another in terms of genomic distances. Therefore, the conventional perspective of genomics has no clarity on the exact mechanism by which these disorders manifest. Admittedly, our knowledge of how these disorders arise has been growing at a phenomenal rate. However, the genomic explanations for such pathologies are missing. One of the reasons for this lacuna is that we don't have a model of genomic architecture. If we have a reasonably good model of genomic architecture, it should be possible to define systemic properties of a genome and correlate them with these disorders.

While we will omit the specific details in this discussion, we will limit ourselves to provide a general schema of how this model could be used to demonstrate that these disorders are indeed halting problems. As mentioned above, we will discuss the specific examples in the succeeding monograph. Presently, we will look at the general arguments. The halting problem in the case of genetic disorders can be demonstrated if we can define three parameters.

1. A specific gene or a well-defined operon is involved in a given disorder. This disorder arises because of the improper gene expression of the concerned gene.

2. However, there is no immediate or proximate cause for this improper gene expression.
3. The manifestation of this disorder finds a significant statistical correlation with a remote but well-defined DNA sequence.

Under such conditions, it is possible to construct a topological model of the genome and demonstrate a specific type of genomic rearrangement without any changes in the DNA sequence that can lead to the manifestation of this genetic disorder. Once this is established, it is possible to define the topological changes that can cause the onset of the pathology arising from that genctic disorder. The key point is that we don't need to define a particular topology of the genome beforehand. Once we have significant statistical correlations, the necessary topologies can be generated using a standard protocol. Admittedly, this appears to be wishful thinking rather than a scientific hypothesis. However, at this stage, we are trying to understand the basic features of genomic architecture. Therefore, this general schema should suffice. In the preceding sections, we have covered a variety of different topics of gene expressions and their ontology. Therefore, in the next section, we will try to summarize these discussions.

4.21 Conclusion

In the preceding 20 sections, we looked at the various aspects of a possible genomic architecture. We discussed certain semantic and structural ambiguities of the conventional perspective of genomics. In order to resolve these ambiguities and unify some of the different facets of functionalities and structuralism, a topological model of genomic architecture was outlined. In order to simplify these nuances, a pointwise summary is presented in this section.

1. In congruence with our phylogenetic perspective, a notional entity called genomic singularity is proposed. This entity represents two semantic connotations.
2. Firstly, it represents a primordial genome from which all the subsequent genomes could have emerged. This conception is analogous to the conception of LUCA (Last Universal Common Ancestor). This singularity could be a notional entity or a material entity existing in the distant past. Its physical existence is not critical, but its semantic primacy is.
3. Secondly, the term genomic singularity stands for all the manifest and potential functionalities of a genome which are somehow coded into the structuralism of the genome. This is also a notional entity which acts as a repository of all the functionalities of a genome.
4. The topological model of a genome proposed here conceptualizes the genome as an ensemble of molecules spread over multiple dimensionalities. Each dimensionality represents different features.
5. A genome may operate from any of the dimensionalities it occupies. It can change its dimensionality through an operator of involution.

6. Even in its nascent state, different dimensionalities of a genome are connected to one another through the operator of involution.
7. This operator of involution can be visualized as an inward folding of one of the dimensions of the genome onto its remaining dimensions. This inward folding results in the information transfer from the dimension undergoing involution to the remaining dimensions. Thus, every dimensionality has different types of information content, and therefore, it represents different functionalities of the genome.
8. The highest dimensionality of this representation is named as the genomic singularity. The next highest dimensionality represents two different facets of functionalities and structuralism of the genome. These can be thought of as submanifolds of the parent manifold representing genomic singularity.
9. As we further reduce the dimensionality, different types of modularities manifest themselves. The order of their emergence and their relative topological placements are depicted in Figs. 4.1 and 4.2.
10. Each modularity contains different types of modules which occupy the same dimensionality. For instance, according to this model, different DNA sequences are spread over different chromosomes. Therefore, they occupy the same dimensionality and each chromosome represents a different module.
11. The relationship between different modularities are defined by the operator of involution. This operator represents a unitary mechanism. Thus, all dimensionalities are connected to one another by a unitary mechanism. As a result, the functioning of the entire genome can be formalized as an algorithm in which each step consists of a single operation which is formalized as the operator of involution.
12. Thus, the process of information content from one dimensionality to another in a fixed manner gives rise to different types of structural complexity and different types of functionalities. This unitary approach justifies why the process of natural selection is domain independent. It also provides a semantic foundation of phylogenetics.
13. The operator of involution brings about information transfer from the dimension undergoing involution to the remaining dimensions. This information content could be either in the form of the genomic details or in the form of the structural template of spacetime.
14. This model can also be applied to genomic evolution as well. We can think of biological evolution as inward transfers from the environment to the proto living organisms. In that case, the environment (which includes spacetime as well) acts as the parent manifold and the proto-living organisms as its submanifolds.
15. The key feature of this model is that it postulates that functionalities and structuralism are both located in the DNA sequence, but in different dimensionalities. It is just that since the structuralism of the genome is manifest in the three-dimensional space that we are able to comprehend it in the form of the DNA sequence. The corresponding functionalities occupy higher dimensionality and are therefore beyond our sensory perceptions.

16. More importantly, the proposed model places functionalities in the dimensionality that is higher than the dimensionality of the corresponding structural template. Therefore, in the evolutionary perspective, functionalities can influence the variations in the structural template of the genome. However, since there is no one to one influence from a functionality to its corresponding structuralism, this does not imply any teleological processes. At the same time, it explains a variety of features like punctuated evolution or speciation. More importantly, it explains how complexity emerges during biological evolution without resorting to any design principles.

Admittedly, the proposed model is a contrarian model. However, its ability to explain the otherwise inexplicable and incongruent features of genomics, makes it a serious scientific hypothesis. In the following chapters, we will try to apply this model to some of the broader issues of genomics to test its explanatory power.

References

Alcock J (2001) The triumph of sociobiology. Oxford University Press, Oxford

Allison L (1986) A practical introduction to denotational semantics. Cambridge University Press, Cambridge

Appasani K (ed) (2016) Genome wide association studies: from polymorphism to personalized medicine. Cambridge University Press, Cambridge

Arnove A (ed) (2008) The essential Chomsky. The New Press, New York

Bard J (2016) Principles of evolution: systems, species and the history of life. Garland Science, New York

Bonner JT (1988) The evolution of complexity by means of natural selection. Princeton University Press, Princeton

Borrello ME (2010) Evolutionary restraints: the contentious history of group selection. University of Chicago Press, Chicago

Brendon GE (1993) Topology and geometry. Springer, Dordrecht

Bromham L (2008) An introduction to molecular evolution and phylogenetics. Oxford University Press, Oxford

Chhaya P (2020) The universe within: and a view from that within. (Provisionally accepted)

Chhaya P (2022a) The origin and nature of mathematics. (Submitted)

Chhaya P (2022b) The origin and nature of spacetime. (Submitted)

Chhaya P (2022c) The origin of the counterintuitive nature of quantum phenomena. (Submitted)

Chhaya P (2022d) The origin and nature of cognition. (Submitted)

Chhaya P (2022e) The computational universe. (Submitted)

Chhaya P (2022f) A new paradigm for therapeutic genomics. (To be Submitted)

Cohen IR, Miller A (eds) (1994) Autoimmune disease models: a guidebook. Academic Press, San Diego

Cottingham J (2008) Cartesian reflections. Oxford University Press, Oxford

Cottingham WN, Greenwood DA (2013) An introduction to standard model of particle physics. Cambridge University Press, Cambridge

Cretoiu D, Xiao J, Das S (2020) MicroRNA signaling. Frontiers Media, Lausanne

Delisle RG (ed) (2021) Natural selection: revisiting explanatory role in evolutionary biology. Springer, Dordrecht

DeVos J, Pluth E (eds) (2015) Neuroscience and critique: exploring the limits of neurological turn. Routledge, London

Flew A (2017) Darwinian evolution. Routledge, London

Fritz A (2014) Elucidating chromosome territories organization during the cell cycle and in breast cancer cells. Proquest Publishing, Ann Arbor

Grene M (1986) Dimensions of Darwinism: themes and counter themes in the twentieth century evolutionary theory. Cambridge University Press, Cambridge

Hamilton I (1989) An evaluation of the "mitochondrial eve" hypothesis: theory, methodology, validity and relationships with the regional-continuity model of human emergence. Simon Fraser University Press, Burnaby

Hartle DL, Clark AG (1989) Principles of population genetics. Sinauer Publishing, Sunderland

Herken K (ed) (1995) The universal Turing machine: a half century survey. Springer, Berlin

Herrlich H (2006) Axiom of choice. Springer, Dordrecht

Hodge J, Redick G (eds) (2009) Cambridge companion to Darwin. Cambridge University Press, Cambridge

Howard K (ed) (2013) RNA interference: from biology to therapeutics. Springer, Dordrecht

Howard S, Abell C (2015) Fragment based drug discovery. The Royal Society of Chemistry, Cambridge

Jackendoff R (2002) Foundations of language: brain, meaning, grammar and evolution. Oxford University Press, Oxford

Kimura M (1983) The neutral theory of molecular evolution. Cambridge University Press, Cambridge

Kline P (1990) Psychology and Freudian theory: an introduction. Routledge, London

Kresina TF (ed) (1998) Immune modulating agents. CRC Press, Boca Raton

Laudal OA (2021) Mathematical models in science. World Scientific, Singapore

Lozano NA (2017) The influence of epistasis, epistatic mutability and pleiotropy on evolvability. California State University Press, Long Beach

Lynch M (2007) The origins of genome architecture. Oxford University Press, Oxford

Margulis L (1970) Origin of eukaryotic cells. Yale University Press, New Haven

Mazza ME (2007) Homeobox evolution: insights into Homeobox clusters in the basal metazoan Nematostella vectensis. Boston University Press, Boston

Miller JH, Reznikoff WS (1980) The operon. Cold Spring Harbour Laboratory Press, New York

Montani G, Battisti MV, Benini R, Imponente G (2011) Primordial cosmology. World Scientific, Singapore

Pevsner J (2015) Bioinformatics and phylogenetics. Wiley, Hoboken

Popper K (1963) Conjectures and refutations: the growth of scientific knowledge. Routledge, London

Portner PH (2005) What is meaning: fundamentals of formal semantics. Wiley, Hoboken

Press SJ, Clyde CM (2003) Subjective and objective Bayesian statistics: principles, models and applications. Wiley, Hoboken

Provine WB (2001) The origin of theoretical population genetics. University of Chicago, Chicago

Reavey C (2013) Analysis of transcription activating distance as a polygenic trait in Saccharomyces cerevisiae. Harvard University Press, Cambridge

Roy K (2015) Quantitative structure activity relationships in drug design, predictive toxicology and risk assessment. IGI Global Publications, Hershey

Shi C (1989) The applications of molecular mechanics methods and models to the study of intermolecular phenomena. Illinois Institute of Technology Press, Chicago

Shmulevich I, Dougherty ER (2014) Genomic signal processing. Princeton University Press, Princeton

Smith E, Morowitz HJ (2016) The origin and nature of life on earth: the emergence of the fourth geosphere. Cambridge University Press, Cambridge

Sober E (2015) Ockham's razor. Cambridge University Press, Cambridge

Soltis PS, Soltis DE (2012) Polyploidy and genome evolution. Springer, Berlin

Stamenov M (ed) (1992) Current advances in semantic theory. John Benjamins Publishing Company, Amsterdam

Sussman RW, Cloninger CR (eds) (2011) Origins of altruism and cooperation. Springer, Dordrecht
Szabo A, Ostlund NS (1989) Modern quantum chemistry: an introduction to advanced electronic structure theory. Dover Publications, Mineola
Theng BKG (2019) Clay mineral catalysis of organic reactions. CRC Press, Boca Raton
Weinzierl ROJ (1999) Mechanism of gene expressions: structure, function and evolution of the basal transcriptional machinery. Imperial College Press, London
Winter Y (2016) Elements of formal semantics: an introduction to the mathematical theory of meaning in natural languages. Edinburgh University Press, Edinburgh
Yarus M (2010) Life from an RNA world: the ancestor within. Harvard University Press, Cambridge

Nature of Developmental Processes in Mammals

5

Abstract

There is an old dictum that phylogeny recapitulates the ontogeny. In this chapter, we will revisit this dictum using a topological model of genomic architecture described in previous chapters. For this purpose, we would pick up mammalian development as a typical example. In particular, we will try to deconstruct the functionalities of the homeobox genes using the formalism of the involuted manifold. It would be demonstrated that the functional and the structural separations of genes present in the homeobox can be formally represented in the formalism of the involuted manifold. This successful formalization of homeobox vindicates the topological model of genomic architecture proposed in previous chapters. More importantly, it paves the way for new therapeutic approaches to several congenital disorders.

5.1 Introduction

One of the most appreciated, but least understood features of developmental biology is that developmental stages of several species often mimic the evolutionary journey of that species (Gould 1985). This encapsulation of the evolutionary past into the developmental stages is loaded with semantics which are not fully understood. The old dictum that the phylogeny recapitulates the ontogeny needs to be reinterpreted in the genomic context. This is important for two reasons. Firstly, though we know homeobox variation across the species (Mazza 2007), the evolutionary significance of these variations is yet to be unraveled. This hasn't been possible because we don't know the placement of homeoboxes in the overall genomic architecture. Therefore, we know which variations bring about which changes in gene expressions, but we don't know the genomic context of these variations. Secondly and more fundamentally, developmental strategy, as embodied in homeobox, as a genomic functionality

© The Author(s), under exclusive license to Springer Nature Singapore Pte Ltd. 2023

P. Chhaya, *The Topological Model of Genome and Evolution*,
https://doi.org/10.1007/978-981-99-4318-0_5

must have been retained by natural selection. Therefore, homeobox, as a genomic module, must be a unit of selection. However, this perspective is not fully articulated in literature.

Both these aspects of homeobox and its genomic context have not been articulated in the absence of any model of genomic architecture. However, one such model has been articulated in previous chapters. Therefore, we will apply this model to define the genomic context of homeobox. Prima facie, there are two legitimate inferences that we can draw from this perception that the recapitulation of ontogeny by phylogeny, encapsulates some fundamental insights into the Darwinian semantics. Firstly, the functional template of the genome (as distinct from the structural template of the DNA sequence) is transferred, in toto, to each organism Therefore, it is the functionality that is memorized in the form of this encapsulation. In other words, if Homeobox is taken as a higher level module of genomic architecture, what is conserved during natural selection is not just the individual genes present in Homeobox, but the relationships between these individual genes. Secondly, it is also legitimate to think that the regulatory framework of the genome operates as a single functional unit and it influences the structural template of the genome that is spatially distributed across the DNA sequence of a genome. This inference leads to another inference that the genome is not really a simple collection of genes but a functional unit endowed with a well-defined architecture. Our current understanding of genomics does not justify both these inferences, though, it doesn't deny such a scenario.

In previous chapters, a topological model of genomic architecture using the formalism of the involuted manifold is articulated. However, it is imperative that such speculative models be tested for empirical evidence. Therefore, we will pick up the processes of developmental biology, as a genomic module and seek some congruence between the proposed model and the known features of developmental biology. For this purpose, we will look at the Homeobox genes and their functions. As mentioned above, the developmental stages of any species carry its evolutionary legacy. However, this legacy is not easy to discern. Fortunately, in the case of mammalian developmental biology, there exists a definite and localized structural template governing different developmental stages. The Homeobox genes are well-characterized and their individual and collective roles are also well-documented (Duboule 1994). In addition, it is possible to obtain an evolutionary perspective of the Homeobox as it has been investigated across the spectrum of mammalian species (Mazza 2007). Therefore, the Homeobox genes are a natural choice for validating the model of genomic architecture proposed in the accompanying papers.

In this chapter, we will try to deconstruct the structural and functional templates of the Homeobox genes using the formalism of the involuted manifold. For this purpose, we would restrict ourselves to human Homeobox genes and their expressions. We would first develop a topological model of the developmental strategies of human Homeobox in a higher dimensional model. Having done that, we would invoke the operator of involution to formalize the influence of these functionalities on the gene expressions. For the sake of simplicity, this chapter has been further divided into 16 sections. Section 5.2: Developmental Strategy in

Mammals, Sect. 5.3: HOX and Homeobox, Sect. 5.4: Functional Strategy of Homeobox, Sect. 5.5: How Strategy Gets Translated into Functional Template, Sect. 5.6: Structural Template of Homeobox, Sect. 5.7: Evolutionary Perspective of Homeobox, Sect. 5.8: Variations of Structural Template of Homeobox, Sect. 5.9: Variations in Functional Template of Homeobox, Sect. 5.10: Nature of Relationships Between Functional and Structural Templates of Homeobox, Sect. 5.11: Static Model of the Relationship, Sect. 5.12: Dynamic Model of the Relationship, Sect. 5.13: The Proposed Mode of Homeobox, Sect. 5.14: Semantics of Topological Separation, Sect. 5.15: Conventional Perspective Versus the Proposed Model, Sect. 5.16: Therapeutic Possibilities According to the New Model, Sect. 5.17: Conclusion.

5.2 Developmental Strategy in Mammals

Prima facie, it is intuitively clear that the mechanism of developmental strategy of multicellular organisms must have evolved during the pre-Cambrian era when multiplicity of body plans suddenly emerged from the unicellular organisms (Cabej 2020). It seems reasonable to think that different morphological features could have evolved by random mutations and the subsequent gene fixation through genetic drifts. However, what remains unclear is whether any strategy of coordinating different genes to give rise to different body plans was already present in the genomes. It is difficult to decide whether the strategy (in the form of genomic architecture) was already in place or whether it evolved post facto after different morphological features surfaced as a result of random mutations. This ambiguity could have been resolved if we knew genomic architecture. In the absence of any such model of genomic architecture, we can deconstruct the course of evolution of multiplicity of multicellular body plans by revisiting the Darwinian paradigm. It is possible to argue that even unicellular organisms had reasonably organized genomes. Admittedly, these genomes didn't have to provide for intercellular organization, they still had to formalize the internal structure within a single cell. The key point is that the principle of spatial and temporal organization within a single cell must have evolved when Life evolved into the Eukaryotic stage. Therefore, a strategy for spatial and temporal placement of different organelles must be present somewhere on the genome. Therefore, the question arises whether this strategy evolved into the strategy for developing different body plans in multicellular organisms or the strategy of multicellular organisms evolved de novo after different body plans emerged because of mutations.

This is where the conventional Darwinian paradigm offers a plausible explanation. Considering that Nature juggles around with existing features to arrive at new features (including exadaptation), it seems reasonable that Nature must have tinkered around with the strategy of unicellular organisms to arrive at the strategy of evolving different body plans. In other words, there must be some kind of structural template within the genomes of the unicellular organisms that was capable of changing into templates for different body plans of the multicellular organisms. Apparently, the structural template present in the genomes of the unicellular organisms couldn't have been in the form of its DNA Sequence. It had to be present in some abstract manner. This is precisely what functionalities are supposed to be. A functionality per se, is a property of the genome that influences the chemical and structural changes in gene expressions. Moreover, the emergence of different body plans must have been caused by the changes in this functionality of spatial and temporal organization. More importantly, these changes need not have been reflected in the underlying DNA sequences.

This possibility points toward a fundamental semantic proposition of the Darwinian paradigm. It implies that functionalities also undergo parallel natural selection along with the evolution of structural templates of genomes. Admittedly, this has always been implicit in the conventional wisdom, particularly behind the domain of phylogenetics (Bromham 2008, see Chapter 5). If it has not been acknowledged or articulated, it is only because it raises our old fears of revival of the design principles or even teleological arguments. Just as our current understanding of genomics has taught us that we can abandon the dogma of the one-way influences from DNA to RNA and then to proteins (Robert 2004), we need to overcome the fear of design principles and seek the possibility of natural selection of genomic functionalities independent of the genomic structural template. Just as epigenetic processes did not revive Lamarckian theory, independent changes in functionalities will not revive any teleological theory.

Thus, it seems reasonable to think that natural selection operates at two levels of structuralism and functionalities. After all, it is customary to define units of selection differently for different types of natural selection (Okasha 2010, see Chapter 2). If we can think of individual genes and individual members of a species as different units of selection, there is no reason why structuralism and functionalities could be separate units of selection. If the changes in the structuralism (say genotypes) are influenced by the changes in the environment (including mutations), then according to the Darwinian semantics, the corresponding changes in the functionalities too must be influenced by the environment. However, this is hardly possible. This is where the proposed model offers a way out. It says that when we think of the environment, we conventionally think of ecological parameters. However, according to this model, we can extend the definition of the environment to include spacetime itself. Therefore, the structural template of spacetime (as defined in Physics) too could influence the functionalities. The fine structure of spacetime can be thought of as abstractions of the underlying Mathematics. Therefore, different types of mathematical templates, say, topological organization, could mold different types of genomic architecture or its structural elements. Thus, different types of body plans could have arisen from the structural template of body plan of the unicellular organisms out of mathematical or topological compulsions. If mathematical rules of association can define the details of the fundamental particles in physics, they can surely define different mathematical constructs behind different body plans of the multicellular organisms. Thus, invocation of the finer details of the structural template of spacetime in shaping the biological functionalities can help us to deconstruct the evolutionary perspective of the spatial and temporal controls of multicellular organisms. Admittedly, it will be next to impossible to connect the fine structure of spacetime to the spatial and temporal control elements present in the genomes. However, we can always employ the underlying mathematical commonality between these two domains to confirm such a scenario. In the preceding monograph dealing with the nature of spacetime (Chhaya 2022b), this model has been employed to define the fine structure of spacetime. Therefore, if we can demonstrate that the same mathematical construct of the involuted manifold can be successfully employed in deconstructing the genomic architecture, it is intuitively clear that we must accept that spacetime is an integral part of the environment that shapes biological evolution and natural selection.

With this background, let us look at the developmental strategy in mammals. Before we discuss the mammalian developmental strategy, it is necessary to outline a few caveats. These caveats refer to the general approach of this model as well as to abstract principles of any development of body plans. Since the proposed model is based on the premise that the information transfers within a genome are not confined to chemical transformations only, we need to define a topological notion of proximity for the long-range influences of the stereochemical forces to be passed on. Conventionally, we think of the genomic controls in the language of intermolecular interactions brought about by the physical interplay between the concerned molecules, say, an initiator and an upstream nucleotide (Weinzierl 1999, see Chapter 2). However, as we move to long-range influences in genomes, we employ the notion of the chromosome territories (Fritz 2014). However, the nature of interactions is still the same as that observed in a normal initiation of gene expressions. Therefore, presently, we will continue to assume that these "nonchemical" information transfers occur using the stereochemical proximity brought about by higher dimensional rearrangements of the genome. In other words, at this stage, we will not alter the basic mechanism of information transfer but instead postulate a new mode of information transfer. Later on, when we have gone through a detailed description of the proposed model, we will add a new mechanism as well. However, at this juncture, we adhere to intermolecular van der Waal forces and their manifestation in genomics.

The second caveat refers to the body plans and the kind of controls it necessitates. Since we are articulating this model of genomic architecture for the first time, it is necessary to lay down the general principles of controls. Later on, when we have a more detailed description of the proposed model, we will try to apply it to each body plan separately. For the present, we will describe the general schema of controls during developmental stages of a generic body plan. With these caveats in place, let us begin with the nature of higher dimensional rearrangements that can bring about the changes in the degree of proximity between otherwise widely separated two DNA sequences. As mentioned above, once brought into proximity, these two DNA sequences representing, say, a gene initiator and the upstream nucleotides, they would interact as conventionally understood.

Prima facie, such higher dimensional rearrangements are difficult to be conceptualized because we intuitively mistake such higher dimensions as identical with spatiotemporal dimensions that we are familiar with. Since we are biologically endowed with the four-dimensional perspective of spacetime, we fail to comprehend any conception of a higher dimensional perspective of spacetime. In a way, this cognitive compulsion to limit our understanding of reality to the four-dimensional perspective is necessary because the entire kinetic and thermodynamic paradigm of energy transfers which define the biochemistry of genomics is manifest in the four-dimensional spacetime only. However, it is important to keep in mind that the postulate of higher dimensional perspective of the genome ultimately translates into the kinetic and thermodynamic perspective of gene expressions. In other words, the proposed higher dimensionalities ultimately devolve into our familiar four-dimensional spacetime and give rise to routine biochemical processes. Therefore, the proposed model does not replace the conventional perspective of genomics, it merely augments it. In other words, the proposed higher dimensional rearrangements of the genomic architecture eventually translate into the conventional perspective of genomic functionalities. Therefore, the point of using higher dimensional entities to deconstruct the genomic functioning is to deconstruct the otherwise implicit dynamics of the genome and its functioning. At the same time, the postulated higher dimensional entity is not to be taken as a hermeneutic device. The higher dimensional model of the genomic architecture proposed here refers to the higher dimensional nature of spacetime and therefore, it is as real as the four-dimensional spacetime. It is just that since we can't comprehend the nature of higher dimensional features of spacetime, that we need to translate these higher dimensional configurations into their projections onto the four-dimensional spacetime.

Prima facie, irrespective of the body plans, there are two types of controls required to execute different body plans. These are spatial and temporal controls. Therefore, let us discuss how a higher dimensional rearrangement can give rise to spatial and temporal separation. Admittedly, it is intuitively clear that the eventual body plan requires a fine tuned sequence of spatial control and temporal control. However, this fine tuning can be thought of as a kind of algebra of these basic elements of spatial and temporal control elements. Therefore, we will overlook the compositionality of these control elements here and instead, focus on the individual spatial and temporal control elements. For the sake of simplicity we will name the element of spatial control as spation and the element of temporal control as chronon. Now, let us see how the proposed model defines the mechanisms by which these two elements of control operate. Let us begin with the chronon. According to this model, the function of chronon is to delay the next gene expression. Admittedly, it is not easy to conceive how a genome controls the gene expressions of thousands of genes. However, for the sake of simplicity we will assume that different genes express themselves in a fixed sequence. Therefore, the sequence in which genes are expressed needs to be explained. However, as discussed below, this can be defined using the elements of spatial control spation. In other words, spation determines not just the physical notion of distance, but it also determines the distance in terms of the number of nucleotides that separates any two given genes. (This is somewhat similar to the notion of kilobase pairs as a unit of genomic separation, except that spation also denotes stereochemical notion of distance, something that is missing in the conventional notion of kilobase pairs.)

Returning to the mechanism by which a chronon operates, there are three prerequisites that a genome must possess. Firstly, it must possess a certain number of higher dimensional configurations that it may occupy at any given time. Thus, when the genome changes its configuration in a given higher dimensionality, it activates a chronon. Secondly, these different higher dimensional configurations are defined by topological constraints analogous to the stereochemical constraints that manifest in the four-dimensional spacetime. However, the stereochemical constraints are translated into kinematics of the DNA sequence of the genome. On the other hand, these higher dimensional configurations are not governed by thermodynamic considerations. They are governed by topological factors. We can understand this scenario better if we compare it with the quantum superposition state. A quantum superposition state, as discussed in the preceding monograph (Chhaya 2022c, see Chapter 1), exists in higher dimensional spacetime wherein there are no time-like and space-like distinctions among the dimensions of the higher dimensional spacetime. It is only when a quantum superposition state reverts back to the four-dimensional spacetime that the thermodynamic forces take over. Similarly, in the case of the genome, its higher dimensional configurations are devoid of any thermodynamic consequences. It is only when the changes in these higher dimensional configurations result in the activation of a chronon that the thermodynamic consequences manifest in the form of stereochemical forces. Thirdly, this activation of chronon results in the changes in the conformational energies in the three-dimensional arrangements of the genome. These changes in the conformational energies set up an energy gradient which results in the delay in gene expressions of the DNA sequence undergoing conformational changes.

We will discuss the semantic nuances of this model in the following sections and in the following chapters. However, the key point is: The higher dimensional configurations are decided by the fine structure of spacetime and thereby making spacetime an active element of the environment that shapes biological evolution and natural selection. Another key point is that since the stability of the higher dimensional spacetime is decided by the fine structure of spacetime, quantum mechanics plays an important part in defining different higher dimensional configurations. This ensures that there is inherent nondeterminism in defining different higher dimensional configurations. This ensures that there are no design principles involved in gene expressions. At the same time, the underlying topology is fixed. Therefore, we can have only certain types of higher dimensional configurations of a given genome. This ensures that the subsequent emergence of complexity during biological evolution and natural selection follow certain patterns of complexity. It is this coexistence of quantum randomness and topological constraints that manifests itself as punctuated evolution (Gould 2007). Admittedly, this exposition is rather vague and we will refine it as we go along. However, before elaborating these details, it is necessary to define the element of spatial control, viz., spation.

If a chronon is thought of as a mechanism for delaying gene expressions, a spation must be thought of as a mechanism for eliminating genomic distance during simultaneous gene expressions. This might sound contradictory. The proposed model postulates a control element to increase the time lag and yet postulates a control element to decrease the space gap between two genes. Therein lies the most fundamental semantic proposition of genomic architecture. The basic premise of genomic architecture is that it must create a three-dimensional object from a linear set of instructions. Therefore, it must create a spatial simultaneity from a linear set of instructions and yet it must phase out different simultaneities. Thus, a good genomic architecture is aimed at preventing gene expressions at a wrong time and at the same time ensuring simultaneity over a distance. The operating principle is not to speed up but delay gene expressions. In order to operate this delaying tactics genomic architecture must ensure that genomic distances do not come in the way of controlling the timings of gene expressions. Therefore, the conception of spation is secondary to the conception of chronon. This is best understood by visualizing a living organism as a spatiotemporal system. The temporal dimension has only one degree of freedom but the spatial dimensions have three degrees of freedom. Therefore, in order to maintain the temporal sequence it is necessary to compromise on spatial degrees of freedom. The property of spation of eliminating the genomic distances must be viewed as a device to reduce the spatial degrees of freedom. However, to achieve this objective, it is necessary to stagger the rate of gene expressions. This is precisely what a chronon does. It helps a spation to execute its strategy by delaying gene expressions.

Admittedly, this is a simplistic scenario of a control mechanism of genomic functioning. In reality, when a genome contains hundreds or thousands of genes, this mechanism needs to be refined. However, the basic structure is identical. This explains the evolution of modularity. An identical template is repeatedly used in different contexts. This results in modular architecture. It is important to keep in mind that this strategy of using a primitive template in different contexts is consistent with the conventional Darwinian paradigm. After this rather long interlude, it is time to go back to the present discussion of the mammalian developmental strategy. In the next section, we will briefly outline two theoretical constructs of HOX genes and the corresponding mammalian construct of Homeobox. This will be followed by deconstruction of Homeobox using the proposed model.

5.3 HOX and Homeobox

The discovery of a group of genes working in tandem as a functional unit in the form of HOX genes should be considered as a landmark in the domain of genomic architecture (Duboule 1994). The eventual realization that this group of genes has been naturally selected and passed along in several evolutionary branches has made HOX an emblematic identity of genomic architecture (perhaps rivaled only by the conception of operon (Miller and Reznikoff 1980)). The structuralism of HOX, its evolutionary perspective, and its phylogenetic studies are extensively reported in literature. Therefore, we will assume this literature as being read and instead focus on the principles on which it is built and how it could have evolved. In order to refresh

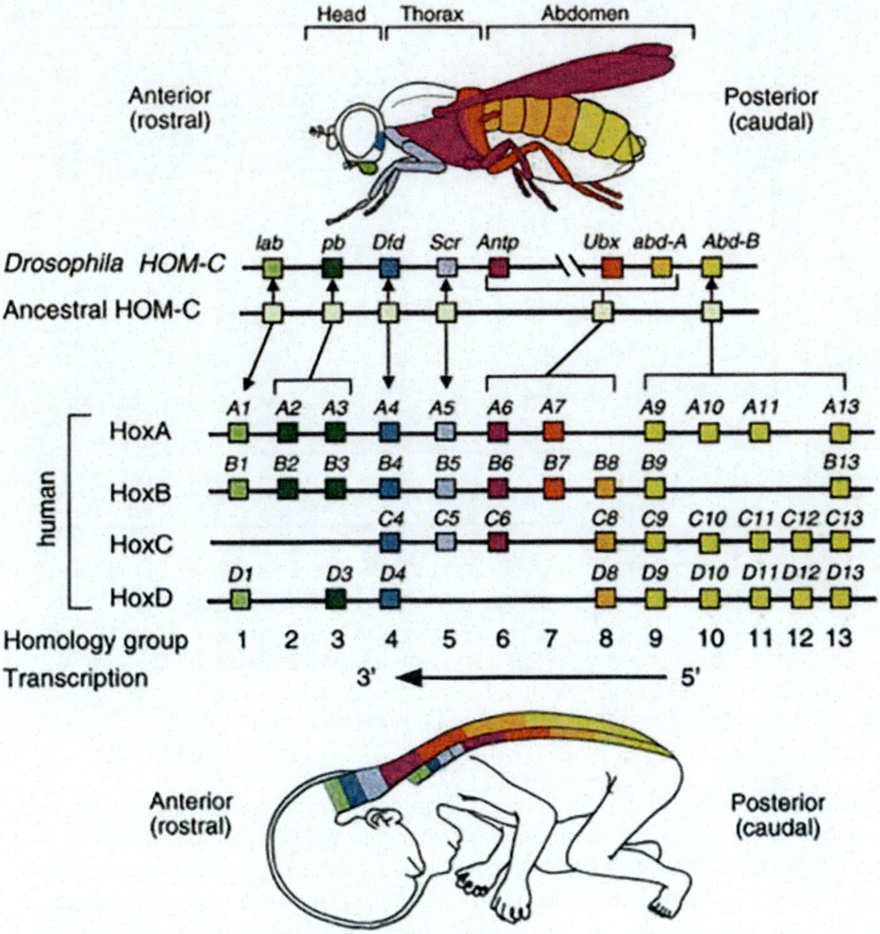

Fig. 5.1 HOX genes. (Image courtesy Mark et al. (1997))

our understanding of HOX genes a couple of diagrams available in the public domain are presented here. It has to be admitted that there are better and more detailed representations of HOX available in literature. However, for the present discussion, the accompanying diagrams Fig. 5.1 (Image courtesy Mark et al. (1997)) and Fig. 5.2 (Image courtesy Pick (2015)) ought to suffice. This is because we are not going to discuss any molecular biology of these genes, but only their abstract relationships among themselves. In this section, we will also discuss the evolutionary perspective of Homeobox and its relationship with HOX genes. This will be followed by more detailed discussion on Homeobox in the following sections.

Even while discussing the evolutionary perspective of HOX genes, we will not discuss the conventional perspective of the kinds of mutations/insertions/deletions that these genes have undergone. This is because these changes have been discussed

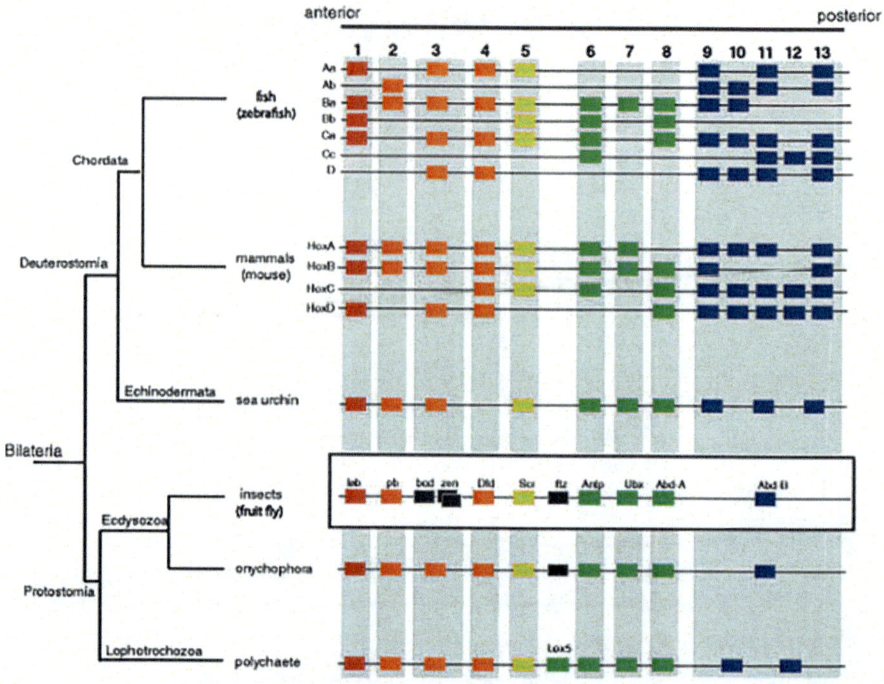

Fig. 5.2 Evolutionary perspective of HOX genes. (Image courtesy Pick (2015))

in great detail in literature. Moreover, as mentioned above, we are not going to look into molecular biology of these genes. Rather, our focus will be on the structural imperative of the control elements. Even if we were to disregard the proposed model based on the constructs like chronon and spation, it is intuitively clear that if genomic architecture per se, was the unit of selection, it must have certain structural imperatives. Natural selection can, in principle, improve upon the efficacy of these control elements. However, it cannot create these control elements de novo. Admittedly, this is a broader problem of the Darwinian paradigm. It is often overlooked that natural selection is not synonymous with biological evolution. Natural selection merely improved survival once the living organisms came into existence. This is not to belittle the primacy of Darwinian thought in modern science. Rather, it is to point out that there exists a semantic ambiguity about where to draw the line between biological evolution and natural selection. Biological evolution has certain structural imperatives. Once a system, a living organism, achieves a certain degree of complexity, the functionalities of Life would manifest axiomatically. Our endeavor to formalize the RNA world hypothesis (Yarus 2010, see Chapter 2) must be evaluated in this context. Any molecule that has certain stereochemical complexity would inevitably be capable of carrying information. The same logic applies to the duplicating ability of such a molecule. The functionality of replicative memory is not a random coincidence. It is axiomatic that any molecule having a certain degree of complexity will exhibit the functionality of replicative memory.

What is random is the identity of such a molecule and what kind of memory it could replicate. Therefore, exobiologists never rule out noncarbonaceous Life (Schulze-Makuch and Irwin 2008, see Chapter 6). The key point is this: there is a certain degree of finality about the relationship between structuralism and functionalities. Therefore, it is not subject to natural selection. What is subject to natural selection is the presence of such an entity in such an environment. Once the entity having the capabilities of replicative memory (say, RNA molecules) is present in a given environment (say, a prebiotic soup in which Life must have evolved), then natural selection takes over and improves upon the functionality of replicative memory. Admittedly, we don't know where to draw a line between biological evolution and natural selection. However, there is no denying that such a line must exist. Therefore, once we accept that any structural unit of genomic architecture, say, HOX genes together, is naturally selected, we must also accept that this unit was selected on the basis of its functionality of controlling the body plans. However, at the same time, we must also accept that this natural selection must also be based on a structural template of control. It is this aspect of HOX genes that we must try to deconstruct here. HOX genes among themselves must possess some relationships that define the functionality of control. More importantly, these relationships must be in addition to their molecular biological functionalities. A hint to such a possibility is available in the conventional perspective, only it is yet to be articulated and formalized.

It is possible to argue that this attempt to define individual control elements before even the articulation of the overall genomic architecture is like putting a cart before the horse. However, a little exposure to the field of genomics, particularly phylogenetics, ought to convince everyone that this argument is fallacious. Just as the genome is not a series of genes strung together, genomic architecture is not a series of modules clubbed together. There are semantic and structural overlaps between the genome and genes, just as there are overlaps between different control elements and genomic architecture. The question that needs to be analyzed is what possible structural imperatives define HOX? Once we accept this scenario, it is intuitively clear we would arrive at either with the conception of chronon and spation or something similar to it.

The key point is that the conventional perspective of genomics hasn't done anything about it. More importantly, this lacuna is based on the misconception that such structural imperatives will undermine the most fundamental semantic proposition of the Darwinian paradigm, viz., natural selection is nondeterministic. However, as discussed in the preceding chapters, this fear of the return of teleological arguments or the design principles is unwarranted. Having taken a contrarian view of HOX genes, let us return to the conventional perspective and briefly look at the relationship between HOX and Homeobox. It would be correct to say that there exists a considerable confusion about the relationship between HOX genes and Homeobox genes. This is partly because of the manner in which scientists found them and partly because of scientists' tendency to create acronyms. It was while discovering HOX genes that scientists discovered a DNA sequence that was very well conserved. This sequence was named as Homeobox domain and the protein encoded in that DNA sequence as Homeoprotein. The Homeoprotein functions as a regulator of gene expressions. Subsequently, it was discovered that Homeoprotein and its corresponding Homeodomain are present in a larger number of genes, most of which have nothing to do with the functions of the HOX genes. Thus, it will be fairly correct to say that Homeobox genes are those genes which contain Homeodomain. More importantly, of these Homeobox genes, only a few are HOX genes whose function is to regulate the body plan. In other words, Homeobox genes are those genes which carry within themselves, a mechanism for controlling their gene expressions in the form of Homeodomain. Similarly, HOX genes are those Homeobox genes which control the body plan during the developmental stages. Apart from the overlapping nomenclature, the presence of Homeodomain at several places in a genome points toward a common template. There are two ways to interpret this. Firstly, it points toward an architectural template that is domain agnostic. In other words, the same mechanism is employed in a diversely different genomic context. Secondly, it also points toward Nature's tendency to redeploy the same tools in different evolutionary contexts.

Returning to the architectural perspective, it is necessary to keep in mind a few general observations. Firstly, Homeoprotein can act on the gene expression of the very gene it is present. This is possible because of the RNA interference mechanisms. Secondly, a genome may contain other DNA sequences doing something similar to what the Homeoprotein does. Secondly, the expressions of HOX genes can also be controlled by mechanisms other than the one involving Homeoprotein. The most common example of this is the role played by maternal RNAs present in ovules. These facts are germane to the present discussion on several counts. Firstly, it is possible to view the Homeobox mechanism as an example of a more general class of controls of gene expressions including RNA interference. Secondly, there is something unique about the Homeobox mechanism that has helped its conservation during biological evolution. In fact, it is one of oldest mechanisms present and conserved. This leads to a possibility that there is something inherently important about the mechanism of Homeobox that has enabled its conservation.

Admittedly, we do not know the exact evolutionary path of evolution and the conservation of Homeodomain. We do not know whether Homeodomain evolved within HOX and later on, passed on to other genes or whether Homeodomain evolved first and HOX genes merely exploited it. In spite of this lack of knowledge, it is necessary to deconstruct the relationship between HOX genes and Homeodomain. Even in the absence of knowledge about the evolutionary perspective of the relationship between HOX and Homeobox, it is possible to deconstruct this relationship purely by looking at their present modes of interactions. There are two facts about them that we know. Firstly, Homeodomain is a general strategy that Nature has evolved for controlling gene expressions. On the other hand, HOX genes, collectively, have a limited role of setting up a foundation for body plans. Thus, we have two architectural features here. One of them happens to be general and nonspecific, while the other element is limited and specific. Therefore, from the architectural perspective, it is necessary to think about which evolutionary scenario should be more beneficial for survival.

Let us look at both these scenarios from the architectural perspective. Let us begin with the scenario wherein Homeodomain was naturally selected and eventually incorporated into HOX genes. From the minimalist perspective, this seems to be a natural choice. It is minimalist in the sense that it doesn't presuppose any specific architecture of genomes. It could have evolved, at least in principle, after a single gene had evolved. It doesn't require any particular architecture for its evolution. All that is required for the evolution of Homeodomain is a presence of a comparable DNA sequence in any of the genes of the primitive living organisms. Therefore, this scenario is congruent with the conventional Darwinian paradigm, particularly because it doesn't require any design as a prerequisite. Therefore, in this scenario, it is intuitively clear that having evolved, Homeodomain would have found its way into several genes, including HOX genes.

Now, let us look at the alternative scenario wherein Homeodomain could have evolved within HOX and later on passed on to different genes. Purely from the structural perspective, this scenario is complex. This is because it presupposes that HOX genes and their architecture was already present before Homeodomain could evolve. Therefore, it is also incongruent with the conventional wisdom. In addition, we know that HOX genes were present even in the simple eukaryotic organisms (Pourquie 2009, see Chapter 2). Therefore, their functionalities of organizing body plans weren't operational in these primitive genomes. Therefore, it seems least likely to have occurred. The reason behind outlining both these scenarios is to use them as test cases for the proposed model. This is because the proposed model postulates that an entity named here as genomic singularity is a repository of all the genomic functionalities. Therefore, during biological evolution, these functionalities became manifest. Thus, according to this model, architectural design was built into genomic singularity and biological evolution merely enabled unfolding of these hidden genomic functionalities. Therefore, if the proposed model is valid then, it is the second scenario that is congruent with the proposed model. Before we discuss the details of the second scenario according to this model, let us summarize the difference between these two scenarios and thereby summarize the differences between the conventional wisdom behind the Darwinian paradigm and the proposed model.

Prima facie, the first scenario which is congruent with the conventional wisdom is preferable because it requires fewer presumptions. (cf. Okham's razor (Sober 2015, see Chapter 1)). On the other hand, the second scenario which is consistent with the proposed model, requires a few semantic propositions which are in addition to the ones behind the first scenario. Therefore, under normal circumstances, we should prefer the first scenario. However, there are two observations which are pertinent in this case. Firstly, if any additional presuppositions were to offer an insight into some prevalent semantic ambiguities or they were to offer a window of opportunity to experimentally verify the new model, these presuppositions must be welcomed. In this particular case, the conventional wisdom which supports the first scenario, is ambiguous about the origins and nature of biological functionalities. This aspect has been discussed in the preceding chapters. Therefore, if we were to make additional presuppositions to unravel the origins of biological functionalities, they should be analyzed, provided they are verifiable. It is important to keep in mind that the conventional wisdom behind the Darwinian paradigm doesn't reject the idea that biological functionalities do not play any roles in biological evolution and natural selection. On the contrary, the conventional wisdom insists that it is these biological functionalities that are units of selection. It is just that the conventional wisdom has chosen to remain agnostic about the relationship between biological structuralism and the corresponding functionalities. Therefore, in the following sections, we will try to evaluate the second scenario and the manner in which the proposed model formulates this scenario.

For the sake of convenience, we will adhere to the conventional nomenclature of Homeobox to denote genes carrying Homeodomain within their DNA sequences. Admittedly, this nomenclature is ambiguous because logically it should contain all the individual genes having Homeodomain sequence. However, we will discuss the group of genes working together as a functional unit. Even among the groups of genes, we will focus on HOX genes. While the first caveat is due to ambiguous nomenclature, the second caveat is due to the lack of space that every book faces. However, since we are not going to discuss molecular biology of these genes in this monograph, the second caveat is pardonable. We will discuss the structural logic behind the organization of HOX, that logic must be valid for all the groups of genes having Homeodomain and functioning as units. Therefore, in the next section, we will try to deconstruct the relationship between different genes of Homeobox and their relationships with Homeodomain. Admittedly, we will try to employ the topological perspective of the proposed model to deconstruct the evolutionary perspective.

5.4 Functional Strategy of Homeobox

As mentioned above, we will call all the genes containing Homeodomain as Homeobox. Therefore, in this and the following sections, we will discuss the relationship between the parent genes and the Homeodomain within them. Moreover, since we are discussing HOX genes, we will refer to HOX genes as Homeobox. Wherever necessary, we will refer to individual genes or groups of genes by their respective names. Unless mentioned specifically, the term Homeobox refers to a group of genes and Homeodomain refers to the DNA sequence. Let us begin with deconstructing the strategy of Homeobox and the role of Homeodomain in it. Prima facie, it is intuitively clear that keeping aside the question of how the regulatory framework of Homeobox genes could have evolved and what kind of molecular mechanisms employed by it, we would look at the strategy of Homeobox as a purely logical schema. Once we reduce the functioning of Homeobox to a logical schema or an algorithm, it is intuitively clear that the architecture of Homeobox must consist of simple feedback loops (possibly of a single type) arranged in a nested hierarchy. In other words, Homeobox can be viewed as a network problem. Such attempts have been reported in literature (Davidson 2006, see Chapter 1). The problem however has been that such an approach doesn't succeed. This is because the network theory as applied to decision-making trees, is incapable of the functionality of self reference. In a network model (Caldarelli 2007, see Chapters 3 and 4), the network represents a set of instructions (or a set of logical operations). None of the models employed in Network theory are capable of formalizing self reference. In every case, logical operators are different from their arguments. However, as mentioned above, in this case DNA sequence is both an algorithm and an operand. Therefore, the applications of Network Theory to genomics have met with a limited degree of success (Davidson 2006).

In order to overcome this dilemma, the most intuitive way to resolve it is by using "stacks" that we employ in computation. Instead assuming that the DNA sequence of Homeobox is a one-dimensional mathematical object, we should treat it as a stack spanning several mathematical dimensions. It must be kept in mind that from the perspective of molecular biology, any DNA sequence is a three-dimensional object. However, from the mathematical perspective, the DNA sequence can be viewed as a one-dimensional object. Therefore, for the present discussion, when we assign higher dimensional status to the DNA sequence, it is only in the mathematical sense. Admittedly, there is a definitive relationship between molecular dimensions and these mathematical projections. We will return to this topic in the following chapters. Presently, we will assume that the DNA sequence occupies a higher dimensional mathematical space.

Once we do that, it is possible to translate the standard Network theory models into their corresponding topological models. Now each shelf of this stack represents one dimensionality. Using this model, now we can define self reference in a classical manner. When an instruction is directed toward itself, it can be shown to apply to a different shelf in the same stack. Thus, now we can imagine how Homeobox can connect to Homeodomain and vice versa. Therefore, all that we need to do is to place Homeobox and Homeodomain in different dimensionalities. Therefore, Homeobox can communicate with Homeodomain (in the form of a signal to begin its transcription) by reducing its own dimensionality. Similarly, Homeodomain can communicate with Homeobox by increasing its dimensionality. Thus, a single DNA sequence can express self reference. Moreover, this is not as outlandish as it appears at the first sight. After all, DNA sequence in any chromosome is organized in variable conformations. Thus, by merely changing the degrees of coiling and uncoiling, it is possible for Homeobox to communicate with Homeodomain and vice versa. Admittedly, it is simplistic to equate topological dimensionality with the stereochemical configurations. However, we will return to this topic in the following sections and the following chapters. However, there is one caveat to this scenario. To propose that different dimensionalities are connected to one another by a well-defined architecture and by well-defined functional changes is questionable. We need to provide a definitive semantic explanation for such a proposal. We will discuss this topic in Sect. 5.14. In the next section we will discuss how this strategy gets translated into the functional template of Homeobox.

5.5 How Strategy Gets Translated into Functional Template

As discussed in the previous section, in order to create self reference it is necessary to remodel the decision-making trees and to organize different elements of control in a hierarchy, it is necessary to define a dimensionality based topological model. It was suggested that we can formalize the relationship between Homeobox and Homeodomain in a similar manner. While conceptually, this strategy might sound appealing, it will be of little practical value unless we can demonstrate that it is actually present in genomes, particularly in Homeobox. This translation of an abstract strategy into molecular language need not be seen as an attempt to introduce a design principle into genomic architecture. Rather, it is aimed at demonstrating that mathematical constructs (a hierarchy of dimensionalities in this case) find uncanny congruence in a variety of natural phenomena. Admittedly, from the metaphysical perspective, this "unreasonable effectiveness of Mathematics" is troublesome (Wigner 1960). However, this is a general and perhaps a more fundamental problem of science and not that of the Darwinian paradigm alone. In fact, the conventional perspective of the Darwinian paradigm has yet to provide a cogent explanation for the emergence of complexity during biological evolution and natural selection. Therefore, it is just the complexity per se and not genomic architecture that also must be deconstructed with a minimum number of propositions. If this deconstruction leads to a congruence with some mathematical constructs, then it is sensible to not to imply any such design principles.

Returning to the present discussion, let us see how this strategy gets incorporated into the molecular paradigm. It was mentioned in the previous section that the topological perspective of this strategy need not necessarily be reflected in the stereochemical conformations of genomes. Therefore, we should not think that the interactions between Homeobox and Homeodomain will always be in the form of the conformational changes in Homeobox whereby it activates the transcription of Homeodomain sequence. Admittedly, it will always result in the transcription of the Homeodomain sequence, but only as a consequence of the changes in the dimensionalities as mentioned above. The transcription per se is not the strategy, but it is the consequence of this strategy. Therefore, let us see what the proposed model suggests on this topic. There are two propositions of this model that help us to define how an abstract strategy gets translated into a molecular strategy. Firstly, according to this model, spacetime itself exists in multiple dimensionalities simultaneously. Therefore, even the fundamental particles (and by implication, even molecules) must exist in multiple dimensionalities simultaneously. Therefore, according to this model, Homeobox (and by implication, Homeodomain) too would be spread over multiple dimensionalities simultaneously. Therefore, just as the higher dimensionalities of spacetime get projected onto the four-dimensional spacetime, the higher dimensional manifestation of the molecules also get projected onto the four-dimensional layout of these molecules. Similarly, just as the details of the higher dimensionalities of spacetime upon projection onto the four-dimensional spacetime, give rise to time-like features, the higher dimensional manifestations of the molecules upon being projected to the four-dimensional layout of the molecules give rise to kinematics of these molecules. As discussed in the preceding monograph dealing with the nature of spacetime (Chhaya 2022b), this scenario is synonymous with the conventional perspective of kinematics. Therefore, this proposition, per se, doesn't add anything new to the conventional perspective of the relationship between Homeobox and Homeodomain.

However, the second proposition of this model provides a clue to what is missing from the conventional perspective of molecular interactions. This refers to the structural restrictions on the projections from the higher dimensionalities to the four-dimensional manifold. According to this model, whenever a higher dimensionality gets projected onto a lower dimensionality, the projection is not random, but it is defined by the rules of the involutive algebra. Admittedly, there is never any single outcome of such a projection. Similarly, there is never any one single dimensionality that is projected onto the four-dimensional spacetime. Therefore, while there are multiple outcomes of the projection from every single dimensionality (and there are multiple dimensionalities that get projected onto the four-dimensional spacetime), the number of possible outcomes is not infinite. It is these limited numbers of outcomes that are responsible for the interactions between Homeobox and Homeodomain. Let us see how.

According to this model, different projections from different dimensionalities are governed by the group theoretical framework. Therefore, the number of possible ways in which Homeobox interacts with Homeodomain is limited and sporadic. If we could map these interactions, it would resemble an atomic spectral distribution. Most of these interactions are sharply defined and bunched together. It is important to keep in mind that a fraction of these interactions manifest as conformational orientations. The remaining interactions have never been investigated. These conformational changes are manifest because they entail thermodynamic and kinetic changes. However, the remaining interactions that do not get translated into thermodynamic and kinetic changes remain unobserved. It is legitimate to question the very notion of some molecular interactions which don't result in any thermodynamic and kinetic changes. After all, if these interactions do not reflect the changes in thermodynamic and kinetic energy transfers, they must be fictitious.

However, it is important to keep in mind one very strange but experimentally verified feature of biomolecules. It is a verified feature of protein synthesis in a biological cell that the resulting protein acquires a correct tertiary structure without much problem (Admittedly, when this strategy occasionally fails, the wrongly folded protein molecules are ushered into proteasomes for degradation.) (Wetlaufer 2019, see Chapter 2). From the perspective of thermodynamics and Kinetics, this highly efficient method of folding of molecules is unexplainable. There exists millions of possible conformations which are permissible for a given protein molecule. Even if Nature were to try out these conformations rapidly, the time taken to arrive at a correct conformation would be more than the age of the universe. But, Nature does it in a matter of seconds. Our computational simulations are based on thermodynamic and kinetic principles. Therefore, they throw up an infinitely large number of possible conformations. Nature, on the other hand, has found an effective algorithm to arrive at the correct conformation. The origin of this discrepancy between our thermodynamic and kinetic formalization of conformational analysis and Nature's way of computing it, lies in a simple fact. Our thermodynamic and kinetic computation strategy focuses only on the four-dimensional spacetime. Nature, on the other hand, uses higher dimensionalities as computational devices to arrive at the correct conformations of the proteins. Nature can do so because spacetime itself consists of multiple dimensionalities, each having different types of information contents.

It is important to keep in mind that even Nature's algorithm fails occasionally. Therefore, there are no design principles behind this efficiency. It is only a better algorithm. This algorithm arises from the participation of the higher dimensionalities of spacetime. Something similar happens in the case of the interactions between Homeobox and Homeodomain. Some of the pathways consist of the conventional favorable conformational states. However, there will always be some pathways that are not stereochemical in nature. It is these pathways that give rise to Darwinian randomness. While there is no direct evidence of this model, there is enough indirect evidence of this. One of the most intriguing features of developmental biology is that an apparently linear code of instructions gives rise to three-dimensional bodies of living organisms. Anyone familiar with computation theory would know that writing a program for three-dimensional objects is a challenging task, particularly when the intended three-dimensional object is intricately designed. It requires a highly nested architecture of commands.

Nature has accomplished this task rather efficiently. Admittedly, the process of natural selection acts as a debugging protocol for this software development, the resulting success is still amazing. Till date, there is not much attention being paid to this computational perspective of developmental biology (Bower and Bolouri 2001). More often than not, we try to deconstruct these intricate patterns using mathematical equations. For instance, tiger's stripes have been shown to be generated by the Turing machine model of diffusion of the products of different gene expressions (Sekimura et al. 2003). However, this is a case of missing trees for wood. It is one thing to demonstrate that spatial and temporal separation of different morphological features obey any particular set of equations and quite another to trace it all the way back to genomic architecture. The key point is that we need to deconstruct the mechanism by which a linear set of nucleotides can be made to give rise to these spatial and temporal separations. Admittedly, a partial answer to this question is available from the kinematics of gene expressions. It is possible to correlate different diffusion constants of different proteins that are synthesized, to these spatial and temporal separations (Gilbert and Barresi 2020). After all, the history of Homeobox is a testimony to this approach. However, the conventional perspective misses one crucial point of these separations.

While Homeodomain proteins can translate the differential rates of gene expressions (and the subsequent different rates of diffusion of the resulting proteins) into spatial and temporal control elements, the distribution of different genes within Homeobox plays a critical role in shaping the body plans. In fact, it was the experiment to insert one of the genes from the Homeobox to different locations that led to the discovery of the importance of Homeobox (cf. Antennapedia)

(Bermingham 1989). From the evolutionary perspective, what is inexplicable is that this group of genes were placed in a correct sequence by random shuffling of DNA sequences during the course of biological evolution. Conventional wisdom suggests that this random shuffling of DNA sequences was a purely structural event. The resulting functional changes, particularly the emergence of body plans, was an accident. This view is justified because according to the conventional perspective, there is no definitive relationship between structuralism and functionalities. While such a view was acceptable in the pre genomics era, it cannot be taken at a face value in the present time.

As discussed in the preceding chapters, if genomes are also subject to natural selection, it is axiomatic that genomic functionalities are actually phenotypes and genomic architecture is a new genotype. Once we accept this proposition, it is intuitively clear that the shuffling of DNA sequences cannot be random anymore. It is the different genomic functionalities that skew this random shuffling. To understand this, we must go back to the days of population genetics (Provine 2001, see Chapter 5). This statistical model rests on the semantic proposition implicit in classical genetics that the units of selection are discrete in nature (Darden 1991). (It must be kept in mind that it was Mendel's fortuitous choice of attributes of pea that gave birth to genetics. Had he chosen a different set of features of the plant which were polygenic, Mendel may not have succeeded!) . This approach breaks down if the units of natural selection happen to be nondiscrete entities. Even if we were to think of statistical analysis per se, say any event not related to biological evolution, it is intuitively clear that such an analysis succeeds only if the outcomes are discrete by themselves. Therefore, the moment we introduce any internal linkages between these outcomes, the notion of randomness becomes an artifact of the methodology employed.

Therefore, once we concede that genomes, like their constituent genes, are also subject to natural selection, the conventional wisdom behind the evolution of Homeobox by random shuffling of different DNA sequences doesn't make sense. The conventional strategy when faced with such a situation is to employ Bayesian logic and work out the conditional probabilities (Press and Clyde 2003). However, this cannot be applied here for two reasons. Firstly, Bayesian logic also requires a set of discrete outcomes. Apparently, this is not the case with Homeobox because there would always be some intermediate stages of the evolution of Homeobox when one or more genes will be connected to one another. Therefore, the functionalities of these intermediate stages cannot be formalized as discrete elements but only as a continuum. Therefore, we can employ a Bayesian model only if the constituent genes of Homeobox had evolved separately and then brought together by the shuffling. Such a scenario is incongruent with the conventional perspective of Homeobox.

The second reason why we can't apply the Bayesian approach here is that this approach is essentially meant to address the effect of one outcome on the remaining possible outcomes. However, in the case of Homeobox genes, this is not the case. Individual genes of Homeobox do not compete with one another (at least not directly). Therefore, the evolution of Homeobox cannot be rationalized using

Bayesian logic. Therefore, there are two possible scenarios of the evolution of Homeobox. Firstly, we must concede that there exists a separate level of functionalities which is linked to DNA sequences. Therefore, while DNA sequences undergo random changes due to physical forces like mutations, deletions, insertions and migrations, the resulting functionalities have their own rules of associations to give rise to different but new functionalities. This reasoning is analogous to the reasoning behind the selection and conservation of tertiary structures of proteins and not that of underlying DNA sequences (Bryson and Vogel 1965, see Parts III, IV, and V). Thus, the functionality of body plans arises when the constituent genes are brought together by the above-mentioned physical processes. This can be said to represent the conventional view of the origin of the functionalities involving multiple genes like Homeobox. Admittedly, the conventional perspective doesn't postulate a separate level of functionalities of each of these DNA sequences. However, barring this, this scenario is congruent with the conventional wisdom. Moreover, the reason why the conventional perspective hasn't postulated such a separate level of functionalities is that it is afraid of introducing some design principles through the backdoor.

In fact, it is intuitively clear that it is this reluctance to postulate a separate level of functionalities that has created a semantic lacuna in the conventional perspective. It cannot explain why natural selection requires separate units of selection and inheritance. As discussed in the preceding chapters, the evolutionary need to have dualities of genotype /phenotype, structural templates /functionalities and DNA /RNA remains unarticulated in the conventional perspective of biological evolution and natural selection. However, in the case of Homeobox, there is an additional problem with the proposed scenario as well as with the conventional perspective. Whether there exists a separate level of functionalities or not, the problem with the evolution of Homeobox is that it gives rise to a design. A body plan, irrespective of its perfection, embodies a certain design. This could not have arisen from coming together of different genres of Homeobox. There are no records of individual organisms, or ancestors having single structural features which could have arisen from the individual constituent genes of Homeobox. In fact, the fact that HOX genes are also found in unicellular organisms excludes the possibility of a gradual coming together (Pourquie 2009).

It is important to distinguish between the absence of individual structural elements of a body plan and the conventional scenario based on comparative morphological features. After all, it was this comparative morphology that led Darwin to articulate the theory of descent with modification. However, this is not the same (or even analogous to) presence of individual structural features of a body plan. Most of the phenotypic features of morphology are the products of the combined activities of several genes. Therefore, there exists a prior grouping of genes before the emergence of any given morphological feature. Natural selection merely refines these morphological features, obtained by the actions of multiple genes, depending on the environment. Natural selection does give rise to different morphological features, it merely tinkers with them, once they have evolved. It is for this reason why the conventional theory of natural selection cannot explain phenomena like the pre-Cambrian explosion of morphological diversity (Cabej 2020). The notion of punctuated evolution merely formalizes this ergodic phenomenon without explaining its origin (Gould 2007). It is possible to argue that the fact that we can obtain the manifestation of wrongly placed organs by tinkering with the sequence of genes of Homeobox demonstrates that the individual genes can operate independent of one another (After all that is how we discovered the *antennapedia* gene.) (Bermingham 1989). However, this argument is fallacious. Firstly, it is nobody's case that these genes cannot operate separately. What is being postulated here is that their functional independence is not the same as their independent evolution. The fact that they can act independently is not synonymous with the postulate that they evolved separately from one another. It is less problematic to think that these genes evolved together and then separated out. This is where the second scenario comes into the picture.

The second scenario that can explain how a body plan can evolve is based on the premise that emergence of functionalities (and their underlying structural complexities) is in fact, manifestation of what is latent in genomes. The model postulates that there exists a notional entity called genomic singularity which is a source of all the subsequent manifestations of functionalities. As discussed in the preceding chapters, our conventional perspective concedes the possibility of there being a common ancestry of all the living organisms. This has been often named as LUCA (Last Universal Common Ancestor) (Bard 2016, see Chapter 9). However, the general consensus is that this is a notional entity. There may or may not be a single organism from which all the living organisms came into existence. At the same time, it is possible to back-project all the features of living organisms to a single source. Therefore, the conventional perspective has taken a pragmatic view and postulated a notional existence of LUCA. (Interestingly, a similar conception of Mitochondrial Eve (Margulis 1970) has gained popularity in literature. The logic is the same as the one behind LUCA. There may or may not be any single woman from whom all the subsequent human beings were born. However, if we take different genes present in mitochondria, it is possible to trace it all the way back to a single woman.)

In a similar manner, we can postulate a genomic singularity which acts as a source of all the subsequent genomic functionalities. It is important to keep in mind that at least at this stage, this entity called here as a genomic singularity is a notional entity. Whether there existed a physical entity like a genomic singularity can only be justified if we were to find evidence of it. Presently, a genomic singularity must be taken as a notional entity from which all the genomic functionalities could have arisen. Admittedly, the problem with this proposal of genomic singularity is not much about its conception, but it is about its semantic implications. It is legitimate to be skeptical about something that is postulated to be omniscient. This genomic omniscience (of knowing what future holds for biological evolution) smacks of transcendental arguments or some form of theism. Historically, this is precisely what Darwin fought against.

However, the proposed model offers a naturalistic deconstruction of this supposedly omniscient entity. The genomic singularity derives its omniscience not from any transcendental agency, but from the nature of spacetime itself. Let us step back and think about the most conservative interpretation of the Darwinian paradigm (Grene 1986). Even in these most puritanical interpretations, there is a consensus that the environment plays a critical role in natural selection. All that is required is to add two propositions to this view. Firstly, we must accept that the notion of what constitutes an environment must include the nature of spacetime. This is not a farfetched demand. Spacetime would influence all natural phenomena, including biological evolution. Therefore, it is axiomatic that the nature of spacetime would also influence the course of biological evolution. This is not difficult to explain. Just imagine that biological evolution has taken place on a planet whose gravity was twice as strong as that of Earth. Naturally, the size of animals produced on such a planet would be smaller than those found on Earth. Thus, even the most conservative interpretation of the Darwinian paradigm is congruent with the notion that spacetime would influence the course of evolution.

Now, let us add a second proposition. The proposed model postulates that spacetime doesn't merely influence biological evolution, but spacetime gets woven into Life. Thus, spacetime is not merely a passive restrictive agent in biological evolution and natural selection, but rather, it is an active participant in biological evolution. Once we accept this possibility, it is intuitively clear that structural complexities of living organisms arise from the fine structure of spacetime itself. This is not as outlandish as it appears to be at the first sight. We can go back to quantum chemistry to understand this proposition (Szabo and Ostlund 1989, see Chapters 2 and 3). In quantum chemistry, the core argument is this: all the chemical properties of all the possible molecules can be sourced from the wave function of these molecules. The next logical step is to ask what is the constituent of a wave function of any molecule. The answer is the distribution of quantum fields (of the constituent fundamental particles of the given molecules) and spacetime. Therefore, all the conceivable chemical properties are decided by the fine structure of spacetime. (Interestingly, if we accept the string theoretical perspective, quantum field and spacetime are one and the same.) The next logical step in this scenario is to accept that DNA is like any other molecule. Thus, the fine structure of spacetime would be woven into the structural template of DNA. Therefore, the fear that the postulate of genomic singularity is tantamount to the acceptance of omniscience is unfounded. Just as incorporation of spacetime in molecules does not lead to any determinism, the incorporation of spacetime in genomes (in the form of genomic singularity) doesn't introduce any determinism or teleological principles. Having dealt with the broad scenario, let us get back to the specific example of Homeobox and its mechanism.

According to this model, the basic schema that operates in genomics is that at a higher dimensionality, there are some functionalities which remain in their latent forms. However, at any dimensionality that is lower than this, these functionalities become manifest. Thus, all the possible genomic functionalities remain latent at the highest dimensionality of a genome. This highest dimensionality of a genome is defined as a genomic singularity. Therefore, whenever the dimensionality of a genomic singularity is reduced, one or more functionalities become manifest.

Since according to this model, a genome may exist in multiple dimensionalities simultaneously, it is possible to envisage several operations which reduce the dimensionality of the genomic singularity, each operation leading to different dimensionalities and thereby activating different functionalities.

The same schema can be applied to Homeobox genes collectively. As a group of genes, Homeobox behaves like a genomic singularity wherein the functionalities of the constituent genes remain latent. When the dimensionality of Homeobox, as a single unit, is reduced, the functionalities of individual constituent genes become manifest. Thus, Homeobox can be viewed as a hierarchy of the constituent genes, each occupying a characteristic dimensionality. This arrangement allows different member genes of Homeobox to express themselves in two ways. Firstly, these genes can express themselves in a particular order depending on the dimensionalities they occupy. It is apparent that in such a scenario, a wrong topological placement of genes can give rise to various types of pathologies. The case of *antennapedia* (Bermingham 1989) mentioned above fits into this category. The only difference being that this change in hierarchy is artificially created under laboratory conditions.

The second way that Homeobox can operate is through the Homeoproteins. This is essentially a conventional mechanism whereby the Homeoproteins regulate the rate of different gene expressions. There are two key questions that need to be addressed: how is the dimensionality of a genome altered? How does a change in dimensionality induce a particular gene to initiate its expression? This is where the direct participation of spacetime comes into the picture. Let us begin with the first question. It is always tempting to be misled by clever ideas because their novelty lulls our critical analysis. This logic applies to the proposed model as well. There is a reasonable chance that the notion of dimensionalities of genomes and their relationships with different functionalities are such clever constructs. Therefore, it is imperative that these concepts and their underlying semantics must be verified using the empirical evidence available in literature. Therefore, let us assume that prima facie, a genome exists in multiple dimensionalities simultaneously. In other words, let us assume that the proposed model is a true representation of genomes. Then, it is obligatory for the model to define what these dimensionalities are and how a genome switches from one dimensionality to another. This is where the direct participation of spacetime comes into the picture.

As discussed in the preceding sections and in the preceding chapters, the conventional wisdom suggests that large molecules, including DNA, can have different conformations. In the case of DNA, the conventional perspective employs these conformational changes routinely to explain different activities of genomes. This includes the conformational changes during cell division as well as the conformational changes during transcription, say, in the form of chromosome territories. Therefore, it is tempting to extend this conformational perspective to the proposed higher dimensionalities and formalize using similar mathematical constructs. However, this is not necessary, rather, it will be counterproductive. It is important to keep in mind that this conventional perspective of conformational changes is grounded in the Newtonian paradigm wherein spacetime, albeit as separate entities, remains passive during these conformational changes. In contrast, the proposed model seeks to employ the relativistic paradigm wherein spacetime is an active agent of all the thermodynamic and kinetic processes. (It is necessary to place a caveat to this invocation of the relativistic perspective of spacetime here. This invocation is not meant to imply that the relativistic paradigm in any way leads to or is congruent with the proposed model. The invocation of the relativistic paradigm is based only on one of the semantic propositions of the relativistic paradigm, viz., spacetime is a physical entity capable of interacting with other material entities like molecules, including DNA.)

The proposed model postulates that spacetime interacts actively with DNA molecules. Since according to this model, spacetime itself can be formalized as having multiple dimensionalities simultaneously, all the molecules, including DNA, are also spread across the different dimensionalities of spacetime. Therefore, what we experimentally observe, say, different conformations of DNA, is a fraction of the spread of these molecules. In other words, the conventional perspective of molecular conformations is a subset of the changes in the shape of the molecules that occur across the dimensionalities of spacetime. However, this is not an escapist argument. This does not imply that the changes in the shape of the molecules in higher dimensionalities are beyond scrutiny and therefore, we must accept some kind of "holistic" approach. On the contrary, the proposed model provides a way to link the changes in the shape of molecules in different dimensionalities with one another. Thus, according to this model, it is possible to expand the formal description of conformational changes to include the corresponding changes in the shape of molecules that occur in higher dimensionalities. It is just that it is not necessary to go into the mathematical formalism of the link between the changes in the shape of the molecules and their respective dimensionalities.

The key argument about this linkage between the changes in the shape of the molecules and the dimensionality in which these changes occur is this: at higher dimensionalities of spacetime, there is no distinction between the time-like and the space-like dimensions. Therefore, the changes in the molecular shapes do not appear as conformational changes in the strict sense. In fact, according to this model, spacetime itself consists of several time-like and space-like dimensions. More importantly, in each case, a time-like dimension occupies a higher dimensionality vis a vis its corresponding space-like dimensions. It is only when these time-like dimensions get involuted into the lower space-like dimensions that we obtain our four-dimensional spacetime. This leads to two important semantic consequences. Firstly, our conventional perspective of conformational changes is actually a composite perception arising from multiple blending of time-like and space-like dimensions. Therefore, it is a subset of all the changes in the shape of a molecule. Secondly, since the blending of time-like and space-like dimensions is governed by a single mechanism, the resulting changes in the shape of the molecules appear to be random only when viewed from the perspective of the four-dimensional spacetime. According to this model, the blending of time-like dimensions with the space-like dimensions through involution occurs continuously (actually from the perspective of the higher dimensionalities, these blendings are permanent). It is only from the perspective of the four-dimensional spacetime, that these blendings appear to be continuous. Thus, to answer the question, we can say that the changes in dimensionalities occur ceaselessly from the perspective of the four-dimensional spacetime. It is just that our observational capacity is limited to the four-dimensional perspective that we cannot perceive all the changes in the shape of the molecules, including DNA. Therefore, there is no way to verify these higher dimensional changes in the shape of molecules, including DNA.

However, this scenario can explain a couple of hitherto unexplained phenomena of molecular biology. In transcription of DNA sequences, it has always been mysterious to observe that the resulting amino acid sequences, more often than not, fold correctly (Wetlaufer 2019). What is mysterious about this very efficient manner of folding is that from the computational perspective there exists an infinitely large number of conformations possible for a given protein. If Nature were to employ a random method of folding, it would take an infinitely long time (perhaps longer than the age of the universe) to arrive at the correct conformations of proteins. It seems reasonable to think that Nature employs some unknown algorithm to fold a given sequence of amino acids into the desired conformations of protein. However, till now, we do not know how Nature does this apparently mysterious task. According to this model, this now can be explained without invoking any transcendental arguments. Nature employs the changes in the dimensionalities of spacetime to reduce the computational time for folding these proteins. Because Nature employs changes in the higher dimensionalities to limit the number of possible conformations, it manages to obtain a correct conformation in a reasonably short time. Incidentally, according to this model, even the changes in the shape of molecules are determinable even in the higher dimensionalities. In these higher dimensionalities, there are multiple pathways available. It is just that they are fewer in number. Therefore, there are times when Nature too ends up with wrong conformations of proteins. This forces Nature to employ the ubiquitin governed destruction of the wrongly folded proteins (Mayer et al. 2006). This rare failure of Nature also supports a naturalistic explanation and eliminates any transcendental or holistic explanation of Nature.

Now let us look at the second question about why the changes in the dimensionalities should result in different functionalities of genomes. In order to answer this question, let us look at two analogous phenomena, viz., chemical reactivity of different stereoisomers (Patel 2000) and different absorption spectra of different vibrational modes (Lightner and Gurst 2000). Both these phenomena are well documented and adequately formalized. For instance, there are a host of enzymatic reactivities, catalytic efficiencies and reactivities in biochemistry and molecular biology wherein one particular stereochemical orientation of a given molecule is more efficient than the others. This feature is attributed to the three-dimensional shape of the molecule. This in turn can be understood in terms of Kinetics and thermodynamics of the chemical reaction under investigation. It so happens that we are accustomed to think in terms of energy transfers in the four-dimensional spacetime. Therefore, we can visualize how different three-dimensional arrangements of a molecule can give rise to different chemical properties. If we could visualize higher dimensional configurations of different objects, say, molecules, then perhaps we could see that topological changes in dimensionalities ought to have similar changes in the molecular reactivities. It is just that these higher dimensional configurations of molecules do not entail any time-like dimensions. Therefore, these changes in the high dimensional configurations do not get translated into the corresponding kinetic or thermodynamic changes and therefore, they remain unnoticed.

Similarly, in the case of the absorption spectrum of a given molecule, it is universally accepted that different vibrational modes absorb different wavelengths of light. This happens because each vibrational mode is assigned a specific energy transfer. Of course, it is possible sometimes to flood the molecules with excess electromagnetic radiation and obtain an overlap of different vibrational modes. However, under normal laboratory conditions, it is possible to observe different vibrational modes by attenuation. By analogy, we can think of different dimensionalities as different states that require energy transfers to switch from one dimensionality to another. Therefore, the key point is to establish that molecules exist in multiple dimensionalities simultaneously. Once we can establish that, it follows axiomatically that the changes in dimensionalities would result in energy transfers triggering the thermodynamic and kinetic consequences.

It is possible to argue that if any scientific phenomenon remains beyond experimental verification, for all practical purposes, it doesn't exist. Therefore, even if the above-mentioned scenario is valid and if it doesn't alter the kinetic or thermodynamic properties of the molecules, it needn't be taken seriously. It could be a fanciful but irrelevant conception. However, upon a little reflection, it is intuitively clear that this is not the case with this scenario. Firstly, once we accept that there exists higher dimensional configurations of molecules, it is axiomatic that the changes in these higher dimensional configurations would lead to some changes in the chemical properties. Therefore, the key point is how to establish that higher dimensional configurations of molecules exist? Admittedly, there is no way to establish the existence of such higher dimensional configurations. However, we can take inspiration from other domains which employ similar strategies. String theory is perhaps the best example of this approach (Conlon 2016, see Chapter 14). There is no evidence that there are ten or eleven dimensions of spacetime (which is what string theory models require for their formal description, depending on which particular model we wish to develop). Still, this lack of experimental evidence hasn't prevented string theorists from developing higher dimensional models of spacetime in order to explain quantum phenomenology. The key point is that if a postulate of higher dimensional perspective helps us to deconstruct the otherwise inscrutable phenomena, there is no harm in developing such models. It is possible that higher dimensional configurations could just be a hermeneutic device. Even then, its utility justifies such modeling.

The second reason why the proposed scenario is not a fanciful nonsense is that it is not invoked to explain just this particular aspect of genomics. As discussed in the accompanying chapters, the proposed model involving higher dimensional perspective is capable of deconstructing a whole lot of otherwise enigmatic features of biological evolution and natural selection. In fact, as discussed in the preceding monographs, this approach is applicable to a wide range of natural phenomena like quantum reality, cognitive processes and linguistics. Therefore, the balance of evidence is in favor of the proposal, even if it is not presently verifiable.

Returning to the present discussion, let us assume that when a dimensionality of a genome is changed it results in expression of different types of functionalities. The key point is how this comes about. In order to understand this all that we need to do is to postulate that the changes in the higher dimensional configurations of genomes translates into subtle changes in the three-dimensional configurations of genomes. What is implied by subtle changes here is that these changes fall within the range of thermodynamic energy levels of the underlying DNA sequences. To begin with, at present, we still don't know how a genome switches from one configuration to another during the cell cycle. True, we know how different kinesins provide the necessary energies for folding and unfolding of genomes. However, we do not know how this is regulated. Therefore, it seems reasonable to think that these subtle changes in the three-dimensional configurations of genomes (which are triggered by the corresponding changes in the higher dimensional configurations) are actually governed by some control mechanism involving kinesins. Thus, it is possible to argue that these changes in higher dimensional configurations, through the agency of kinesins, give rise to subtle changes in the three-dimensional configurations of genomes. These different three-dimensional configurations can legitimately give rise to different sequences of gene expressions, thereby giving an appearance of different functionalities. In other words, different genomic functionalities are essentially the changes in the sequence of gene expressions. This perhaps explains the phenomenon of chromosome territories (Fritz 2014). While this scenario looks plausible, it still needs to be fleshed out with more details. Therefore, in the next section, we will look at the structural template of Homeobox itself and see whether it supports the proposed scenario.

5.6 Structural Template of Homeobox

In the previous section, we looked at how the strategy of organizing different gene expressions into a particular sequence can be seen as a functional template of Homeobox. Therefore, the next logical step would be to find out how this functional template actualized in the structural template of Homeobox. In this section, we will try to deconstruct the physicality of the functional template of Homeobox. Before going into the details, it is necessary to place one caveat. Normally, at least in reference to genomics, structural templates ought to refer to the DNA sequence and its ensembles with chromatin and other protein molecules. This is correct because different conformations of DNA sequences of a genome are controlled by the proximity of chromatins and other proteins, particularly those necessary to build transcription complexes (Weinzierl 1999). Therefore, any changes in the functionalities of a genome would be decided not just by the DNA sequence, but also by these proteins.

However, we will eschew this approach in this section for two reasons. Firstly, we are interested in decoding the higher dimensional configurations of Homeobox as a strategy for spatial and temporal control elements. Therefore, it can be argued that whatever the interactions that a given DNA sequence may have with these proteins are dictated by these higher dimensional configurations. The conventional three-dimensional or stereochemical interactions between the DNA sequence and these proteins arise as a consequence of the changes in the higher dimensional configurations of DNA sequences and can be studied later on, after the higher dimensional configurations have been deconstructed. Admittedly, we can extend this approach to these proteins as well. However, till such time when this approach has been verified experimentally, it will be presumptuous to expand this approach to include these proteins.

The second reason why we will eschew the conventional approach lies in the deeper semantic content of the proposed model. According to this model, as mentioned above and in the preceding monographs, the notion of discreteness is a derivative of the dimensionality from which we make our observations. Thus, what appears to be an ensemble of discrete entities can appear to be a single entity in a higher dimensional perspective. Thus, according to this model, fundamental particles, in general, can be viewed as "crystallizations" of the spatiotemporal universe itself. The same logic applies to atoms and molecules. They all are localized manifestations of the cosmic singularity. Similarly, once we accept the notion of genomic singularity, it is intuitively clear that not just the DNA sequences, but also these proteins are local manifestations of the genomic singularity. Thus, when a higher dimensional genomic singularity devolves into lower dimensionalities, it "fragments" into different molecules. Thus, according to this model, it is legitimate to think about the DNA sequence and its partner proteins as conjoined twins which are united at higher dimensionality, but appear to be different entities at the four-dimensional spacetime. Admittedly, such a scenario is fraught with the perils of design principles. However, we will discuss in the following chapters why this fear is misplaced. Presently, as discussed above, it is clear that there are semantic imperatives that necessitate this eschewal of the conventional perspective of the structural template of Homeobox.

Having outlined this caveat, let us deconstruct the structural template of Homeobox. It is legitimate to question that in the absence of any discussion of the structuralism of the DNA sequence, what other structural template we can conceptualize. After all, every DNA sequence is defined by its three-dimensional configurations and all its functionalities can be traced back to these three-dimensional configurations. However, we will take an unorthodox, if not heretical, view of Homeobox. We will assume that it possesses a multidimensional configuration in which several dimensionalities coexist. Admittedly, at the three-dimensional spatial perspective, this structural template will be congruent with the conventional stereochemical perspective. Thus, there is no contradiction between the proposed model and the conventional perspective. The difference between these two perspectives lies in what kind of structuralism Homeobox can possess at higher dimensionalities. Admittedly, according to the conventional wisdom, the Homeobox doesn't have any structure of its own in the dimensionality of higher than three. If we wish to be liberal with the conventional interpretation of genomics, we can conceptualize a four-dimensional perspective of Homeobox (or any other biocatalysts, for that matter). We can at least in principle, map the catalytic functionalities of Homeobox in its four-dimensional structuralism. Admittedly, this approach is never pursued seriously, but we will be justified in doing so even while remaining within the conventional perspective of genomics. However, the conventional perspective doesn't allow us to theorize the structural template of Homeobox beyond the four-dimensional perspective.

Therefore, we will pursue this unorthodox approach, if only to find out what the proposed model offers. As discussed in the preceding chapters and in the preceding sections, we will postulate that in some of these higher dimensionalities, there are no atoms or molecules. More importantly, there are no time-like and the space-like dimensions as well. Similarly, while the proposed model puts no restrictions on the numerical values of the dimensionalities of any object, for the sake of simplicity, we will postulate that the Homeobox model will consist of dimensionalities whose numerical values are natural numbers. Thus, the proposed model of Homeobox will take into consideration only the dimensionalities with integral numerical values. Even here, we will confine ourselves to the dimensionalities up to ten. It must be kept in mind that neither the proposed model, nor the principles of genomics deny the possibility of Homeobox having fractal dimensionality or dimensionality greater than ten. It is only for the sake of manageability of the formal description of Homeobox that we choose to restrict ourselves to integral dimensionalities up to ten.

As a next step, we will arbitrarily assign several functionalities of Homeobox to different dimensionalities. Admittedly, both the choice of functionalities and their ascription to different dimensionalities are arbitrary and ad hoc. However, as we deconstruct this model further, we can arrive at some thumb rules of assigning different dimensionalities to different functionalities. However, at present, this choice of dimensionality of each of these functionalities is arbitrary, albeit governed by implicit semantics of genomics. For this purpose, we will select the following functionalities of Homeobox and assign them different dimensionalities. The details are given in Table 5.1. We will discuss each of these five functionalities from the higher dimensional topological perspective.

We will divide this discussion in two parts. Firstly, we will deconstruct the first three functionalities as a group. This will be followed by the second group of functionalities (functionalities 4 and 5 in Table 5.1). This segregation is necessary because the functionalities numbered 1–3 are well understood from the conventional perspective. Moreover, the proposed model offers a parallel explanation of their mechanisms. Admittedly, the proposed model offers additional insights into these functionalities, but by and large, the mechanism of these functionalities proposed by this model is analogous to the one available in the conventional perspective. Therefore, we will discuss these functionalities as a separate group. Our focus will be on the new insights available from the proposed model.

Having done that, we will discuss the functionalities 4 and 5 as a second group. This is because not only we will employ the insights made available from the

Table 5.1 Topological hierarchy of Homeobox functionalities

#	Functionality	Dimensionality	Explanation
1.	Initiation of gene expressions	Four	Three-dimensional stereochemical arrangement along the time dimension
2.	Promotion of gene expressions	Four	Three-dimensional stereochemical complex formation in the time dimension
3.	Suppression of gene expressions	Four	Three-dimensional stereochemical complex formation in the time dimension
4.	Activation of RNA complexes in protein synthesis	Five	Four-dimensional nucleosome conformational changes in the time dimension
5.	Interference of RNA complexes in protein synthesis	Five	Four-dimensional nucleosome conformational changes in the time dimension

functionalities 1–3 to deconstruct the functionalities 4 and 5, but also because the conventional perspective has nothing much to offer by way of mechanism of the functionalities 4 and 5. In fact we would discover that the ability of the proposed model to offer mechanistic insights into the functionalities 4 and 5 arises from the new insights made available from the deconstruction of the functionalities 1–3 by the proposed model. We will end this section with a brief discussion why it is necessary to replace the conventional perspective with the proposed model. This comparison between the conventional perspective and the proposed model offers a new semantics of the higher dimensional models of genomes.

Let us begin with a brief description of the conventional perspective of the functionalities 1–3. According to the conventional perspective, the functionalities 1–3 arise from a stereochemical or rather, conformational contours of the proteins synthesized previously with the target sequence of the gene expression under investigation. Admittedly, we had to deploy sophisticated experimental devices to capture the fine details of the stereochemistry of these molecules. However, purely from the theoretical perspective, there is nothing much that is additionally required to explain these functionalities (Weinzierl 1999). Of course, we end up marveling at the fine tuning of these molecules and their stereochemistry. However, given the stereochemical congruence between the previously synthesized proteins and the target DNA sequence of the gene under investigation, it is axiomatic that these functionalities would manifest themselves.

Secondly, the conventional perspective offers some tantalizing hints about how the sequence of the gene expressions could be arranged in such a way that the right initiator or suppressor of a given gene is always available. While the conventional perspective doesn't have any definitive guide on how a genome switches on the correct genes at the correct time, it has no framework to formalize the principles of sequence of gene expressions. Finally, the conventional perspective is totally incapable of explaining how the preferred sequence of gene expressions could have evolved. Thus, the conventional perspective can be divided into three types of explanations, viz., the well-defined stereochemical framework, implicit and poorly understood mechanisms of the choice of the sequence of gene expressions, and unknown mechanism for the evolution of the sequence of gene expressions. As discussed in the preceding chapters and in the preceding sections, this is largely because the conventional perspective restricts itself to the three-dimensional stereochemical perspective of a genome.

Therefore, in this section, we will avoid discussion of the first type of functionalities which are very well explained by the conventional perspective. Instead, we will focus on the second type of functionalities like the mechanism by which the sequence of gene expressions is executed in genomes, particularly by Homeobox. The third aspect of the evolution of this mechanism will be discussed in the next section. As mentioned above, the conventional perspective gives some tantalizing hints about the mechanism by which the sequence of gene expressions is executed in genomes. However, the mechanism, by itself, remains beyond the scope of the conventional perspective. Therefore, in the following paragraphs we will discuss a possible mechanism for deciding which genes must be expressed in which sequence. More importantly, we will discuss how Homeobox plays a role in this determination of the sequence of gene expressions. Admittedly, we will discuss this topic from the perspective of the proposed model. It must be kept in mind that the scenario outlined below is purely speculative and no evidence to support it will be provided. At the same time, it must also be kept in mind that the proposed mechanism is consistent with the conventional perspective. It is just that the proposed model ventures into a speculative thought without any inhibitions.

Before going into the specific details of how Homeobox could influence the sequence of gene expressions, let us look at an abstract template of the proposed mechanism. The schema outlined here is generic in nature and can be applied to any phenotypic expressions involving multiple genes acting in tandem.

Admittedly, Schema 5.1 is a simplistic depiction of how topological transformations of genomes and their components like Homeobox bring about the manifestation of different types of functionalities. However, it is illustrative of the basic principles of the proposed model. There are three aspects of this schema that are germane to the present discussion. We will merely mention them here. Having done that, we will discuss these aspects in reference to the functioning of Homeobox in little more detail. The relevant aspects of Schema 5.1 are: at lower dimensionalities, the proposed model is congruent with the conventional perspective. Therefore, there is no need to redefine the conventional perspective or its stereochemical details (functionalities 1–3 in Table 5.1); at higher dimensionalities, the proposed model postulates new topological transformations and correlates them with different functionalities; the proposed model defines a universal operation of the changes in the dimensionalities (defined as the operation of involution) that connects the proposed model to the conventional perspective of genomics. The advantages of such an approach are obvious. Firstly, there is no need to replace or overhaul the current paradigm of genomics. Secondly, this approach connects the current paradigm of genomics with a novel topological perspective using a single mathematical operator of involution. This allows us to invest new semantics to genomics without having to replace the old paradigm. Finally, the newly introduced

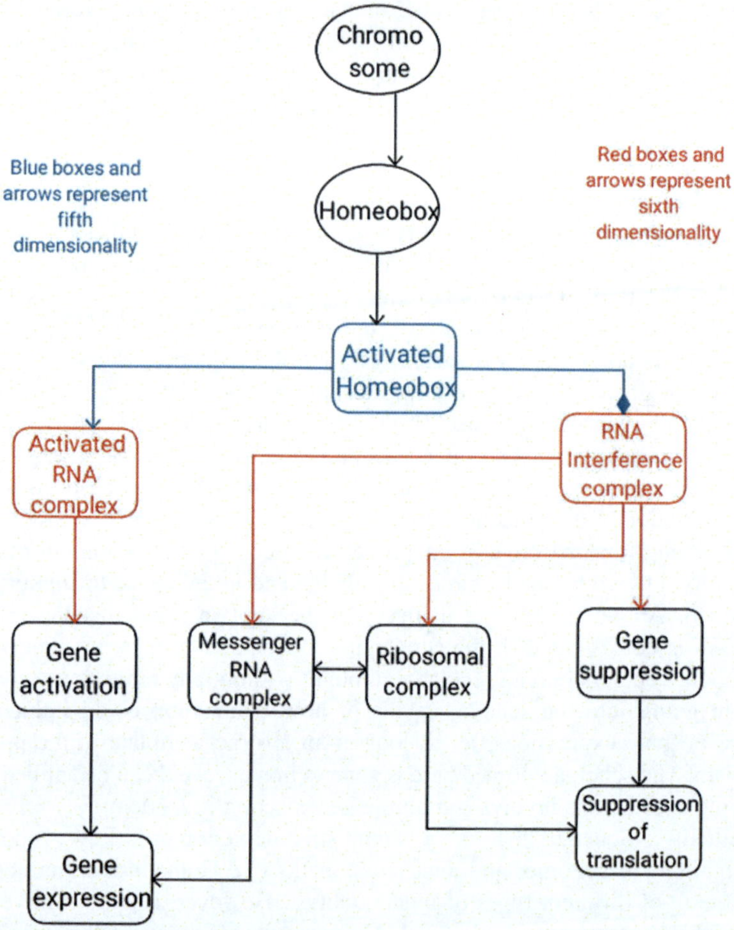

Schema 5.1 Generic template of controlling the sequence of gene expressions

semantic propositions embellish the current understanding of genomics without undermining the principle of natural selection as articulated in the Darwinian paradigm.

It is possible to be skeptical about such semantic interpolations, particularly because they might introduce some kind of design principles surreptitiously into the Darwinian paradigm. More importantly, in the absence of any experimental evidence, such supplementary semantics may well turn out to be a speculative anthropism. Most of such apprehensions have been addressed in the preceding monograph and in the preceding chapters. However, we will try to allay such misapprehensions here with the specific deconstruction of the functioning of Homeobox. If the proposed model can offer a cogent interpretation of the functioning of Homeobox, it will be legitimate to accept the proposed model as a serious scientific template, albeit subject to experimental proof. Therefore, in the remaining part of this section, we will try to deconstruct the functioning of Homeobox, particularly with respect to the topological nuances of the distinction between higher and lower dimensionalities.

Therefore, let us try to reconfigure Homeobox as a topological manifold with different dimensionalities representing different functionalities. Admittedly, the proposed model of Homeobox is arbitrary in the sense that it is not based on any molecular perspective. However, it is based on the intuitive conception of the control element that Homeobox seems to exemplify. A brief outline of a topological model of the evolution of Homeobox is given in Schema 5.2 and the topological separation of its functionalities is given in Schema 5.3. At present, we will not discuss the exact nature of the higher dimensionalities and why it should influence the functioning of Homeobox. This topic is discussed in later chapters along with other details of

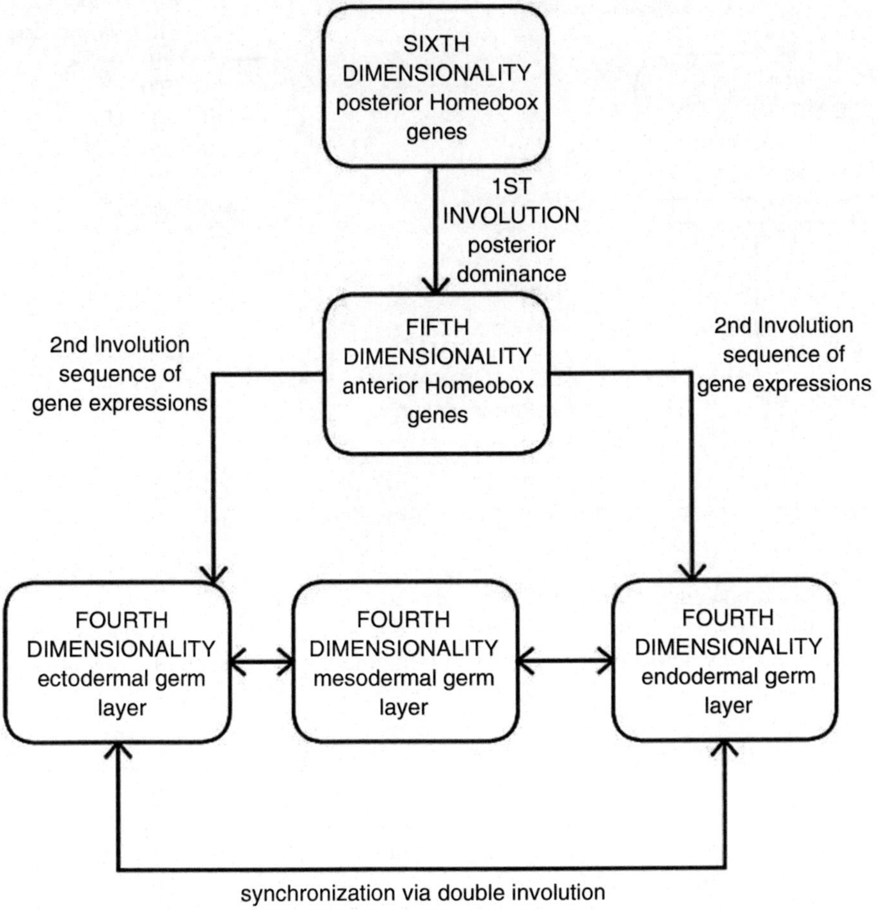

Schema 5.2 Topological model of Homeobox

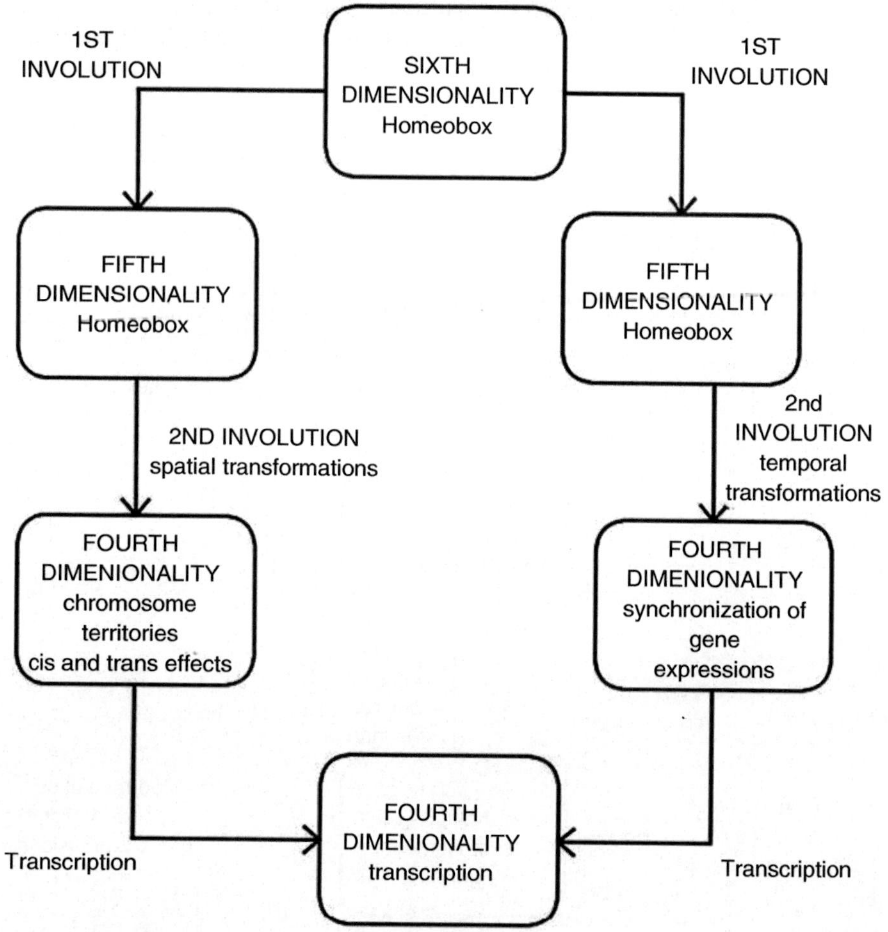

Schema 5.3 Topological model of the functionalities of Homeobox

genomic architecture. The emphasis here is on the possibility that different functionalities of Homeobox are related to one another and more importantly, they can be selectively activated by merely changing the dimensionality of Homeobox.

5.7 Evolutionary Perspective of Homeobox

Literature is replete with reports on evolution of Homeobox during the course of evolution and speciation (Mazza 2007; Duboule 1994). Therefore, instead of reproducing it here, we will focus on a rather neglected aspect of the evolution of Homeobox. The reason why we are sidestepping the main thrust of the evolutionary perspective of Homeobox and looking at its hitherto overlooked aspect, lies in the very conception of natural selection in the proposed model. The proposed model postulates that all potential genomic functionalities that found their expressions were present in their incipient forms in an entity named here as a genomic singularity. Therefore, according to this model, all the genomic functionalities simply manifest themselves during the course of biological evolution. The role of natural selection is simply to enable these incipient functionalities to find their expressions. In other words, natural selection merely explicates what was implicit in genomic singularity. Admittedly, it is possible to argue that such an entity is nothing but teleology in disguise, and therefore, it must be rejected. In the preceding chapters, this topic has been adequately discussed and this fear of teleological arguments has been allayed. The proposed model merely provides an explanation for the emergence of complexity during biological evolution and natural selection. It doesn't predict which kind of complexities can arise.

However, what is germane to the present discussion is whether this scenario (of having something like a genomic singularity) would reflect itself in the evolutionary trajectory of Homeobox. Therefore, having an enormous amount of literature on the evolutionary perspective of Homeobox would provide us with raw material to build up a case for the genomic singularity. Thus, instead of reproducing the previous literature, we will develop an argument of what kind of evolutionary course Homeobox could have taken if there was indeed something like a genomic singularity in the past. Having looked at such a possibility, we will try to find any support for this scenario from the published literature. Therefore, we will take the present literature on the evolutionary perspective of Homeobox as read and use it only to prove or reject the scenario outlined below.

In view of the extensive literature on this topic, we will focus on three critical features of Homeobox and its evolution, viz., the nature of modularity within Homeobox, the evolution of the modularity of Homeobox in the scenario beginning with the genomic singularity, and the role of mutations in the evolution of the modularity of Homeobox. The choice of these three features is based on a hunch that any conception of modularity would be greatly influenced by the presence or absence of the entity named here as genomic singularity. It must be kept in mind that the notion of modular architecture of genomes is generally accepted (Pevsner 2015; Peter and Davidson 2015), though the evidence for it doesn't match its popularity. Some of the general concepts of modular architecture of genomes and their nuances in the conventional perspective as well as in the proposed model are discussed in the following chapters. Therefore, in this section, we will restrict ourselves to the conventional conception of modularity in Homeobox.

Let us begin with the nature of modularity of Homeobox. In continuation with the arguments presented in the preceding chapters, we can think of Homeobox modularity in terms of temporal and spatial control of the gene expressions of genes under the influence of Homeobox. This separation of temporal and spatial influences is well documented in the literature and therefore, it seems reasonable to think that the modularity of Homeobox functionalities can be mapped on axes of temporal and spatial separations. The most remarkable aspect of this modularity, particularly the control of temporal and spatial influences, is that there is no mechanism proposed in literature. If we were to take a very simplistic model of these control elements, the conventional perspective suggests that the temporal control manifests itself in the form of the sequence of gene expressions and the rate of diffusion of the products of these gene expressions. In fact, it is possible in a more detailed description to ascribe the spatial control also to the quantitative difference between different rates of diffusion of these products of gene expressions (Bothma 2013). However, there is no explanation for a particular sequence of gene expressions and its evolution. Therefore, the conventional explanation of the origin of the modularity of Homeobox remains vague. As discussed in the following sections, the proposed model offers a single mechanism for temporal and spatial control exercised by Homeobox.

The reason why the proposed model can offer such a unified explanation of the origin of the modularity of Homeobox lies in the higher dimensional topological perspective of the proposed model. In consonance with the quantum mechanical perspective, the proposed model postulates that spacetime exists in multiple dimensionalities simultaneously. Therefore, at the higher dimensionality (higher than the four-dimensional spacetime), there is no distinction between the time-like and the space-like dimensions. Therefore, according to this model, if we can formalize the molecular perspective of Homeobox in higher dimensionalities, it is possible to think of a unitary mechanism for temporal and spatial controls arising from Homeobox. While our present computational capabilities do not provide us to formalize the higher dimensional configurations of Homeobox, the proposed model offers several general guidelines on the nature of the higher dimensional configurations. These guidelines can be verified by looking at the evolution of Homeobox using phylogenetic studies. Before we look at this, let us look at what the proposed model says about the role of mutations. This is important because the proposed model begins with an entity named above as a genomic singularity. Therefore, let us see how genomic singularity relates with the process of mutations. This is important because our current understanding of evolution, speciation and genetic drifts is based on the random nature of mutations.

Prima facie, it appears that if the conception of a genomic singularity is valid, then it must imply some design principles which are antithetical to the effects of random mutations. Therefore, it is necessary to accommodate the random nature of mutations within the semantics of genomic singularity. We will discuss this topic in greater detail in the following chapters. Presently, we will simply look at the gist of this argument. There are three propositions attached to the conception of a genomic singularity that are germane to the present discussion. Firstly, genomic singularity doesn't imply that all the functionalities of a genome are present in their totality in the genomic singularity. Secondly, though genomic functionalities do enjoy ontological primacy over genomic structuralism (due to the fact that they occupy higher dimensionality as compared to the genomic structuralism), there is no one to one relationship between functionalities and structuralism. Therefore, a given genomic functionality can devolve into multiple structural templates. Similarly, multiple functionalities can devolve into a single structural template. This is important because a genome functions at more than one dimensionality simultaneously. Therefore, within a range of changes in both functionalities and structural templates, there is no preordained design principle. In other words, the genomic functionalities are not present in toto in a genomic singularity. Rather, they are present in a notional sense. Thirdly, as a result of this incipient presence of genomic functionalities in a genomic singularity, the genomic singularity is not omnipotent, but it has potential to become one.

When viewed collectively, these three propositions offer a new perspective on the evolution of any module, including Homeobox. Since according to this model, a higher dimensional configuration can devolve into multiple lower dimensional configurations, it cannot decide the outcomes of lowering the dimensionality. Similarly, whenever a lower dimensional configuration undergoes changes like mutations, its linkage to higher dimensional configurations alters, but not necessarily. It is this many to many linkages between higher dimensional and lower dimensional configurations that ensures that there exists no predetermined structuralism. It must be kept in mind that at the same time, this scenario does restrict the range of changes in functionalities as well as those of structuralism. Therefore, the main function of genomic singularity is to restrain the number of variations both in genotypes and phenotypes. A more intuitive way to understand this distinction between the changes in the higher dimensionalities and the mutations in the four-dimensional perspective is to think of the former as changes based on topological compulsions and the latter as thermodynamic compulsions. Interestingly, according to this model both these types of changes are analogous because in the higher dimensionalities there are no time-like features. Therefore, there are no thermodynamic consequences of these changes. However, when the topological compulsions force the system to decrease its dimensionality, the time-like features emerge, thereby giving rise to thermodynamic consequences. Thus, it is the changes in the nature of dimensions in different dimensionalities that gives rise to analogous compulsions for changes.

It is possible to argue that if the postulate of genomic singularity serves such a limited role in explaining biological evolution and natural selection, why use it. In any case, several finely nuanced interpretations of the Darwinian paradigm are reported in literature (Grene 1986). Moreover, these interpretations are adequate for explaining major features of natural selection. Thus, using the principle of parsimony (Sober 2015), it seems tempting to reject the notion of a genomic singularity. However, this argument is fallacious. As discussed in the following chapters, there are several fundamental dilemmas about the Darwinian paradigm. Foremost among them is that the Darwinian paradigm is totally agnostic about the origin of Life. While natural selection is adequate to explain the original Darwinian theme of descent with modification, it fails to explain the finer details of the course of biological evolution. As discussed in Chaps. 1 and 2, these issues need to be addressed sooner than later.

More importantly, as discussed in the following chapters, to think that conception of a genomic singularity serves just the purpose of limiting the range of changes in phenotypes and genotypes is rather simplistic. The conception of a genomic singularity aligns biological evolution with the remaining natural phenomena. Just as the cosmic singularity due to its peculiar symmetry breaking gives rise to only a certain number of phenomenology of spacetime, a genomic singularity ensures that only a certain number functionalities arise during biological evolution. Thus, just as the cosmic singularity provides a common ontology of a diverse range of phenomenology, a genomic singularity provides a common ontology of a diverse range of genomic functionalities. At a still deeper level, the cosmic singularity acts as a source of all the information content that is present in the manifest universe. Similarly, a genomic singularity acts as a source of all the information content that is present in biota. As we know from quantum field theory (Lawrie 2013, see Chapter 5), the units of information content undergo different algebraic associations to give rise to different phenomenologies of spacetime. However, all the information content of the unified quantum field and even their rules of algebraic associations originate from the cosmic singularity. Similarly, not only the information content of all the phenotypes, but the rules of their structuralism arise from a genomic singularity. Thus, the conception of a genomic singularity is the fountainhead of all the semantic and syntactic framework of genomic architecture.

Returning to the specific example of modularity of Homeobox, let us briefly see how it could have evolved. To begin with, it is important to keep in mind that the modular nature of Homeobox is not evolved from a combination of several genes either in one go or in a stepwise manner. It is true that the constituent genes of Homeobox can be transplanted in different places of a genome. Such experiments have been reported in literature (Bermingham 1989). In fact, it was this approach that led to the discovery of Homeobox. However, it doesn't imply that Homeobox could have evolved by combination of these individual genes. In fact, there are no reports of partial evolution of Homeobox wherein only a few of these genes are combined to give rise to partly functional Homeobox which later on further evolved into a fully functional Homeobox. At the same time, there is enough evidence of Homeobox (or rather its individual genes) undergoing translocation, duplication and further mutations. Therefore, it is possible to find a specific example wherein there are multiple copies of Homeobox (either identical or mutated) in a single genome (Mazza 2007).

Therefore, the conventional perspective remains silent on the evolution of Homeobox or its functionalities. The proposed conception of a genomic singularity helps us to understand the evolutionary context of Homeobox. In fact, it explains the known facts about the evolutionary changes in Homeobox in an intuitive way. Let us assume that a genomic singularity exists. In that case, a modular cluster like that of Homeobox can be thought of as a higher dimensional configuration of genomes. As mentioned above a single higher dimensional configuration can devolve into multiple four-dimensional configurations, each representing a different body plan. More importantly, different four-dimensional configurations can legitimately represent the same body plan. This scenario opens up a room for mutations giving rise to anomalies like the one observed in the experiments leading to the discovery of the gene *antennapedia* (Bermingham 1989). Moreover, multiple four-dimensional configurations can also explain different roles played by the identical (or near identical) Homeodomain sequence in different stages of developmental biology. Before we elaborate the topological perspective of Homeobox as defined by the proposed model, let us look at some of the known variations in the structural details and the corresponding functionalities of Homeobox.

5.8 Variations in Structural Template of Homeobox

It is customary to investigate the changes in the DNA sequences of single genes while studying for the evolution of any particular gene. These changes include single nucleotide polymorphism (SNP) or insertions/deletions (INDELS). However, while studying the evolution of a group of genes or a genomic module, it is necessary to study changes caused by translocation, duplication and splicing of the group of genes under investigation. The rationale behind this change in approach is obvious. What we are looking for in the evolutionary trajectory of a module like Homeobox is not necessarily the functioning of individual genes, but rather the relationship among the constituent genes of Homeobox (Mazza 2007). This is necessary because the functioning of the modularity of Homeobox (or any other modules) lies in the intergenic regions of the genome. Since we don't know the exact genomic architecture, this comparative study becomes the only source of our knowledge of the exact mechanism of modularity. This conventional perspective is all the more critical to the proposed model because the proposed model lays the stress on the high dimensional configurations of genomes which are definitely influenced by intergenic regions of the genome.

However, in the conventional perspective the separation between different genes of a module are less important. This is because the conception of the temporal and spatial influences of a genomic module is defined by the particular sequence of gene expressions and the comparative diffusion constants of different proteins that are synthesized by the genes present in the given genomic module. Apparently, the changes in the length of intergenic regions would play no role in deciding the magnitude of the diffusion constants of different proteins. Therefore, the changes in the length of intergenic regions can play a role only in influencing the sequence of gene expressions of a given module. In comparison, the situation changes if the proposed model is valid. This is because according to the proposed model, the higher dimensional configurations of genomes also include the time-like and the space-like features of spacetime. Therefore, whenever there is a change in the dimensionality of a genomic module, it influences both thermodynamic and kinetic profiles of the genomic module. Therefore, the surest way to confirm that the proposed model is valid, is to correlate the length of intergenic regions of various species and the degree of spatial and temporal control elements. Admittedly, this will provide a rough correlation, but it would provide a definitive answer. Of course, once we figure out the details of higher dimensional configurations, it would be possible to have a quantitative description of the mechanism by which higher dimensional configurations influence the outcome of gene expressions.

In order to formalize the relationship between the lengths of intergenic regions with a postulated higher dimensional configuration, we can think of several classes of these lengths and postulate that each such class arises from a unique higher dimensionality. Thus, even without knowing the exact template of various higher dimensional configurations, it is possible to create a hierarchy of higher dimensionalities simply by classifying different lengths of intergenic regions. Once we assign different dimensionalities to different lengths of intergenic regions, the next step would be to correlate different dimensionalities to different rates of diffusion. This would, at least in principle, provide a way to define spatial and temporal control elements on par with one another. In the case of Homeobox genes, the situation is slightly complicated by the fact that in this case, the control elements, in the form of Homeobox genes, are present in the close vicinity. However, since we have an enormous amount of data available on the details of which of these Homeobox genes, individually and collectively, influence different gene expressions (Duboule 1994), it would require computer simulations to arrive at the topological perspective of these genes. However, as a first step, it is possible to visualize how different genes were segregated into different chromosomes. A schematic representation of one such template is given in Schema 5.4. Admittedly, this is a very primitive representation, but it serves our present attempt to formalize the topology of Homeobox. The actual template may be derived by employing some computer simulations by incorporating numerous genes and their expressions that are known to be governed by Homeobox.

Even if we accept this primitive framework at a face value, it is imperative that we should try to correlate it with the corresponding functional template of Homeobox. This is imperative because the proposed model asserts that functional and structural templates enjoy a definitive relationship. Therefore, in the next section we will look at what kind of functional template of Homeobox is possible.

5.9 Variations in Functional Template of Homeobox

As mentioned above and in the preceding chapters, the conventional perspective of functionalities in general is derived indirectly from the corresponding structural template. Therefore, within the confines of the conventional perspective, it is difficult to conceptualize any functional template without any reference to the corresponding structural template. Admittedly, the proposed model denies any independent existence of any functional template without the underlying structural template. However, according to this model, functionalities have their own template which is independent of the underlying structural template. Of course, according to this model, both these templates enjoy a definitive relationship. Therefore, in this section, we will try to deconstruct a functional template of Homeobox which is distinctly different from the underlying structural template of Homeobox, even when it is formally connected with it.

From the conventional perspective, functionalities of genomes are essentially catalytic in nature. In other words, while it is possible to differentiate different

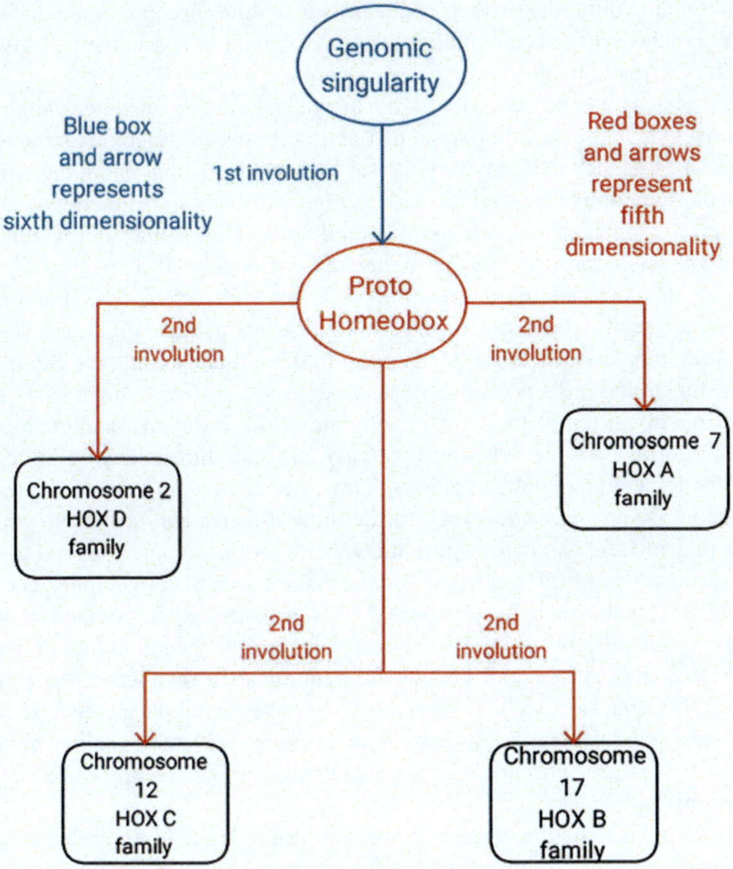

Schema 5.4 Topological representation of Homeobox genes

genomic functionalities into several categories, in essence, they are just catalytic agents. The proposed model also concedes the primacy of catalytic functions from which different genomic functionalities are derived. However, the proposed model goes one step further and postulates that these different functionalities acquire their distinctive characteristics from the parent catalytic functionality by a fixed set of rules. More importantly, these rules are defined by the inherent structuralism of the higher dimensional spacetime in which they manifest. Thus, there is a subtle difference between the nuances of these two approaches. The conventional perspective defines catalytic functionality as a primordial functionality from which we can define different genomic functionalities. However, according to this model, all the genomic functionalities enjoy ontological primacy. The catalytic functionality is merely an abstraction of these functionalities. Secondly, according to this model, different functionalities derive their characteristic features because of the dimensionality that they occupy. It is the underlying structuralism of spacetime that gives rise to plurality of functionalities.

With these caveats in place, let us try to understand the functional template of Homeobox. One of the most important distinctions between structural and functional templates of Homeobox (or any module for that matter) is that while the structural template of Homeobox is spread the wide genomic distances, its functional template is confined to a smaller genomic architectural span. This feature is implicit in the proposed model because it assigns higher dimensionality to a functionality with respect to its corresponding structural template. Therefore, the influence of functionalities seem spread out to the lower dimensionalities which are occupied by structural templates. In fact, it is tempting to think that by using the extent of the spread of functional influences, it is possible to assign different dimensionalities to different functional templates. Thus, even among the different functionalities, the functionalities like Homeobox must occupy highest dimensionality, if only to account for its wide range of influences. Thus, just as we can employ the measure of intergenic distances as a yardstick for defining dimensionalities of the structural template of a module, we can employ the yardstick of topological spread to define the dimensionalities of different functional modules in a given genome. Thus, it is intuitively clear that even in the absence of any formal description of the topology of the genome, it is possible to infer some rough features of genomic topology purely from the conventional wisdom behind the genomics. In that sense, the proposed model merely explicates what is implicit in the conventional perspective. If we find the proposed model counterintuitive, it is because we are mistaken about the equivalence between absence of any structuralism of natural selection and nondeterminism.

The next step is to define functional primacy within Homeobox genes. While our conception of genomic architecture is still ambivalent about the nature of discreteness within the genome, we still expect the elements of any genomic architecture to be discrete. This is analogous to our conception of genetics. Admittedly, we have abandoned the belief that a genome consists of a series of genes strung together. However, we still try to define genes as a discrete entity. This ambivalence arises perhaps from our cognitive compulsions. Therefore, even while we accept Homeobox as a module, the fact that it contains discrete entities in the form of individual genes needs to be accounted for. Just as now we concede that the relationship between a genome and its constituent genes needs to be defined in order to understand the overall architecture, we also need to accept that Homeobox, as a module, must enjoy a definitive relationship among its constituent genes. Thus, the notion of modularity need not be limited to the relationship between the modules, but must be extended to the relationship among the constituent genes of Homeobox as well. This is analogous to our understanding of the units of selection (Okasha 2010, see Chapter 2). As discussed in the preceding chapters, the definition of the units of selection is flexible and perhaps contextual. Similarly, the notion of modularity too must be taken as flexible and contextual.

Once we accept this reasoning, it is intuitively clear that even the relationship among the constituent genes of Homeobox too must be defined by some parameters. Since the proposed model employs a topological perspective, it is inevitable that the relationship among the constituent genes of Homeobox too must be defined in some topological parameter. As mentioned above, we can use the length of intergenic regions as a yardstick for structural template and the yardstick of topological spread for functional template. Therefore, upon a little reflection, it is intuitively clear that the correct yardstick to define the relationship among the constituent genes of Homeobox must be that of dimensionalities. Thus, according to this model, Homeobox (or any module for that matter) must occupy multiple dimensionalities simultaneously. There are two reasons for this. Firstly, to the extent the genes present in Homeobox are functionally discrete, they would be influencing different genes in expressing themselves. Since the DNA sequence of a genome is well spread out, the influence of different genes of Homeobox must travel different distances. This is a natural corollary from the concept of using intergenic distances as a yardstick. Thus, once we accept that different lengths of intergenic regions arise because of a particular ontology of the genomic architecture, it is inevitable that in order to influence different genomic distances, the genes present in Homeobox must be placed in different dimensionalities. This provision ensures that the nature of long-range influences (irrespective of their hitherto undefined mechanism) is unitary in nature. In other words, it is the difference in the dimensionalities of the genes present in Homeobox that ensures that their influences travel different genomic distances.

The second reason why we must accept that Homeobox (like any other module) must occupy multiple dimensionalities simultaneously lies in the very conception of biological evolution beginning with the genomic singularity. Once we accept that different functionalities (including modularity) arise from the genomic singularity, it is intuitively clear and perhaps semantically inevitable that the mechanism by which different functionalities separate out of the genomic singularity must be unitary. The reason why we have different functionalities is not because there are different mechanisms of natural selection. Rather, we have different functionalities because of different types of information content present in these functionalities. According to the proposed model, the underlying mechanism of natural selection is unitary. What differentiates these functionalities is their inherent information content and its organization. The notion of a unitary mechanism of natural selection is implicit in the contextual definition of the units of selection. Therefore, we are simply extending it to the modularities as well. Thus, now we have three parameters to define the template of Homeobox.

1. Measure of intergenic distances for defining the structural template of Homeobox.
2. Measure of topological spread of the influence of the constituent genes of Homeobox.
3. Measure of the range of dimensionalities in which the constituent genes of Homeobox exist.

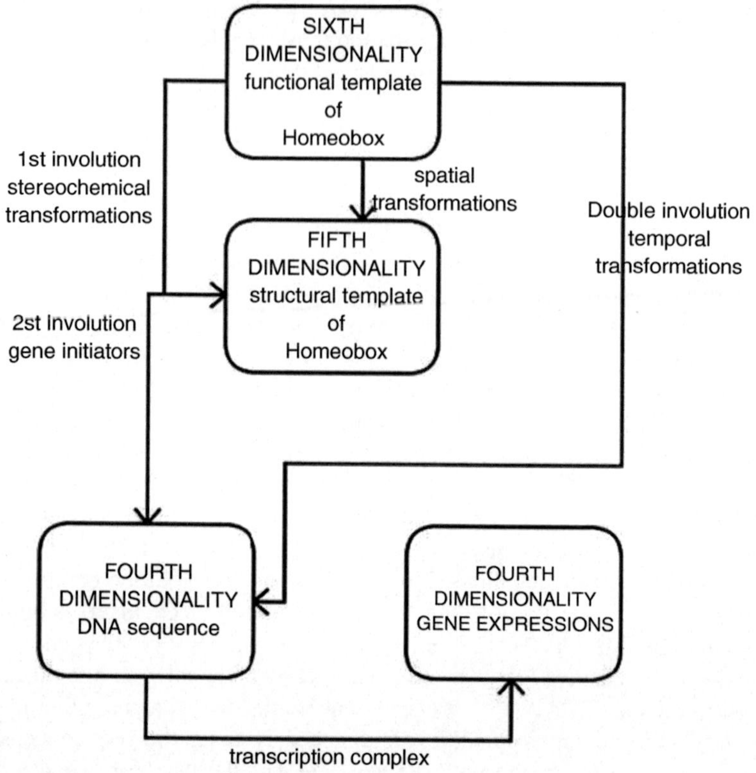

Schema 5.5 Topological model of molecular biology of Homeobox

It is important to keep in mind that these yardsticks are not dependent on any particular mechanism. They are defined by the conception of the genomic singularity and the subsequent topological compulsions. Without going into the details of the individual genes present in Homeobox and their molecular biological operations (which are extensively reported in literature) a tentative topological model is presented in Schema 5.5. Having looked at the structural and functional templates of Homeobox, in the following section, we will look at the possible relationship between them.

5.10 Nature of Relationship Between Functional and Structural Templates of Homeobox

One of the objectives of describing the variations in structural and functional templates of Homeobox in the preceding sections was to lay a foundation for defining the exact relationship between functional and structural templates. This is necessary for two reasons. Firstly, the conventional perspective doesn't acknowledge that functionalities can have their own templates. Therefore, the conventional perspective normally uses the structural template as a basis for formalizing functionalities. However, according to this model, structuralism and functionalities can have their own templates. Therefore, we will use our discussion from the preceding sections to demonstrate that a functional template exists independent of the structural template. More importantly, both these templates enjoy a definitive relationship between them. Secondly, we wish to discern the reason why Nature requires two different templates for structuralism and functionalities of genomes. In order to understand this aspect, Homeobox is perhaps the best example. It is suggested that the need for having separate templates for structuralism and functionalities arises from the difference between the static and dynamic models of genomic architecture. It is the contention of this monograph that the conventional perspective has focused only on the static model of Homeobox (and by implication that of genomes). Therefore, an attempt would be made to demonstrate that the proposed model is complementary to the conventional perspective of genomics.

In the conventional perspective, we assume that the template of structuralism is adequate to explain the functionalities. Thus, as discussed in the preceding sections, we define temporal and spatial influences in terms of the sequence of gene expressions and the rates of diffusion. Since both these parameters are dynamic in nature, the conventional perspective finds it difficult to formalize them as it is based on the static model of DNA sequences. Admittedly, we can revert back to biochemistry to formalize rates of diffusion (Bothma 2013), but there is no way to formalize the sequence of gene expressions. In contrast, the proposed model offers a common framework for both these parameters. This is because it postulates that there exists higher dimensional configurations which include the temporal and spatial influences (in the form of the fine structure of spacetime). Therefore, both these parameters, viz., the sequence of gene expressions and the rates of diffusion, are parameterized in these higher dimensional configurations. Admittedly, till such time when we can formalize these higher dimensional configurations, the determination of both these parameters would be of little practical value. However, the key point is that the proposed model offers a semantic foundation for parameterization. As discussed above, these parameters are tentatively named as chronon and spation. While we do not have any formal definition of these two parameters, the underlying semantic content can help us to deconstruct hitherto unknown organization perspectives of Homeobox. This is precisely what we will discuss in this section, viz., the nature of relationships between functional and structural templates of Homeobox.

Let us, at least in principle, accept that the conception of higher dimensional configurations of Homeobox (or genomes) is valid. In such a scenario, it is intuitively clear that it would influence the lower dimensional properties as well. Therefore, if we can demonstrate that at least some of the unexplained control elements of Homeobox are amenable to comprehension using the conception of a higher dimensional configuration, then we can consider the proposed model as a serious scientific template, albeit subject to experimental proof. For this purpose, we will discuss three control elements of Homeobox which have been extensively dealt with in literature (Mazza 2007) and yet have defied any causal explanation. These features are (1) The position of Homeobox vis a vis the genes under its influence, (2) The copy number variations of Homeobox and its influence on the body plans, (3) The evolution of proto Homeobox into Homeobox and related clusters. Admittedly, these three topics are widely discussed in literature, therefore, we will merely point out the shortcomings of the conventional perspective and demonstrate how these shortcomings can be addressed by the proposed model. Some of these broad objections are discussed in the preceding chapters. Therefore, we will confine ourselves to these shortcomings in reference to Homeobox genes collectively.

It is abundantly clear from literature that Homeobox, in its most primitive form, appeared as Proto Homeobox in cnidarians and bivalves (Fusco 2008, see Chapter 10). Eventually, through a series of mutations, duplications and translocations it split into several clusters in different lineages including chordates. From chordates to mammalian species the journey of this cluster of genes is fairly well documented (Mazza 2007). However, the three topics mentioned above do not find any articulation in literature. Perhaps, as discussed in the preceding chapters, this ambiguity is generic in nature and may have arisen from the lack of knowledge of genomic architecture. However, there are several reviews on the evolution of Homeobox and related clusters (Mazza 2007; Duboule 1994). It is apparent from these studies that while the functionalities of Homeobox are discovered by changing the places of individual genes of Homeobox (cf. Antennapedia (Bermingham 1989)), the exact evolutionary scenario of the corresponding functionalities is not available. Therefore, we will try to understand these three topics by assuming that there exists some higher dimensional principles (not necessarily those postulated in the proposed model).

Let us begin with the first feature of the position of Homeobox vis a vis the genes under its influence. As mentioned above, according to the conventional perspective, there is no explicit relationship between a control element and the genes under its influence. Admittedly, factors like prior expressions of initiators and facilitators or relative rates of diffusion or temporary proximity (as implicit in the concept of chromosome territories (Fritz 2014)) are cited as possible causes for genomic controls. However, there is no systemic articulation of any of these phenomena. Thus, if chromosome territories happen to control the sequence of gene expressions, it is purely accidental (albeit naturally selected after its chance occurrence). Similarly, if some initiators and facilitators are synthesized prior to their employment, it is just an evolutionary accident, albeit exploited subsequently by natural selection. This type of reasoning is congruent with the implicit randomness of the Darwinian paradigm. However, as discussed in the preceding chapters, this is a category mistake. We can retain the randomness in natural selection even after a well-defined architecture of genomes is established.

In order to understand this, let us look at what the proposed model offers on this topic. According to this model, the genomic distance between a control element and the genes under its influence is defined by the number of changes in the dimensionalities of the control element and the genes under its influence. Admittedly, when this distance is measured by the conventional unit of kilobase pairs of nucleotides, there is no generic relationship discerned. Therefore, no such model has been proposed. However, the proposed model replaces the notion of genomic distances (as defined by kilobase pairs of nucleotides) by a topological unit of dimensionality. Irrespective of its merits, the question is how this measure of genomic distance can be measured? It is all the more difficult to measure this unit because the definition of dimensionality in the proposed model refers to the dimensionality of spacetime and not some information theoretical parameter. It is our cognitive inability to observe Nature in different dimensionalities that is the reason for our failure to perceive and measure this unit of genomic distance. However, our cognitive inability cannot be an alibi to accept this model. There ought to be some tangible evidence of this measure for us to accept this model. Fortunately, such evidence is available, but we have overlooked it.

The evidence lies in the fact that we have overlapping DNA sequences of different genes (Sabath 2009). Such an overlap represents different functionalities existing in different dimensionalities projecting onto the same DNA sequence. In fact, this type of DNA overlap remains unexplained in the conventional perspective of genomics. Even when we look at this overlap from the perspective of natural selection, there is no justification for such overlap. At best, we can think of it as one of the quirks of biological evolution. However, according to this model, this overlap must occur. This is because the number of dimensionalities in which biological functionalities can exist is limited. Therefore, when these different higher dimensionalities devolve into the four-dimensional DNA sequence, it is inevitable that there will be some overlap between the DNA sequences of different genes. There is an exact analogy for this in physics. As discussed in the preceding monograph (Chhaya 2022b), a fundamental particle possesses multiple properties like spin, mass, charge, etc. In the conventional quantum mechanical perspective there is no explanation why a fundamental particle should possess different properties at the same time. Albeit, in physics, we have symmetry principles which partly explain in which combination these different properties can manifest themselves in a single particle. However, in field theoretical perspective, there is no explanation for a single particle possessing these properties simultaneously. This feature is taken as a priori. However, as discussed in the preceding monograph, the proposed model explains this phenomenon. It assigns different dimensionalities to different properties. When each of these properties devolve into the four-dimensional spacetime, they manifest themselves as different properties. Moreover, because these different dimensionalities devolve into a single point of the four-dimensional spacetime, they appear to us as particles with different properties. Similarly, we can think of different biological functionalities existing in different dimensionalities. When they devolve into the four-dimensional spacetime, we get this overlap of DNA sequences representing different genes. The only difference between the fundamental particles and genes is that because quantum mechanical functionalities occupy different ranges of dimensionalities vis a vis the range of dimensionalities occupied by biological functionalities. Therefore, in the case of the former, the devolvement results in convergence to a single point of spacetime. In the case of the latter, the devolvement converges to a larger sliver of the four-dimensional spacetime which contains the DNA sequence.

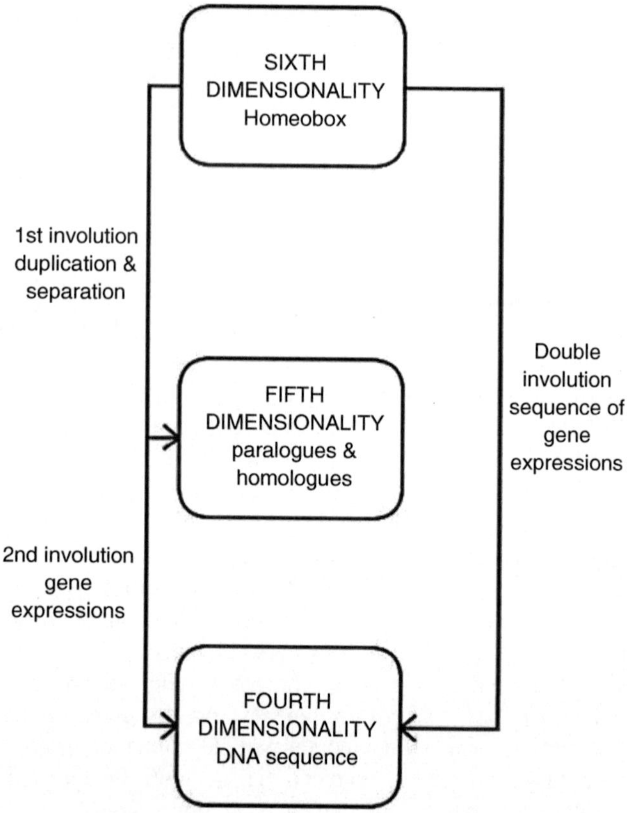

Schema 5.6 Topological model of structural and functional templates of Homeobox

It is tempting to think a host of other features of genomics can be viewed from this perspective. For instance, phenomena like RNA splicing (Stamm et al. 2012), RNA interference (Cretoiu et al. 2020; Howard 2013), and sense/antisense transcriptions (Mostovoy 2014) can very well be explained by the symmetry breaking processes which devolve the higher dimensional functionalities into the four-dimensional spacetime where different biomolecules exist. Returning to the specific example of Homeobox, a tentative topological model representing the relationship between the functional and structural templates of Homeobox is given in Schema 5.6. For the sake of simplicity, we have used a generic nomenclature of functionalities. Therefore, as shown in the Schema 5.6, the first involution leads to a structural segregation of paralogs and homologs. It is important to keep in mind that according to the proposed model, at the higher dimensionality (sixth dimensionality as shown in the diagram) is a single entity which segregates out after the first involution into separate templates. The second involution leads to the individual gene expressions. It is important to keep in mind that while this pathway gives the mechanism of Homeobox functioning, the process of synchronization is enforced by double involution leading to the exact sequence of gene expressions. Admittedly, this is a simplistic depiction. However, the underlying logic is consistent with the conventional wisdom behind the functioning of Homeobox. In the following two sections, we will try to deconstruct this separation of functionalities and structuralism into static and dynamic models of Homeobox, as implicit in the conventional perspective. Following that, we will try to understand why the conventional perspective fails to account for the relationship between the static and dynamic models of Homeobox. Finally, we will outline a new model of Homeobox as defined by the proposed model.

5.11 Static Model of the Relationship

At the outset, it is necessary to distinguish between what is labeled here as static and dynamic models. This is necessary because in the conventional perspective, the emphasis is on the static model of Homeobox. This largely consists of the DNA sequences of the genes present in Homeobox and the transcription machinery (in the form of protein complexes). The corresponding dynamic components (which are labeled here as the dynamic model) are not viewed as a template. Rather they are taken as operations of functional genomics. As mentioned above, this is in congruence with the conventional perspective which delinks structuralism and functionalities. According to the conventional perspective, functionalities do not have any framework of their own. Functionalities arise as a consequence of thermodynamic and kinetic propensities of the concerned biomolecules. However, according to the proposed model, functionalities, collectively, possess a framework of their own. Moreover, this framework incorporates spacetime as an integral element. Therefore, according to this model, what is considered as incidental thermodynamic and kinetic propensities, are in fact, consequences of the direct participation of spacetime. Therefore, this is labeled here as a dynamic model of Homeobox.

With this caveat in place, let us look at the static model of Homeobox. By and large, the proposed model is congruent with the conventional perspective of what constitutes a static model of Homeobox. However, there are two aspects of the proposed model which differ from the conventional perspective of Homeobox. They are the relationship between the constituent genes of Homeobox and the role of protein complexes necessary for transcription. Let us look at these differences. Let us begin with the relationship between the constituent genes of Homeobox. As mentioned above, there exists a voluminous literature on the evolutionary perspective of how the constituent genes of Homeobox have evolved (Mazza 2007; Duboule 1994), either individually or in relation with the remaining members of Homeobox. Therefore, the conventional wisdom of the evolution of Homeobox needs to be accepted as such. Admittedly, there are lacunae in the finer details of how Homeobox, either as a whole or as an ensemble of genes, has evolved. However, these lacunae will eventually be filled as more and more evidence is gathered. The proposed model doesn't question this conventional wisdom. Rather, it differs on one key point about assembling the constituent genes to create a module of

Homeobox. This refers to the origin of this modularity. As mentioned above, the earliest appearance of Homeobox or HOX genes is reported in Cnidarians (Duboule 1994). Later on, during the course of evolution and natural selection, it underwent several indels, duplications to its present form in *Homo Sapiens.* Therefore, the conventional perspective is silent on the ontology of Homeobox. As mentioned above, there is no evidence of partial evolution of Homeobox. It appears with its major functionalities in toto in the form of proto Homeobox (Fusco 2008, see Chapter 10). However, according to the proposed model, this total arrangement of the constituent genes of Homeobox (as it emerged in Cnidarians) arose from one source, viz., the genomic singularity. While this proposition can be dismissed as it is unverifiable and therefore unscientific. However, the reason why it must be taken seriously is that once we accept the original arrangement of different constituent genes within Homeobox (as it appeared proto Homeobox), it helps us to anticipate what kind of changes it could have undergone to its final modularity in *Homo Sapiens.*

Therefore, according to the proposed model, the earliest template of Homeobox arose from the genomic singularity due to the topological compulsions of the symmetry breaking processes working on the genomic singularity. Moreover, once the primitive template of Homeobox emerged (in Cnidarians), the same topological compulsions provided several further modifications. However, due to the inherent nature of changes in the environment, only a few of these all the possible modifications were selected. It is important to keep in mind that this scenario does not bring in any design principles through the backdoor. It merely says that at each stage, the number of modifications available are available for natural selection. It is just that the number of modifications is limited by the topological compulsions of the genomic singularity. Even this scenario can be dismissed as a superfluous assumption. However, if it can be demonstrated that these topological compulsions (which are expressed in their algebraic rules of the involutive algebra) indeed tally with our phylogenetic studies of Homeobox, then we must consider the postulate of genomic singularity as a valid scientific hypothesis.

Alternatively, if we can corroborate this scenario with the available data from functional genomics, then the conception of genomic singularity becomes a credible scenario. This is where the second difference between the proposed model and the conventional perspective lies. This refers to the role of protein complexes in transcription. The conventional perspective takes the formation of the individual proteins of the transcription complexes as a priori. It is considered a priori because in the conventional perspective, there is no explanation why these proteins must be synthesized prior to gene expressions. In other words, in the conventional perspective, there is no causal link between the prior syntheses of these proteins and the gene expressions in which these proteins participate. This reasoning is applicable to the genes present in Homeobox. These genes express themselves and in turn, they control the gene expressions of the target genes. Admittedly, the conventional perspective rests on the sound logic of thermodynamic kinetic processes that govern the influence of these transcription complexes. Therefore, the ontological question of why and how these protein complexes are synthesized prior to their employment remains unaddressed in the conventional perspective. As suggested above, this lacuna arises because we are apprehensive of reintroducing some design principles if we attempt to introduce the ontological perspective of these protein complexes.

Before we look at the reasons why this apprehension is misplaced, let us look at what the proposed model offers on this topic. In continuation with the reasoning given above, according to this model, the protein complexes and the target DNA sequences are causally unconnected with one another. The causal processes set in only after the protein complexes interact with the target sequence of nucleotides and not prior to it. Thus, there is a certain degree of commonality between the conventional perspective and the proposed model. However, there is a huge difference between these two approaches as far as the ontological perspective is concerned. According to the proposed model (unlike the conventional perspective), there exists a higher dimensional representation of genomes wherein these protein complexes (albeit through their parent genes like those present in Homeobox) and the target DNA sequences are linked to one another. However, in these higher dimensionalities, there are no time-like features and therefore, there are no causal linkages between them. It is only when these higher dimensional configurations of genomes devolve into the four-dimensional spacetime that spacetime becomes a blend of time-like and space-like dimensions, thereby giving rise to the thermodynamic forces and the scenario implicit in the conventional perspective unfolds.

At first sight, this explanation doesn't make sense. This is because from the perspective of the four-dimensional spacetime (in which we test our scientific theories), both these approaches are undistinguishable. Moreover, the proposed model requires an additional unverifiable postulate to justify itself. Therefore, it is natural to reject such a scenario. However, there is one aspect of the proposed model which qualifies it to be taken seriously as a scientific hypothesis. This refers to the topological imperative. Let us briefly understand how this topological imperative arises. According to the conventional perspective, the entire catalytic properties of these protein complexes (which are the foundations of control of gene expressions) rests on the stereochemical congruence between these protein complexes and the target DNA sequences (along with). In the conventional perspective, there is no explanation why this stereochemical congruence between these proteins and DNA sequences arise. This is a serious lacuna because it would be wrong to apply the paradigm of natural selection here. It is wrong because natural selection operates on genes (and perhaps on the entire genome), but not on the shapes of these molecules. More importantly, if we accept that natural selection operates on shapes of these molecules (Bryson and Vogel 1965), we will have to abandon our belief that natural selection doesn't have any structural template. It is possible to argue that there is no reason why natural selection cannot operate on the three-dimensional configurations of individual molecules. To be honest, it can. After all, as mentioned above, even the tertiary structures of proteins are conserved during natural selection. Therefore, there is no reason why the stereochemical configurations of these protein complexes couldn't have arisen due to natural selection. However, there is a flaw in this reasoning. Once we accept that natural selection operates on three-dimensional configurations of molecules present in genomes, we are introducing a generic mechanism for natural selection. This structural generic mechanism of natural selection would reintroduce the design principles back into the Darwinian paradigm. However, the proposed model saves natural selection from being deterministic. According to the proposed model, these higher dimensional configurations of genomes can devolve into multiple templates in the four-dimensional spacetime. The above-mentioned topological imperative demands that any higher dimensional configuration must devolve into a set of predefined lower dimensional configurations. Thus, the topological compulsions of this model are such that they force these higher dimensional configurations to devolve into the four-dimensional spacetime, but in multiple configurations. As mentioned above, the types and the number of these configurations are defined by the parameter of spation. This ensures that there is an ontological explanation for the functionality of controlling gene expressions without succumbing to any deterministic design principles.

There is an additional advantage of this model. It connects different thermodynamic and kinetic properties of these protein complexes among themselves and with those of the target DNA sequences through a formal description. Thus, the stereochemical congruence between these protein complexes and the target DNA sequences is causal in nature due to inherent structuralism of the higher dimensional spacetime. It is legitimate to wonder how such a connection between thermodynamic and kinetic profiles of these diverse entities can be established merely on the basis of higher dimensional perspective. In order to understand this, we will look at the dynamic model of Homeobox. It must be kept in mind that we have not looked at the molecular perspective of either Homeobox genes or their targets in this section. This is largely because the arguments presented above are fundamental and apply to a whole host of genomic functionalities. However, we will discuss a schematic representation of static and dynamic models of Homeobox in Sect. 5.13.

5.12 Dynamic Model of the Relationship

In the preceding chapters and in the preceding sections, it was suggested that the conventional perspective of genomics (and the biochemical science in general) tries to deconstruct the conception of functionalities on the basis of their underlying structural templates. In other words, it is never thought necessary to conceptualize functionalities as having templates of their own. On the other hand, the proposed model insists that both structuralism and functionalities have their own templates and more importantly, these templates enjoy a definitive relationship between them. However, as discussed in the preceding sections, in the case of Homeobox, it was suggested that the dynamic model of Homeobox, in the form of its control mechanisms, cannot be discerned separately. More importantly, it was suggested that this hypothetical dynamic model eventually leads to a stereochemical perspective which is identical to the one available from the conventional perspective of thermodynamics and Kinetics. Under these circumstances, it is natural to doubt the very conception of a dynamic model of Homeobox. If any new theory gives identical perspective as that of an old theory and if the new theory demands additional assumptions, it would be legitimate on the grounds of parsimony (Sober 2015) to discard the new theory. Therefore, in this section, we try to understand the reason why such a new theory is legitimate and deserves a serious investigation.

Let us begin by understanding what exactly constitutes a dynamic model of Homeobox according to this model. As mentioned above, we can think of the constituent genes of Homeobox and even the protein complexes generated from these genes as a static model of Homeobox. As mentioned above, according to the conventional perspective, this is all there is for the structuralism of Homeobox. Its functionalities of controlling various gene expressions is merely a chance convergence of the stereochemical congruence between these protein complexes and the target DNA sequences. Admittedly, this control is executed through thermodynamic and kinetic effects. However, neither the stereochemical orientation, nor the subsequent thermodynamic/kinetic effects, either by themselves, or collectively, can be thought of as a template for functionalities of Homeobox. The stereochemical configurations of these proteins and their resulting thermodynamic/kinetic effects are taken as separate phenomena in the conventional perspective.

The reason behind this conception of the functionalities of Homeobox lies in our conception of thermodynamics and kinetics which is rooted in the Newtonian paradigm. One of the founding propositions of the Newtonian paradigm is that spacetime is an inert background on which the energy transfers occur. Therefore, according to the Newtonian paradigm, the thermodynamic and kinetic properties of molecules (in this case the protein complexes and the target DNA sequences) are solely dependent on their individual molecular structure. This implicit semantics of spacetime becomes explicit when we try to define chemical functionalities (including the catalytic functionalities of these protein complexes) using quantum chemistry. Apart from the fact that these computations become intractable for the molecules containing more than a handful of atoms, the reason why we can't define chemical functionalities using quantum chemistry (Szabo and Ostlund 1989) is that spacetime is no longer an inert background. Once we accept that spacetime plays an active role in thermodynamic and kinetic processes, it is intuitively clear that the earlier assumption that stereochemical configurations and the thermodynamic/kinetic properties are separate entities needs to be abandoned. We need a framework wherein the stereochemical configurations and the thermodynamic /kinetic properties of molecules can be unified through the agency of spacetime.

It is possible to argue that even if this reasoning is valid, we have successfully employed the conventional perspective which is simpler. However, it must be kept in mind that after using this conventional perspective for all these years, we have not been able to formalize genomic architecture (let alone Life) in a purely algebraic sense. Therefore, it is worth trying out new approaches, if only to obtain a better understanding of the complex architecture of genomes. More importantly, the proposed model offers a way to unify spacetime and matter in a single framework. More importantly, it provides a way to define a relationship between spacetime and its segregation into time-like and space-like features. Therefore, it is possible to formalize spacetime as an active medium through which the stereochemical configurations and the thermodynamic/kinetic properties of these protein complexes can be interconverted. Let us, for the sake of argument, accept the rationale given here as valid. However, semantic implications apart, this approach must offer a definitive mathematical formalism which helps us to define a dynamic model of all the genomic functionalities, including those modularized as Homeobox. Admittedly, the exact details of the computational perspective of the proposed model are not available. However, it is possible to obtain the operative logic behind this formalism.

According to the proposed model, spacetime exists in multiple dimensionalities simultaneously. Of these multiple dimensionalities, only some of the lower dimensionalities manifest feature temporarily. In other words, the lower dimensionalities (including the four-dimensional spacetime that we are accustomed to) manifest passage of time. Therefore, it is possible, at least in principle, to define thermodynamic/kinetic energy exchanges by defining them in terms of the changes in the dimensionalities of spacetime. Similarly, the proposed model postulates that spacetime and matter are actually isomorphs. Thus, when domains of higher dimensional spacetime get involuted to give rise to the four-dimensional spacetime, it segregates out as conjoined twins of the four-dimensional spacetime and matter. Thus, our initial expectation of spacetime acting as a link between the stereochemical configurations and the thermodynamic/kinetic properties of molecules can be fulfilled by using the model of spacetime existing in multiple dimensionalities simultaneously. All that we require is to define an operator for changing the dimensionalities. The proposed model offers one such operator.

As discussed in the preceding monograph (Chhaya 2022a), a topological model of modified involuted manifold was described. Conventionally, we define an operator of involution as the relationship between a parent manifold and any of its submanifolds. However, the proposed modified operator of involution defines involution as the relationship of the parent manifold with itself. This operator can be visualized as an inward folding of one of the dimensions of a manifold into its remaining dimensions. As discussed in the preceding monograph, this operator when applied to spacetime gives us a model of spacetime possessing multiple dimensionalities simultaneously. Thus, it is possible to extend this approach to define thermodynamic/ kinetic effects discussed above (Chhaya 2022b). The advantage of using this model to define long-range influences is that we can define higher dimensional configurations of genomes. Since matter and spacetime are considered as conjoined twins in this model, the higher dimensional configurations of genomes would axiomatically involve the participation of spacetime. Therefore, when we try to reduce the dimensionality of any of these higher dimensional configurations of spacetime, the biomolecules and spacetime separate out into two different realms. Moreover, since spacetime has its own structural template in the form of its fine structure, this separation leads to only a few of all the possible thermodynamic/ kinetic properties of these biomolecules. Thus, the postulated higher dimensional configurations of genomes containing both biomolecules and spacetime, ensure that the stereochemical configurations of these molecules in the lower, four-dimensional spacetime, are congruent with their underlying thermodynamic/kinetic properties which are responsible for their catalytic functionalities.

Thus, entwining of spacetime and biomolecules at higher dimensionalities gives rise to a definitive relationship between structuralism and functionalities of genomes as well as Homeobox. There are two comments that are pertinent in the present discussion. Firstly, this model explains why this particular assembly of transcription complexes arise out of nothing. In the conventional perspective, there is no explanation for these molecules to develop stereochemical congruence if natural selection operates only on phenotypes. This is because phenotype in this case is not these proteins individually, but their intricate assembly. In a wider context, this argument is not valid because phenotype is different from genotype and it is produced at the end of the transcription. However, in this case the transcription machinery itself is a phenotype and can't be separated from transcription itself. Purely from the semantic perspective, this scenario refers to one of the drawbacks of the Darwinian model of natural selection (Grene 1986). The Darwinian paradigm is essentially focused on natural selection after Life had evolved. It can't be applied to biological evolution per se. The proposed model extends the Darwinian paradigm to include biological evolution as well.

The second comment that is relevant here refers to the fact that this scenario, in spite of employing higher dimensional configurations, doesn't entail any design principles. The postulate of higher dimensional entwining of spacetime and biomolecules of genomes merely restricts the number of possible conformations of these biomolecules and their subsequent catalytic functionalities. Therefore, it is always possible that this devolvement from higher dimensionalities to the four-dimensional spacetime may lead to wrong conformations of proteins and may stem further transcriptions. Thus, there is no preordained intelligent design principle being implied in this model. If at all, the proposed model restricts the number of possible outcomes during natural selection. However, this kind of punctuated evolution is within the confines of Darwinian semantics (Gould 2007). Moreover, higher dimensional configurations influencing lower dimensional configurations can also explain two different types of biochemical enigmas. As discussed in the preceding chapters, natural selection tends to preserve the tertiary structures of proteins. Conventionally, this is inexplicable because it is intuitively clear that a given protein molecule may be naturally selected. However, it is difficult to explain why a particular stereochemical orientation could be selected. However, the proposed model offers a cogent explanation for this. The tertiary structures of proteins are conserved because natural selection operates using a certain mechanism which involves higher dimensionalities. Therefore, it is the topological compulsions that decides the natural selection of tertiary structures of proteins.

The second example refers to the post transcriptional modifications of proteins and their folding (Wetlaufer 2019). Purely from the stereochemical perspective, any large polymer like a protein can possess an infinitely large number of conformations. However, misfolded protein in vivo is an exception. The proposed model postulates that the correct folding of newly synthesized proteins occurs via its higher dimensional configurations. This argument is strengthened by simple empirical evidence. If a large molecule were to undergo random transitions from one conformation to another to arrive at the correct conformation, at room temperature, the time taken to arrive at the correct conformation will be longer than the age of the universe. Thus, if Nature is avoiding this tortuously long process routinely, then it is axiomatic that it cannot be a chance occurrence. Nature must employ some mechanism to arrive at the correct conformations. Purely from the computational perspective, we manage to predict correct foldings of proteins only when we employ artificial intelligence coupled with massive computational capabilities (Plummetti 2022). Therefore, it makes sense to think that Nature also employs some unknown algorithm to fold a given sequence of amino acids into the desired conformations. The proposed model points toward one such model. The proposed model offers a naturalistic explanation for this otherwise inexplicable phenomenon. So far, we have discussed static and dynamic models of genomic architecture in general. Admittedly, we have instantiated this discussion with reference to Homeobox as given in Schema 5.6. However, the focus has been on the general theoretical perspective of genomic architecture. In the next section we will look at a more detailed model of Homeobox according to this model.

5.13 The Proposed Model of Homeobox

In this section, we will try to expand the details given in Schema 5.6. For this purpose, we employ three semantic propositions that have emerged from the above discussion. These three propositions are: the genes present in Homeobox express themselves in a particular sequence due to the higher dimensional configuration of Homeobox; the position of Homeobox genes vis a vis their targets reflects the difference between the dimensionality of Homeobox genes; the rates of diffusion of the products of these gene expressions is defined by topological distances between the respective Homeobox genes and their targets. While it is possible to question these propositions, they are amenable to experimental verification. Admittedly, such a verification is beyond the scope of this monograph, but this possibility of being verified experimentally puts these propositions in the category of scientific hypotheses. Before we expand the Schema 5.6, let us understand the implications of these three propositions.

Let us begin with the first propositions, viz., the genes present in Homeobox express themselves in a particular sequence due to the higher dimensional configuration of Homeobox. Admittedly, the sequence of gene expressions of the constituent genes of Homeobox is a well-researched phenomenon (Duboule 1994). Without going into any quantitative analysis of this phenomenon, it is intuitively clear that to the extent the old dictum that phylogeny recapitulates the ontogeny is valid (Gould 1985), the genes present in Homeobox would follow a particular sequence of gene expressions. Therefore, this points toward a conservation of functional template of Homeobox during the course of natural selection. Admittedly, the conventional perspective doesn't acknowledge the existence of an independent template of functionalities of Homeobox. However, it doesn't deny the possibility of such a template either. Therefore, the key point is that the proposed model simply tries to define such a framework of Homeobox functionalities in the form of topological spread into different dimensionalities. It is important to keep in mind that the proposed model assigns physical dimensionality of spacetime to define the functionalities of Homeobox. However, even if eschewing this stronger version of physical connotation of dimensionalities, from the modeling perspective (Laudal 2021), it makes sense to assign different dimensionalities to different control elements. Thus, to the extent the control elements of Homeobox are discrete and in correspondence with the genes present in Homeobox, this assignment of different dimensionalities to different functionalities of Homeobox is a mainstream modeling exercise. Of course, the proposed model confers the physicality of spacetime dimensionalities to these postulated dimensionalities. We will return to the reasons why these topological dimensionalities must be taken as the dimensionalities of spacetime itself in the following chapters.

Presently, let us look at the second proposition, viz., the position of Homeobox genes vis a vis their targets reflects the difference between the dimensionalities within the Homeobox genes. In other words, if we map the distance between a Homeobox gene and its target DNA sequence, say, by using the number of nucleotides (kilobase pairs), then according to this model, it provides a measure to define a topological distance in terms of different dimensionalities. This proposition is necessary because according to the conventional perspective, the placement of homeoboxes in the overall genomic architecture is not governed by any design principles. In fact, the conventional perspective replaces any structural template with the kinetic processes like diffusion of the products of gene expressions of these Homeobox genes. Therefore, it is legitimate to question the need for an additional postulate on the grounds of parsimony. However, if as suggested above, we can analyze quantitative description of different rates of diffusion and find a fixed pattern, then it legitimizes this proposition. Even prior to any such verification, there exists a valid semantic reason for such a proposition. Let us see what that semantic imperative is. Once we accept that Homeobox has largely been conserved during natural selection, we need to accept that it is its functionality that is conserved. Admittedly, the process of natural selection operates on the morphological features that arise from body plans as shaped by the control elements of Homeobox. However, the bottom line is that it is the functionality of Homeobox that is selected. Therefore, if it is possible to think of the resulting body plans in the language of dimensions, then it is axiomatic that we can formalize the corresponding control elements of Homeobox in the language of dimensionalities that are higher than the four-dimensional spacetime in which body plans manifest. While the conventional perspective ascribes the origins of body plans to the different rates of diffusion of the products of different genes within Homeobox, the proposed model simply connects the differences between these rates of diffusion to a topological mapping. This is one of the reasons why the proposed model insists that the dimensionalities postulated in the proposed model represent spacetime dimensionalities. This physical connotation of dimensionality represents the link between the stereochemical orientations and the resulting kinetic properties like the rates of diffusion. It is important to keep in mind that if a body plan can be defined using a discrete number of dimensions, it is intuitively clear that the corresponding mapping of the functionality of controlling the sequence of gene expressions must be defined using higher dimensionalities.

Now let us look at the third proposition, viz., different rates of diffusion of the products of different genes within Homeobox are themselves a measure of different dimensionalities in which their respective genes are present. If the first two propositions were unorthodox in their semantic implications, they at least in principle be reconciled with the conventional perspective. This is because while these propositions were unproven, they are not excluded from being possible candidates. The third propositions of different diffusion rates being themselves a measure of topological spread of the parent genes, sounds not just orthodox, but even heretical. Let us understand why. Our conception of thermodynamics and Kinetics suggests that the diffusion constants of any molecule in a given medium is defined by the electrons present in the highest occupied molecular orbital of that molecule and physical properties of the medium through which the candidate molecule is diffusing. In this case, for the sake of simplicity, we will assume that the physical properties of the medium (say cytoplasm) remains fixed for different molecules diffusing through it. In other words, the difference in diffusion rates of different molecules must be solely dependent on their structural differences between different diffusing molecules. Therefore, according to the conventional perspective, the antecedents of any molecule (in terms of where its parent gene is located) do not decide the magnitude of its diffusion constant. This is in sharp contrast to this third proposition. Let us see how these two mutually exclusive approaches can be reconciled.

Even in the conventional perspective, it is intuitively clear that the magnitude of a rate of diffusion of a given molecule is decided by the number of electrons being present in its outermost molecular orbital (HOMO cf. Quantum chemistry (Szabo and Ostlund 1989)). Admittedly, the shape and size of the molecule orbital also play a crucial role in determining the magnitude of its rate of diffusion. However, to eliminate any ambiguities, we normally employ a parameter of electron density at each point on the outermost molecular orbital. Therefore, in an approximate sense, the number of electrons present in the outermost molecular orbital can be used as a thumb rule for correlating different rates of diffusion. Now, according to the proposed model, the number of electrons present in the outermost molecular orbital can be viewed as an ensemble of entangled electrons. This entangled state of multiple electrons can be represented in the proposed model with different dimensionalities depending on the number of electrons. This is largely because our conventional modeling methodology (Laudal 2021) employs the notion of dimensionality as a measure of information content in the form parameters. Thus, a system having a certain number of parameters would be assigned a dimensionality based on the number of parameters. Therefore, even in the conventional perspective, it is the number of parameters (which in this case exists in the form of the number of electrons) that decides the dimensionality of a system. Therefore, while representing a molecular orbital of a given molecule, the number of electrons present in its highest molecular orbital would decide the dimensionality. Admittedly, the proposed model insists that the notion of dimensionality used here is that of spacetime itself and not a theoretical conception of dimensionality. However, this physical connotation of dimensionality used here provides us with an added justification for correlating the magnitude of the rate of diffusion with dimensionalities. Let us understand how.

We normally associate the notion of inertia to the fact that spacetime is a continuum. Upon a little reflection, it is possible to refine this proposition to think that the degree of inertia would also depend on the degree of "viscosity" of the medium through which an object moves. Now, according to this model, each dimensionality of spacetime would have its own degree of complexity. If we accept the notion of indestructibility of information content of spacetime, then it is axiomatic that any increase or decrease in dimensionality would also bring about corresponding changes in the "density" of the information content of spacetime at different dimensionalities. This is implicit in the definition of modified involution. This operator can be visualized as an inward folding of one of the dimensions of spacetime into the remaining dimensions of spacetime. Since this operation results in decrease in dimensionality with the accompanying increase in the "density" of information content in the resulting involuted manifold. Once we accept this reasoning, it is intuitively clear that the degree of inertia would increase with the decrease in dimensionality.

As a result, the higher dimensional configurations of genomes (or any other molecules) would experience lesser resistance from spacetime at those higher dimensionalities. Now it is easy to visualize how different rates of diffusion of the products of different constituent genes within Homeobox can be related to the dimensionality of the parent genes. In higher dimensionalities, the distribution of information content between spacetime and molecules is more even. However, after involution, the information content of spacetime increases due to the fact that now it has two types of dimensions, viz., the time-like and the space-like dimensions. This means that spacetime would generate more inertia for molecules moving around in the lower dimensionalities. All we have to do is to visualize a scenario wherein different higher dimensionalities undergo different numbers of involutions. Therefore, if different constituent genes within Homeobox are present in different dimensionalities, then it is axiomatic that to give rise to different proteins, they would have to undergo different numbers of involutions. Therefore, the products of these gene expressions (viz., the constituent genes of Homeobox) would have different diffusion constants. This is precisely what these three propositions discussed above imply. Admittedly, these three propositions are very difficult to be accepted as such. However, if we could quantitatively compare different rates of diffusion of the products of different gene expressions of different genes within Homeobox, it can validate these propositions. Till such time, we can at least qualitatively postulate a rough topological model of the genes present in Homeobox. One such model is given in Schema 5.7. It is difficult to accept such a radical perspective in the absence of any empirical evidence. However, until such evidence is available, it is necessary to deconstruct the semantics of the new model of Homeobox. Therefore, in the next section, we will try to deconstruct the semantics of the topological separation of Homeobox into different dimensionalities.

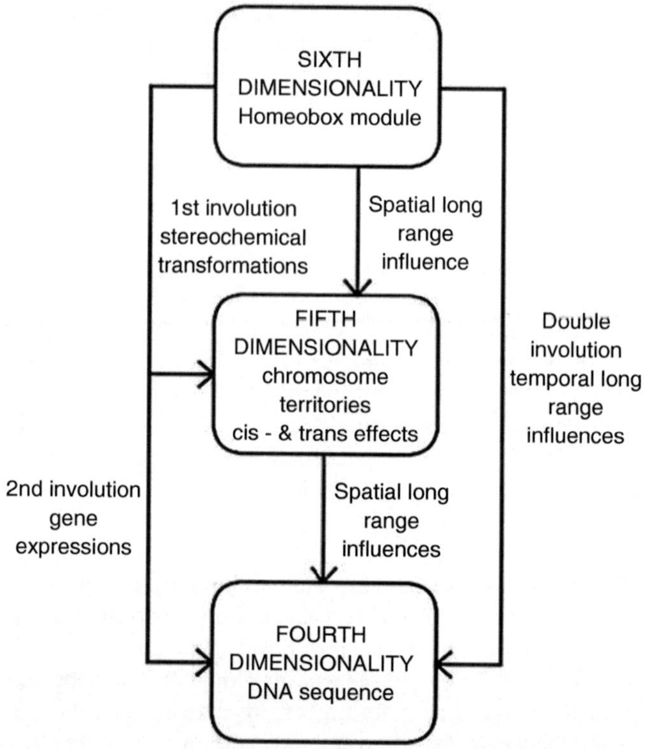

Schema 5.7 Topological map of long-range influences in Homeobox

5.14 Semantics of Topological Separation

From the conventional perspective, it is legitimate to be skeptical about something as intangible as unobservable higher dimensionalities. Apart from the radical nature of this model, there is a far deeper reason why we are skeptical about it. This refers to the relationship between mathematical constructs and the physical reality that we can directly comprehend. While we no longer concede the transcendental nature of mathematical constructs (as implicit in the Cartesian paradigm (Cottingham 2008)) we are not quite convinced about the origin of mathematics. This is reflected in the fact that we, as scientists, employ mathematics by taking it as a priori. At best, we subscribe to a version of mathematical naturalism based on the analyzability of mathematics. (cf. Quine (Quine and Ollian 1978)). Therefore, we are averse to any model that is based on the stronger version of mathematical realism. This is precisely what the proposed model rests upon. Having said that, it is imperative that we must seek semantic consistency of the type of mathematical realism implicit in the proposed model. Therefore, in this section, we will try to deconstruct what this conventional separation of structuralism and functionalities (or at different levels, the separation between genotype and phenotype as well as the separation between DNA and RNA) using the language of dimensionalities. Conventionally, this translation of biological separation into topological separation would have been a little unorthodox, but not unthinkable. However, since the proposed model employs a physical connotation of dimensionality, it is imperative that we deconstruct this biological separation using spatiotemporal dimensionalities.

One of the reasons why the proposed model postulates the separation of structuralism from functionalities on the basis of dimensionalities is that it seeks to modify the emergence principle conventionally employed in molecular biology (Smith and Morowitz 2016, see Chapter 4). In general, we concede that any reductionist model would lead to unconvincing inferences. Therefore, as a reverse strategy, we think of complexity giving rise to additional features in any system. Thus, a single nucleotide cannot possess replicative functionality, but when present as a constituent of DNA, it can. Similarly, a single neuron cannot possess a functionality of sentience, but when present in a human brain, it can. Thus, we ascribe to the belief that as the system gets more and more organized, new functionalities emerge in such a system. This is a simplistic depiction of the emergence principle. As a corollary to the emergence principle, we subscribe to the belief that the level of organization is reflected in the degree of complexity manifest in the system. Upon a little reflection, it is intuitively clear that this principle is intuitive and therefore, represents our cognitive expectations. The proposed model simply tries to formalize this intuitive understanding into the language of dimensionalities.

The trouble however, lies in the fact that the proposed model insists that the notion of dimensionality is that of spatiotemporal dimensionality. Therefore, the proposed model needs to justify this physical connotation of dimensionality by demonstrating its congruence with biological evidence. Even then, this is not a serious problem because our naturalistic foundation of science allows us to think of morphological features as ensembles of atoms and molecules. Therefore, to the extent molecules represent information content, it makes sense to think that the more complex information content necessarily requires higher dimensional representation. However, according to the conventional perspective, this higher dimensional representation is notional and not physical. This is because, as mentioned above, we think of mathematics as an abstract entity. However, if we wish to subscribe to mathematical realism, it is inevitable that as the complexity increases, we will require higher dimensional representation. Admittedly, mathematical realism is not a well-accepted doctrine, and therefore, it is tempting to dismiss it. However, irrespective of the validity of mathematical realism, there is another reason why we must concede that a more complex system would require higher dimensional representation. This reason arises from our experience in quantum computation (Nielsen and Chuang 2010). In quantum computation, we are forced to concede that information per se is a physical entity. Therefore, to the extent the complexity of a biological system, say, a genome consists of information content (in the form of biomolecules), it is axiomatic that it has to be represented in higher dimensional configurations. The difference however is that this time, we have to accept that these higher dimensional configurations must be physical in nature, if only to account for the physicality of the information content of genomes.

Thus, we arrive at a scenario, wherein higher dimensional configurations are thought of as physical entities and not as some abstract entities. The next stumbling block for this model is its insistence of topological representation rather than geometric representation. Admittedly, if we confine ourselves to a three-dimensional spatial perspective, there is not much conflict between topology and geometry. In fact, it is possible to transform a stereochemical perspective which is essentially a geometric perspective, into a topological perspective. This is precisely what we do while computing molecular models in quantum chemistry (Szabo and Ostlund 1989). The trouble begins when we wish to define a geometric perspective in the higher dimensional topology. There doesn't seem to be any formal way to transform a three-dimensional geometric frameworks into higher dimensional topology. This ambiguity of the relationship between geometry and topology is best illustrated by Riemann conjecture (Mazur and Stein 2016). Without going into the details of the mathematical issues underlying this conjecture, it is intuitively clear that this conjecture cannot be derived formally because we cannot comprehend the formal description of the relationship between geometry and topology. Of course, it is possible to use approximations in translating a three-dimensional geometric construct into a higher dimensional topological model. This is precisely what the Hilbert space model does. It enables us to conjure up higher dimensional configurations resembling our three-dimensional configurations. However, it comes with a cost. We are required to abandon some of the possible configurations (which are called nonorthogonal states) (Nielsen and Chuang 2010). This scenario naturally prompts us to be skeptical about any model that is based on higher dimensional spacetime. Such a model, even if true, would be mathematically intractable.

As if these were not, the proposed model runs into another difficulty. Let us temporarily assume that the arguments presented above are valid and that we can build a higher dimensional model of genomic architecture which is based on higher dimensionalities of spacetime itself. The proposed model postulates another counterintuitive proposition. According to the conventional perspective, if a mathematical model has to be employed, its framework would depend on the amount of information content in the form of some parameters. Thus, higher the amount of information content, the larger the framework. Thus, in the case of a topological model, dimensionality of a model would be decided by the number of parameters required for formalizing a given system. However, the proposed model takes an exactly opposite approach. It postulates that the information content in the form of complexity is inversely related to the dimensionality of the model. This is apparently counterintuitive. Therefore, let us see how this counterintuitive feature arises in the proposed model.

It is important to keep in mind that the proposed model begins with the cosmic singularity (Chhaya 2022b, see Chapter 1). Therefore, it is axiomatic that whatever the information content that any natural phenomenon possesses must be derived from the cosmic singularity, albeit after a large number of operations of involution. This is necessary because we believe that information per se is physical and it can't be either destroyed or created. Therefore, in such a scenario, complexity, either structural or functional, arises from the information content already present in the cosmic singularity. Thus, an operator of involution transforms the preexisting information content of spacetime (or that of the cosmic singularity) into a more complex arrangement. Upon a little reflection, it is intuitively clear that any type of complexity arises only when the dimensionality of the system is reduced by involution. Thus, the more complex system must occupy the lowest dimensionality. It is apparent that this seemingly counterintuitive feature arises quite naturally from the basic postulates of the proposed.

The reason why we find it counterintuitive is that we follow the Cartesian paradigm (Cottingham 2008). Because of the Cartesian split, we think of different mathematical frameworks as abstract entities present in the transcendental realm. Since under the Cartesian influence, we think that our cognitive faculty (or to use a more popular term, our consciousness) is also a part of the transcendental realm. Therefore, whenever we attempt mathematical modeling, we imagine ourselves to be outside the system under investigation. Thus, we develop models from an inherently external perspective. However, according to this model, our cognitive faculty is a natural phenomenon and owes its origin to the cosmic singularity. Thus, functional complexity of our cognitive faculty is represented in lower dimensionalities in this model. Therefore, when we create an external perspective of a given system, we overlook the fact that our perspective is the one available from a few dimensionalities from which our cognitive faculty operates. Thus, what appears to us as an obvious perspective is in fact one of the many perspectives. More importantly, since according to this model, our cognitive faculty resides within the spatiotemporal universe, what we have is an internal perspective. We often overlook that we are observing the universe and its natural phenomena from within and not from outside (Chhaya 2020).

This is where the semantic importance of involution arises. The modified operator of involution represents an internal template of every system including our cognitive faculty. Once we accept that involutive algebra (as modified in this model) represents the universe as it exists then it is possible to reconcile it with our cognitive artifact of external perspective. Returning to the present discussion, it is intuitively clear why we find it difficult to formalize genomic architecture. According to this model, genomes acquired this architecture by a series of operations of involution. Whereas we have been trying to formalize it using external projections from our cognitive faculty. In order to understand this dichotomy between the external and internal perspectives of Homeobox, in the next section, we will deconstruct the conventional perspective using the proposed model. This is necessary because whatever the semantic justification this model offers, its validity rests solely on its verification.

5.15 Conventional Perspective Versus the New Model

For the sake of brevity, we will restrict ourselves to the Schemas 5.6 and 5.7. Admittedly, even Schema 5.6 is based on topological arguments. Therefore, it cannot be said to represent the conventional view. However, Schema 5.6 retains the conventional perspective of the functionalities of Homeobox. Therefore, it retains the essential semantic propositions of the conventional perspective of genomics. On the other hand, Schema 5.7 represents the key semantic proposition of an active role of spacetime in defining the long-range influences. Therefore, it seems reasonable to compare these two schemas. It is important to keep in mind that the implicit topological perspective in Schema 5.6 offers us a tool for comparison. With these provisos in place, let us deconstruct Schema 5.6 in the light of Schema 5.7. For this purpose, we will confine ourselves to three functional features of Homeobox, viz., the chromosomal placement of Homeobox and the sequence of expressions of the constituent genes of Homeobox; the duration of expressions of the constituent genes of Homeobox; the overlap of expressions of the constituent genes of Homeobox. These three features of Homeobox functionalities are semantically loaded functionalities.

For instance, the placement of different copies of Homeobox on different chromosomes is well documented (Mazza 2007). However, according to the conventional perspective, these placements are the outcomes of random events involving duplications and translocations. Thus, it eliminates any ontological perspective. At the same time, the conservation of individual placements during natural selection must have been due to the efficacy of these selected placements in giving rise to phenotypic morphological features. Thus, according to the conventional perspective there is no need to invoke any semantic propositions while explaining the placement of homeoboxes in different chromosomes. The proposed model doesn't question this conventional wisdom. However, it suggests that the process of translocations is governed by the higher dimensional configurations of genomes. Therefore, while duplication of Homeobox according to this model is essentially a random occurrence, once duplicated, the copies of Homeobox translocate according to the higher dimensional configurations. Even then, these topological compulsions don't lead to translocation to any single destination. Rather, it leads to more than one possible destination. Thus, the proposed model suggests that there exists a definitive mechanism for translocation but it leads to multiple possible outcomes. It is intuitively clear that this scenario suggested by the proposed model doesn't introduce any determinism or teleological principles. However, it leads to limitations of the possibility of number of translocations and their destinations.

Similarly, there are subtle differences between the conventional perspective and the proposed model about the duration of gene expressions of the constituent genes of Homeobox. The conventional perspective largely depends on the stereochemical and thermodynamic considerations for justifying different durations of gene expressions. In fact, this reasoning applies to any gene expression. For instance, the conventional perspective doesn't postulate that the gene expressions of the constituent genes of any module have any different mechanisms than the expressions of other genes which are not controlled by any modular influences. Thus, the conception of modularity in the conventional perspective remains nonstructural. This is partly because we discover different modularities by correlating different gene expressions using Bayesian logic (Press and Clyde 2003). If an expression of a gene A can be shown to influence the gene expressions of other genes under different conditions, we hypothesize a certain modularity consisting of gene A and other genes influenced by it. We rarely, if ever, discover modularity on the basis of structuralism of genomes. The proposed model, on the other hand, suggests that the duration of gene expressions of genes constituting a module is governed by different mechanisms than the duration of gene expressions of other genes. The constituent genes of any module are governed by higher dimensional configurations of that module. Therefore, even the duration of expression of the constituent genes of a module are decided by the higher dimensional configurations of that module. The remaining genes, either the genes under control or genes without any control of long-range influences, manifest different durations of gene expressions which are determined by the control elements or by the conventional thermodynamic and stereochemical considerations. Admittedly, according to this model, even these gene expressions are governed by overall higher dimensional configurations of genomes, but the durations of the gene expressions of these genes can be approximated by the conventional perspective.

This brings us to the third proposition of the mechanism of overlap between the gene expressions of different genes. Upon a little reflection, it is intuitively clear that Nature employs the overlap of gene expressions of different genes to give rise to feedback mechanisms which orchestrate the development of body plans during the developmental stages. Admittedly, there are instances of simple feedback loops like the one observed in the case of the *lac* operon (Miller and Reznikoff 1980). However, not all such feedback mechanisms are as straightforward as that. The conventional perspective is silent on the evolution of such feedback. This is true even in the cases of well-studied modules like Homeobox. Therefore, from the perspective of the conventional wisdom, the origins of these overlaps of gene expressions (and the origins of modularity per se) remain unexplained. However, according to the proposed model, this phenomenon of overlap of gene expressions of different genes within any given module is necessitated by the topological compulsions. Therefore, at least in principle, these overlaps are manifestations of higher dimensional configurations of modules and this feature can be used to decipher the higher dimensional configurations.

These three semantic propositions are qualitative and therefore, speculative. In the absence of any quantitative analysis, it is not possible to define a higher dimensional configuration of Homeobox. However, Schema 5.7 depicts a rough sketch of this model, which is lacking in Schema 5.6. It is important to note that even in the absence of any concrete model of higher dimensional configurations of Homeobox, it is possible to review the pathologies arising from Homeobox genes and seek some justification for the proposed model. Therefore, in the next section, we will try to link pathologies reported in literature with the proposed model. It must be admitted that this is simply an indicative introduction to a new approach and not a rigorous analysis of the pathologies arising from the changes in Homeobox.

5.16 Therapeutic Possibilities According to the New Model

Purely from a logical perspective, it is intuitively clear that any module exercising control over multiple genes during developmental stages can also be a source of pathologies in the case of its malfunction. Therefore, Homeobox cannot be an exception to this logic. However, in the case of Homeobox, since it functions during developmental stages of an organism, most of the malfunctions would result in premature abortion. Thus, developmental genetic disorders (Butler and Meaney 2019) have a different etiology than those observed in late-onset genetic disorders (Evers-Kiebooms et al. 2002). In the context of the present discussion, this distinction is important. Furthermore, the genetic disorders originating due to Homeobox malfunction are documented (Duboule 1994), but never investigated from the perspective of Homeobox having a particular functional template. This limits the range of therapies available for these disorders. However, the proposed model outlines a functional template of Homeobox. Therefore, in this section, we will try to understand whether the proposed functional template can offer any new type of therapies. Once again, we take the existing literature as being read and instead focus on a few general strategies that the proposed model offers. Admittedly, these strategies are generic in nature. However, they are amenable to clinical verifications.

Ab initio, we can think of the therapies targeted toward the mutations in Homeobox to operate by modifying the abnormal Homeobox genes. The obvious choice would be that of reverse transcription. However, since Homeobox genes function in utero, any such therapy must work prior to in vitro fertilization. However, the conventional perspective of Homeobox and its functioning cannot provide any information other than the DNA sequence of the abnormal Homeobox genes. This is because the conventional perspective is not based on any higher dimensional configuration of Homeobox. Therefore, the therapies of the type mentioned above would have to be very precise. However, the proposed model offers a simpler strategy. Instead of taking up the task of modifying the nucleotide sequences and then moving upward, the proposed model takes a top-down approach. If we could alter the higher dimensional configurations directly, then the sequence of gene expressions of the constituent genes of Homeobox, their diffusion constants can be automatically adjusted. This is because according to this model, it is the topological distances that decide both the sequence of gene expressions and the diffusion constants of different proteins thus synthesized.

The key point is how to alter the higher dimensional configurations of a mutated DNA sequence of the constituent genes of Homeobox. In the conventional perspective, while altering the nucleotide sequences of mutated genes is cumbersome, once successful, the higher dimensional configurations automatically change. However, since we wish to avoid this cumbersome protocol, we need to think of a direct way to alter the higher dimensional configurations of Homeobox or more specifically. There is one aspect of the proposed model that offers a way to solve this problem. According to this model, every higher dimensional configuration can devolve into multiple four-dimensional configurations. Therefore, it is not necessary to alter the DNA sequence of the mutated constituent genes of Homeobox to repair or restore its functionalities. We can instead achieve the desired higher dimensional configuration of Homeobox by altering the intergenic sequences. This is because given a right lower dimensional configuration, we can arrive at the identical higher dimensional configurations of Homeobox. This obviates the need to alter the DNA sequence of the mutated constituent genes of Homeobox. Admittedly, this strategy requires a prior knowledge of the correct higher dimensional configurations and the rules of their devolvement to four-dimensional configurations. However, since it is easier to alter intergenic DNA sequences by *indels*, it is a more viable strategy than replacing the mutated constituent genes of Homeobox. This is a particularly attractive strategy in the post CRISPER era (Luo 2019). Even from the theoretical perspective, it is feasible to work out higher dimensional configurations of given DNA sequences up to five or six dimensionality configurations, provided we can work out the topological model having a suitable metric. It is legitimate to question the validity of this approach, particularly when we know so little about genomic architecture. For instance, at first sight, it doesn't make sense to think that the products of mutated constituent genes of Homeobox would behave differently just because the neighboring DNA sequences have been altered. After all, the diffusion constants and the stereochemical configurations of the products of these mutated constituent genes of Homeobox would remain unaltered and would have the same deleterious effects. However, if the proposed model is valid then, it is possible that the duration and onset of the gene expressions of these mutated constituent genes of Homeobox would be altered. The products of these mutated constituent genes of Homeobox will obviously have unchanged stereochemical configurations and diffusion constants, but their timings would change. These changes in their timings ought to alleviate the resulting pathologies. This brings us to the end of the chapter. Therefore, in the next section, we will conclude with a brief summary.

5.17 Conclusion

For the sake of simplicity, we will summarize the above discussion in a point-wise manner.

1. The conventional perspective presently lacks a formal description of genomic architecture. This is also evident from the fact that there is no formal architecture of Homeobox as well.
2. In the preceding chapters, a topological model of genomic architecture was outlined. It was argued that spacetime plays an active role in biological evolution and natural selection. Therefore, spacetime must be a structural element of genomic architecture.

3. This chapter outlines a topological model of Homeobox in which spacetime is included in the formal conception of Homeobox.
4. It is proposed that the functionalities of Homeobox of controlling various gene expressions are encoded in chemical properties of the biomolecules as well as in the nonchemical properties of the topological arrangements of these molecules.
5. It is proposed that Homeobox (like genomes in general) is spread across several dimensionalities of spacetime and the four-dimensional perspective of Homeobox is just a local perspective of Homeobox.
6. The functionalities of Homeobox can now be divided into two categories. The conventional functionalities involving initiation, activation and suppression of gene expressions are manifest in the lower dimensionalities of spacetime representing Homeobox.
7. Homeobox functionalities like the sequence of gene expressions of the constituent genes of Homeobox, the magnitude of diffusion constants of different proteins and the placement of homeobox vis a vis the target genes are placed in the higher dimensional configurations of Homeobox.
8. The changes in the dimensionalities of the higher dimensional configurations of Homeobox are defined by two operators called chronon and spation. They help us to define how higher dimensional configurations of Homeobox get translated into lower four-dimensional configurations of Homeobox. These operators represent what is conventionally called long-range influences. Thus, the proposed model replaces the hitherto undefined long-range influences with well-defined operators.
9. Chronon and spation convert the higher dimensional details of spacetime (in which there are no distinctions between the time-like and the space-like dimensions) into the thermodynamic and kinetic properties of these biomolecules. Moreover, since the relationship between the higher dimensional spacetime and the four-dimensional space is defined by topological constraints, the operators of chronon and spation give rise to only a limited number of stereochemical configurations, thereby introducing the catalytic functionalities of these biomolecules.
10. According to the proposed model, there exists a notional entity called genomic singularity from which various genomic modules have evolved. This postulate is consistent with the conventional phylogeny. It is possible to explain the evolution of Homeobox using the proposed entity of genomic architecture.
11. The proposed model suggests that it is possible to define two templates of Homeobox, viz., the structural template and the functional template. The proposed model helps us to define a formal relationship between these two templates.

References

Bard J (2016) Principles of evolution: systems, species and the history of life. Garland Science, New York
Bermingham JR (1989) Structure and expression of Antennapedia gene in Drosophila melanogaster. University of Colorado Press, Boulder
Bothma J (2013) Mechanisms of transcriptional precision in Drosophila embryo. University of California Press, Berkeley
Bower JM, Bolouri H (eds) (2001) Computational modeling of genetic and biochemical networks. MIT Press, Cambridge
Bromham L (2008) An introduction to molecular evolution and phylogenetics. Oxford University Press, Oxford
Bryson V, Vogel HJ (1965) Evolving genes and proteins. Academic Press, New York
Butler M, Meaney FJ (eds) (2019) Genetics of developmental disabilities. CRC Press, Boca Raton
Cabej NR (2020) Epigenetic mechanisms of the Cambrian explosion. Academic Press, London
Caldarelli G (2007) Scale free networks: complex Webster in nature and technology. Oxford University Press, Oxford
Chhaya P (2020) The universe within: and a view from that within. Provisionally accepted by Universal Press, Irvine
Chhaya P (2022a) The origin and nature of mathematics. (Submitted)
Chhaya P (2022b) The origin and nature of spacetime. (Submitted)
Chhaya P (2022c) The origin of the counterintuitive nature of quantum phenomena. (Submitted)
Conlon J (2016) Why string theory? CRC Press, Boca Raton
Cottingham J (2008) Cartesian reflections. Oxford University Press, Oxford
Cretoiu D, Xiao J, Das S (2020) MicroRNA signaling. Frontiers Media, Lausanne
Darden L (1991) Theory change in science: strategies from Mendelian genetics. Oxford University Press, Oxford
Davidson EH (2006) The regulatory genome: gene regulatory networks in development and evolution. Elsevier, Amsterdam
Duboule D (1994) Guidebook to homeobox genes. Oxford University Press, Oxford
Evers-Kiebooms G, Zoeteweij MW, Harper PS (2002) Prenatal testing for late onset neurogenetic diseases. Bios Publishing, Oxford
Fritz A (2014) Elucidating chromosome territories organization during the cell cycle and in breast cancer cells. Proquest Publishing, Ann Arbor
Fusco G (2008) Evolving pathways: key themes in evolutionary developmental biology. Cambridge University Press, Cambridge
Gilbert SF, Barresi MJF (2020) Developmental biology. Oxford University Press, Oxford
Gould SJ (1985) Ontogeny and phylogeny. Harvard University Press, Cambridge
Gould SJ (2007) Punctuated evolution. Harvard University Press, Cambridge
Grene M (1986) Dimensions of Darwinism: themes and counter themes in the twentieth century evolutionary theory. Cambridge University Press, Cambridge
Howard K (ed) (2013) RNA interference: from biology to therapeutics. Springer, Dordrecht
Laudal OA (2021) Mathematical models in science. World Scientific, Singapore
Lawrie ID (2013) A unified grand tour of theoretical physics. CRC Press, Boca Raton
Lightner DA, Gurst JE (2000) Organic conformational analysis and stereochemistry from circular dichroism spectroscopy. Wiley VCH, Weinheim
Luo Y (2019) CRISPR gene editing: methods and protocols. Springer, Dordrecht
Margulis L (1970) Origin of eukaryotic cells. Yale University Press, New Haven
Mark M, Rijli FM, Chambon P (1997) Homeobox genes in embryogenesis and pathogenesis. Pediatr Res 42:421–429
Mayer RJ, Ciechanover MA, Rechsteiner M (eds) (2006) Protein degradation: the ubiquitin proteasome systems, vol 2. Wiley, Berlin

Mazur B, Stein W (2016) Prime numbers and Riemann hypothesis. Cambridge University Press, Cambridge
Mazza ME (2007) Homeobox evolution: insights into homeobox clusters in the basal metazoan Nematostella vectensis. Boston University Press, Boston
Miller JH, Reznikoff WS (1980) The operon. Cold Spring Harbour Laboratory Press, New York
Mostovoy Y (2014) Regulation and function of antisense transcription in Saccharomyces yeast. University of California Press, Berkeley
Nielsen MA, Chuang IL (2010) Quantum computations and quantum information. Cambridge University Press, Cambridge
Okasha S (2010) Evolution and the levels of selection. Oxford University Press, Oxford
Patel RN (2000) Stereoselective biocatalysis. Marcel Dekker, New York
Peter IS, Davidson EH (2015) Genome control processes: development and evolution. Academic Press, London
Pevsner J (2015) Bioinformatics and Functional Genomics. John Wiley and Sons, Hoboken
Pick L (2015) Hox genes, evo-devo and the case of ftz gene. Chromosoma 125:535–551
Plummetti M (2022) Molecular dynamics and complexity in catalysis and biocatalysis. Springer, Dordrecht
Pourquie O (2009) Current topics in biology: Hox genes. Academic Press, London
Press SJ, Clyde CM (2003) Subjective and objective Bayesian statistics. Wiley, Hoboken
Provine WB (2001) The origins of theoretical population genetics. University of Chicago Press, Chicago
Quine WV, Ollian JS (1978) The web of belief. Random House, New York
Robert JS (2004) Embryology, epigenesis and evolution: taking development seriously. Cambridge University Press, Cambridge
Sabath N (2009) Molecular evolution of overlapping genes. University of Houston, Houston
Schulze-Makuch D, Irwin LN (2008) Life in the universe. Springer, Dordrecht
Sekimura T, Noji S, Ueno N, Maini P (eds) (2003) Morphogenesis and pattern formation in biological systems: experiments and models. Springer, Dordrecht
Smith E, Morowitz HJ (2016) The origin and nature of life on earth: the emergence of the fourth geosphere. Cambridge University Press, Cambridge
Sober E (2015) Ockham's razor. Cambridge University Press, Cambridge
Stamm S, Smith S, Luhrmann R (2012) Alternative pre messenger RNA splicing. Wiley VCH, Weinheim
Szabo A, Ostlund NS (1989) Modern quantum chemistry: an introduction to advanced electronic structure theory. Dover Publications, Mineola
Weinzierl ROJ (1999) Mechanism of gene expression: structure, function and evolution of the basal transcriptional machinery. Imperial College Press, London
Wetlaufer DB (2019) The protein folding problem. Routledge, London
Wigner E (1960) The unreasonable effectiveness of mathematics in natural science. Commun Pure Appl Math 13:1–14
Yarus M (2010) Life from an RNA world: the ancestor within. Harvard University Press, Cambridge

Nature of Aging Processes: Genomic Ontology of Aging

6

Abstract

The process of aging, like the embryonic development, must be seen as a polygenic phenomenon embedded at a deeper level of the genomic architecture. Our current focus in gerontology is focused on delaying the aging processes by identifying and altering individual genes and their expressions. However, in order to understand the nature of aging, it is necessary to understand the evolutionary context and the degree of embedding of this functionality in the genomic architecture. In preceding chapters, a topological model of genomic architecture using the formalism of the involuted manifold was articulated. In this chapter, we would employ this model to deconstruct aging. This deconstruction suggests that aging is not just impregnated into the deeper level of the genome, but it is also an inevitable price for the evolution of multicellular organisms. This model suggests that aging is an outcome of shortcomings of several processes like intercellular communications, genomic wear and tear, and telomere shortening. More importantly, aging is not just a side effect of evolution, but an integral objective of evolution.

6.1 Introduction

With the improving standards of public health and diagnostics, the average lifespan of human beings is also increasing. This has left modern society with larger and larger percentages of the population in their postemployment stages of their lives. This demographic shift has increased the burden of the national resources of all the countries. This has resulted in more resource allocation to the research in gerontology. Therefore, the domain of gerontology has witnessed growth in understanding the process of aging and its attendant pathologies. Irrespective of the inherent social relevance, the discipline of gerontology or geriatrics has interesting challenges in the

© The Author(s), under exclusive license to Springer Nature Singapore Pte Ltd. 2023

P. Chhaya, *The Topological Model of Genome and Evolution*,
https://doi.org/10.1007/978-981-99-4318-0_6

context of genomics (Vijg 2007) and molecular biology (Guarente et al. 2008) of aging. From these perspectives, the process of aging is still an enigma. Our present approach has been focused on the wear and tear of transcription and replication of the human genome. In addition, there is a tacit belief that the process of aging also affects the overall genomic functionalities resulting in pathologies related to gene expressions. However, this belief is not really articulated as well as the belief in wear and tear of replicative machinery.

Prima facie, this primacy of belief in the wear and tear theory seems reasonable. However, this may be due to two reasons. Firstly, it is easier to devise experimental set ups for studying individual genes, their expressions and the mechanistic details of their transcriptions and translations. Secondly, as discussed in a preceding monograph on the reinterpretation of the Darwinian theory (Chhaya 2020, see Chapter 8) and in the accompanying chapters of this monograph, molecular biology is historically biased toward analytical tools rather than toward synthetic approaches. Thus, molecular biology is focused on genomic architecture and its deconstruction. Molecular biology has little to offer on the evolution of genomic architecture. While this is understandable, the discipline of genomics (Pevsner 2015) has reached a level of theoretical sophistication wherein it is possible to explicate the tacit belief mentioned above. We have reached the point where we have sufficient semantic depth in the genomics of aging, where we can look for genomic level aberrations influencing the process of aging. In preceding chapters, a topological model of genomic architecture has been outlined. We would employ this model to deconstruct the nature of aging from the genomic perspective.

The process of aging is also enigmatic from the semantic perspective of the Darwinian paradigm (Grene 1986). While it seems reasonable to think that evolutionary need for aging is merely an extension of the popular, though inaccurate in itself, semantic proposition of the survival of the fittest. However, both semantically and functionally, the process of aging seems to be as much an outcome of survival doctrine, as it is of genomic inevitability. Therefore, this chapter would try to disentangle the genomic compulsions of aging from the survival argument. In order to disentangle various facets of the process of aging and the role of genomics in it, this chapter has been further divided into 16 sections. It must be admitted that different facets act in a concomitant and complex manner. Therefore, this segregation of topics is merely a narrative strategy and doesn't reflect on the actual onset of aging. Moreover, we will overlook the clinical and therapeutic approaches to aging, or rather, we will take these aspects as a priori to the content of this chapter. This chapter has been further divided into 16 sections. Section 6.2: Genetics of Aging, Sect. 6.3: Genomics of Aging, Sect. 6.4: Evolutionary Perspective of Aging, Sect. 6.5: Can Genomic Perspective Solve the Problem?, Sect. 6.6: Aging as a Side Effect of Genomic Complexity, Sect. 6.7: Aging as a Genomic Module, Sect. 6.8: Aging as a Tool for Natural Selection, Sect. 6.9: Can Aging Be Prevented?, Sect. 6.10: Should Aging Be Prevented?, Sect. 6.11: The Proposed Model of Genome, Sect. 6.12: Aging According to the Proposed Model, Sect. 6.13: Aging as a Genomic Functionality, Sect. 6.14: Mechanisms of Aging, Sect. 6.15: Topological Model of

Mechanisms of Aging, Sect. 6.16: Possible Therapeutic Approaches, Sect. 6.17: Conclusion.

6.2 Genetics of Aging

There are three areas of interest in the genetics of aging that are germane to the present discussion: oxidative stress and its impact on individual gene expressions; cellular senescence; the failure of apoptosis leading to cancer. Admittedly, these three areas overlap with one another. However, from the evolutionary perspective, they have different origins. For instance, the phenomenon of oxidative stress is a historical accident of Life having originally evolved in anaerobic conditions and then having to adapt to aerobic conditions (Oparin 1968, see Chapter 5). Similarly, cellular senescence arises from mechanical failure of machinery of transcription and translation of DNA sequences (Demaria 2018). For instance, it is intuitively clear that the lengths of telomeres would continually decrease in a given cell as the number of cell divisions increases. This is essentially a mechanical failure arising from peculiar double helix structuralism of DNA. Similarly, epigenetic changes in histone complexes give rise to suppression of gene expressions. Admittedly, these mechanical failures occur in a variety of situations and therefore, it is not easy to discern a commonality among them. However, from the semantic perspective, these plethora of failures share one feature, viz., the mechanical nature. The question that is yet to be answered is are these diverse types of failures of cellular machinery due to inherent flaws of genomic architecture or are they due to simple biochemical wear and tear? Finally, the emergence of apoptosis in the multicellular organisms is by itself, an enigma (Tutar and Tutar 2018). Prima facie, it couldn't have evolved at a level of individual cells. However, if it did evolve only after the emergence of multicellular organisms, the question arises whether it evolved in response to oxidative stress? Or did it arise from the compulsions of genomic architecture?

Admittedly, these are semantically loaded questions and enigmas. As mentioned in the introduction, the general approach is to work at the level of individual genes and the biochemical consequences of their gene expressions. Apart from the experimental feasibility, this approach is guided by the lack of knowledge of genomic architecture. When we think of genomic architecture, we adopt a bottoms up approach. We try to assemble blocks of genomic architecture from the perspective of individual genes. We do not have any theoretical framework of genomic architecture. As discussed in the preceding chapters, this lack of any theoretical perspective of genomic architecture arises from our belief that any such framework would imply some design principles which are antithetical to the Darwinian semantics (Grene 1986). The trouble with any such attempts to formalize genomic architecture starting from individual genes is that it becomes intractable. To begin with, we don't know how many genes exist in a genome. Secondly, even if we accept our current estimate of a few thousand genes in the human genome, the number of ways these genes can be placed in different architectural designs is enormously large, much beyond our computational capabilities.

Thus, genetics of aging remains an unfinished business. It can't complete its objective without any help from the insights into genomic architecture. At the same time, it has to be admitted that our present approach to study concerned genes and their expressions individually has yielded vast amounts of information. However, this impressive gain needs to be integrated into a coherent theory of aging. This is something that is possible if we develop a new perspective of genomics of aging. Therefore, in the next section, we will try to understand what kind of framework this putative theory of genomics of aging ought to possess.

6.3 Genomics of Aging

Ab initio, we expect such a theory of genomics of aging, as distinct from genetics of aging, to have three principal features, viz., the evolutionary logic of aging, functionalities of long-range influences, and phylogenetic evidence to justify itself. At first sight, these three prerequisites suggest themselves as self-evident propositions. However, let us try to understand the semantic imperative of these propositions. Having done that, we will try to understand why our current understanding of the genetics of aging cannot offer to justify these propositions. It is important to keep in mind that our current understanding of the genetics of aging doesn't refute these propositions, but it doesn't support it either. The reason behind this semantic agnosticism, as mentioned above, lies in our mistaken belief that these propositions are the elements of some design principles in disguise. The objective of this chapter is to disambiguate ourselves.

Let us begin with the semantic imperative of having some evolutionary logic that would make the process of aging inevitable, if not desirable. Everyone who has studied biological evolution knows that the most characteristic feature of biological evolution is that it seeks to achieve permanence amidst the incessant flux. It seeks to fix a pattern amidst chaos. However, there is something more to this feature of biological evolution. While seeking permanence, it doesn't reject flux, biological evolution uses it. While trying to fix a pattern amidst chaos, biological evolution doesn't avoid chaos, it employs chaos to obtain a better pattern. This self-contradictory duality sums up the semantics of biological evolution. Biological evolution doesn't aim to create machines, it aims to create self repairing and self generating machines. Therefore, biological evolution embeds a strategy in the process of natural selection which is not confined to any particular organism or any particular environmental niche. In order to free itself from the limitations of individual instances, biological evolution has embedded a domain neutral logical template in the process of natural selection.

When viewed from this perspective of evolutionary logic, it is intuitively clear that senescence is an inevitable byproduct of this strategy. It arises because biological evolution seeks to exploit the molecular flux to achieve a temporary fixation of a pattern. This fixation of pattern has to be temporary because the environment is in a flux and therefore, a permanent fixation of pattern would be detrimental. This constant churning of the environment and the repeated attempts to fix a pattern constitute a Darwinian duality. It is important to keep in mind that senescence doesn't arise because of this duality of fixation amidst flux. It arises only when the Malthusian logic sets in (Flew 2017, see Part III). Historically, the notion of "survival of the fittest" represents this logic. Given limited resources, competitive survival is inevitable. This logic lies in the heart of the Darwinian paradigm. Different species sharing an ecological niche, would always compete, resulting in elimination of a few species. Similar logic applies to different cells in multicellular organisms. The key point is that biological evolution has provided necessary machinery for the elimination of less fit cells. Once we accept this reasoning, it is intuitively clear that the need for senescence arises only in the case of multicellular organisms. Senescence is unheard of in unicellular organisms. Upon a little reflection, it is apparent that in that case, the mechanisms by which senescence and eventual death must be embedded in genomes and they can't be present at the level of individual genes or even individual cells.

Thus, roots of senescence must lie in genomic architecture and specifically in the form of long-range influences. Admittedly, this is antithetical to the conventional wisdom. We know for sure that senescence is essentially a cellular phenomenon arising from biochemical wear and tear of the biomolecules (Demaria 2018). Even the subsequent need for cellular death is affected by the machinery present at the cellular level (in the form of the mechanisms of apoptosis). Therefore, let us now deconstruct the second proposition that senescence is initiated by long-range influences arising from genomic architecture. Prima facie, it is intuitively clear that if senescence sets in primarily because of the biochemical wear and tear of biomolecules, the only possible way to induce this wear and tear is to send inappropriate signals to the cellular machinery. In order to understand how inappropriate signals can give rise to wear and tear of biomolecules, let us perform a thought experiment. Let us assume that there exists a protein P which acts as an enzymatic catalyst for a house keeping metabolic process. For the sake of simplicity, we will assume that this protein P would be synthesized from a single gene whenever there is a biochemical signal available. Thus, the presence of a block signal would initiate the gene expression leading to the synthesis of protein P. Similarly, the protein P would have a limited half-life, after which it would be degraded by normal cellular machinery. Thus, as far as the protein P is concerned, its concentration would vex and wane in a cyclic manner.

With this scenario in place, let us see how improper signals can induce wear and tear. Prima facie, this wear and tear leading to senescence can occur from several different sources. For instance, it could arise from oxidative stress wherein the gene responsible for synthesizing protein P is impaired. Similarly, wear and tear of protein P could also occur because of the excess of metabolic substrate for which the protein P acts as a catalyst. However, we will overlook all the other remaining causes of the wear and tear of the protein P and instead focus on an idealized scenario wherein the wear and tear of the protein P is caused because of long-range influences of genomic architecture. Apparently, these influences irrespective of their nature would have to translate into any of the known biochemical pathways that cause wear and tear of protein P. This is necessary because otherwise we would be forced to concede some kind of transcendental effects like the "vital force." Therefore, it is imperative that irrespective of the nature of these long-range influences, once it reaches a cell, its effects would be in the form of making a choice among the known biochemical pathways. Once we concede this reasoning, it is intuitively clear that the nature of these long-range influences must be in the form of timing the signals rather than in the form of a new type of signal. Thus, genomic mechanisms of aging in the form of long-range influences must be the altered signals of initiations and terminations of a gene expression, which in this case would be the gene coding the protein P.

Thus, once we accept that genomic mechanisms essentially bring about long-range activation or termination of a gene expression, then we can formalize the genomic model of aging. Upon a little reflection, it is intuitively clear that we already know some of these influences in the form of chromosome territories (Fritz 2014) or in the form of cis and trans long-range influences (Donaldson 2000). Thus, without invoking any transcendental arguments and without abandoning our conventional biochemical details, we can formalize the genomics of aging if we can define the mechanism of these phenomena. At present, there is no formal description of chromosome territories which explains what causes chromosomes to acquire a suitable configuration for chromosome territories to influence different gene expressions. Similarly, we have observed cis and trans effects in influencing gene expressions. However, we do not know why chromosomes change their three-dimensional configurations during the various stages of a cell division. Our conventional perspective of the coiling and uncoiling of DNA sequences present in chromosomes is based on the influence of cdk proteins (Morgan 2007). Upon a little reflection, it becomes obvious that even these signals are actually a variation of intercellular long-range influences. Thus, whether these long-range influences are intracellular or intercellular, they simply synchronize the gene expressions. It is often overlooked that beyond the random thermodynamic and kinetic processes, we don't have any explanation of how a genome synchronizes different gene expressions.

However, if we postulate that spacetime is an integral element of genomic architecture, it is possible to create a framework wherein these different configurations of genomes and its constituent DNA, change depending on the nature of participation of spacetime. Thus, different alignments of chromosomes that give rise to the desired chromosome territories or different degrees of coiling of chromosomes can be derived by linking them to the different degrees of participation of spacetime with the genome. As argued in the preceding chapters, this need not lead to any determinism. Moreover, apparent thermodynamic and kinetic effects of activators like cdk proteins could be shown to arise from some higher dimensional configurations of genomes in which spacetime is entwined with the genome. The advantage of postulating higher dimensional configurations of genomes is that it would allow us to link various stages of the cascade of activators into a single framework. For instance, different cdk proteins operate at different stages of a cell division. The synchronization of these cdk proteins cannot be explained by the conventional perspective. However, by defining a higher dimensional framework of genomes, we can explain how these cdk proteins are synchronized. Admittedly, there is no evidence as yet to prove anything about higher dimensional configurations of genomes. Therefore, it is legitimate to be skeptical about it. While we will try to defend this idea in the following sections, in this section, let us look at the nature of evidence we should look for to support the postulate of higher dimensional configurations of genomes. Self-evidently, the best way to look for such evidence is to study phylogenetics.

Therefore, let us try to understand what kind of phylogenetic linkages can convince us about the existence of higher dimensional configurations of genomes and more importantly, whether such evidence is available. Prima facie, since the long-range influences arise from higher level of organization of genomes and not from the individual genes, it doesn't make sense to think these long-range influences would be verifiable by phylogenetic studies. However, as mentioned in the preceding chapters and in the preceding sections, if genomes have any higher dimensional configurations, then it ought to reflect in the length of intergenic distances. Since a genome can be thought of as a linear molecule, it is intuitively clear if we were to organize a genome in more complex configurations in the three-dimensional space, these configurations would be allowed only if the intergenic distances have a certain length. Therefore, it seems logical to extend this analogy to higher dimensional space. After all, we routinely observe the relationship between the degree of coiling and its effects on gene expressions. Therefore, it is not unreasonable to extend the scenario to higher dimensional configurations.

However, there is one problem with this analogy. When we try to correlate different configurations of chromosomes, say, during different stages of cell divisions, we are not changing the nature of spacetime. However, the higher dimensional configurations proposed in this model consist of different degrees of blending of time-like and space-like dimensions. In other words, the nature of underlying spacetime is different for each of these higher dimensional configurations of genomes. More importantly, unlike the conventional perspective, in the proposed model, spacetime plays an active role in genomic architecture. Therefore, it is not feasible to use analogy and try to conceptualize phylogenetic studies to uncover evolutionary evidence of this model genomic role in aging. At the same time, such a genomic model of aging must leave its mark in the form of phylogenetic linkages. The question is what kind of linkages? In the absence of any quantitative description of this model, it is not possible to define such linkages. However, a general rule can be articulated. Since according to this model, different dimensionalities of spacetime consists of different degrees of blending of time-like and space-like features, it is inevitable that when these higher dimensionalities devolve into the four-dimensional spacetime, they give rise to a fixed amount of blending of time-like and space-like features. Therefore, while the temporal information of the mechanism of these long-range influences might not be preserved in the four-dimensional configurations of genomes, the spatial information of the long-range influences would be echoed in the four-dimensional configurations of genomes.

Upon a little reflection, it is obvious that if genomes exist in multiple higher dimensionalities, then each higher dimensionality would devolve differently into the four-dimensional configurations of genomes. Therefore, irrespective of different degrees of blending, each higher dimensionality would have its unique ratio of conversion to the spatial separations of different genes within a genome in its four-dimensional configurations. Therefore, the length of intergenic distances of a given genome would be indicative of the number of involutions that the genome has undergone while devolving from a given higher dimensionality to the four-dimensional spacetime. Thus, while we can't simply measure the intergenic distances to prove the proposed model, the distribution of lengths of various intergenic distances can give us some idea of the possible topological configurations. Given the large number of genes (and therefore, large number of intergenic distances) in a given genome, the distribution of the lengths of intergenic distances for a given genome can definitely help us to check the validity of this model. It is important to keep in mind that this strategy is amenable to phylogenetic studies because we already have a fairly good idea of evolutionary linkages between different species. A word of caution is necessary. Since we don't have a quantitative description of the topology of genomes, we can't predict a particular pattern of distribution of intergenic distances. Therefore, while evaluating a particular distribution of the lengths of intergenic distances of a given genome, we may encounter multiple solutions. This is particularly true in the case of a genome having a few thousand genes. It is interesting to note that this limitation is similar to the problem of rooting in a typical phylogenetic investigation (Bromham 2008, see Chapter 5). It is a matter of speculation whether the problem of rooting is ontologically linked to this problem of distribution of lengths of intergenic distances. There is an additional problem with this approach. It is assumed all along in the discussion presented above that genes are linearly distributed on a given genome. Therefore, this approach may not work in the cases where there are known cases of gene overlaps. This brings us to the end of the discussion on the possibility of genomic mechanisms of aging. In the next section, we will look at the evolutionary perspective of aging as implicit in the conventional perspective.

6.4 Evolutionary Perspective of Aging

In the introduction, semantics of aging in biological evolution was discussed. In the preceding sections, the genomic imperative of aging was outlined. However, the role of natural selection in shaping the exact mechanism of aging needs to be analyzed. It is one thing to accept that aging is an essential ingredient of biological evolution. However, it is quite another to think that the present methods of aging are optional. There are three aspects of this topic that would be deconstructed in this section, viz., what are the possible mechanisms of aging that could have evolved, but didn't; is the current "graceful degradation" of functionalities (Wechsler 1992) a naturally selected mechanism of aging?; does the selection of graceful degradation tell us anything about the nature of natural selection? Prima facie, it is difficult to conceptualize different mechanisms of aging. This is particularly true because the conventional perspective has taken a view that aging is a rather unwelcome consequence of genomic complexity. However, if we take an alternative perspective postulated in the proposed model that aging is not an undesirable outcome of biological evolution, but an integral element of the strategy of biological evolution, then it is necessary to deconstruct the possible alternatives to the graceful degradation of functionalities as a mechanism of aging.

In order to understand this conundrum, let us deconstruct the graceful degradation of functionalities and how it could have evolved. Ab initio, it is intuitively clear that this feature arises from two underlying aspects of genomic architecture. Firstly and perhaps self-evidently, it arises from a certain degree of plasticity of genomic architecture. It arises because there are multiple and overlapping long-range influences within a genome. Polygeny (Reavey 2013) and pleiotropy (Lozano 2017) are testimony to this functional and structural overlap. Secondly, graceful degradation arises because there is a distinction between functionalities and structuralism. Interestingly, this duality of functionalities and structuralism manifests itself in different forms. The semantics of these dualities are discussed in Chap. 1. If we were to take a contrarian view that the genome is simply a string of discrete numbers of genes, apparently, the graceful degradation would not have evolved. The fact that graceful degradation occurs, points toward some kind of nonlinear genomic architecture. Conversely, the phenomenon of graceful degradation must be indicative of a particular architecture of genomes.

Once we accept this reasoning, it is possible to think about different mechanisms by which graceful degradation could have evolved. It is also intuitively clear that there must be two types of opposite processes operating in genomes. First type of process must be governing the cessation of genomic activities. This process can be visualized as the one causing cellular senescence and eventual death. This type of process can be initiated by a variety of factors like oxidative stress or deleterious mutations, etc. The second type of process can be visualized as the one trying to restore functionalities from alternative sources. However, this type of process can only be initiated by genomic compulsions and not by any external triggers. Thus, it is intuitively clear that the phenomenon of graceful degradation is an outcome of two conflicting processes. More importantly, one of these processes is initiated by external or cellular signals affecting individual genes or a cluster of genes. On the other hand, the second process is initiated by internal signals arising from the genome itself. Conventionally, we have focused on the first type of process caused by oxidative stress or deleterious mutations. However, we have overlooked the second type of process which arises from the genomic compulsions. This selective neglect of genomic influences is caused by the absence of any prior knowledge of genomic architecture. More specifically, it is caused by our belief that structuralism and functionalities of genomes are synonymous. However, the proposed model takes a slightly different view. It postulates that structuralism and functionalities have a separate framework of their own. More importantly, it postulates that these two frameworks enjoy a certain degree of overlap without being synonymous. This reasoning tempts us to think that genetic mechanisms and genomic mechanisms enjoy the same relationship that exists between structuralism and functionalities of genomes.

Once we accept this reasoning, it is intuitively clear that natural selection not only operates on genomes as units of selection, but it acts differently on these two frameworks. Thus, in view of this reasoning, we can now think of graceful degradation as having arisen from the different compulsions of structuralism and functionalities of genomes. Natural selection through the environmental impact causes changes in the structural template of genomes at the level of individual genes. However, natural selection changes the functional template of genomes at the higher level of genomic modules. The emergence of graceful degradation symbolizes the different pulls that the structural and functional templates of genomes experience due to natural selection. It is legitimate to wonder what causes the changes in the functional template? After all, in the structural template of genomes, we know that it is the changes in the environment that bring about factors like oxidative stress or deleterious mutations which change the DNA sequences and their expressions. However, there is no way an environment can bring about changes in the functional template of genomes. Admittedly, the structural changes in genomes would be reflected in the corresponding changes in the functional template, but there is no room for the functional template of genomes to undergo changes on its own. The environment cannot possibly alter the functional template without changing the corresponding structural template. Thus, we need to justify the existence of separate functional templates of genomes by offering some mechanism causing changes in the functional template of genomes. This is where the direct participation of spacetime in natural selection comes into the picture. According to the proposed model, spacetime itself becomes the environment for the functional template of genomes. Moreover, according to this model, spacetime exists in multiple dimensionalities. Thus, whenever an object (a biomolecule in the present case) moves from one dimensionality to another, it undergoes changes in its structuralism. Now, according to this model, the functional template of genomes occupy a higher dimensionality vis a vis its corresponding structural template. Therefore, according to this model, both these templates keep changing their dimensionalities. While this change in the dimensionality of the structural template of genomes gives rise to changes that are perceptible from the perspective of the four-dimensional spacetime, the corresponding changes in the functional template of genomes remain imperceptible because they happen in higher dimensionalities of spacetime.

Thus, what appears as the environment for the structural template consists of the details of spacetime at the four-dimensional spacetime. This happens to be visible to our cognitive faculty and therefore, we can formalize it. On the other hand, the environment at the dimensionality, wherein the functional template of genomes exists, consists of spacetime itself. Moreover, at these dimensionalities, spacetime doesn't manifest the distinction between the time-like and the space-like dimensions. Therefore, the changes in the functional template cannot give rise to any perceptible changes which we can observe. The final piece of this jigsaw puzzle is what causes the changes in the dimensionalities of the functional template of genomes? The answer is quantum fluctuations. If we accept that spacetime exists in multiple dimensionalities simultaneously, then it is axiomatic that the fine structures of spacetime in different dimensionalities too would be different. Therefore, due to quantum fluctuations, the higher dimensional configurations of genomes would keep on switching from one higher dimensionality to another. Apart from these inherent changes in the dimensionalities of spacetime at higher dimensionalities, there are also external causes for this. Because of the partial overlap between these two templates of genomes, it is possible that some of the changes in the structural template would alter the dimensionality of the corresponding functional template. This is because structural complexities of biomolecules also spread across different dimensionalities. Thus, an isolated riboprotein would occupy different dimensionality than the one occupied by, say, ribosomal RNA. This scenario explains why we need to postulate separate templates for functionalities and structuralism. More importantly, it also explains why we didn't perceive the existence of a separate template of functionalities.

Thus, it is evident from the above discussion that the phenomenon of graceful degradation arises because of the conflicting tendency of the template of structuralism and the template of functionalities. More importantly, this conflict arises because functionalities have coarser units as compared to the units of structural template. Since the template of functionalities supervenes over the template of structuralism, functionalities overlap over different units of the template of structuralism, resulting in plasticity that gives rise to graceful degradation. This brings us to the next question: does the emergence of graceful degradation tell us about the nature of natural selection? If the reasoning behind the graceful degradation given above is valid, then it suggests that natural selection must have its own structuralism. Upon a little reflection, it is intuitively clear that the scenario described above rests on the topological perspective of the relationship between functionalities and structuralism. Therefore, it is axiomatic that natural selection too must operate on some topological principles. As discussed in the preceding chapters, the possibility of natural selection having its own native structuralism doesn't undermine the Darwinian semantics of randomness (Bonner 1988). Similarly, it doesn't introduce any design principles either. This is because the structural template of natural selection implicit in this scenario doesn't arise from any intelligent preconceived design. It arises from the structural template of spacetime itself. Therefore, to the extent we accept that the fine structure of spacetime didn't arise from any design principles, we must accept that the native structural template of natural selection doesn't represent any design principles. Of course, we can always question why spacetime has such a template of fine structure. To be honest, there is no answer. In fact, as discussed in the preceding monograph (Chhaya 2022, see Chapter 1), our belief in the cosmic singularity being itself a nonstructural entity is inconsistent with the conception of spacetime having such a fine structure. Science hasn't reached a stage where it can question the source of all the structural details of the manifest universe. If the cosmic singularity is the source, then the question is from where did the cosmic singularity derive the structural details? It doesn't help to think that the source of these structural details of the manifest universe lies in mathematics because then, we are required to explain the source of mathematics itself. Thus, we can reasonably assume that the possibility of natural selection having its own native structuralism doesn't imply any design principles, at least not any more than the possibility of spacetime having its own native structuralism does.

This brings us to the last question: if aging as a phenotype is a product of a particular template of genomes, does it eliminate any other phenotypic attributes from being naturally selected? Admittedly, we can employ the technique of modal logic to think of various phenotypes that could have arisen if aging wasn't a *fait accompli* of the genomic architecture. The most obvious and perhaps most desirable phenotype would be eternal youth. However, upon a little reflection, it is relatively easy to demonstrate that such a phenotype would stop the very process of natural selection. More importantly, it would violate the hitherto universally obeyed second law of thermodynamics. As mentioned in the introduction, the guiding principle of biological evolution is how to fix a pattern amidst chaos. The moment natural selection produces an eternally youthful species, it can't improve upon it, and hence, natural selection would cease to operate. Similarly, if there was such an eternally youthful species, it would keep on producing generation after generation of replicas of itself. This would result in creating more and more organized information content which violates the second law of thermodynamics. The phenotype of aging is an inevitable outcome of natural selection and it embodies the semantics behind biological evolution and natural selection.

Instead, it is more fruitful to think about what kind evidence is available from natural selection that justifies the scenario outlined above. If we look at graceful degradation as a phenotype, it is apparent that it arises from the conflict between functionalities and structuralism of the genome. More importantly, it arises because of different types of complexities of the templates of functionalities and structuralism. Therefore, this need not get reflected in the form of graceful degradation only. It would give rise to different features. There are several features of biological evolution that have remained nebulous, viz., the origins of polygeny, pleiotropy, the origin of open reading frame in in vivo synthesis of proteins, and the origin of punctuated evolution (Gould 2007). Perhaps, the scenario discussed above offers an explanation for all these enigmas. Polygeny could arise when a single unit from the functional framework devolves into several units of structural template. This leads to a situation wherein multiple genes contribute to a single functionality. Similar logic applies to Pleiotropy. In the case of punctuated evolution, it would arise when the mechanism of natural selection is topological in nature. In that case, when a functionality exists in a higher dimensionality, the process of natural selection would alter the dimensionality of that functionality to a lower dimensionality. In this process, due to its inherent topology, initially contiguous functionality would now manifest as different and disjointed sub functionalities. Thus, over a period of time, evolutionary course appears to be stratified into various branches, thereby giving rise to a punctuated evolution. Returning to the present discussion, even if we were to accept this scenario at face value, the question arises whether this conception of aging being a genomic strategy and not a genetic strategy can help us to alleviate some of the pathologies arising from aging in general and cellular senescence in particular? Therefore, in the following sections, we will try to find answers to this question.

6.5 Can Genomic Perspective Solve the Problem?

Conventionally, since we believe that aging is essentially governed by individual genes, our perspective of aging has centered around two propositions. Firstly, we believe that aging is inevitable and irreversible. Secondly, in spite of this inevitability and irreversibility, we can minimize the pathologies arising from aging by focusing on the individual genes responsible for aging. This is a pragmatic view and it has evolved after sustained clinical observations. Therefore, the question arises whether this proposition that aging is a genomic functionality can help us to alter our approach to study aging? This is important because even if aging arises from genomic causes, it becomes manifest only at the level of individual genes. Therefore, if we can successfully handle the gene expressions of these genes individually, there is no need for the proposed model irrespective of its validity. Therefore, in this section, we will try to deconstruct how the genomic origins of aging can alter our approach to study pathologies arising from aging.

In that context, it is intuitively clear that we should look for pathways that pass on the genomic functionality of aging to the individual genes that we know are responsible for aging. There are individual genes that give rise to apoptosis cascade. However, these genes are activated by intracellular signals. For instance, presence of free radicals or reactive oxygen species causes damage to the different organelles or even the genes present. However, in that case, apoptosis is necessary and even desirable. This is particularly true for cancers. Therefore, we should look for those pathologies arising from mistimed and improper gene expressions for this purpose. The key point is that if there are these types of faulty gene expressions (and self-evidently, there are), what kind of long-range influences can alter these aberrations. More importantly, can this model offer any help in identifying and altering these faulty gene expressions?

The answer to this question is yes, but only indirectly. Let us understand how. According to this model, the genomic functionalities, including the functionality of aging, is coded in the form of higher dimensional configurations of genomes. Moreover, these higher dimensional configurations of genomes are not independent of the underlying DNA sequences. Therefore, any major changes in the length of DNA sequences would lead to the corresponding changes in the higher dimensional configurations of genomes. Upon a little reflection, it is intuitively clear that we can't alter the length of the DNA sequence of a genome by inserting additional DNA polynucleotides in the places known to contain functional genes. This is because such an insertion would give rise disruption of a functional gene, leading to unwarranted pathologies arising from that genes. Obviously, such an insertion of polynucleotides must be aimed at the intergenic regions. This strategy would increase the length of genomes without disrupting its constituent genes. Once we achieve this, according to this model, the resulting genome would have different higher dimensional configurations. These new higher dimensional configurations would definitely alter the genomic signals for aging, without altering the genetic signals for aging.

However, it has to be admitted that this strategy has several pitfalls. Firstly, if the intergenic regions that we choose to extend are active in microRNA control of gene expressions (Cretoiu et al. 2020), then it would be quite harmful. Secondly, in the absence of any prior knowledge of higher dimensional configurations of genomes and their relationships with the underlying DNA sequences, there is no way to predict what kinds of control elements are present in the higher dimensional configurations. Thirdly, we must ensure that the inserted polynucleotides do not express themselves, thereby adding additional pathologies. However, purely from the theoretical perspective, this strategy is amenable to experimental verification. Moreover, with the present techniques like microarray of gene expressions, this strategy can be easily tested in vitro. In addition, with our current expertise in plasmid insertion methodology, the choice of lengths of plasmids and the sites of insertion can be managed to obtain specific increases in the length of intergenic distances.

6.6 Aging as a Side Effect of Genomic Complexity

Though there are no theoretical models, it is intuitively clear that aging is possibly a side effect of genomic complexity. This belief is consistent with the fact that unicellular organisms do not seem to manifest aging per se. Moreover, since these organisms divide at regular intervals, the phenomenon of cellular senescence is also missing. Of course, deleterious mutations do occur in unicellular organisms. However, they give rise to different species, albeit with varying survivability. Therefore, it seems reasonable to think aging is an outcome of increasing complexity of genomes. Moreover, since genomes of unicellular organisms are also quite complex, it seems reasonable to think that aging could be an artifact of genomic architecture rather than that of genomic complexity per se. More importantly, since aging is manifest only in multicellular organisms, it is reasonable to think that aging arises from those features of genomic architecture that control the synchronization of activities of different cells. Thus, it is tempting to think that aging is a product of natural selection simply because genomic architecture is naturally selected. This inference leads us to two interesting conjectures. Firstly, if aging is a consequence of a particular architecture of genomes, then it must be caused by long-range influences. Secondly, it ought to be possible to minimize age related pathologies by altering the overall genomic architecture rather than altering individual genes involved in these pathologies. This reasoning is consistent with the fact that the onset of age related pathologies varies from individual to individual. Had the onset of these pathologies been governed by wear and tear of replicating machinery or by oxidative stress alone, there would have been a uniformity in the time of onset of these pathologies. From this perspective, it would be interesting to investigate *Progeria* types of genetic disorders.

6.7 Aging as a Genomic Module

In the previous section, two inferences were presented, viz., (1) The process of aging is a by-product of our typical genomic architecture. In other words, had genomes evolved differently, the pathologies arising from aging would have been different. (2) Aging could have arisen from those features of genomic architecture that bring about synchronization among different parts of the organism. However, these inferences, even if true, don't help us to understand the genomic origins of aging. However, as mentioned above, *Progeria* types of genetic disorders provide a hint to the manner in which genomic architecture gives rise to aging. *Progeria* types of diseases are characterized by rapid onset of aging followed by faster progression (Seager 2005). Admittedly, we don't know what actually causes these diseases. However, there is one inescapable inference available from the clinical examination of the patients suffering from these diseases. None of the pathologies arising from these diseases relate to lack of synchronization of different physiological cycles. It is just that every physiological process is speeded up. This points toward a possibility that aging is controlled by a separate module of genomes. It is as if there is a separate biological clock embedded in genomes which supervenes over the remaining modularities. This conjecture is consistent with the fact that we have not been able to locate any particular DNA sequence responsible for progeria from genome-wide association studies. Therefore, in this section, we assume, albeit provisionally, that there exists a separate module which synchronizes different modularities. More importantly, we will assume that *Progeria* types of diseases are caused by dysfunction of this module. For the sake of simplicity, we will name this module as "Synchrone."

Let us examine the possible consequences of having a separate module of Synchrone. Since aging is universal, it seems reasonable to think that it can't be caused by any particular DNA sequence or a functional gene. Had it been otherwise, phylogenetic studies would have located it. By analogy, aging could not be caused by any of the conserved genes or what are colloquially called housekeeping genes. Therefore, it is axiomatic that this functionality must originate from a higher level of organization of genomes. Moreover, synchronization of various physiological processes occurs at various levels, from intracellular level to organismic level. Therefore, it is reasonable to think that these various synchronizations could have a basic mechanism which manifests itself at various levels. In such a scenario, if indeed there was a module which we have named Synchrone, then it must enjoy both ontological and functional primacy. The ontological primacy arises because this module must have evolved along with or just after the evolution of multicellularity and much before any other functionalities. Similarly, in developmental stages of any multicellular organisms, the process of synchronization sets in in the earliest stages. More importantly, it doesn't set in in the case of a fertilized egg or even in the embryo which has undergone one or two cell divisions. This observation suggests that the functionality of synchronization is not a functionality manifest in unicellular organisms. However, there must be some factor which synchronizes different biochemical processes operating within the organisms. Therefore, the functionality of synchronization must be present in some incipient form in the unicellular organisms. The question however, is how did this incipient functionality of synchronization in the unicellular organisms evolve into a full-fledged modularity in the multicellular organisms? It is tempting to think that maybe the underlying mechanisms of synchronization in the unicellular organisms were adapted to function in diverse contexts.

This scenario is consistent with the earlier suggestion that intracellular synchronization too must be governed by the mechanism which governs the intercellular synchronization. In order to deconstruct this ambiguity, we must look at the mechanism of synchronization at the intracellular and intercellular levels. At the intracellular level, synchronization occurs by feedback servo mechanism (Elsasser 2016). There is a signal, either chemical or electrical, which initiates a biochemical process. For instance, sudden release of Ca^{2+} ions can give rise to activation of some enzymes to shift an equilibrium of a biochemical process (Verkhratsky and Toescu 1998, see Part I). Similarly, a messenger molecule may enter the cell through a designated cell receptor to trigger a biochemical reaction (Krauss 2005, see Chapter 5). Using this simple scenario, it is easy to visualize how synchronization can be brought about. It can be brought about by controlling the emergence of chemical (messengers) or electrical (Ca^{2+} ions) signals. This can happen only when there exists a hierarchy of controls which is what is defined as feedback servo mechanism. Now, let us look at synchronization occurring at intercellular levels. The communication between two cells in proximity, say by paracrine mechanism, would also be brought about by the same feedback servo mechanism. However, in this case, the signal is always chemical (even in the case of two neighboring neurons the messenger molecule is always a chemical messenger). However, for this synchronization to occur, it is necessary that the cell producing a messenger molecule must exhibit a cyclic production of the messenger molecule. This is also achieved by feedback servo mechanism. Thus, the mechanism of synchronization at every level from intracellular level to organismic level (which includes paracrine, exocrine, and endocrine) remains the same.

This unitary nature of feedback servo mechanism in both these cases suggests that the reason why unicellular organisms do not show aging must lie in the nature of the messenger molecules. To be more precise, it must lie in the way these messenger molecules are escorted out of the cell producing them. This is because otherwise, these messenger molecules would be confined to control the physiological pathways of the cell producing them. The syntheses and the transport of paracrine and endocrine messenger molecules is an extensively studied topic (Goldbetter 1989). The key step in the transport of these messenger molecules is that of vesicle formation which prevents these messenger molecules from acting in an autocrine manner. Thus, it is the evolution of transport mechanisms of these messenger molecules that was a key to the evolution of multicellular organisms. This also suggests that the efficacy of intercellular synchronization lies not only in the regulation of the synthesis of these messenger molecules alone, but also in the synthesis of lipid bilayers necessary for vesicle formation.

It is tempting to think that if aging is caused by impaired synchronization among different cells in multicellular organisms, then the transport mechanisms also contribute to aging. While we don't know the exact details of the connection between the genes responsible for synthesizing messenger molecules and the genes responsible for synthesizing the lipid bilayers necessary to transport them, it is intuitively clear that this too must require a hierarchy of controls. Thus, from the present discussion, it is reasonable to think that the incipient functionality of synchronization of the unicellular organisms must have evolved into a separate module only when it was integrated with the functionality of synthesizing lipid vesicles. This makes sense because the genes responsible for synthesizing the lipid bilayers too have evolved during the transition from unicellularity to multicellularity (Marsh and Goodle 1994). Therefore, aging must be treated as a genomic module which is ancient in its origin and that it must possess a hierarchy of controls within its own structuralism. It seems reasonable to think that evolution of aging as a genomic module is a consequence of multicellularity. However, this module must have also played a role in natural selection. Therefore, in the next section, we will look at aging as a factor in natural selection of multicellular organisms.

6.8 Aging as a Tool for Natural Selection

Even in the absence of any detailed information about how aging is an inevitable outcome of multicellularity, it is intuitively clear that aging affects the survivability of the individuals. Therefore, it must be a factor in natural selection. However, there is no clarity on the exact role of aging in natural selection. Prima facie, aging ought not to influence the course of natural selection. This is because, like many of late-onset biological aberrations, age-related pathologies cannot be passed on to the next generation. In the case of higher organisms, the embryonic segregation of germ lines from the rest of the embryo ensures that any late-onset pathologies cannot be passed on to the next generation (Marsh and Goodle 1994). However, aging can impair survival of an individual. Therefore, any influence that aging can exercise on natural selection would be in the form of group selection (Borrello 2010). In any given environmental niche, the amount of resources available and the size of population of species dependent on these resources would shape the course of natural selection. Therefore, to the extent aging impairs the survivability of individuals, it would alter the size of groups competing for a given environmental resource. Thus, aging per se doesn't influence the course of natural selection, at least not directly. However, it could influence group dynamics, thereby altering the group selection. It is also possible that there exists a mechanism at a genomic level which decides the pace of aging, thereby changing the longevity of different species. This also could influence the course of natural selection. The possibility of aging being a genomic strategy and hence having influence on the course of natural selection is tenable. However, this conjecture needs to be formalized as an integral element of genomic architecture. Apart from this academic perspective, the possibility of aging being a genomic module deserves to be taken seriously from the clinical perspective as well.

Thus, it is reasonable to think that if we can understand the exact genomic mechanism of aging, we cannot only alleviate some of the pathologies arising from aging, but we can also increase the longevity of human beings by altering the genomic architecture. However, the question is can we do it? Or more importantly, should we do it? We will discuss these topics in the next two sections.

6.9 Can Aging Be Prevented?

The question whether aging can be prevented has two connotations. Firstly, it refers to pathologies arising from aging. Secondly, it refers to romanticism implicit in the idea of eternal youth. In this section, we will focus on the first connotation. The second connotation of the search of eternal youth will be partially addressed in the next section. The conventional perspective on the pathologies arising from aging is based on individual genes and their expressions (Vijg 2007). This is essentially a genetic strategy. However, there is no corresponding genomic strategy available in the conventional perspective. In other words, our current efforts are based on the premise that individual genes operate mostly independently and rarely, if ever, in tandem to give rise to these pathologies. Ideally, systems biology should provide a systemic view of genomic influences on aging. However, in the absence of any formal description of genomic architecture, there is no systemic perspective of aging and the pathologies arising from it. Even within the conventional perspective, there is a fundamental reason why we have not made any remarkable progress in preventing or delaying aging. This is because the primary cause of the pathologies arising from aging is oxidative stress. In the days of ecological evangelism, it is tempting to think that the environment, or rather its deterioration, is responsible for the increase in age-related pathologies. To an extent, this is a valid inference. However, the primary factor causing age-related pathologies through oxidative stress lies in the way biological evolution has proceeded.

Biological evolution began when the environment on Earth was devoid of oxygen (Oparin 1968). Therefore, the basic metabolic processes of unicellular organisms were fine-tuned by natural selection in the context of the anaerobic conditions. The rise of oxygen level in the atmosphere on Earth simply steered natural selection in a different direction. As is her wont, Nature tinkered around with the existing metabolic pathways to survive and thrive in the newly formed aerobic conditions. Thus, our basic metabolism consists of a patchwork. The original framework is that of operating under anaerobic conditions. However, in order to survive, Nature has tried to fill the gap arising from aerobic conditions by doing a cut and paste job. Those familiar with modern day computers and their operating system would immediately understand Nature's predicament. Computer's operating system can be altered by adding new subroutines. However, there comes a time when the patchwork becomes unwieldy and prone to failure. This is precisely what is happening in our bodies. Unfortunately, Nature, unlike companies developing computer's operating systems, can't rewrite the genomic architecture. Therefore, Nature and consequently, we are stuck with these legacy problems. Each cell in our bodies carries such a medley of subroutines embedded in the basic framework of anaerobic metabolism. Therefore, it is tacitly accepted, but never openly stated that aging is a *fait accompli*. At best, we can manage pathologies arising from aging. Therefore, our present focus is twofold. Firstly, we are trying to delay the onset of aging and its attendant pathologies. Secondly, we are trying to alleviate some of the pathologies arising from aging.

With this limited but pragmatic strategy, the question arises whether the conventional choice of focusing on individual genes and their expressions is adequate or not. More importantly, can a different strategy, say the above-mentioned genomic strategy, improve our chances of fulfilling these modest objectives? As discussed in the preceding sections, this genomic strategy rests on a simple premise. A genome exists in higher dimensionalities and most of the genomic functionalities exist in higher dimensional configurations of genomes. More importantly, these genomic functionalities influence the sequence and the duration of individual gene expressions through long-range influences. Since genomes acquire their higher dimensional configurations on the basis of the lengths of their DNA sequences, it ought to be possible to alter the higher dimensional configurations of genomes (and therefore their genomic functionalities) by inserting additional DNA polynucleotides in genomes. Moreover, since we don't wish to alter any of the known genes, these insertions must be restricted to the intergenic regions. Admittedly, this strategy needs a lot of fine tuning about the nature of inserted polynucleotides as well as the places of insertion. Thus, we can at least delay, if not prevent aging provided we know the details of genomic architecture. We will return to this topic in the following sections. Presently, let us discuss the second question about whether we should prevent aging.

6.10 Should Aging Be Prevented?

A new strategy to prevent or delay aging was suggested in the previous section. Assuming that one day, we will be able to really prevent or delay aging, the question arises whether we should do it or not. Prima facie, there are two arguments that can be put forth against any such attempts. The first argument arises from bioethics while the second argument arises from clinical risks that any new technology faces. Therefore, in this section, we will briefly outline these two arguments. Let us begin with bioethics. There are two facets to this bioethical objection. Firstly, it can be argued that Nature has incorporated aging into genomic architecture for some reason. Admittedly, we don't comprehend the wisdom behind Nature's choice, but fact remains that aging is a consequence of biological evolution and therefore, it should not be thought of as a disease or illness (Of course, this argument doesn't apply to pathologies arising from aging, but applies aging per se). However, this argument is debatable because it can be applied to any medical advancement. However, the second facet to this bioethical objection is rather difficult to refute. It can be argued that if any such technical advancement is put in place, we will be changing the demographic distribution of our society. After all, one of the motivations for Darwin to develop the theory of descent with modification was the

Malthusian logic (Flew 2017, see Part III). This issue would raise itself if we succeed in preventing aging. If aging is prevented (or even delayed) the proportion of older people in a society would keep on increasing. In today's world, where we can't manage to feed everyone, the prevention of aging would put an unbearable resource crunch. This would necessarily result in a reduction in the birth rate. This scenario forces us to face a moral dilemma. Should we sacrifice the younger population for the sake of the older population? Moreover, even if we acquiesce to do so, we will be interfering with the process of natural selection. Thus, it seems less objectionable to think about reducing the pathologies arising from aging rather than eliminating aging per se.

Second objection arises from the clinical perspective. To be honest, our knowledge of genomic architecture (and genomics in general) is far from adequate. Therefore, it is possible that any half-baked strategy can lead to unacceptable consequences for the patients. Our recent experience with cloning and CRISPR techniques doesn't inspire confidence (Isaacson 2021). Admittedly, it can be argued that perhaps we would never know everything about genomes to ensure a total success for such novel techniques. It shouldn't prevent us from trying new ideas. Therefore, it seems reasonable to think that we must take a cautious approach about any new techniques. Therefore, it makes sense to take a middle of the road approach. It seems reasonable to focus on the pathologies arising from aging and on understanding the genomic origins of aging. The approach advocated here is congruent with this cautious optimism. Moreover, the proposed model also offers a way to formalize genomic architecture as well. Therefore, it offers us a way to evaluate origins of aging in the context of overall genomic architecture. Thus far, we have discussed some broad issues about aging. Therefore, now it is time to look at more specific issues related to aging and how it is represented in the proposed model. In the next section, we will summarize the genomic architecture as implicit in the proposed model. The details of this model are discussed in various chapters of this monograph. Therefore, we will try to pick up the features of genomic architecture that are germane to the present discussion in the next section.

6.11 The Proposed Model of Genome

For the sake of simplicity, we will summarize the proposed model of genomes in a point-wise manner.

1. Genomic architecture is spread over multiple dimensionalities of spacetime simultaneously. This is in contrast to the conventional perspective in which genomes exist in four-dimensional spacetime.
2. Genomic architecture exists in multiple dimensionalities simultaneously and supervenes over the four-dimensional DNA sequence and its different ensembles with chromatin.
3. Genomic architecture consists of two parallel frameworks of structuralism and functionalities.

4. Of these two frameworks, the framework of functionalities occupies higher dimensionality as compared to the dimensionality occupied by the structural framework.
5. Genomic architecture consists of several modules. Thus, just as the structural framework of genomes consists of chromosomes, gene clusters and intergenic regions, the functional framework also consists of different functionalities.
6. Just as the structural unit of genomes is defined as genes, the functional unit of genomes is operon.
7. While the structural framework of genomes appears to be spread over different chromosomes in the four-dimensional spacetime, the functional framework exists in higher dimensionalities in different units.
8. In other words, genomes exist in multiple dimensionalities simultaneously and the distinction between functional framework and the structural framework rests on the dimensionality. There is only one framework, but it behaves as a structural framework at some of the lower dimensionalities and as a functional framework in higher dimensionalities.
9. There exists a single mechanism by which dimensionality of genomes can be altered. This mechanism has been formalized by a mathematical operator called involution.
10. Since genomes exist in multiple dimensionalities simultaneously, the operator of involution lowers the dimensionality from which genomic expressions take place. Thus, at the four-dimensional spacetime, the genomic expressions brought about by lowering the dimensionality using the operator of involution appear to be gene expressions.
11. According to the proposed model, similar expressions occur in different dimensionalities of genomes. Thus, gene expressions are the special case of genomic expressions.
12. At higher dimensionalities, genomic expressions consist of changes in the higher dimensional configurations of genomes. These changes appear in the four-dimensional spacetime as long-range influences.
13. This long-range influences at the four-dimensional spacetime appear to be nonlocal because of the topological compulsions of the operator of involution.
14. Since according to this model, genomes have evolved from a hypothetical entity named here as genomic singularity, different modules have evolved at different stages of biological evolution.
15. As discussed above, the module responsible for aging must have evolved at the time when multicellularity evolved.
16. According to the proposed model, the dimensionalities of different modules are decided by their ontological primacy.
17. Therefore, according to this model, the module of aging occupies the highest dimensionality among other modules of multicellular organisms.
18. Therefore, the functional influences of the module of aging would be routinely passed on to all the other modules of multicellular organisms.

Using this point-wise description of genomic architecture, we will try to deconstruct the nature of aging in the following sections. In the next section, we will describe how aging, particularly the pathologies arising from it, can be passed on to different functionalities.

6.12 Aging According to the Proposed Model

Using the description of the genomic architecture given above, we will try to deconstruct aging according to this model. We will continue with the proposition that aging has arisen from multicellularity. Therefore, in this section, we will look at the possible scenarios prior to and after the emergence of the aging module. Having done that, we will try to speculate some details about the nature of the module of aging. Admittedly, these scenarios would be brief and simplistic. However, they are adequate for the present discussion because this monograph deals with the ontology and epistemology of genomes and not with molecular biology of genomes.

Let us begin with the scenario prior to the emergence of the module of aging. As discussed in the preceding chapters, according to this model, all the genomes must have arisen from the putative proto genome named here as genomic singularity. The semantics and structuralism of genomic singularity will be discussed in the following chapters. Presently, we will assume that during the course of evolution, genomic singularity had differentiated into a typical genome possessing the functionalities of eukaryotic organisms. According to this model, this transformation of genomic singularity into a genome of a typical eukaryotic genome was brought about by a series of involutions each leading to the lowering of dimensionalities and the differentiation of singularity into multiple modules. It is this modular genome that must have given rise to multicellularity. The topic of multicellularity and its evolution has been discussed in literature (Niklas and Newman 2016, see Part V). At present, we will maintain an agnostic stance on this topic and focus on the emergence of the module of aging. It is apparent that multicellularity per se requires evolution of mechanisms of synchronization of intercellular communications. Therefore, we will assume that such a mechanism was already in place before the module of aging separated from the genome as a separate module. Upon a little reflection, it is intuitively clear that this putative mechanism for synchronizing intercellular communications would be generic in nature. This is because when multicellularity eventually gave rise to different body plans and to separation of biological functionalities into different organs, this mechanism must be already in place. Therefore, it makes sense to think that separation of the module of aging must have preceded the modularization leading to the evolution of different organs. This scenario is depicted in Schema 6.1.

While there is no definitive evidence for supporting the scenario outlined in Schema 6.1, it is one of the most likely scenarios. This is because it requires a least number of assumptions. Moreover, it is possible to justify it by the known physiological pathways of different types of organs and different types of body plans. For instance, according to this scenario, the mechanism for synchronizing

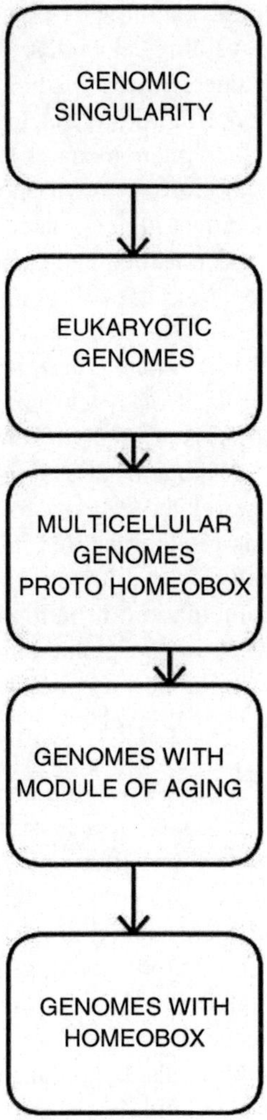

Schema 6.1 Evolution of the module of aging

intercellular communications evolved prior to its modularization leading to the module of aging. Therefore, the same mechanism ought to operate in different physiological conditions present in different organs. This is self-evidently true. It is important to keep in mind that while the physiological consequences of this unitary mechanism may vary from organ to organ, the mechanism is the same. For this purpose, we will look at two examples from human pathologies arising from aging due to the impairment of three different organs, viz., cataract, osteoporosis, and dementia. It has to be admitted at the outset that these pathologies could arise from different physiological causes. However, we will assume that the nature of the trigger is not relevant to the present discussion, but how aging responds to this trigger is. In other words, we will consider the idealized scenario in these cases and assume that there are no comorbidities. The basic approach would be to think of these pathologies as having arisen from the lack of synchronization of different physiological cycles. We will assume that it is the impairment of homeostasis that triggers these pathologies in the individuals who do not have any other comorbidities. It is important to keep in mind that homeostasis, per se, is a far wider phenomena and aging is one of the many types of impaired homeostasis. Therefore, we will not discuss the mechanism of homeostasis here, except as manifest in age-related pathologies.

Let us look at cataract as an age related pathology. Conventionally, there are two generic causes of cataract, viz., oxidative stress and diabetes. However, as mentioned above, we will not consider both these causes. Instead, we will try to think of physiological consequences of oxidative stress only. Here again, we will focus on the lack of synchronization leading to cataract formation. For instance, oxidative stress is normally taken care of by a redox system consisting of glutathione (Whelan and Nugent 2009, see p. 83) operating around crystallins. Therefore, in the age-related cataract formation it is the lack of synchronization of the synthesis of this glutathione that must be thought of as a trigger for cataract formation. Admittedly, in real life situations, cataract formation is very complex and arises from the combination of multiple triggers. However, we will look at the simplified scenario wherein the glutathione synthesis is impaired in the eyes. This simplification is justified because we are looking for genomic origins of aging and how it manifests at a local level.

From the physiological perspective, we can think of cataract formation as a result of protein crystallin and how it is packed in lenses. The key to maintain the optimal refractivity of a lens is to achieve a certain range of water molecules within the packing of different crystallin molecules. The oxidative stress produces an oxidation of free SH groups of crytalin to give rise to S-S sulfide bonds. This crosslinking of adjacent crystallin molecules gives rise to impervious packing of crystallin in the lens, thereby causing dehydration and subsequent decrease in refractivity (Yorio et al. 2008). In a healthy individual the regular synthesis of a glutathione that restores free SH groups from S-S sulfide prevents cataract formation. However, in old individuals, the feedback servo mechanism to control the synthesis of a glutathione is impaired, leading to cataract formation. Upon a little reflection, it is intuitively clear that cataract formation is not a one step process, but a gradual process. Thus, in a young individual, the refractivity of the lens remains within the permissible range without any impairment of vision. However, over time, there is an accumulation of S-S sulfide bonds and a gradual decrease in refractivity. This accumulation of sulfide bonds is aggravated by oxidative stress and lessened by the glutathione produced by homeostatic forces. With the progressing age, this homeostasis mechanism of synchronization between the increase in sulfide bonds and the synthesis of the glutathione gets impaired. This sums up the physiology of cataract formation. Therefore, we will focus on the glutathione and the regulation of its expression as a typical example of feedback servo mechanism and see how the proposed model reconfigures this cycle of feedback. A brief sketch of this mechanism is given in Schema 6.2.

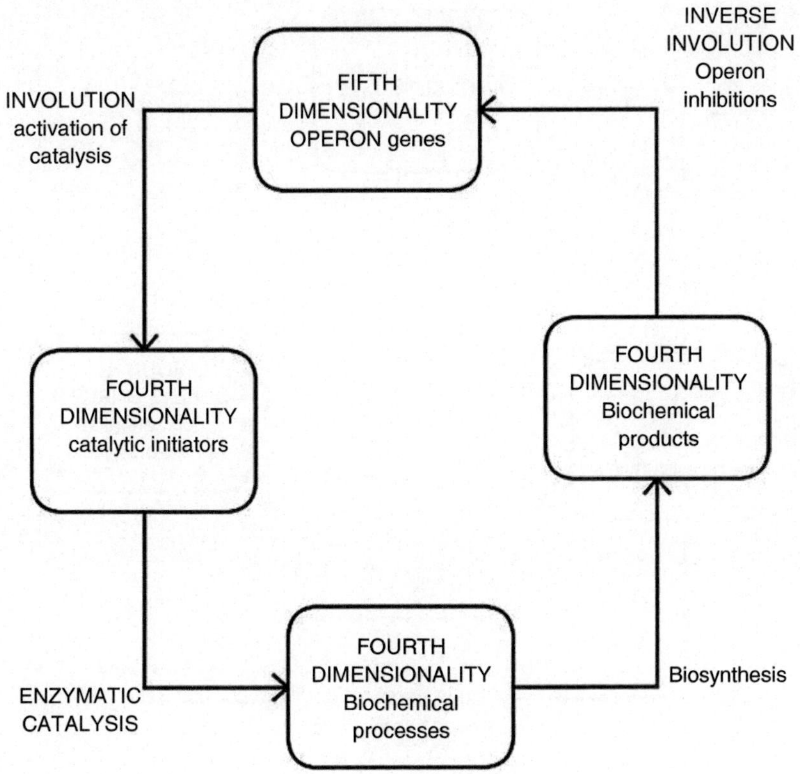

Schema 6.2 Topological model of feedback servo mechanism

Now let us look at the age-related pathology of osteoporosis. Just as in the case of cataract formation, onset of osteoporosis is governed by various triggers (Leder and Wein 2020). Conventionally, availability of vitamin D and estrogen is reported to be a causal factor for osteoporosis. At a deeper level, osteoporosis has been shown to be a result of equilibrium between the populations of osteoblasts and osteoblasts. However, we will sidestep this etiology and look at the cellular causes of osteoporosis. Clinically, it has been demonstrated that cellular senescence of osteoblasts causes osteoporosis. Moreover, the number of senescent osteoblasts increases with age, thereby establishing a cellular level mechanism of age-related osteoporosis. At a genomic level, this mechanism of senescence of osteoblasts is reported to be through the classical wnt signaling pathways (Yang 2019, see Chapter 2). Therefore, we will try to deconstruct this mechanism using the proposed model. A brief description of this interpretation of wnt signaling is given in Schema 6.3.

Now, let us look at the third type of age-related pathology, viz., non-Alzheimer types of dementia. The conventional perspective of dementia has been based on the belief that neurons do not regenerate in adults. However, with the discovery of neural stem cells (Mobley 2019), the situation has changed. This observation points toward

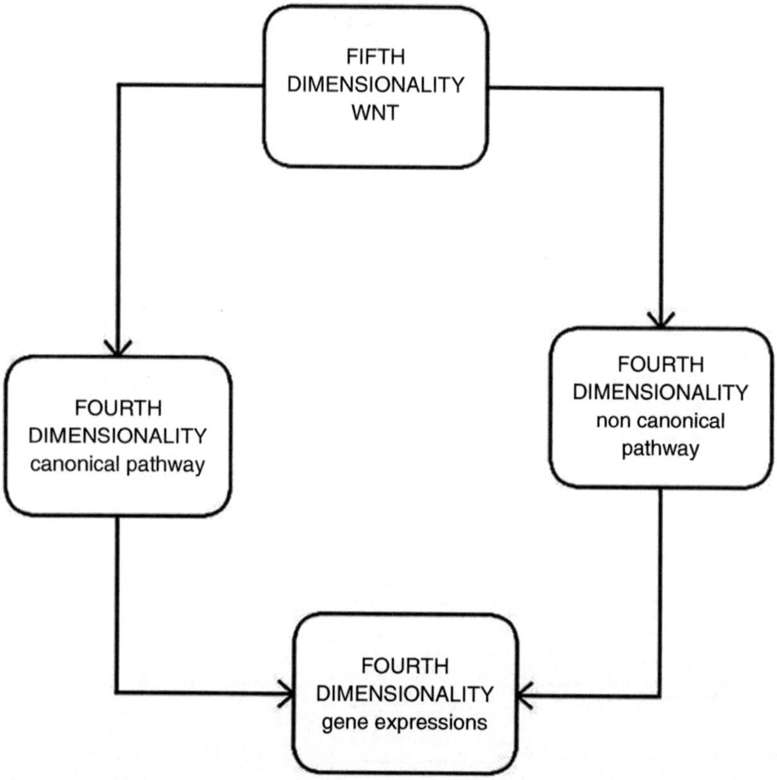

Schema 6.3 Topological model of wnt signaling

a possibility of new therapeutic approaches. Among the known age-related pathologies, dementia is perhaps the most pervasive pathology. Clinically, two major types of dementia have been extensively reviewed, viz., Alzheimer disease (Haass 2005) or vascular dementia (Meyer et al. 2001). To some extent, both these dementias are age-related pathologies. However, the primary cause in each of these dementias is essentially not related to aging. For instance, in the case of Alzheimer disease, it is accumulation of protein (Tau protein) that gives rise to dementia (Al Baghlani 2013). Admittedly, the excess synthesis of this protein must be due to lack of synchronization during the feedback mechanism. However, this type of dementia is not caused by aberrant or impaired synaptic junctions. Therefore, we would omit this pathology from our discussion. Similarly, in the case of vascular dementia, the primary cause of dementia is the inadequate vasculature and not the neuronal dysfunction. Admittedly, improper vasculature does affect the repair and maintenance of neurons. However, dementia in this case does not arise primarily due to neuronal dysfunction. Therefore, we will omit this topic as well. Instead of looking at these most common types of dementias, we will focus on dementia arising from neuronal dysfunction. It is important to keep in mind that the approach outlined here is applicable to these two types of dementia, but with the necessary modifications.

When viewed from this perspective, it is intuitively clear that dementia per se ought to arise from the lack of synchronization between the genes responsible for generating and repairing synaptic junctions. Therefore, it is possible to think of these pathologies as having arisen from the impairment of genomic controls rather than having arisen from the impairment of gene expressions of the concerned genes. We will discuss these pathologies while discussing the mechanisms of aging in Sect. 6.14. However, before that, let us look at the suggestion that aging is not an outcome of the dysfunction of individual genes. Rather, aging is an outcome of failures of genomic controls.

6.13 Aging as a Genomic Functionality

In the previous section, we looked at three typical age-related pathologies, viz., the cataract formation, osteoporosis, and dementia. Admittedly, as discussed in that section, the onset of these pathologies is triggered by multiple factors. However, we have chosen only the cellular level causes of these pathologies and its genomics. It is important to keep in mind that these cellular level mechanisms of these pathologies are well known, and the proposed model has nothing new to offer, except for a new perspective. This is important because it is legitimate to argue that if the mechanisms of these pathologies are well established and the proposed model doesn't improve upon them, then it is not necessary to employ the proposed model because it doesn't alter the therapeutic approaches. At best, the proposed model could be treated as a hermeneutic device, nothing more.

This argument is based on the misconception that genetics and genomics of aging are one and the same. The cellular mechanisms discussed above are based on genetics. They deal with the individual genes and their expressions. The feedback mechanisms implicit in these gene expressions are governed by the cytoplasmic conditions. There is no mention of genomic controls in regulating the gene expressions of these genes. However, according to this model, there exists a definitive feedback mechanism which is genomic by definition. For instance, in the case of wnt signaling (Yang 2019), we know lots of details of how it alters gene expressions. However, very little is known about what controls the initiation of the wnt cascade. It is the contention of this model that there exists another level of control which manipulates the initiation of the wnt signals. More importantly, this control exists in the form of configurations of genomes themselves. These higher dimensional configurations of genomes bring about long-range influences that activate wnt cascade. In other words, there exists a genomic module, named here as Synchrone which controls the process of synchronization of all the other modules. Of course, it is possible to dismiss this claim. However, as discussed in the following sections, this model offers credible therapeutic alternatives which are amenable to clinical verification. Therefore, it is necessary to scrutinize the proposed model in the course of scientific inquiry. In order to facilitate such an inquiry, in the next two sections, we will outline a genomic mechanism of aging and its topological representation.

6.14 Mechanisms of Aging

As discussed in the preceding sections, aging arises from several different physiological sources, viz., oxidative stress and cellular senescence. However, the mechanisms by which aging is actually initiated remain genetic in nature. For instance, oxidative stress can induce damage to different genes. Similarly, different stages of cell divisions can be impaired due to regular wear and tear of the replicating machinery. In such cases the ultimate cause is impaired replacements which arise from the gene expressions of the genes responsible for synthesizing proteins necessary for replicative machinery. Alternatively, repeated replication leads to telomere shortening which again impair the replication. The fact that the ultimate cause of aging lies in genes and their expressions, should be comforting to those who are looking for therapeutic solutions for preventing or delaying aging. This is because if we can repair genetic aberrations, aging wouldn't manifest, just as it doesn't in the case of unicellular organisms. There is nothing in the biochemical processes per se that causes aging. It is only the program that controls these biochemical processes that becomes dysfunctional leading to aging.

It is legitimate to wonder that if the reasoning given above is valid, then why we have not been able to prevent or delay aging. The reason for this failure lies in the fact that we have so far focused on the individual genes and their expressions. It is important to keep in mind that this approach has worked reasonably well because we have managed to cope with the age-related pathologies. However, geriatrics is waiting for a paradigm shift. Our conventional approach has reached a level where we obtain an incremental increase in therapeutic knowledge of geriatric. Upon a little reflection, it becomes obvious that this approach would lead to an impasse. The paradigm shift that can dramatically benefit the discipline of gerontology or geriatrics must come from genomics. It is only when we are to understand genomic influences on the expressions of individual genes that we can hope to prevent or delay aging (Albeit, as discussed in Sect. 6.6, it is a separate question whether we should prevent aging). Therefore, in this section, we will try to distinguish between genetic mechanisms and genomic mechanisms of aging. More importantly, we will try to understand why genomic mechanisms are more equipped for delaying (if not preventing) aging than the genetic mechanisms are.

Before we look at the specific mechanisms of aging, let us understand the distinction between a genetic mechanism and a genomic mechanism of any biochemical processes. A typical genetic mechanism for controlling a given biochemical process, can be conceptualized as a feedback servo mechanism. This has conventionally been idealized as operon (Miller and Reznikoff 1980). The concerned gene synthesizes a protein which catalyzes the given biochemical process. The product of this biochemical process, in turn, controls the timing of the gene expression of the gene responsible for synthesizing the required catalyst in the form of a protein. Admittedly, this is a rather simplistic scenario, but it is adequate for the present discussion. Now let us look at the corresponding genomic mechanism. Admittedly, there are some known phenomena like chromosome territories (Fritz 2014) which can prima facie be thought of as genomic mechanisms. However, in the conventional perspective, there is no framework for their formal descriptions. Therefore, instead of picking up an actual example, we will take a hypothetical example of a genomic mechanism. Admittedly, this model is based on the proposed model, but it doesn't employ any topological arguments to justify itself. In that sense, this hypothetical example is just what the name suggests, hypothetical. However, this example is consistent with the conventional perspective except for the fact that it hasn't been proposed, nor has it been experimentally tested.

This hypothetical genomic mechanism could be a long-range influence wherein a noncontiguous DNA sequence somehow influences a gene expression of a distant gene. This could happen by some changes in the conformations which bring these otherwise noncontiguous DNA sequences, viz., the influencer sequence and the sequence whose expression is being influenced. Once these two sequences are in close proximity to one another, as we know in the case of chromosome territories, the gene expression of one of these DNA sequences would be influenced by the second DNA sequence. The question that is germane to the present discussion is this: does this influence differ significantly from the influence that is formalized as an operon? We will overlook the fact that this influence occurs between two DNA sequences which are some distance away from one another. Instead, we will focus on the timing and the duration of gene expression. In the case of operon type of control, timing is decided by cytoplasmic conditions like the quantity of the products of the biochemical reaction under the control of the operon. It is also intuitively clear that in this case, the timing and the duration of gene expression of either a catalyst or an inhibitor are governed by the cytoplasmic conditions. There is no discretion available for an operon to decide whether to activate or not. In other words, it is a closed system. At a certain level of the desired product, operon would be switched on, and at some different concentration of the desired product, the operon would be switched off.

The situation in the case of a genomic mechanism is slightly different. Admittedly, even in this case, there is a degree of fixity. The moment two spatially segregated DNA sequences come in proximity, the expression of one of the DNA sequences would be controlled by the second DNA sequence. However, when these two sequences would come to gather remains uncertain. Therefore, it provides some flexibility in deciding the timing and the duration of the gene expression. It is this flexibility that separates a genomic mechanism of control from the genetic mechanism of control. Admittedly, it is possible to argue that unless we define the exact nature of genomic mechanisms, this scenario cannot be taken seriously. After all, the example of a genetic mechanism, viz., the operon mechanism, is very defined and even verified. Therefore, now we will limit ourselves to the proposed model. This is because the conventional perspective of long-range influences is not well defined. However, the proposed model offers a mechanism for this flexibility of controls of gene expressions. It must be admitted that the scenario described here is unproven and therefore, speculative. However, if the proposed mechanism of genomic controls is found capable of explaining the hitherto unexplained long-range influences, then this mechanism must be accepted as a valid scientific hypothesis.

According to the proposed model, genomes exist in multiple dimensionalities of spacetime simultaneously. While from the perspective of the four-dimensional spacetime, genomes appear as a DNA sequence impregnated with chromatin, at the higher dimensionality of spacetime, genomes appear in the form of molecular orbitals. According to the proposed model, different molecular orbitals possess different numbers of electrons and therefore occupy different dimensionalities. [It is important to keep in mind that the conventional quantum chemical perspective (Gupta 2016) also assigns different numbers of electrons to different molecular orbitals.] However, in the conventional perspective, different molecular orbitals possess identical dimensionality. However, in the proposed model, each molecular orbital, depending on the number of electrons present in it, would occupy different dimensionalities. The justification for this postulate is outlined in the preceding monographs. Moreover, according to this model, the metrics of different molecular orbitals would also be defined by their dimensionalities.

Long-range influences according to this model arise by the interactions between different molecular orbitals of a given genome. These interactions are formalized as involutions wherein a higher dimensional molecular orbital devolves into a lower dimensional molecular orbital. This operation of involution results in nonlocal influences which from the perspective of the four-dimensional spacetime appear to be long distance influences. Thus, it is because of the topological compulsions of the operator of involution that gives rise to long-range influences. In addition to these spatially nonlocal effects, this scenario also explains the order of gene expressions. This is because different spatially segregated genes are influenced simultaneously because they are generated by a single operator of involution. In fact, according to this model, it should be possible to define genomic architecture by working backward from the order of gene expressions and the lengths of intergenic distances between the genes expressing simultaneously.

With this scenario in place, let us look at the aging process. Once again, we will overlook the age-related pathologies arising from oxidative stress. Instead, we will focus on two biological processes, viz., the cycle of cell divisions (Morgan 2007) and transcriptions during mitosis (Weinzierl 1999). Admittedly, these two processes are connected with one another. For the sake of simplicity, we will assume that they operate independent of one another. Of course, the scenario outlined below works as well even if we assume that these two processes operate in tandem. Therefore, the assumption that these processes operate separately is only a hermeneutic device. The underlying logic remains the same in both these cases. The cycle of cell divisions is triggered by the activation of cdk proteins. Admittedly, the trigger for this activation can be either intracellular or intercellular. The intercellular signals can be thought of as genomic mechanisms and the intracellular signals can be thought of as genetic mechanisms. Thus, we can think of the cycle of cell divisions as being under simultaneous control of genetic and genomic mechanisms. With this perspective in place, in the next section, we will depict a topological model of genetic and genomic mechanisms that explains why these two mechanisms are harmonized leading to delay in the onset of aging. More importantly, under what circumstances this synchronization is disrupted during aging. However, before we look at the model, let us look at the process of transcription during mitosis.

Prima facie, transcription during mitosis appears to be a consequence of signals available from cdk proteins. Therefore, conventionally, the transcription is considered to be dependent on the cycle of cell divisions. However, there is one aspect of transcription that is dependent on genomic mechanisms. This refers to the expression of telomerase genes. The centrality of telomeres and telomerase in aging and even in cancers has been extensively dealt with in literature (Mattson 2001). It is generally conceded that onset of cancers is the price that we pay for our longevity. At the other extreme of this belief is our gradual realization that cancers are the outcomes of Nature's way to search for eternal life. Between these profound semantic propositions lie the details of mechanisms of transcription. It is important to keep in mind that the importance of telomeres lies in the fact that DNA molecules exist in double stranded helix configuration. Had DNA sequence existed as a circular strand, there would have been no need for shortening of DNA sequence at the end of transcription and there would have been no need for either telomeres or telomerase. Therefore, there would have been no instances of aging or cancers. But then, there would have been no higher organisms, including human beings.

Returning to the present discussion, we will try to deconstruct two specific processes leading to initiation of cdk protein cascade and telomerase synthesis. It is important to keep in mind that once both these processes are initiated, they are followed by complex chains of reactions. Therefore, we will focus on the first step in both these cases, viz., initiation of cdk protein cascade and the synthesis of telomerase. Since both these initial points are singular in nature, it would be easier to conceptualize them as switches controlled by genetic as well as genomic mechanisms. Therefore, in the next section, we will describe how the proposed model offers a correct topology of these competing mechanisms and how they initiate both cdk protein cascade and the gene expression of the gene encoding telomerase.

6.15 Topological Model of Mechanisms of Aging

In the preceding chapters, a topological model of genomic architecture has been outlined. Admittedly, it is based on the premise that genomes occupy multiple dimensionalities of spacetime simultaneously. It was suggested that these dimensionalities are not to be taken as abstract entities representing the information content of genomes. Rather, these dimensionalities were postulated to be physical dimensionalities. Similarly, in the preceding sections, it was mentioned that the higher dimensional configurations of genomes must be viewed as molecular orbitals occupying different dimensionalities. Admittedly, both these concepts are contrarian. However, in this section, we will overlook the underlying semantics of these two concepts and instead accept this scenario at face value. Moreover, according to this model, different modules must occupy different dimensionalities and therefore, there would be some topological hierarchy among the modules of a given genome. We will also overlook this aspect as well. In this section, we will focus on the module of aging by itself, without considering its relationships with the remaining modules. However, as discussed in the preceding sections, it is important to keep in mind that according to this model, the module of aging must have evolved along with multicellularity. Therefore, from the evolutionary perspective, the module of aging must enjoy ontological primacy over other modules of genomes. Therefore, the module of aging, which we have named Synchrone, also influences the remaining modules. This ensures that the nature of genomic mechanisms (or long-range influences) remains unchanged in different types of synchronization. With this perspective in place, let us look at the proposed topological template of the module of aging.

In the light of the discussion presented above, it is intuitively clear that the genomic mechanisms of aging must occupy higher dimensionality vis a vis the genetic mechanisms. At the same time, within each category of these mechanisms, there must be some internal hierarchy. Thus, within DNA, transcription during mitosis might have different signals to abort transcription and initiate apoptosis. Thus, these different signals leading to apoptosis would constitute a hierarchy among themselves. Similarly, during the cycle of cell division, there exists a nested hierarchy of cdk proteins and their activations (Morgan 2007). We will retain these nested hierarchies in this model. It is important to keep in mind that the semantics of these nested hierarchies is quite different in the proposed model than the one implicit in the conventional perspective. For instance, the sequential activation of different cdk proteins, according to the conventional perspective, depends on different activation energies and different rates of diffusion. However, from the semantic perspective, these chemical properties represent the temporal information. However, in the proposed model, the temporal information is represented in the form of dimensionalities and their devolutions to the four-dimensional spacetime. As discussed in the following chapters, both these interpretations are synonymous. However, a word of caution is necessary because the semantic synonymy has not been articulated in the language of mathematics, at least not as yet.

In order to simplify the scenario, we will restrict the diagrammatic representation of the inherent topology of the module of aging to only a few broad categories. However, it is possible to incorporate more genetic and genomic mechanisms in the diagram given here. For this purpose, we will select four genomic and three genetic mechanisms. These mechanisms are listed in Table 6.1 and their topological model is given in Schemas 6.4 and 6.5.

It is intuitively clear from the Schemas 6.4 and 6.5 that topological representation of both these types of mechanisms, viz., the genetic and genomic mechanisms, that this topological perspective is not merely a hermeneutic device. Rather, it represents hitherto unexplored and perhaps a more fundamental aspect of genomic architecture. At the same time, it has to be admitted that this approach is rather tentative and needs experimental corroboration. The exact strategies for verifying the proposed topological paradigm are beyond the scope of this monograph. However, it is possible to

Table 6.1 Mechanisms of aging

#	Category	Name	Function	Cellular context
1	Genomic	Epigenetic changes	Inhibitions of cdk proteins	Cellular senescence
2	Genomic	Operon	Incorrect gene expressions	Apoptosis
3	Genomic	Spatial long-range influences	Incorrect gene expressions	Apoptosis
4	Genomic	Temporal long-range influences	Loss of synchronization	Cellular senescence
5	Genetic	Mutations	Faulty genes	Cellular senescence
6	Genetic	Telomere shortening	No transcription	Cellular senescence
7	Genetic	No paracrine signals	No cell divisions	Cellular senescence

predict some of the therapeutic approaches based on this model. Therefore, in the next section, we will briefly point out some possible therapeutic possibilities.

6.16 Possible Therapeutic Approaches

As discussed in the preceding sections, there are several moral and ethical issues involved in devising therapies for aging. Therefore, we will try to restrict this discussion to the age-related pathologies rather than on the aging per se. Even here, there are topics that are related to individual lifestyles. These topics too would be overlooked. We will mainly focus on those features of cellular senescence that lead to clinical manifestation of pathologies. With these limitations in place, we will focus on three typical age-related pathologies, viz., cataract formation, osteoporosis, and dementia arising from aging other than that arising due to Alzheimer disease. These pathologies are included here because they are not only widely prevalent in the older population, but also because they offer insights into the underlying genomic influences. Some of the details are discussed in the preceding sections. We will revisit these details, if only to understand the finer nuances of the underlying physiology.

Let us begin with cataract formation (Yorio et al. 2008). As mentioned above, the onset of cataract can be traced back to the lack of synchronization between glutathione concentration and the extent of S-S sulfide bond formations which alter the water permeability of the lens (Whelan and Nugent 2009). It is important to keep in mind that cataract formation is not a one-time accident, rather it is a continuous lack of synchronization between the S-S sulfide bond formations and the gene expressions of the gene encoding two enzymes, viz., glutamate cysteine ligase (GSL) and glutathione synthetase (GS). These two enzymes are ATP-dependent enzymes which synthesize glutathione in two steps with the first step catalyzed by GSL

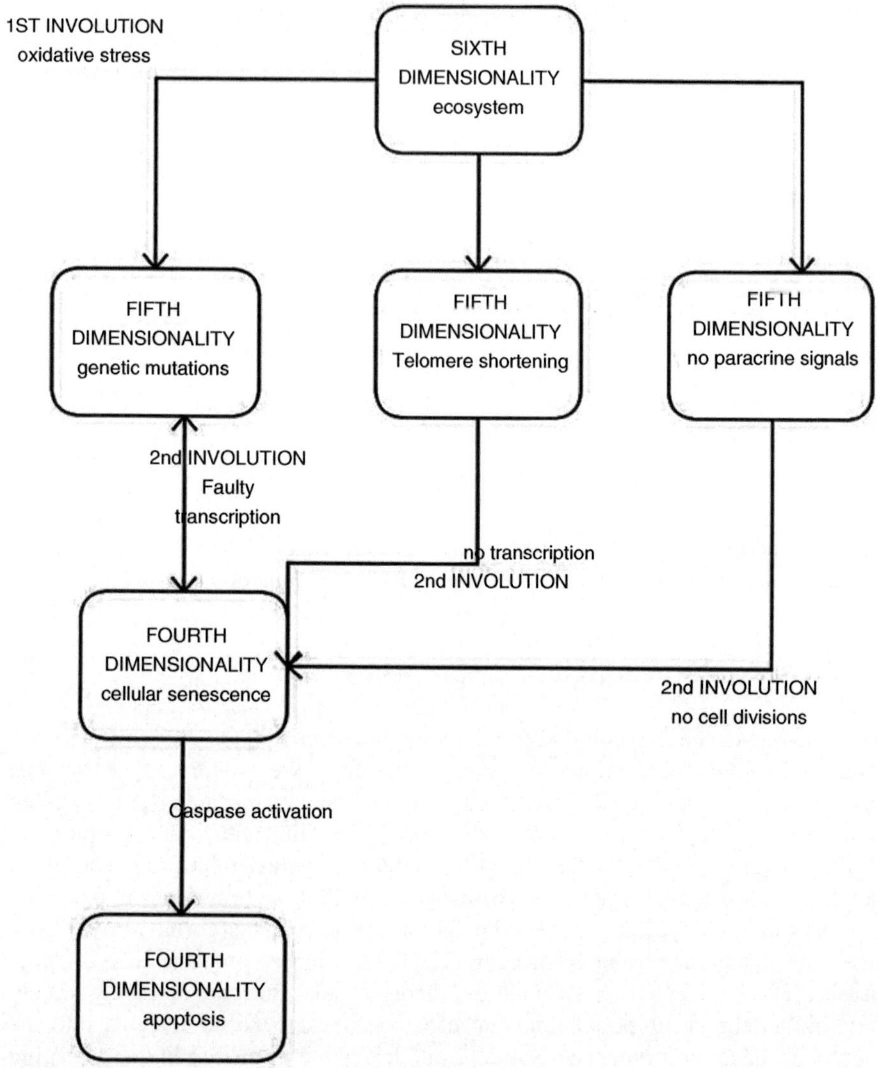

Schema 6.4 Topological model of genetic mechanisms of aging

being a rate-limiting step. Glutathione, thus synthesized, is eventually degraded by usual ubiquitin mediated pathways.

It is reasonable to think that the synthesis of these two enzymes is controlled by a typical feedback servo mechanism which employs a mechanism for sensing either the extent of sulfide bonding or the changes in the concentration of glutathione. Thus, in the case of any changes in these parameters could either switch on the gene expression of the genes encoding the GSL and GS or turn them off. Without going into the identity of the sensor or its mechanism, this system can be idealized as a

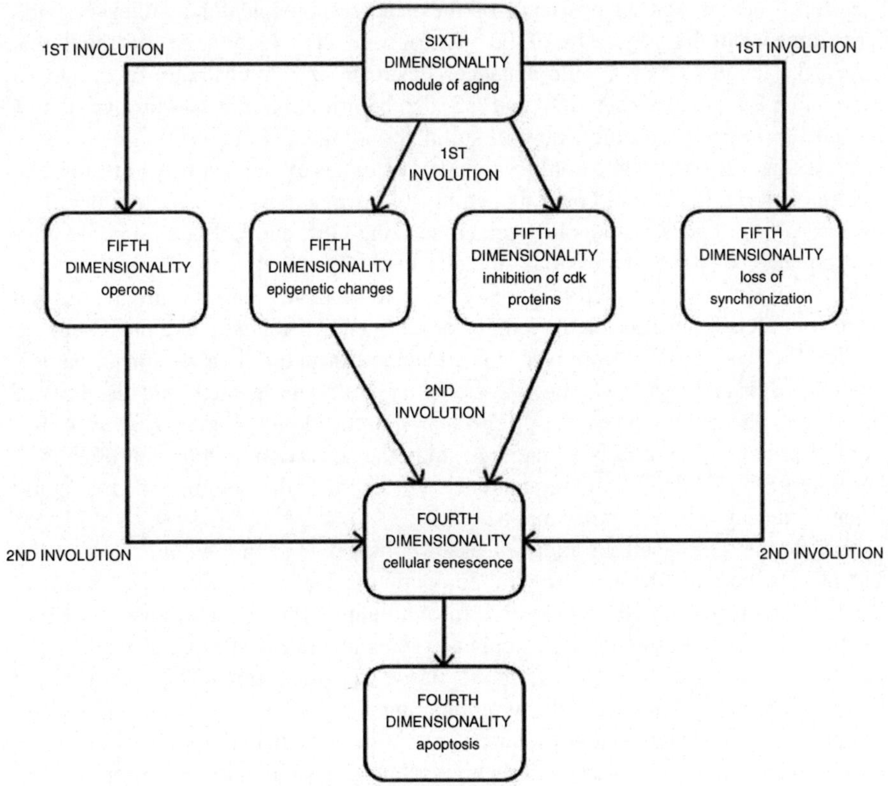

Schema 6.5 Topological model of genomic mechanisms of aging

typical operon. The relevant point is that in the conventional perspective, this feedback system can malfunction due to the failure of the sensor to attach itself to the DNA sequence upstream of the gene encoding the enzymes GSL and GS. Therefore, it is intuitively clear that according to the conventional perspective, onset of cataract formation begins because of the mutations in the DNA sequence upstream of the concerned genes.

Accordingly, the conventional perspective of genomics cannot offer any specific strategy to delay the onset of the cataract formation. In some sense, this pathology is a *fait accompli*. Of course, it is possible to alleviate the problems of low glutathione concentration by using several nutrients. However, bioavailability of glutathione is poor in the case of oral administration of glutathione (Buonocore et al. 2016). On the other hand, the proposed model offers a possible way to mitigate this condition. To begin with, it suggests that the initiation of the gene expression of the gene encoding GSL and GS is controlled by genomic mechanisms and not by the conventional feedback servo mechanism. Admittedly, this feedback servo mechanism is necessary. However, according to this model, there exists an additional layer of control of this gene expression. This layer consists of a higher dimensional configuration.

Therefore, mere mutation upstream of the gene responsible for synthesizing the glutathione is not the only cause of the onset of cataract formation. According to this model, the lengths of intergenic distances upstream and downstream of the genes responsible for synthesizing GSL and GS also contribute to the synchronization of the gene expression with the extent of sulfide bonding.

Therefore, the best way to confirm this proposition is to define a way to restore the synchronization. If we could alter the lengths of upstream or downstream intergenic distances of the genes responsible for synthesizing GSL and GS, then it could alter the frequency of gene expressions, thereby restoring the synchronization. Admittedly, this is not a very attractive strategy because we have yet to acquire the desired degree of precision in altering the lengths of intergenic distances using the technique of CRISPR (Luo 2019). Moreover, since cataract formation is a late-onset pathology, it is a debatable issue whether we should try to alter these intergenic distances of the individuals who are prone to suffer other age-related pathologies. However, as a proof of principle, it seems reasonable to study the activation of gene expressions of these genes individually or sequentially by altering the lengths of intergenic distances during in silico experiments.

Now let us look at the problem of osteoporosis (Leder and Wein 2020). In the case of osteoporosis, there are several competing influences at work. For instance, there are two types of cells responsible for maintaining bone mass density (BMD), viz., osteoblasts and osteoclasts. Osteoblasts generate bone tissue and osteoclasts, as their name suggests, break down bone tissue via resorption. Thus, BMD is a dynamic parameter maintained by competing functionings of osteoblasts and osteoclasts. This dynamic homeostatic cycle is in congruence with the semantics of biological evolution. As discussed above, Nature seeks fixation of pattern amidst chaos and permanence amidst transience. The duality of osteoblasts and osteoclasts exemplifies this semantic proposition. However, this simple dichotomous process is further complicated by several intervening factors. These include the role of vitamin D receptors, the role of calcitonin and RANKL nuclear factor which is produced by osteoblasts to suppress the activity of osteoclasts. Admittedly, these intricate and interlinked mechanisms of maintaining the desired value of BMD makes it very difficult to devise any genomic or genetic therapies. As a result, the therapeutic focus has been on the nutritional micro nutritional supplements augmented by surgical intervention.

However, from the genomic perspective, the fact that there exists a hierarchy of controls in maintaining BMD within the desired range of values opens up a challenge to think of a strategy to coordinate these different pathways by manipulating the long-range influences. The reason why this approach has not been pursued lies in the fact that we don't know how the long-range influences operate at a genomic level. Since this monograph deals with a particular model of genomic architecture, it is necessary to deconstruct this hierarchy of controls in maintaining the values of BMD within the desirable range. Admittedly, we don't know much about the mechanistic details which fine-tune this hierarchy of controls. However, it is possible to visualize a topological model of this hierarchy of controls as depicted in Schema 6.5.

This brings us to the third age-related pathology, viz., non-Alzheimer types of dementia. As mentioned above, the choice of non-Alzheimer types of dementia is based on the rationale that these dementias do not arise from any accumulation of protein in neurons, but they arise from the lack of synchronization of different gene expressions leading to the changes in the synaptic junctions. Admittedly, even in Alzheimer disease signal transduction across the synaptic junctions is also disrupted. However, this happens as a consequence of the accumulation of tau protein (Al Baghlani 2013), but it is not the primary cause of dementia in the patients suffering from Alzheimer disease. Since we wish to explore the possibility of long-range influences on the synchronization of different gene expressions, it makes sense to exclude Alzheimer disease from our discussion and instead focus on the other types of dementias. In addition, there exists a major type of dementia called vascular dementia (Meyer et al. 2001). This type of dementia ought to be included in our discussion. However, the problem of improper or insufficient vasculature is generic in nature, and they manifest in different types of pathologies across the organs. Since vasculature is a highly organized module in the genomes of the multicellular organisms, we will omit this lack of synchronization from our discussion. However, it must be kept in mind that the logic developed here is applicable to the module responsible for vasculature.

With this perspective in place, let us look at how a typical dementia can arise from aging. For this purpose, we will define dementia as an impaired ability to recall memories. This definition excludes the pathologies arising from impaired memory formation. This exclusion is necessary because memory formation is a poorly understood process. Of course, we believe that amygdala plays a crucial role in memory formation (Ferry 2017). However, the exact mechanism of memory formation still eludes us (and so does the mechanism of memory storage). Therefore, it seems reasonable to think that memories are available to an individual, but its recall is impaired in dementia. Though we don't know how memories are stored in our cognitive faculty, it seems reasonable to think that it must be stored in a distributed manner among various neurons. Thus, recall of memories ought to be viewed as some kind of a network operation wherein different neurons contribute different elements of memory. Admittedly, this presupposes that the final outcome, in the form of memory recall, is an algebraic combination of several separate details of memory. While there is no evidence of such a model, the proposed model is at least consistent with our current understanding of sensory signal processing which rests on distributed networks (Zhaoping 2014, see Chapters 5 and 6).

Even if we were to reject this scenario, it has to be admitted that any recall must entail interactions between various neurons and it doesn't entail activation of any single neuron. To invoke popular parlance, there is no "grandmother" neuron which is activated while remembering her! Thus, once we accept that every memory recall involves participation of several neurons, some kind of network model is inevitable. The model proposed here is essentially that, a simplest network model without any unnecessary semantic propositions. The key point in the context of the present discussion is that any network model of memory recall requires synchronization of activation of several synaptic junctions. Therefore, it is intuitively clear that if there is any impairment of memory recall, it must be due to the impairment of synchronization of different synaptic junctions and their activations. It is possible to argue that this impairment of synchronization between different synaptic junctions could also arise from impairment of the individual synaptic junctions and not necessarily from the impairment of synchronization. However, our understanding of cognitive plasticity (Tracy et al. 2015) suggests that even if individual synaptic junctions were impaired, our cognitive faculty would develop alternative network configurations to achieve memory recall. Admittedly, the fidelity of such a memory recall is questionable, but the subjectivity of memory recall is universal and not usually considered pathological.

Thus, it is reasonable to think that dementia is an outcome of the lack of synchronization between the activation of different synaptic junctions. It is important to note that this mechanism is generic and could manifest in any type of dementia. Thus, the types of dementia that we have excluded from this discussion, viz., Alzheimer disease or vascular dementia, would become operational because of this mechanism. Thus, there could be different pathological causes of dementia, and the operative mechanism must be either the one outlined here or its variant. This brings us to the key point. Can the proposed model of synchronization being a genomic functionality alter the therapeutic approaches? More importantly, can the proposed topological template help us to devise any workable therapeutic approaches? Broadly speaking, the answer to these questions is yes. After all, our experience in developmental biology suggests synaptic junctions are continuously remodeled and more importantly, they seem to be under genomic controls. However, this doesn't lead us to any specific strategy. Any such specific strategy would have to wait till we obtain a clear picture of how memories are stored. At the same time, it must be noted that once we accept dementia essentially results from the lack of synchronization of different synaptic junctions and that this synchronization is governed by genomic mechanisms, it is axiomatic that the cure for dementia must be thought of by devising a way to alter the intergenic distances between the concerned genes. As we know from conventional neurophysiology, the best way to influence the large number of synaptic junctions and to synchronize their activations is to employ the classical "amine fountains" (Samardzic 2018). In view of our experience with neurotransmitters like serotonin, it seems attractive to devise a way to alter the gene expression of the gene responsible for synthesizing serotonin by altering the intergenic distances, both upstream and downstream, of that gene (Chattopadhyay 2007).

This brings us to the end of this chapter. Therefore, in the next section, we will summarize the topics discussed in the preceding sections.

6.17 Conclusion

For the sake of simplicity, we will summarize the above discussion in a point-wise manner.

1. Research on aging has several dimensions to it. These include public health, social, economic, ethical and epidemiological dimensions. However, purely from the scientific perspective, it is the genetic and clinical perspectives that have been the focus of research.
2. The conventional perspective of research on aging has centered around genetics. It is only in the last few decades that we have turned to genomics to understand the process of aging.
3. However, this approach has been limited to genome wide association studies. This is because of the absence of any formal description of genomic architecture. This monograph outlines a topological model of genomic architecture.
4. When viewed from a genomic perspective, it is necessary to understand the origins and semantics of aging. The proposed model offers new insights into these topics.
5. The proposed model suggests that aging is not an unwelcome consequence of genomic complexity. Rather, it is an integral element of natural selection.
6. According to the proposed model, aging is programmed into genomic architecture. Therefore, the proposed model assigns a separate module of aging in the genomic architecture. This module has been named as Synchrone.
7. According to the proposed model, from the evolutionary perspective, Synchrone must have arisen immediately after the advent of multicellularity, but prior to different body plans.
8. Synchrone, as a module, synchronizes different competitive and complementary biochemical processes. Therefore, Synchrone can influence all the remaining modules.
9. The modular design of aging necessitates that there must be some kind of long-range influences that supervene the process of cellular senescence.
10. According to the proposed model, long-range influences necessary for modularity of aging arise from the direct participation of spacetime in the genomic architecture. It is proposed that a genome exists in multiple dimensionalities simultaneously. The four-dimensional configurations represent the conventional molecular biological framework of genomes which includes the DNA sequence, chromatin molecules arranged in the well-established stereochemical configurations.
11. However, according to this model, a genome also possesses higher dimensional configurations in different higher dimensionalities of spacetime. These configurations would consist of highest occupied molecular orbitals of these biomolecules. The corresponding quantum chemical perspective has been discussed in the preceding monograph.
12. According to the proposed model, different genomic modules can be assigned different dimensionalities. Therefore, the long-range influences arising from different modules devolve into the four-dimensional spacetime via the reduction of dimensionalities.
13. As a result of this nested hierarchy of genomic modules, genomic functionalities remain unobservable from the four-dimensional spacetime from which we gather evidence.

14. According to the proposed model, the changes in the length of DNA sequences lead to the changes in the higher dimensional configurations of genomes. Thus, there exists a definitive relationship between the length of DNA sequences and their functional templates.
15. Therefore, in principle, it is possible to alter the functional template of genomes by altering the length of their DNA sequences. However, in view of the phenomenon of open reading frames, it is not possible to locate each gene on a given genome. Therefore, it is risky to alter the length of genomes.
16. Therefore, it is proposed that it is safer to insert a few standard sequences of nucleotides in the intergenic regions. This would increase the length of genomes without disrupting its constituent genes. At the same time, the increase in the length of genomes would result in altering the higher dimensional configurations of genomes, thereby changing the functional template of genomes.
17. Using this strategy, the pathologies arising from aging like cataract formation, onset of osteoporosis, and non-Alzheimer types of dementia have been discussed.

References

Al Baghlani R (2013) Targeting the role of tau protein in Alzheimer disease. Florida Institute of Technology Press, Melbourne, FL

Bonner JT (1988) The evolution of complexity by means of natural selection. Princeton University Press, Princeton, NJ

Borrello ME (2010) Evolutionary restraints: the contentious history of group selection. University of Chicago Press, Chicago, IL

Bromham L (2008) An introduction to molecular evolution and phylogenetics. Oxford University Press, Oxford

Buonocore D, Grosini M, Giardina S et al (2016) Bioavailability study of an innovative orobuccal formulation of glutathione. Oxidative Med Cell Longev 2016:3286365

Chattopadhyay A (2007) Serotonin receptors in neurobiology. CRC Press, Boca Raton, FL

Chhaya P (2020) The universe within: and a view from that within. Universal Press, Irvine, CA

Chhaya P (2022) The origin and nature of spacetime

Cretoiu D, Xiao J, Das S (2020) MicroRNA signaling. Frontiers Media, Lausanne

Demaria M (2018) Cellular senescence: methods and protocols. Springer, Dordrecht

Donaldson KM (2000) Relationships between chromosome structure and long distance regulation of gene expressions. University of California Press, San Diego, CA

Elsasser WM (2016) The physical foundation of biology: an analytical study. Elsevier, Amsterdam

Ferry B (2017) The amygdala: where emotions shape perception, learning and memories. Intechopen, London

Flew A (2017) Darwinian evolution. Routledge, London

Fritz A (2014) Elucidating chromosome territories organization during the cell cycle and in breast cancer cells. Proquest Publishing, Ann Arbor, MI

Goldbetter A (1989) Cell to cell signaling: from experiments to theoretical models. Academic Press, London

Gould SJ (2007) Punctuated evolution. Harvard University Press, Cambridge, MA

Grene M (1986) Dimensions of Darwinism: themes and counter themes in the twentieth century evolutionary theory. Cambridge University Press, Cambridge
Guarente LP, Partridge L, Wallace DC (2008) Molecular biology of aging. Cold Spring Harbour Laboratory Press, New York, NY
Gupta VP (2016) Principles and applications of quantum chemistry. Academic Press, London
Haass C (2005) Molecular biology of Alzheimer disease: genes and mechanisms involved in amyloid generation. Harwood Academic Publishers, Amsterdam
Isaacson W (2021) The code breaker. Simon and Schuster, London
Krauss G (2005) Biochemistry of signal transduction and regulation. Wiley VCH, Weinheim
Leder BZ, Wein MN (2020) Osteoporosis: pathophysiology and clinical management. Springer Nature, Cham
Lozano NA (2017) The influence of epistasis, epistatic mutability and pleiotropy on evolvability. California State University Press, Long Beach, CA
Luo Y (2019) CRISPR gene editing: methods and protocols. Springer, Dordrecht
Marsh I, Goodle J (1994) Germline development. John Wiley & Sons, Hoboken, NJ
Mattson MP (2001) Telomerase, aging and disease. Elsevier, Amsterdam
Meyer J, Rauch G, Lechner H, Loeb C (2001) Vascular dementia. John Wiley & Sons, Hoboken, NY
Miller JH, Reznikoff WS (1980) The operon. Cold Spring Harbour Laboratory Press, New York, NY
Mobley AS (2019) Neural stem cells and adult neurogenesis. Academic Press, London
Morgan DO (2007) The cell cycle: principles of control. Oxford University Press, Oxford
Niklas KJ, Newman SA (2016) Multicellularity: origin and evolution. MIT Press, Cambridge, MA
Oparin AI (1968) Genesis and development of life. Academic Press, New York, NY
Pevsner J (2015) Bioinformatics and functional genomics. John Wiley & Sons, Hoboken, NJ
Reavey C (2013) Analysis of transcription activating distance as a polygenic trait in saccharomyces cerevisiae. Harvard University Press, Cambridge, MA, p 8
Samardzic J (2018) GABA and glutamate: new developments in neurotransmission research. Intechopen, London
Seager AL (2005) Molecular and cytogenic analysis of premature aging disease: Hutchinson-Gilford Progeria Syndrome (HGPS). University of Wales, Swansea
Tracy JI, Hampstead BM, Sathian K (2015) Cognitive plasticity in neurological disorders. Oxford University Press, Oxford
Tutar Y, Tutar L (2018) Current understanding of apoptosis: programmed cell death. Intechopen, London
Verkhratsky A, Toescu EC (1998) Integrative aspects of calcium signaling. Plenum Press, New York, NY
Vijg J (2007) Aging of the genome: the dual role of DNA in life and death. Oxford University Press, Oxford
Wechsler H (1992) Neural networks. Academic Press, San Diego, CA
Weinzierl ROJ (1999) Mechanism of gene expression: structure, function and evolution of the basal transcriptional machinery. Imperial College Press, London
Whelan J, Nugent J (2009) Human cataract formation. CIBA Foundation Systems. Symposium. Pitman Publishing, London
Yang H (2019) Osteogenesis and bone regeneration. Intechopen, London
Yorio T, Clark AF, Wax MB (2008) Ocular therapeutics: eye on new discoveries. Elsevier, Amsterdam
Zhaoping L (2014) Understanding vision: theory, models and data. Oxford University Press, Oxford

Nature of Genomic Evolution: Its Imprint in Cancer

7

Abstract

In preceding chapters, a topological model of the genomic architecture is described. This model is characterized by top to bottom control elements which are formalized as involutions. While such a model explains the modular design of the genome, it still needs some empirical evidence to be taken seriously. In this chapter, we would try to offer a legacy argument to suggest that the model proposed here is indeed a serious candidate for genomic architecture. It is generally conceded that the manifestation of cancer is a legacy of the evolution of multicellular organisms. However, there are no reports of structural templates for such a belief. In this chapter, we will look at one such template for modularization and the resulting asynchronization of the functionalities. The topological representation of cancer as modular asynchronization is intuitive and can be used for future therapeutic advances.

7.1 Introduction

In genomics, there are two aspects which have remained beyond formalization. These aspects are genomic architecture and an evolutionary account of such a genomic architecture. In the preceding chapters, a topological model of genomic architecture using the formalism of the involuted manifold was discussed. This model is based on the principle of modularity and is congruent with the Darwinian paradigm. However, the proposed mode lacks any empirical evidence for its validity. In a preceding chapter, its confirmation was sought to be obtained by deconstruction of the Homeobox genes using this model.

In this chapter, we will seek an additional confirmation of this model by deconstructing some of the genomic legacy. For this purpose, ontology of cancer has been chosen. This choice is based on several reasons. Firstly, there are several

© The Author(s), under exclusive license to Springer Nature Singapore Pte Ltd. 2023
P. Chhaya, *The Topological Model of Genome and Evolution*,
https://doi.org/10.1007/978-981-99-4318-0_7

citations in the literature (Dellaire et al. 2014, see Chapter 9) which suggest that emergence of cancer is linked to the evolution of functional complexity of the genome. It seems reasonable to think that it is the complexity of the genome which was acquired during the time when multicellular organisms evolved for the first time, that is the source of risk of cancer. Although these reports provide cogent arguments in support of this proposition, there are no reports of any mechanistic details for such an evolutionary scenario. Therefore, in this chapter, we would try to provide a structural template which can explain the emergence of cancer.

Secondly, the molecular mechanisms of onset and metastasis of cancer (Pecorino 2012) suggest that cancer, as a group of diseases, is a pathology arising from impairment of control elements of the genome which influence gene expressions. Therefore, it seems reasonable to think that these control elements represent the features of genomic architecture rather than being independent operators. The model proposed here claims to formalize the genomic architecture including these control elements. Therefore, it makes sense to employ this model to formally represent these control elements and investigate whether such a topological perspective corresponds to the empirical evidence of cancer pathologies.

Thirdly, this model postulates that these control elements responsible for cancer pathologies must be influenced by the overall architecture of the genome. Therefore, it is possible to influence these elements by altering overall topology of the genome. If this postulate were to be true, it opens up new avenues of cancer therapy. Therefore, we would try to outline topological features of the genome which are likely to participate in the manifestation of cancer pathologies. In order to address these three perspectives, this chapter is further divided into 16 sections. Section 7.2: Evolution of Genomic Complexity, Sect. 7.3: Relationship Between Complexity and Cancer, Sect. 7.4: Genomic Complexity and Cancer, Sect. 7.5: Control Elements of the Genome, Sect. 7.6: Cancer as a Pathology of Aberrant Control Elements, Sect. 7.7: Conventional Perspective of Control Elements, Sect. 7.8: Evolution and Cancer, Sect. 7.9: Shortcomings of the Conventional Perspective, Sect. 7.10: The Proposed Model, Sect. 7.11: Complexity According to the Proposed Model, Sect. 7.12: Control Elements According to the Proposed Model, Sect. 7.13: Topological Model of Regulatory Genome, Sect. 7.14: Cancer According to the Proposed Model, Sect. 7.15: Evolutionary Perspective of Cancer According to the Proposed Model, Sect. 7.16: Therapeutic Possibilities of the Proposed Model, Sect. 7.17: Conclusion.

7.2 Evolution of Genomic Complexity

As discussed in the preceding chapters, the Darwinian paradigm is characterized by two mutually incompatible semantic propositions. On the one hand, the Darwinian paradigm provides a completely naturalistic explanation of Life. As Dobzanski (1982) famously said "Nothing in biology makes sense unless viewed from the perspective of Darwin's theory," it sums up the centrality of the Darwinian paradigm. The explanatory power of this paradigm is so exceptional that it has been employed to explain totally unrelated phenomena like the fluctuations in stock market, programming paradigm, cognitive processes, etc. (Dennett 1995). In contrast to this exceptional quality this paradigm is characterized by a lack of predictive power. We can never use the theory of natural selection to predict outcomes. Admittedly, the semantics of the Darwinian theory (Bonner 2013) rests on the inherent randomness. Therefore, if the theory had any predictive power, it would be self-contradictory. However, we expect a scientific theory to possess explanatory and predictive powers. Therefore, we are faced with a dilemma while dealing with Darwin's theory. The reason why a scientific theory must possess predictive and explanatory powers lies in the domains of epistemology and even metaphysics. Normally, a scientific theory consists of a structure. This structuralism rests on several semantic propositions which endow the theory with its explanatory power. Similarly, the structural template of a theory encodes a certain mechanism which endows the theory with predictive power. However, the Darwinian paradigm is an exception to this rule. The reason why it is an exception has been discussed in the preceding chapters. However, there is one aspect that needs to be discussed in this section.

This refers to the conception of complexity, particularly biological complexity. As mentioned above, the Darwinian paradigm centers around the semantic proposition of randomness. Therefore, historically, it has been very difficult to justify how a random process can give rise to structurally and functionally complex organisms. Conventionally, this emergence of complexity has been justified using a phase space argument (Bonner 1988, see Chapter 7). It is argued that the earliest living organisms must be the simplest or least complex entities. Therefore, the process of natural selection would inevitably lead to improvements in the survivability of these primitive organisms. Therefore, over a period of time, this would ensure that the successive organisms turn out to be more and more complex. Thus, it is the inherent power of the process of natural selection that justifies the emergence of complexity during biological evolution. The problem with this argument is that it assumes that the earliest living organisms were least complex. Admittedly, they were less complex than the later organisms that emerged as the products of natural selection. However, these earliest organisms were very complex, particularly when viewed from the molecular biological perspective. It is not a coincidence that the theory of natural selection is silent on biological evolution. It can be used to explain how different forms of Life emerged during the course of natural selection. However, we can't convincingly employ the theory of natural selection to explain the origin of Life. We normally employ a hypothesis called the RNA world which postulates that Life began on Earth with RNA molecules (possibly as RNA proteins; Yarus 2010, see Chapter 2). However, that hypothesis leaves out more questions than it answers.

The key point is that we have two options. Either, we accept that complexity was already present before natural selection began its operations, or we accept that the process of natural selection has a certain mechanism which generates complexity. It might appear that these two options are mutually exclusive. However, they are not. Let us see how. As mentioned above, the Darwinian paradigm extends beyond biological processes (Bonner 1988). It can be employed to deconstruct a whole lot of otherwise enigmatic phenomena. Therefore, it is possible to argue Darwin's theory can be applied to biological evolution as well. However, unlike Life, other natural phenomena which can be rationalized using the Darwinian paradigm possess structural templates. Therefore, we will have to accept that Life, like any other natural phenomena, must be amenable to predictable consequences. However, the Darwinian semantics (Bonner 2013) prevents us from accepting such a scenario.

In that case, we will have to accept the second option, viz., natural selection has certain mechanisms that engendered certain types of complexities. In that case, it can be argued that the process of natural selection alters the nature of complexity in every natural phenomenon, and if Life appears to be unlike any other natural phenomena, it is because Life has a different type of complexity. However, in this scenario, it is necessary that the process of natural selection must possess a structural template which alters the degree of complexity in every natural phenomenon it operates upon. As mentioned above, this appears to be inconsistent with the essential randomness implicit in the Darwinian paradigm. However, as discussed in the preceding monographs, this apprehension is misplaced. It is possible to postulate a structuralism of the process of natural selection that still leads to random outcomes.

Returning to the present discussion, what we wish to deconstruct is the nature of genomic complexity, its evolution and its connection with pathologies of cancer. According to the proposed model, as discussed in the preceding chapters, genomes have evolved from a hypothetical entity named here as genomic singularity. Moreover, genomes are also units of selection on which natural selection operates. The conception of genomic singularity ensures that different elements of genomic architecture arise as outcomes of natural selection. Moreover, genomic singularity also ensures that different structural elements of genomic architecture remain modular. Thus, the conception of genomic architecture as a hierarchy of different modules arises naturally once we accept the genomic ontology in genomic singularity. At present, the conception of genomic singularity must be taken as a hermeneutic device, just as entities like LUCA (Last Universal Common Ancestor; Bard 2016, see Chapter 9) and mitochondrial Eve (Hamilton 1989) are. However, as discussed in the following chapters, in the case of genomic singularity, it may be more than a hermeneutic device. Thus, we now have two reasonably good postulates to understand the origins of genomic complexity, viz., a common ontology in the form of genomic singularity and a structural mechanism of natural selection. Using these two postulates, it is possible to deconstruct the nature and origin of genomic complexity.

Conventionally, the evolution of genomic complexity per se has remained nebulous. Admittedly, since we don't know the formal description of genomic architecture (Lynch 2007), it is difficult to conceptualize how different features of genomic architecture could have evolved. However, there are certain features of genomes that are generally conceded. Moreover, these features point toward a basic plan of genomic architecture. It is accepted conventionally that there exists a continuity of features of genomes. Admittedly, in the case of individual genes, this structural (and even functional) continuity is well articulated from phylogenetic studies (Bromham 2008, see Chapter 5). The same level of clarity doesn't exist in the case of genomes. All the same, the continuity of genomic features has always been implicit in the conventional perspective and even semantically, it seems a reasonable proposition. However, beyond this tacit acceptance of genomic legacy, the conventional perspective is silent on the exact details of genomic evolution. This is possibly because the conventional perspective doesn't find any evidence of any retention of higher levels of organization of genes. In fact, because of the widespread phenomena of translocations of genes (Zhang 2012), it is felt that there are no higher levels of organization of genes in any given genome.

Thus, the proposed model starts from where the conventional perspective stops. It accepts the legacy argument of genomic complexity and postulates that complexity of a certain type has been selected by the process of natural selection. This selectivity in allowing only certain types of complexities to evolve is possible only if natural selection has a template of its own. Once we accept this reasoning, it is axiomatic that this structuralism of natural selection must be singular in nature and that it must be applied repeatedly. This is not apparent at first sight. However, if we concede the legacy argument of genomic complexity, this legacy would not be maintained if the structuralism of natural selection were to change halfway through biological evolution or there were to be multiple structural templates of natural selection. Thus, once we accept the legacy of genomic complexity, it is implicit that the legacy can manifest only if there is continuity of forms. To reiterate, the legacy of genomic complexity implies constancy of natural selection. The nature of natural selection cannot be multifarious or variable, though its outcomes can be. This inference is consistent with the conventional perspective. It is just that the conventional perspective, unlike the proposed model, doesn't ascribe structuralism to the process of natural selection.

Once we accept that natural selection has its own singular mechanism, it is intuitively clear that the phylogenetic principle must be extended to phylogenomics. This conception of genomics makes it easy to visualize how a repetitive process of natural selection can lead to pathologies. This happens because natural selection operates at multiple levels. While we know how natural selection at the genetic level leads to pathologies like cancer, we are just beginning to appreciate that similar phenomena occur at the genomic level. The operating principle is simple. Natural selection operates on different units of selection in parallel. Thus, it operates on individual genes which over a period of time leads to aberrant genes, thereby inducing pathologies. Similarly, natural selection operates on a genome. However, in this case the individual genes are not directly affected, but their expression patterns are. This happens because the aberrations due to natural selection at a genomic level results in the aberrations of control elements of a genome. This in turn, leads to aberrant patterns of gene expressions. The scenario outlined here is not heretical in the sense that it conflicts with the conventional perspective. However, it has to be admitted that this scenario is a speculative scenario at best. For it to be taken seriously, this scenario needs to deconstruct the exact relationship between genomic complexity and cancer. More importantly, it must demonstrate that cancers also arise from genomic influences as separate from the cancers arising from the influence of the individual genes. It is important to keep in mind that the proposed model doesn't predict new types of cancers (at least not necessarily). It merely provides a new ontology and perhaps a new therapeutic approach. Therefore, in the next section, we will look at the relationship between complexity and cancer from the conventional perspective and from the perspective of the proposed model.

7.3 Relationship Between Complexity and Cancer

Before we look at the relationship between complexity and cancer, let us summarize what constitutes cancer. Admittedly, there exists a humongous amount of literature on the classification, nomenclature and linkages between different types of cancer (Swanton et al. 2017). Similarly, there are different ways in which we can classify different types of cancer (Mohammed 2018). We can employ types of pathologies, types of genes involved, types of molecular mechanisms involved to differentiate between different types of cancers. We will sidestep these details and instead focus on the generic definition of cancer. This simplistic description is justified because we are not looking at the molecular biology behind cancers, but their evolutionary context. Operating details do not define the operational logic, they merely confirm it. To cite the most famous example, Darwin wasn't privy to genetics, let alone molecular biology, when he developed the theory that bears his name. (Admittedly, there is some ambiguity about Darwin's access to Mendel's work. However, there is no explicit acknowledgment of it.) Yet, Darwin's theory was consistent with these later paradigm shifts. In a very abstract way, we can define cancer as a pathology arising from three causes. Firstly, cancer can arise from uncontrolled growth. Secondly, it can arise from the lack of cellular differentiation after a cell division. Lastly, cancer can arise from the failure of apoptosis. Admittedly, these three causes arise from a variety of molecular contexts. In fact, volumes have been written about each of these three causes. However, as mentioned above, we will sidestep molecular and genetic details in this section.

What is germane to the present discussion is that if this is how we conceptualize cancer, how can it be related to complexity ? This is where the conventional perspective comes into play. The conventional perspective (Vlachakis 2019) accepts the proposition that cancer is related to the complexity of the interactions between different genes and the resulting lack of synchronization between different gene expressions. The key point is that the conventional perspective admits that cancers are related to complexity, but to the complexity of interactions between different genes. As discussed in the previous section, because there is a lack of any theoretical models of genomic architecture, the conventional perspective has confined itself to the interactions between different genes. Admittedly, in the last couple of decades, with the advent of functional genomics (Pevsner 2015), there is an increasing awareness that higher organizational factors of genomes could be playing a role in the pathologies arising from cancer. However, since our current genomics places emphasis on individual genes at the core of its semantics, the conventional perspective is forced to think about cancer in the context of the complexity of the interactions between different genes. To be honest, genomics, particularly the one focused on cancer, accepts that there exists a hierarchy of genes which tries to minimize if not prevent the manifestation of cancer. The case of gene p53 is the classic example of the implicit hierarchy of genes (Zambetti 2005). However, at the same time, genomics, in its present avatar, has yet to develop a theoretical framework of the hierarchy of genes. This is best exemplified by the fact that we employ genome wide association studies to locate genes responsible for a given pathology (Appasani 2016).

Leaving aside the present status of genomics, ab initio, we can think of cancer as a pathology arising from the lack of synchronization between different control mechanisms. For instance, three causes mentioned above are actually instances of lack of synchronization between different control mechanisms. Thus, the case of uncontrolled growth of a tissue arises from the lack of messages (mostly intercellular) to stop cdk signaling and thereby stopping the cycle of cell divisions. Similarly, the lack of functionalization at a cellular level can arise from intercellular signals. In the same way, failure to initiate apoptosis arises from the failure of the signaling mechanism. Therefore, even at the cost of being accused of oversimplification, it seems reasonable to think that it is the functional complexity, either at the intracellular level or at the intercellular levels that prevents cancer from manifesting. This functional complexity, as the conventional perspective rightfully maintains, arises in the form of interactions between different genes and their expressions. Therefore, it is time to take one step further and try to define the complexity at the level of genomes. Therefore, in the next section, we will try to articulate this inchoate understanding of the complexity of the interactions between different genes into a more structural framework.

7.4 Genomic Complexity and Cancer

To be honest, the conventional perspective tacitly accepts the possibility of higher levels of control elements in genomes. It is just that it can't formalize any such higher level of organization of genomes. As discussed in the preceding chapters, the basic aversion to any such description of higher level of organization of genomes arises from the apprehension that it might turn out to be antithetical to the randomness implicit in the Darwinian paradigm. In addition, with the advent of the molecular biological perspective of genomics, it would have been reasonable to expect such a higher level of organization of genomes. In fact, different degrees of coiling and uncoiling of DNA sequences could have been ideal for such a model. However, as mentioned above, autonomous functionings of genes even after their translocations, makes it difficult to conceptualize any higher level of organization of genomes.

However, as discussed in the preceding chapters, it is possible to formalize a topological model of higher level of organization of genomes. Admittedly, it rests on some unconventional (and rather contrarian) propositions. Firstly, it postulates that spacetime itself exists in multiple dimensionalities simultaneously. Secondly, it postulates that biomolecules also exist in multiple dimensionalities simultaneously. While the lower dimensionalities represent the conventional stereochemical configurations (including the different degrees of coiling and uncoiling of DNA sequences). However, according to this model, the higher dimensionalities of spacetime are occupied by the highest occupied molecular orbitals of these biomolecules. Since these higher dimensionalities of spacetime are not differentiated into the time-like and the space-like dimensions, the kinetic and thermodynamic effects of these higher dimensionalities would not manifest in the four-dimensional spacetime. Therefore, we can never perceive these higher dimensional configurations of biomolecules, and therefore, they remain beyond measurements and subsequent formalizations.

Traditionally, science is skeptical about any framework that remains beyond experimental verification. Therefore, the proposed model ought to be rejected on the ground that it can never be proven wrong. In the post-Popperian era (Popper 1963), falsifiability is the yardstick of a good scientific theory and the proposed model cannot be exempted from it. Therefore, it is imperative that the proposed model must at least provide explanations for diverse and possibly hitherto intractable features of genomics. It is in this context that we try to understand the nature of genomic complexity (as distinct from the genetic complexity implicit in the conventional perspective) using the proposed model. One of the key points about this model is that it postulates that different higher levels of complexity of genomes are interlinked to one another. In other words, the higher dimensional configurations of genomes form a hierarchy among themselves. More importantly, different higher dimensional configurations of genomes are interconvertible by a singular mechanism consisting of the changes in the dimensionalities. In other words, just like the lower dimensional configurations (the stereochemical configurations and different degrees of coiling of DNA sequences), the higher dimensional configurations also undergo transitions from one configuration to another. In the case of the changes in the lower dimensional configurations of DNA sequences, there are always accompanying energy transfers. However, as mentioned above, in the higher dimensional configurations of DNA sequences, there are no energy transfers (because there is no distinction between the time-like and the space-like dimensions resulting in thermodynamic states). According to the proposed model, energy transfers are replaced by information transfers in the case of higher dimensional configurations. Thus, what becomes measurable in the lower dimensional configurations is the energy transfers. Similarly, what becomes measurable in the higher dimensional configurations is the information transfers. More importantly, these putative information transfers are governed by quantum fluctuations. This ensures that these transitions from one higher dimensionality to another higher dimensionality are random and governed if at all, by symmetry principles.

Admittedly, this scenario is internally consistent and externally congruent with quantum chemistry. However, it still fails because it must be capable of explaining the genomic complexity in the language of functional genomics. Therefore, let us see whether this scenario can give us a template of long-range influences of genomic architecture. It is important to keep in mind that even if there is any higher level of organization of genomes, it must translate into a lower level of individual genes. Therefore, we will assume that this devolvement of higher dimensional genomic complexity would manifest in the form of long-range influences which control the sequence and duration of the expressions of individual genes. Moreover, the conventional perspective has remained ambiguous about the exact nature of long-range influences. For instance, the conventional perspective concedes that the phenomena like chromosome territories (Fritz 2014) and cis and trans effects (Donaldson 2000) are essentially long-range influences. Moreover, the conventional perspective ascribes these phenomena to manifest on some general stereochemical transitions. However, the conventional perspective has no causal theory to predict where and when these long-range influences would manifest. Therefore, it seems reasonable to think that if the proposed model can offer a framework wherein different spatial and temporal long-range influences are defined, it would make the proposed model a serious scientific hypothesis. This brings us back to the question of the nature of genomic complexity and its relationship with cancers. Before we articulate this relationship, let us define the temporal and spatial long-range influences of genomes using the proposed model. Therefore, in the next section, we define control elements of genomes according to this model.

7.5 Control Elements of Genome

As discussed above, the conventional perspective admits that there are several long-range influences that activate gene expressions. However, it fails to formalize them as control elements. Similarly, the conventional perspective apart from the feedback mechanisms for controlling gene expressions, accepts higher levels of organization of genomes. This includes operons (Miller and Reznikoff 1980) and modules (Peter and Davidson 2015). Therefore, ideally these features should have been conceptualized into a generic framework of genomic architecture. However, this hasn't happened. As mentioned above, there are two reasons for our failure to place these structural units in a framework. Firstly, genes often translocate themselves during cell divisions. More importantly, the functionalities of these translocated genes often remain unchanged. This eliminates any possibility of a higher structural template that can act as a template for genomic architecture. Similarly, distribution of genes on chromosomes and the number of chromosomes do not substantially alter the genomic functionalities. Therefore, the conventional perspective has justifiably refrained from formulating higher level genomic architecture. Secondly, there is an unspoken apprehension that any such framework of genomic architecture might be in conflict with the randomness implicit in the Darwinian paradigm.

Both these factors can be accounted for if a model of genomic architecture used a different framework of spacetime itself. After all, the primary objective of conceptualizing genomic architecture is to formalize spatial and temporal controls of gene expressions of individual genes. Thus, if we had a model of genomes in which spacetime itself was a constituent, then it can interact variably with different parts of genomes to exercise some spatial and temporal influences. The trouble with this reasoning is that spacetime is uniform and more importantly, passive during biochemical transformations which define genomics. Not only does spacetime have an inherent symmetry (gauge symmetry) that ensures that the properties of spacetime are identical at every point, but spacetime also has a uniform blending of time-like and space-like features at every point. Therefore, from the conventional perspective, it is impossible to think of any mechanism that would enable spacetime to react differently with different parts of genomes. Even from the perspective of theoretical physics, this is a valid inference. This is because in physics, spacetime remains a passive witness to energy and mass transfers in our day to day experiments (which has been formalized in Newtonian mechanics; Capuzzo-Dolcetta 2019, see Chapter 2). According to physics, spacetime becomes an active participant only under extreme conditions like an object moving at a speed comparable to that of light or under extreme gravitational pull (these instances are formalized in the general theory of relativity; Schutz 2009, see Chapters 7 and 8). However, living organisms do not operate in these conditions. Therefore, it is justified to employ the passive framework of spacetime (as formalized in the Newtonian paradigm) while formalizing biological phenomena.

However, as discussed in the preceding monographs, spacetime can be formalized as a topological manifold having multiple dimensionalities simultaneously. Therefore, using this conception of spacetime, it is possible to formalize spacetime differently. It is possible to unify the Newtonian mechanics and the relativistic mechanics into a single framework wherein the passive and active facets of spacetime coexist in different dimensionalities. More importantly, according to the proposed model, matter and spacetime are isomorphs. Therefore, what appears to be a duality of atoms and molecules embedded in spacetime is actually a unified entity at the highest dimensionality of spacetime. Therefore, it is axiomatic that molecules, including DNA sequence of a genome, would be spread over multiple dimensionalities simultaneously. Another feature of the proposed model is that spacetime manifests different textures at different dimensionalities. Thus, at lower dimensionalities, there are two types of dimensions, viz., the space-like and the time-like dimensions. However, at the higher dimensionalities, such a distinction ceases to exist. Thus, whenever the dimensionality of spacetime is reduced, due to topological compulsions, the devolvement leads to nonlocal influences. Admittedly, this view is not only contrarian, but also a mathematically complex view. The details of this model are available in the preceding monographs. However, in the context of the present discussion, we will take this model as a priori and try to outline a brief schema of genomic architecture spread across different dimensionalities. The key point is that this model gives rise to nonlocal influences which can be employed in defining higher levels of organization of genomes. More importantly, since the proposed model employs spacetime as a constituent of genomic architecture, the above mentioned problems of gene translocations and the distribution of genes among different chromosomes, are taken care of. According to this model, the long-range influences do not arise from a particular distribution of DNA sequences in a given genome, but arise from the higher dimensional configurations of genomes.

Ab initio, we can think of long-range influences to be of two types, viz., spatial and temporal. Of course, within each category, there would always be different degrees of nonlocality. Thus, some spatial long-range influences could be operating at intermediate distances, say within a genomic module. Similarly, there would always be some long-range spatial influences that operate at long distances, say, across the genome. Similar situation would be present in the case of temporal long-range influences. In the conventional perspective, even if we were to conceptualize such a scenario, we would be required to provide different mechanisms for different ranges of distances. This is because we are dealing with thermodynamic/kinetic processes which are scale dependent. However, in a topological model such as this one, distances do not matter. Let us understand why. According to the proposed model, both these types of long-range influences arise when a higher dimensional configuration of genomes devolves into the four-dimensional spacetime. Therefore, different degrees of distances at which these influences are felt are dependent on the magnitude of the difference between dimensionalities. Thus, hypothetically, let us assume that there are two higher dimensional configurations of genomes which exist in the fifth and sixth dimensionalities. Both these configurations devolve into the four-dimensional spacetime to give rise to long-range influences. In such a scenario, the influence arising from the devolvement of the six-dimensional configuration of a given genome would be felt at a greater distance than the distance of the corresponding long-range influence arising from the devolvement of the five-dimensional configuration of that genome. Thus, the inherent topological compulsions ensure different degrees of nonlocality of the long-range influences. We don't have to postulate any thermodynamic/kinetic gradients for this. Thus, a topological model of genomes in which genomes occupy different dimensionalities simultaneously can formalize long-range influences simply by defining a mechanism for the changes in the dimensionalities of these genomes.

The second feature of this topological model is that we don't have to define two separate categories of long-range influences. Let us see how. According to this model, atoms and molecules don't occupy spacetime, rather, they are spacetime. It is only at the four-dimensional perspective that they manifest separately as atoms and molecules and the four-dimensional spacetime. Thus, at higher dimensionalities, there is no distinction between matter and spacetime. They are one and the same. Similarly, at some of the higher dimensionalities, say, intermediate range of dimensionalities, there exists a distinction between atoms and molecules, but not between the time-like and the space-like dimensions in spacetime. Thus, at these intermediate range of dimensionalities, what is manifest is matter and spacetime with its undifferentiated dimensions. In such a scenario, wherein genomic configurations present in these intermediate range of dimensionalities devolve into the four-dimensional spacetime, the resulting long-range influences would split up into two types of long-range influences, viz., the spatial and temporal. This is because when the spacetime from these intermediate range of dimensionalities devolves into the four-dimensional spacetime, it would separate into two types of dimensions. Thus, the inherent symmetry considerations ensure that both these types of dimensions (and therefore, both the types of long-range influences) separate out in a fixed manner. Thus, in the proposed model, there is no need to define two separate long-range influences. They arise from the inherent fine structure of spacetime.

When we combine both these postulates, viz., the numerical values of dimensionalities decide the extent of nonlocality of long-range influences and the inherent nature of the fine structure of spacetime giving rise to two types of long-range influences, we get a detailed description of diverse long-range influences. It is intuitively clear from the above discussion that the conception of genomic architecture based on this particular topology is capable of representing various methods of synchronization between different gene expressions, thereby perpetuating healthy life. As a logical corollary, it is intuitively clear that cancers, as a class of pathologies, must arise when there is an impairment in the transition from any of these higher dimensionalities onto the four-dimensional DNA sequence. However, it is necessary to deconstruct the exact nature of origins of cancer according to this model. Therefore, in the next section, we will discuss the conventional perspective of cancer as an outcome of aberrant control elements.

7.6 Cancer as a Pathology of Aberrant Control Elements

As discussed in the previous section, we can formalize different types of long-range influences using a single topological framework. While this scenario is consistent with the conventional perspective (Swanton et al. 2017) in the sense that the proposed model, like the conventional perspective, defines cancer as arising from lack of synchronization between different gene expressions. Upon a little reflection, there is an important difference between these two approaches. Although both these approaches lay emphasis on the lack of synchronization between different gene expressions, the underlying causes in both these approaches are different. While the conventional perspective puts emphasis on the individual genes and their aberrant behaviors, the proposed model puts emphasis on the overall configurations of genomes. This distinction between genetic and genomic mechanisms has major implications on our understanding of cancer and on the future therapeutic approaches. Therefore, it is necessary to understand the distinction between genetic and genomic mechanisms. Admittedly, there is an overlap between these two mechanisms; however, there is a significant degree of disagreement between the two mechanisms. However, in this section, we will try to understand the distinction between genetic and genomic mechanisms.

Prima facie, this distinction arises from the manner in which long-range influences are defined in these two mechanisms. As discussed in the preceding chapters and in the preceding sections, genetic mechanisms do not encompass long-range influences. Therefore, these influences are deemed to have arisen from the thermodynamic and kinetic parameters like diffusion constants of different messenger molecules, different permeabilities across lipid bilayers etc. (Weinzierl 1999). At first sight, these parameters seem obvious and undisputed. Moreover, as mentioned above, they are congruent with the randomness implicit in the conventional perspective. The key point is not about the validity of these parameters, but about their origins. These parameters can explain individual instances of how and when a given gene expresses itself. However, when we are dealing with thousands of genes and their expressions, there ought to be a systemic explanation of the preferred sequence of gene expressions. Let us use an analogy to understand this. Just think of a binary switch [which is what the idealized notion of an operon (Miller and Reznikoff 1980) is]. Its logic is self-evident. However, when we think of a microprocessor, it contains a few millions of such binary switches. However, no one seriously thinks that a microprocessor operates on the basis of random collective outputs of these millions of binary switches. There is an inbuilt logic about how these binary switches are connected to one another. Therefore, there is no reason why thousands of gene expressions should not be governed by some higher level mechanisms.

Conventionally, such arguments are rejected because they seem to imply some design principles (Fodor and Piattelli-Palmarini 2011), if not a designer. However, the proposed model differs from these design principles. The higher level mechanisms proposed in this model, didn't evolve through natural selection subsequent to biological evolution. Rather, these higher level mechanisms owe their origins to the nature of spacetime itself. Therefore, these higher level mechanisms do not represent any design principles (nor do they represent any designer), at least not in the deistic sense. No one discards the standard model of fundamental particles by imputing any intelligent design principles (Cottingham and Greenwood 2013). The key point is this: there would always be some pattern or the other in any natural phenomenon. These patterns do not represent preconceived (or teleological) plans. These patterns arise from the nature of underlying reality. In the case of the standard model, it is the nature of underlying cosmic singularity that creates the pattern implicit in the standard model. Similarly, if we postulate that it is the inherent fine structure of spacetime that allows only some of the higher dimensional configurations of genomes, it doesn't introduce any design principles.

Instead, the postulate that spacetime is actively woven into the structure of genomes must be viewed as a refinement of loosely defined thermodynamic and kinetic parameters. What should concern us is whether this postulate improves our understanding of long-range influences or not. If the predictions of this model turn out to be true, there is no reason to reject it, at least on the philosophical grounds. Returning to the specifics, let us look at the areas where genetic and genomic mechanisms are synonymous and the areas where they are not. It is intuitively clear that the effects of genomic mechanisms of synchronization would be identical with the genetic mechanisms as far as biochemical processes are concerned. This is because genomic mechanisms would eventually devolve into genetic mechanisms. Therefore, it seems reasonable to assert that biochemistry of genomic and genetic mechanisms is identical. This brings us to the areas where genomic mechanisms are not synonymous with genetic mechanisms. There is one critical feature of this model that distinguishes between the conventional thermodynamic/kinetic perspective of gene expressions and the proposed model. According to the proposed model, different higher dimensionalities upon devolvement into the four-dimensional spacetime give rise to different magnitudes of nonlocal long-range influences. Therefore, by implication, there are two verifiable consequences of this scenario. Firstly, different messenger molecules responsible for initiating gene expressions would possess different magnitudes of the parameters mentioned above. More importantly, these different magnitudes of these parameters would form a distinct pattern of distribution. For the sake of argument, let us suppose that a given genome exists in five-, six-, and seven-dimensional configurations. Apparently, when these configurations devolve into the four-dimensional spacetime, the parameters like diffusion constants of different messenger molecules would accordingly depend on the higher dimensional configurations from which the synthesis of these messenger molecules were triggered. Thus, when we tabulate the parameters like diffusion constants or permeabilities, they should be grouped into well-defined classes, say in three different classes in the above example. Admittedly, in reality, there would be a larger number of higher dimensional configurations and therefore, the distribution of these parameters would not be as simple as that of the example cited above. However, the principle remains the same. Therefore, the key difference between genomic and genetic mechanisms is that these parameters which represent thermodynamic/kinetic perspective would not be randomly distributed, but they would behave as well-defined groups.

This brings us to the last and perhaps the most germane point. Is it possible to link these mechanisms with the origin of cancer ? While the relationship between genetic mechanisms and cancer is extensively reported in literature (Pecorino 2012), the corresponding relationship between genomic mechanisms and cancer is relatively less explored. There are three questions that need to be answered before we can develop a common framework of the relationship between complexity and cancer, viz., (1) Do genetic mechanisms represent a far more general relationship between complexity and cancer? (2) Can genomic mechanisms be derived from this general relationship between complexity and cancer? (3) What is the relationship between complexity and synchronization ? Admittedly, these questions refer to evolutionary genomics and therefore cannot be answered from the conventional perspective, at least till genomic architecture is defined. On the other hand, the proposed model begins with the proposition that genomes are the units of selection of natural selection and therefore, genomic architecture too is a product of natural selection. Before we look at genomic architecture as implicit in the proposed model and how it can formalize the relationship between complexity and cancer, let us try to deconstruct these questions from the conventional perspective.

Let us begin with the first question: whether genetic mechanisms of cancer represent a far more general relationship between complexity and cancer ? Prima facie, this question can be answered if we can demonstrate that different genetic mechanisms arise from a single template. Once we demonstrate that, we can use that template to search for the corresponding genomic mechanisms. Upon a little reflection, it is intuitively clear that according to the conventional perspective, the only commonality between different genetic mechanisms of cancer is the impaired synchronization. For instance, mutations are inherent outcomes of cell divisions. It is the inability of mutated cells to trigger repair the mismatch between two strands of DNA that causes cancer (Madhusudan and Wilson 2013). Similarly, in the case of oxidative stress and the resulting biochemical disruption, it is the lack of feedback mechanisms that gives rise to pathologies (Spitz et al. 2011). This is also true in the case of failure to trigger apoptosis after tumor formation (Zambetti 2005). Finally, in the case of metastasis (Welch and Fisher 2016), it is the lack of signaling that allows malignant cells to migrate.

Thus, the conventional perspective explicitly and implicitly accepts that different genetic mechanisms of cancer arise due to lack of synchronization. As mentioned above, the only problem with the conventional perspective is that it fails to create a framework wherein different elements of synchronization can be integrated. For instance, in the case of mismatch between two strands of DNA, the conventional perspective accepts the role of gene p53 in repair. The conventional perspective also describes the conditions under which gene p53 is expressed and used in repair. However, the conventional perspective fails to define conditions under which gene p53 fails to be activated or its expression is delayed. In fact, this and other instances of feedback mechanisms have not been unified into a general framework. Similar situation exists in the case of oxidative stress. There is a feedback mechanism to quench reactive oxygen species that arise from mitochondrial stress. This feedback mechanism also ought to be a part of a framework for general feedback mechanisms. However, this aspect has never been explored in the conventional perspective. There is one simple reason why this hasn't been done. The conventional perspective assumes that spacetime per se plays no active role in homeostasis. Once we accept that spacetime plays an active role in homeostasis, it is possible to develop a general framework for feedback mechanisms and as a logical corollary, that of synchronization.

Let us now look at the second question: can genomic mechanisms of cancer be inferred from such a generic framework of genetic mechanisms? Apparently, according to the conventional perspective, there is no clarity on this topic. On the one hand, the conventional perspective admits that different genes coordinate with one another to give rise to a fixed pattern of gene expressions. On the other hand, the conventional perspective doesn't postulate that there exists a definitive or generic connection between genetic and genomic mechanisms. Thus, the conventional perspective accepts that there are genes or groups of genes who routinely influence the pattern of gene expressions of several genes. However, there is no theoretical rationale that links different genes and their expressions. The key point is that the conventional perspective is ambivalent about this topic. It doesn't expressly forbid any such generic processes which link genetic mechanisms with genomic mechanisms. At the same time, it doesn't offer any theoretical framework from which we can formalize such a generic process. If at all, the conventional perspective is reluctant to accept such a possibility, only out of fear that it might take it back to design principles or teleological arguments. This ambivalence is best illustrated by systems biology (Alon 2021, see Chapter 2). Systems biology, as the name suggests, is a study of systemic features of biological processes. However, there are no theoretical frameworks which can link the expression patterns of different genes, at least beyond the standard examples like Homeobox (Duboule 1994) or wnt signaling pathways (Goss and Kahn 2011).

However, the proposed model offers a way to conceptualize such generic processes which link genetic mechanisms and genomic mechanisms. Admittedly, this is possible because the proposed model, unlike the conventional perspective, begins with a notional entity called genomic singularity. It assumes that different genomic modularities and different genomic functionalities have evolved from genomic singularity. The key semantic proposition of the proposed model is that all the genomic functionalities never arose de novo during the course of biological evolution and natural selection. Each of these genomic functionalities (including its modularity) were present in incipient forms in the genomic singularity. The role of natural selection (that of environment) has been to allow expressions of these incipient functionalities. The conventional perspective can't accept this reasoning because it rightfully questions the nature of incipient functionalities. After all, genomic functionalities whether incipient or manifest, must be encoded in the putative genomic singularity. There is no evidence of any such information in the earliest living organisms, say, Archaebacteria (Garrett and Klenk 2007, see Chapter 3). The proposed model explains this riddle by postulating that the information content necessary for converting incipient forms of functionalities into manifest forms of functionalities comes from the fine structure of spacetime itself. According to the proposed model, the fine structure of spacetime is woven into the architecture of genomes (just as it is woven into atoms and molecules). Moreover, this information content in the form of the fine structure of spacetime is capable of undergoing changes through the changes in the dimensionalities of spacetime. Thus, biological evolution and natural selection, like any other natural phenomena, are constructed by the way different dimensionalities of spacetime interact with each other.

Admittedly, it is possible to argue that even if this scenario is valid, it is purely a metaphysical scenario which has nothing to do with genomic functionalities. However, this scenario leads to predictable consequences in the form of genomic architecture itself. Thus, if we could expand our conception of phylogenetic methods (Bromham 2008) to include phylogenomics, there would be visible patterns of changes in genomic architecture during the course of biological evolution. More importantly, these changes can be anticipated using the mathematical formalism which is at the heart of the proposed model. It is important to keep in mind that this mathematical formalism doesn't predict any particular pathway of the changes in genomic architecture, but it anticipates several possibilities.

Returning to the present discussion, the proposed model postulates that since genomes are spread over multiple dimensionalities simultaneously, different modules would occupy different dimensionalities. Therefore, different genomic mechanisms say, arising from different modularities, would arise from different dimensionalities. Therefore, whenever they devolve into the four-dimensional configuration of genomes, they would give rise to different types of long-range influences depending on the dimensionality from which they arose. Since the genetic mechanisms invariably manifest in the four-dimensional spacetime, different long-range influences would manifest different degrees of temporal and spatial influences on a given genetic mechanism. Admittedly, since we don't know the exact details of different modularities and their dimensionalities, it is not easy to predict the exact degrees of temporal and spatial long-range influences. However, since the different modularities occupy fixed dimensionalities, it is possible to work backward. This is possible because different dimensionalities are directly related to the magnitude of intergenic distances. Thus, it is possible, at least in principle, to construct a topological model by deconstructing distances of all the intergenic distances. Considering that we are dealing with thousands of genes in a typical genome, it is not possible to construct a topological model manually. However, given the present computational capabilities, it is possible to construct such a model. Having looked at cancer as pathologies arising from the aberrations of control elements, let us look at the types of control elements implicit in the conventional perspective.

7.7 Conventional Perspective of Control Elements

As discussed in the previous section, cancer, as a class of pathologies, has been conventionally thought of as outcomes of aberrant control elements. Admittedly, the conventional perspective doesn't have any generic description of control elements, say, in the form of genomic architecture. However, it has certain implicit templates of different types of control elements. Therefore, the conventional perspective tacitly accepts a possibility of different control elements being part of genomic architecture. It is just that it can't formalize genomic architecture for a variety of reasons discussed above. The point that is germane to the present discussion is whether it is possible to link different control elements accepted by the conventional perspective into a common framework. This is important because once such a framework is available, it is possible to take the next logical step and define a schema of genomic architecture which is implicit in such a common framework of different control elements. Such a conception of genomic architecture would also face lesser resistance from the scientific community than a totally radical plan of genomic architecture would. Therefore, in this section, we will try to link up different control elements accepted by the conventional perspective. Using this framework, in the next section, we will try to deconstruct cancer and its evolutionary perspective. In Sect. 7.9, we will discuss why this attempt to link up these conventional perspectives of control elements or its putative framework cannot explain even this conventional evolutionary perspective of cancer.

Let us begin with the control elements conventionally held responsible for cancer pathologies. Ab initio, we can think about cancer arising as a result of failure to differentiate into a functional state that a newly divided cell is expected to reach (*anaplasia*) (Coleman and Tsongalis 2010), failure to terminate cell division cycles (*hyperplasia*) (Coleman and Tsongalis 2010), and the failure to initiate apoptosis (*neoplasia*) (Sirica 1989). The corresponding control elements and their mechanisms are well established. Therefore, we will start with these three control elements responsible for correct differentiation, correct cessation of cell divisions and correct termination of lifespan of an aberrant cell. For this purpose, we will select three well-established control mechanisms and see whether they have any commonalities, viz., the wnt signaling pathways for cellular differentiation (Goss and Kahn 2011), cdk proteins for controlling cell division cycles (Morgan 2007), and p53 pathways for failure to initiate apoptosis (Zambetti 2005). It must be mentioned that these three mechanisms are fairly complex and even interlinked. However, for the sake of simplicity, we will treat them as independent of one another. Moreover, the inherent complexity of these mechanisms ensures that their influences are fine-tuned and they give rise to subtle changes in physiological pathways leading to different degrees of pathologies. However, for the present discussion, we will employ a simple scenario: think of them as binary switches. In other words, we will think of their influences in either being present or absent. Admittedly, this strategy might appear to be grossly inaccurate to those working in this domain. However, the objective of this discussion is to point out the overall framework of long-range influences. It is possible that once the bigger picture is available, it is possible to formalize these finer details of each of these mechanisms in the corresponding topological details. However, initially, it is desirable and even imperative that we formalize the general plan sans any molecular details.

Let us begin with the first cause of cancer, viz., *Anaplasia*. The wnt signaling pathways which we talked about in the preceding chapters, is an ideal candidate. Apart from the fact that wnt signaling has been extensively studied, its role in carcinogenesis is also extensively reported in literature (Goss and Kahn 2011). As mentioned above, under normal circumstances, onset of *anaplasia*, sets into motion the apoptosis. Therefore, manifestation of cancer pathologies due to *anaplasia* is also due to the failure of apoptosis. However, we will think of *anaplasia* itself as a pathology and treat the failure of apoptosis as a cause of aggravating pathology arising from *anaplasia*. Even within the wnt signaling, there are multiple pathways leading to different types of differentiations. However, purely from systems biological perspective, these different pathways and different types of cellular differentiations are consequent to the initiation of wnt signaling pathways. From the genomic perspective, what is germane is how wnt signaling is initiated and more importantly, how it is suppressed. As mentioned above, the conventional perspective restricts itself to the molecular triggers. Therefore, we will select a schematic representation of wnt signaling for the discussion. This schema is presented in Schema 7.1 (reproduced from Haseeb et al. 2019).

As discussed above, wnt signaling is a sequence of messengers forming a network which operates as a cascade. It is important to note that each step of this schema has one-way signaling. Therefore, the changes in the cellular fate or the degree of specialization in this schema are irreversible. This irreversibility is emblematic of natural selection in the sense that each step toward specialization closes the other options for the cell and therefore its survivability. This irreversibility can be eliminated under certain circumstances. This is where the importance of stem cells comes to the fore. Thus, when we think of *anaplasia* as a pathology, it is in the context of lack of specialization accompanied by *hyperplasia*. In other words, lack of specialization per se is not pathological, but if it is accompanied by uncontrolled cell divisions, it is. Stem cells remain nonfunctionalized, but that is not a pathology. It is only when *anaplasia is* accompanied by *hyperplasia*, pathologies manifest. However, as mentioned above, we will treat each type of pathologies as independent of one another. Therefore, we will not discuss the link between *anaplasia* and *hyperplasia*. Our objective is to demonstrate that the conventional cascade of messengers is a surface level phenomena and genomic architecture provides deeper linkages

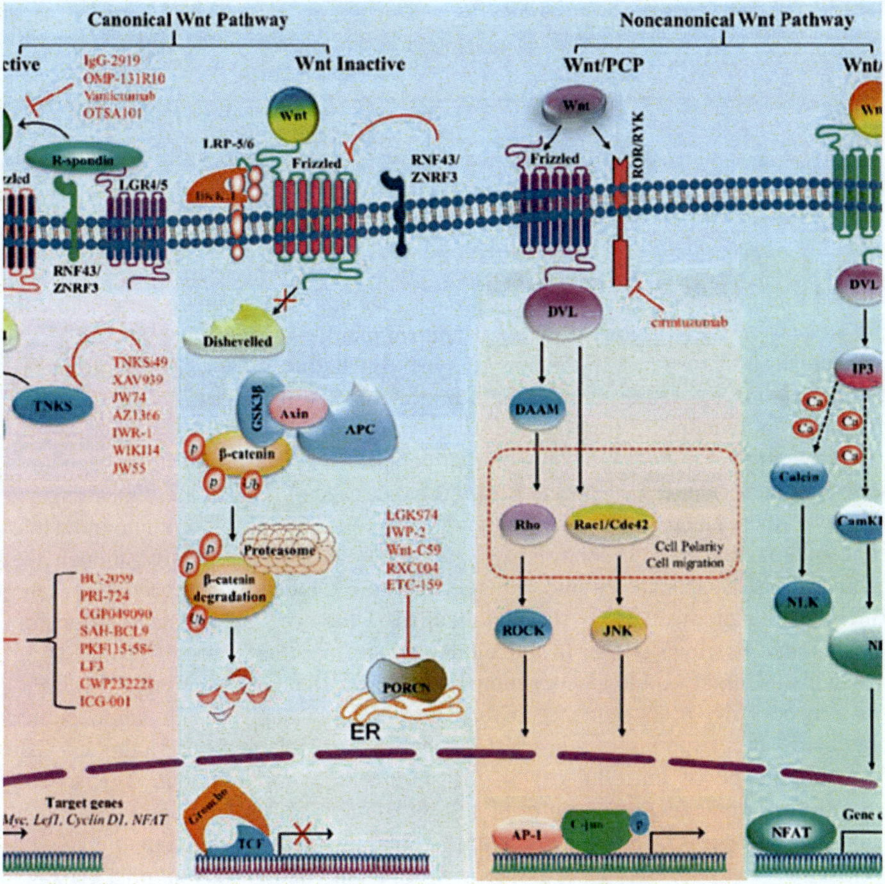

Schema 7.1 Wnt signaling pathways: conventional perspective

between these three types of pathologies. When viewed from this perspective, it is intuitively clear that if the trigger of wnt signaling pathways was somehow linked to genomic architecture, then it should be possible to define formal descriptions of different *anaplasia*, *hyperplasia* and *neoplasia* can be interlinked.

Let us now look at the second cause of cancer, viz., hyperplasia. In a healthy organism, the cycle of cell divisions is intricately controlled by intracellular and intercellular signals. This is essentially an example of a feedback mechanism. However, what we will discuss here is an example of an intracellular signal which disrupts the cyclic expressions of genes encoding cdk proteins. Each stage of cell divisions is controlled by a particular cdk protein. The details of these cdk proteins and their activities are presented in Table 7.1. One event that disrupts this cycle of cell divisions is connected with the proofreading mechanisms during the process of transcription. Whenever there is an error in transcription, it is imperative that either the error must be corrected or the process of cell divisions must be halted. Under

Table 7.1 Cdk proteins and their activities

Type of Cdk	Type of cyclin	Activity
Cdk 1	B	M phase
Cdk 2	E	G1/S transition
Cdk 3	A	S phase, G2 phase
Cdk 4	C	G1 phase
Cdk 5	D	G1 phase

normal circumstances, the relevant feedback mechanism becomes operational and the process of cell divisions stops. *Hyperplasia* manifests itself when this feedback mechanism fails resulting in the lack of synchronization. A typical example of the cell division cycles, roles of different cdk proteins and the onset of *hyperplasia* is presented in Schema 7.2 (reproduced from Finn et al. 2016).

As mentioned above, the conventional perspective once again limits itself to a series of chemical messengers and the timings of the syntheses of these messengers. However, there are a couple of evolutionary contexts that must be kept in mind while conceptualizing genomic controls over this cycle as a whole. Firstly, though the cellular division is an ancient mechanism from the evolutionary perspective, it is the emergence of multicellularity that has resulted in the need to halt the cell division (Niklas and Newman 2016). In unicellular organisms, the failure of proofreading mechanisms might lead to a new genome. Therefore, the manifestation of *hyperplasia* is necessarily a consequence of bad proofreading happening in multicellular organisms. It is tempting to think that maybe it is possible to devise a therapy that takes advantage of unicellularity in the multicellular environment. The second evolutionary context of the control of cell divisions is that if the above scenario is valid, we must focus on the intercellular signaling pathways for our therapeutic approaches. Leaving aside these issues, it is apparent that the conventional perspective can't accommodate the genomic controls in the cycle of cell divisions, particularly under the conditions prevailing prior to the onset of *hyperplasia*.

Let us now look at the third cause of cancer, viz., the failure to initiate apoptosis. The earlier discussion on *anaplasia* and *hyperplasia* referred to specific mechanisms that lead to cancer. There are two relevant aspects of these two types of pathologies. Firstly, they refer to specific causes of cancer. Admittedly, this specificity doesn't imply specificity of the resulting cancers, but the onset depends on specific aberrations. Secondly, these mechanisms essentially deal with the onset of cancer and not prevention of cancer. There is nothing special about these mechanisms that would stop or arrest the onset of cancer. In comparison, the third cause of cancer, viz., the failure to initiate apoptosis (Tutar and Tutar 2018), is generic and invoked when the physiological manifestation of pathologies has set in. In other words, this mechanism operates continuously in a healthy cell. Cancer sets in only when this mechanism fails to be activated. Moreover, this mechanism is activated under a variety of aberrations. Therefore, this is a generic mechanism which causes cancer only when it fails to be initiated. Admittedly, apoptosis is a generic protocol programmed to cause death of the concerned cell. It can be initiated even in the

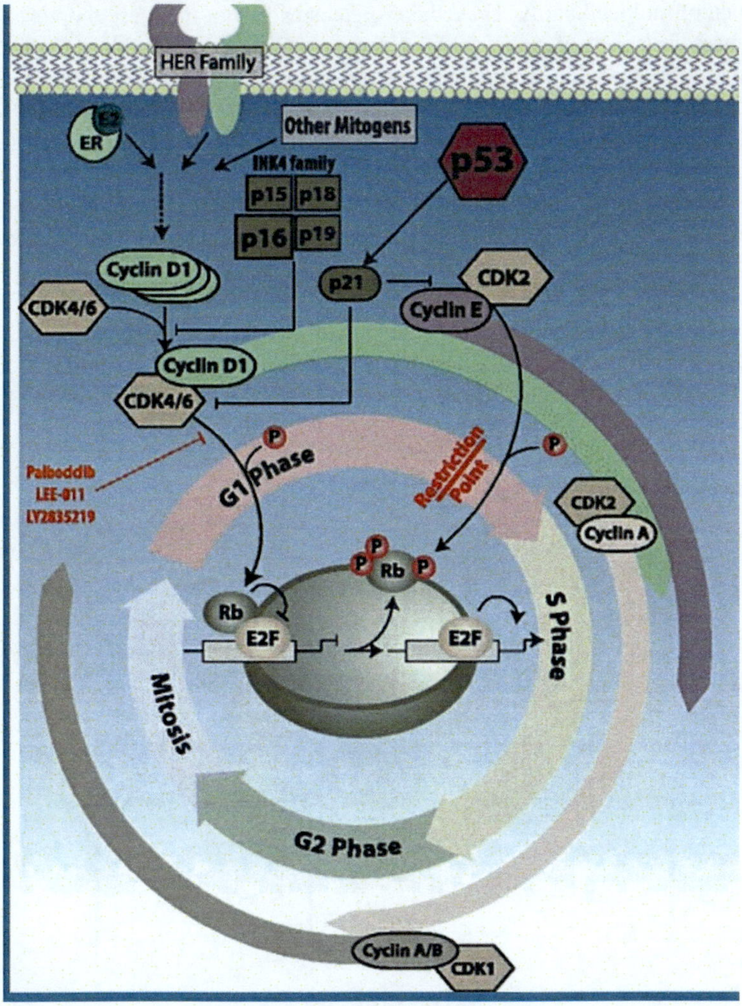

Schema 7.2 Cycle of cell division and onset of *hyperplasia*

absence of cancer, provided there are serious physiological aberrations in the cell. This is particularly true in the case of oxidative stress. Oxidative stress can cause damage to the internal functioning of a cell. If it causes damage to the DNA sequence of the cell's genome, then naturally it will give rise to cancer pathologies and the process of apoptosis would be initiated. However, apoptosis would set in even if there are no cancer pathologies, but cellular housekeeping functions are impaired by oxidative stress.

This fine distinction between the first two causes of cancer and its third cause is necessary because it points toward a fundamental semantic proposition of biological evolution. As discussed in Chap. 6, apoptosis must be viewed as a consequence of the emergence of multicellularity. Unicellular organisms do not manifest apoptosis. Similarly, cancer, as a class of pathologies, is also a consequence of multicellularity. Thus, according to this model, it is intuitively clear that multicellularity, during the course of biological evolution must have differentiated into two different types of control elements, one leading to differentiation of cellular functionalities and another to keep check on this ability to segregate different functionalities. In such a scenario, it is intuitively clear that the functionality of apoptosis must have arisen immediately after the emergence of multicellularity, but prior to aging and the segregation of cellular functionalities. This explains why apoptosis arises not just in the case of cancer, but also in the case of aging and other aberrations caused by oxidative stress. This scenario is presented in Schema 7.3. The exact details of how apoptosis arises are provided in Schema 7.4 (reproduced from Chowdhury et al. 2006). The key point about the

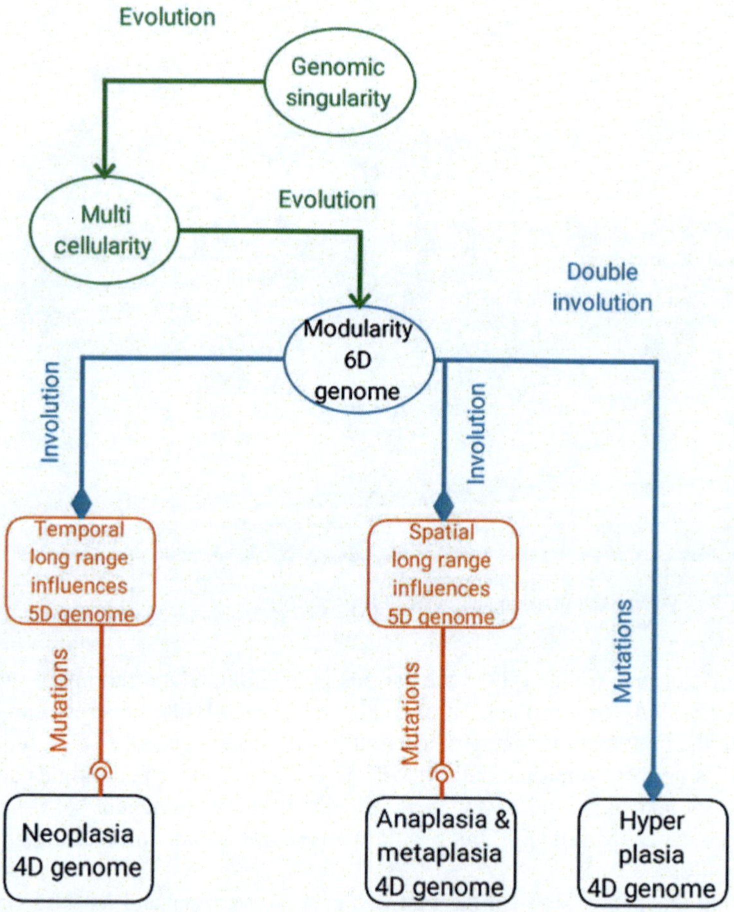

Schema 7.3 Evolution of different mechanisms causing cancer

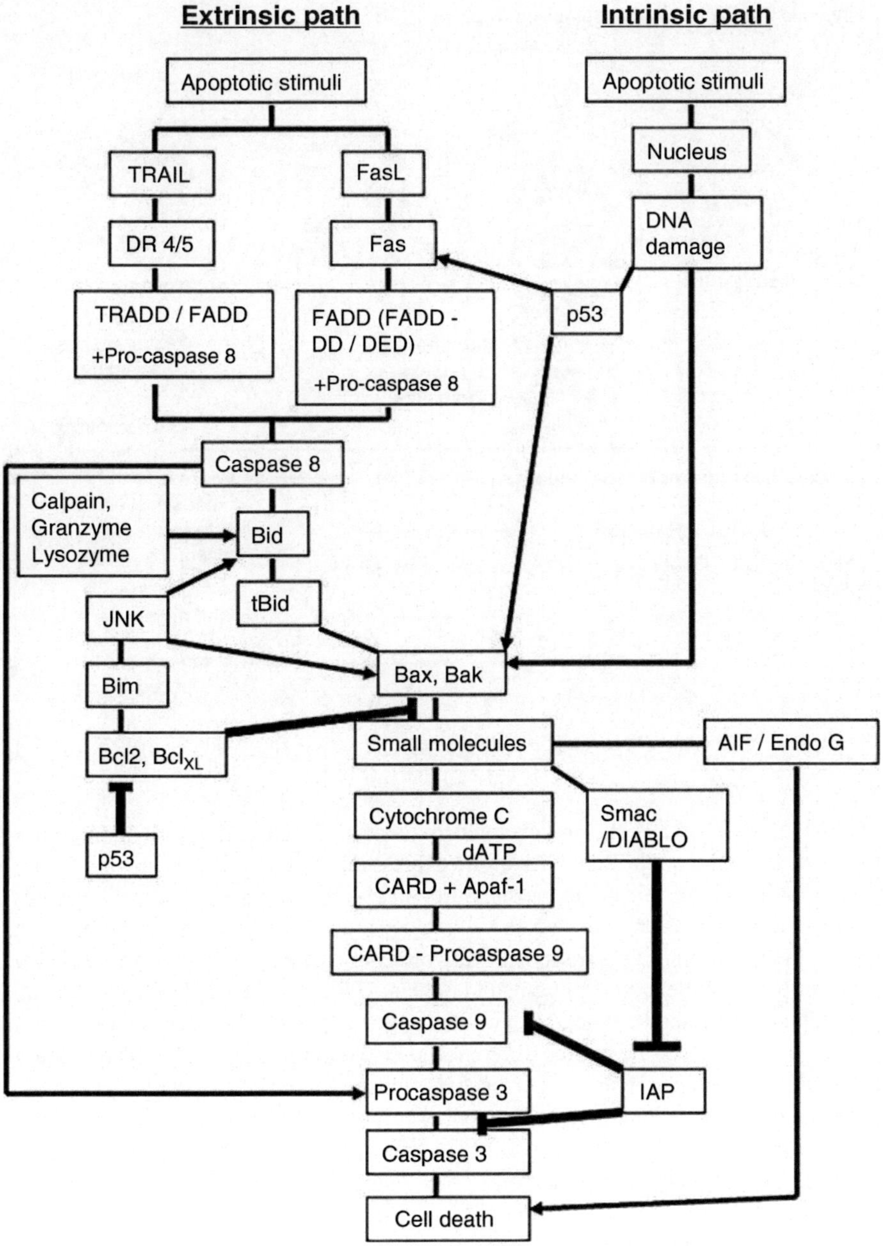

Schema 7.4 Initiation of apoptosis. (Reproduced from Chowdhury et al. 2006)

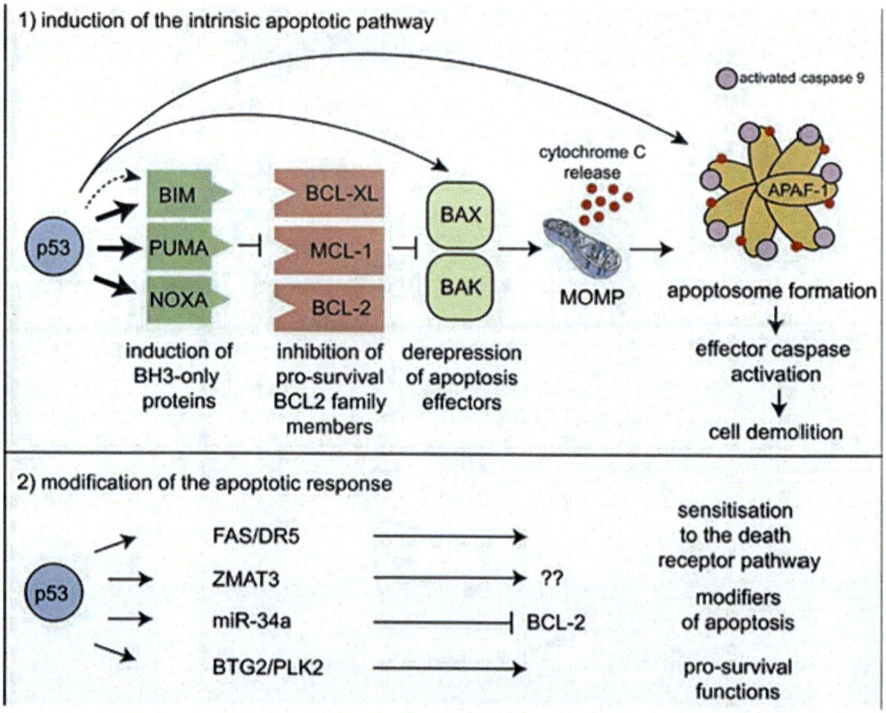

Schema 7.5 Relationship between p53 and apoptosis. (Image courtesy Aubrey et al. 2018)

initiation of apoptosis is the involvement of gene p53. There exists a humongous amount of literature on the centrality of the expression of gene p53 and how its failure leads to the failure to initiate apoptosis. Surprisingly, the origin of gene p53 predates multicellularity. Therefore, it is necessary to deconstruct the relationship between gene p53 and the initiation of apoptosis. A tentative scenario of this relationship is provided by the proposed model and is presented in Schema 7.5 (reproduced from Aubrey et al. 2018). It is important to note that the conventional perspective doesn't provide a genomic perspective of these three different causes of cancer. Therefore, it is necessary to find out why the conventional perspective omits a genomic perspective. However, before doing so, let us look at the evolutionary perspective of cancer. Therefore, in the next section, we will discuss how cancer is an inevitable, although unpleasant consequence of the evolution of complexity.

7.8 Evolution and Cancer

As discussed in the introduction, cancer arises from the failure of the control elements of the genome. This is the reason why the prevalence of cancer in the older population is higher. Due to aging, these control elements suffer wear and tear, leading to cancer. At the same time, if these control elements have undergone changes earlier, cancer can affect younger individuals as well. It is important to keep in mind that cancer is never inherited, but the vulnerability to cancer is. This is because manifestation of cancer pathologies depends on impairment of several genes. Therefore, each one of us may be carrying some of these defective genes. It is the last couple of mutations that separates healthy individuals from cancer patients. Thus, to assert that cancer is a natural consequence of genomic complexity is a tautological assertion. Ever Since the advent of molecular biological perspective, this truism is generally accepted. Therefore, we will not dwell on this topic directly. Instead, we will try to deconstruct the evolutionary context of complexity and its linkage with cancer. Therefore, we will discuss in this section three specific aspects of the connection between genomic complexity and cancer, viz., (1) Is there any specific type of complexity that is prone to give rise to cancer? (2) Did natural selection favor these complexities over other types of complexities that are less likely to lead to cancer? (3) Is it the inherent mechanism of natural selection that allows the complexities prone to give rise to cancer to be selected ?

It is possible that there are no precise answers to these questions. However, by raising these questions and by seeking their answers, we will be able to deconstruct the semantics of natural selection and understand how genomic architecture could have evolved. These questions are also important because the conventional perspective of genomics has no obvious answers to these questions, but the proposed model has something to offer. Therefore, while trying to find answers to these questions will help us to evaluate the proposed model vis a vis the conventional perspective of genomics. Let us begin with the first question: is there any specific type of complexity that is prone to give rise to cancer ? While the conventional perspective concedes that there is some correlation between genomic complexity and cancer (Dellaire et al. 2014), it has no definitive answer. This is because as discussed in the preceding chapters, we don't have any formal description of genomic architecture. Therefore, there is no way to categorize different types of complexities of genomes and link them to cancer. At the same time, there are several hints about such a relationship between complexity and cancer in the conventional perspective. Of course, it is possible to formalize complexity in a mathematical framework and impose it on the data of several genomes that we have. However, instead of taking this approach, we will try to deconstruct molecular biology of known mechanisms discussed above and try to find out what kind of complexities are prerequisites for the onset of cancer. Even without going into the molecular details of these mechanisms, it is possible to think of these mechanisms in terms of spatial and temporal control elements. This is presented in Schema 7.6. Admittedly, we have combined all the three mechanisms in a single map. However, this helps us to understand the generic nature of complexities of these mechanisms instead of their molecular details. Moreover, we have employed the genomic architecture implicit in the proposed model to assign different dimensionalities to genomes and link them to different types of asynchronization. Admittedly, this is purely a speculative scenario, but it leads to cogent explanations.

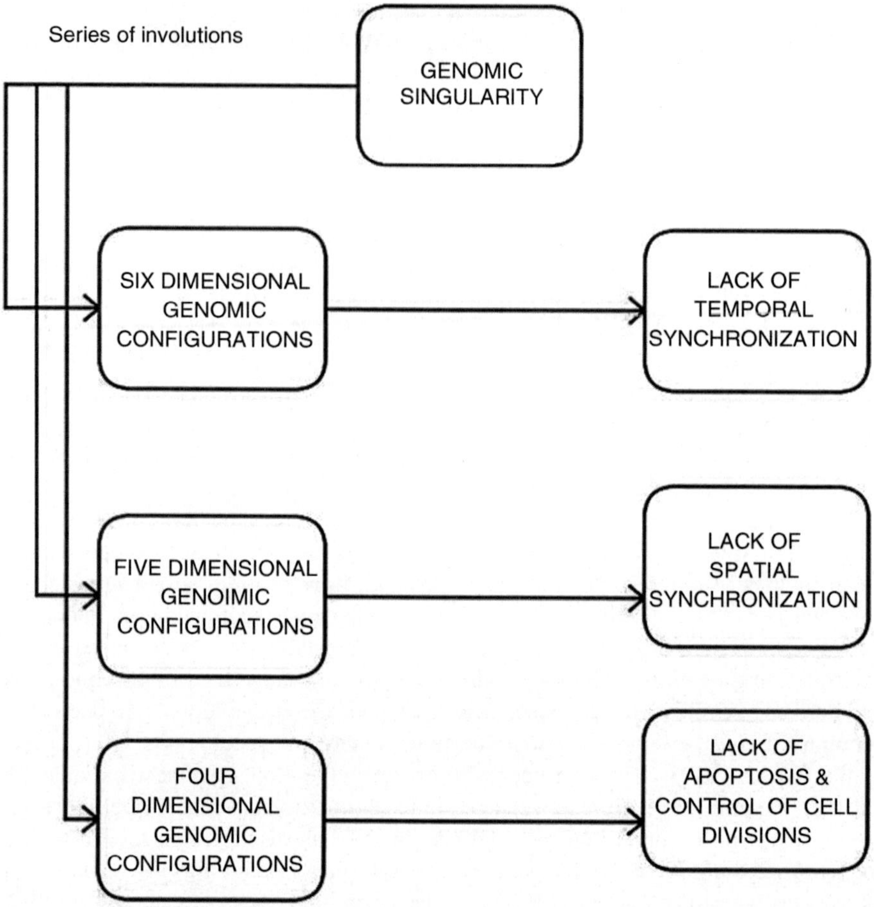

Schema 7.6 Spatiotemporal map of different mechanisms of cancer

When we look at this schematic representation of these mechanisms, few propositions become self-evident.

1. The molecular details of these three mechanisms represent a common logical schema.
2. This logical schema actually represents the key features of spacetime itself.
3. These spatiotemporal features actually offer a blueprint of genomic architecture.
4. Cancer pathologies arise not just because of faulty genes, but they also arise from faulty gene expressions.
5. By altering the spatiotemporal parameters, it should be possible to prevent faulty gene expressions, thereby preventing the onset of cancer pathologies.
6. However, this requires a topological model of genomic architecture.

Admittedly, we need a suitable topological model to verify the last proposition (proposition 6). We will return to this topic in the following sections. Presently, let us look at the second question: did natural selection favor the cancer prone complexities over the remaining types of genomic complexity ? As mentioned above, the conventional perspective is ambivalent about these topics. Therefore, we will try to answer these questions ab initio. Hypothetically speaking, natural selection would not select anything that threatens the survival of the species. Therefore, we can begin this discussion with the assumption that while natural selection might have chosen one type of genomic complexity over the remaining types of genomic complexities based on their respective survivabilities, the cancer pathologies must be treated as a collateral damage. This is perhaps implicit in the conventional perspective (Dellaire et al. 2014). Moreover the fact that germ lines are segregated out during the developmental stages of embryo, suggests that this isolation of germ lines ensures that later genetic damages are not passed on to the succeeding generations. Thus, the conventional perspective, taking a conservative approach, has remained ambivalent. However, since we are trying to deconstruct this topic using the out of the box thinking, the possibility of natural selection preferring one type of genomic complexity over the remaining types of genomic complexities can be achieved only if natural selection employs a structural template while making choices. Thus, the question of natural selection preferring one type of complexities over the remaining types of complexities is linked to the question whether natural selection is essentially a structural mechanism which makes choices based on its own structuralism.

As discussed in the preceding chapters, our reluctance to postulate any structural prerequisite for natural selection is based on the apprehension that it might compromise the essential Darwinian randomness. As it has been argued in the preceding chapters, this apprehension is misplaced. Therefore, let us look at what kind of structural template natural selection can possess and more importantly, what is the origin of this template. Conventionally, this possibility of natural selection having its own structuralism is not taken seriously because the only possible source of the structural template of natural selection will have to be something transcendental. Since any such transcendentalism influences will undermine the naturalistic foundation of the Darwinian paradigm, this approach was never contemplated. However, the proposed model offers a way to avoid any such transcendental influences, thereby retaining the naturalistic foundation of the Darwinian paradigm. According to the proposed model, spacetime itself is a constituent of genomic architecture. Therefore, it is axiomatic that like any other natural phenomena, natural selection operates on the structural template of spacetime itself. Once again, conventionally, the structural template of spacetime was defined in such a way that it remained a passive witness to all natural phenomena. It is true that there are two types of natural phenomena, viz., quantum phenomena and relativistic effects, in which spacetime is supposed to play an active role in these phenomena (Chhaya 2022a). For that purpose, a specific structural template of spacetime is normally employed. However, natural selection (for that matter, biological evolution) cannot be shown to involve either quantum or relativistic processes. Therefore, even if we wish to employ the structural template of spacetime to derive the native structuralism of natural selection, there is no such template of spacetime available to us.

However, as discussed in the preceding monograph (Chhaya 2022a, see Chapter 2), the proposed model offers a different perspective on the structural template of spacetime. More importantly, this template can be used to derive the structural template of natural selection. According to the proposed model spacetime itself exists in multiple dimensionalities simultaneously. Each dimensionality possesses its own metric. Moreover, each dimensionality of spacetime is connected with the remaining dimensionalities by a universal operator of involution. The key point is that each dimensionality possesses a unique blend of the time-like and the space-like features. (Incidentally, according to this model, some of the dimensionalities of spacetime do not possess any distinction between the time-like and the space-like features.) However, all these dimensionalities devolve into the four-dimensional spacetime, but with different degrees of influences. Thus, it is intuitively clear that if our belief that the process of natural selection is a universal protocol and that its biological manifestation during evolution is a specific instance, then the native structuralism of natural selection must be in the form of the changes in the dimensionalities of spacetime. Thus, in any given natural phenomenon that entails the changes in the dimensionalities, the process of natural selection would operate in the form of the changes in the dimensionalities. This provides a way to extend the scope of natural selection to a host of natural phenomena.

In the context of the present discussion, it is intuitively clear that natural selection during the course of biological evolution would consist of the influence of different higher dimensionalities of genomes on the working of individual genes at the four-dimensional spacetime. As postulated in the preceding chapters, genomes are spread over multiple dimensionalities of spacetime simultaneously. Therefore, from each higher dimensionality of genomes, there would be some influences that are imposed on individual genes which exist in the four-dimensional spacetime. However, not all these higher dimensional influences would be compatible with the spatiotemporal influences that arise from the higher dimensionalities of spacetime itself. Thus, we have two sets of influences arising from the higher dimensional configurations of genomes as well as from spacetime. Therefore, natural selection consists of allowing only those higher dimensional influences of genomes that are compatible with the higher dimensional influences of spacetime itself. This scenario establishes two postulates. Firstly, spacetime plays an active role in biological evolution and natural selection. Secondly, the process of natural selection has a structural template that selects only those higher dimensional genomic configurations which are compatible with the structural template of the four-dimensional spacetime.

Once we accept this reasoning, it is intuitively clear that the types of complexities of genomes that are preferred by natural selection will be mainly those that involve long-range spatial and temporal control of gene expressions. This is because the degree of nonlocality of these long-range influences would depend on the higher dimensionalities from which they originate. Thus, at the four-dimensional spacetime, there will be multiple long-range influences (both spatial and temporal) each having different lengths at which they influence gene expressions. Thus, if we think of cancer as an outcome of the lack of synchronization, it is intuitively clear that this lack of synchronization must arise when the long-range influences of genomes are incongruent with the long-range influences of spacetime itself. Thus, the proposed model offers a distinctly different view of how cancer could have arisen during biological evolution. Therefore, it is necessary to compare this perspective with the conventional perspective of how cancer evolved. Therefore, in the next section, we will look at the conventional view of the evolutionary perspective of cancer.

7.9 Shortcomings of the Conventional Perspective

As mentioned above, the conventional perspective of the evolutionary perspective of cancer is based on the individual genes, their expressions and their evolutionary trajectories. Prima facie, this is a conservative, but legitimate perspective. This is particularly so since the conventional perspective lacks any theoretical models of genomic architecture. Therefore, conventionally, it is conceded that there could be genomic influences (in contrast to genetic influences) that also could result in the onset of cancer pathologies. However, in the absence of any clarity on the nature of genomic influences (possibly in the form of long-range influences), the conventional perspective has chosen to focus on the individual genes and their expressions. This pragmatic approach has been remarkably successful. However, partly due to improvements in public health and the changing demographics, the proportion of older individuals in society has been continuously increasing. Since even according to the conventional perspective, cancer is directly linked to human longevity, it is imperative that we must augment the conventional perspective of cancer with some new and if necessary, out of the box thinking. One such model was outlined in the preceding sections. Therefore, it is necessary that we evaluate the conventional perspective in the context of the ideas presented above.

The key point is: whether by avoiding the notion of genomic architecture the conventional perspective is missing out on any key features of the onset of cancer. Therefore, in this section, we will deconstruct the conventional perspective using three propositions, viz., (1) the absence of genomic architecture, reduces the predictive power of the conventional perspective, (2) Genomic influences play a distinct role in the onset of cancer which is different from the conventional perspective which is based on the molecular biology of individual genes and their expressions, (3) A genomic perspective of the onset of cancer, in principle, offers a better therapeutic approach.

Let us begin with the first proposition: The absence of genomic architecture reduces the predictive power of the conventional perspective. Prima facie, it is axiomatic that if we know the exact nature of genomic architecture, it would help us to improve our ability to predict the onset of cancer. However, the point is that the conventional perspective cannot and doesn't accommodate the conception of a fixed template of genomic architecture. This is because, as mentioned above, the conventional perspective can't accept this notion because of the apprehension that any such fixity of genomic architecture would undermine the essential randomness implicit in the Darwinian paradigm (Bonner 2013). Similarly, the conventional perspective doesn't find sufficient evidence of any such fixed pattern of the distribution of genes in genomes. However, with the increasing computational capabilities, it is at least feasible to test various types of models for genomic architecture. Given the fact that we already have complete genomes of several species, it is worth trying several techniques of mathematical modeling. Broadly speaking, mathematical modeling would be in the form of trial and error. This is because there is no link between mathematics and biology. However, the proposed model bridges this gap. As discussed in the preceding monographs, the proposed model offers a common ontology of mathematics and Life. More importantly, it postulates a universal computation template for all the natural phenomena including biological evolution. Therefore, it makes sense to employ this model. We will discuss the details of this approach in Sect. 7.16.

Let us now look at the second proposition: Genomic influences play a distinct role in the onset of cancer which is different from the molecular biological perspective of the individual genes. Prima facie, according to the conventional perspective, the onset of cancer pathologies arises from the mutations of individual genes (Madhusudan and Wilson 2013) and the genetic damages caused by oxidative stress (Spitz et al. 2011). While the mutations in genes occur either due to errors in transcription during mitosis or due to oxidative stress, including radiations. Oxidative stress, on the other hand, acts in multiple ways. It can either damage genes due to reactive oxygen species. It can also influence epigenetic silencing. Alternatively, it can disrupt biochemical feedback mechanisms and disturb homeostasis (Hagen 2021, see Chapter 11). From these various causes of the onset of cancer, disruption of homeostasis and the accompanying lack of synchronization are the causes where genomic mechanisms play an important role. As discussed in the preceding sections, synchronization of different gene expressions, per se, is a less understood phenomenon. Therefore, it is axiomatic that if there exists a model of genomic architecture which explains the exact mechanism of synchronization of different gene expressions, then such a model can provide a better predictive power to oncologists. What is debatable is whether such genomic architecture exists or not. More importantly, whether such an architectural design is amenable to comprehension or not. As discussed above, the conventional perspective has a very reason to doubt the very existence of genomic architecture. The scenario described in the preceding sections, is an attempt to justify the existence of such a genomic architecture that corresponds to the known molecular biology of cancer and at the same time, it resolves the semantic ambiguities of the conventional perspective.

Now let us look at the third proposition, viz., the genomic perspective of cancer, at least in principle, offers better therapeutic approaches. Prima facie, this proposition is true. However, its scope is limited. As discussed above, the majority of the causes of cancer are manifest at the level of individual genes. Therefore, the influence of genomic mechanisms could be limited to suppress the onset of cancer, rather than elimination of cancer. Even in the conventional perspective of cancer, we define two types of genes involved in cancer pathologies, viz., oncogenes

(Ross 1998, see Chapter 3) and tumor suppressor genes (Polinski 2007). Therefore, the role of genomic mechanisms must be limited to prevent the expressions of oncogenes and to initiate the expression of tumor suppressor genes. Prima facie, no amount of clarity on the nature of genomic architecture can help us to alter the molecular biology underlying the cancer pathologies. Molecular biology of genes must be accepted as *fait accompli*. All that we can hope for is to find a way to harness molecular biology to prevent the onset of cancer. This brings us to the end of the discussion on the shortcomings of the conventional perspective.

7.10 Proposed Model

In the preceding sections, we discussed the shortcomings of the conventional perspective of cancer. All along, it was suggested that if there is any genomic architecture, its formalization can help us to understand, predict and treat cancer pathologies, albeit in a limited way. During these discussions, various features of the proposed model were hinted at, but without any formal description of genomic architecture implicit in the proposed model. Therefore, in this section, we will try to articulate the details of this model. In the preceding chapters, this model has been discussed in a variety of contexts. Therefore, in this section, we will articulate this model of genomic architecture in the context of the emergence of complexity and its relationship with cancers. Therefore, there are two parts of the model that would be discussed here. Firstly, we will discuss the emergence of complexity during biological evolution per se. In the second part, we will try to demonstrate that the types of complexities naturally selected can give rise to only some types of cancers. Since some of the details of the proposed model are discussed in the preceding chapters, we will selectively highlight the features that are germane to the present discussion on cancer.

As mentioned in the preceding chapters, the emergence of complexity during biological evolution remains enigmatic phenomena in the Darwinian paradigm (Grene 1986). Given the centrality of randomness and the absence of design in the Darwinian semantics (Bonner 2013), the emergence of complexity is prima facie incongruent with the Darwinian paradigm. Of course, it can be argued that since the earliest living organisms were least complex entities, any improvement in their survivability would introduce complexity. However, the problem with this argument is that biological evolution has given rise to only a certain type of complexity. Even here, it can be argued that since the Darwinian paradigm rests on the principle of descent with modification, the legacy of complexity must follow axiomatically. Therefore, it is possible to argue that if biological evolution has led to the emergence of only certain types of complexities, it is because of the underlying continuity of structuralism. If an ancestor has a certain kind of complexities, it is intuitively clear that its successors would have the types of complexities that are derivable from the complexity of the ancestor.

This argument has two drawbacks. Firstly, it assumes that at the start of biological evolution, the earliest living organisms were least complex. Admittedly, these organisms were less complex than their successors. However, purely from the structural perspective, these earliest organisms were very complex nonetheless. Therefore, the earlier assumption that complexity emerged during biological evolution is a cognitive artifact. It will be more accurate to postulate that Life requires a certain degree of complexity ab initio. The process of natural selection merely refashioned this native complexity. This reasoning makes sense because when we look back, Darwin's theory is about what might have happened after living organisms came into existence. It is rarely, if ever, discussed how Life originated in the first place. Once we accept this reasoning, the second drawback suggests itself. If the process of natural selection merely refashioned the type of complexity in order to improve the survivability, it couldn't have done so unless it had a definitive structural template of its own. In other words, the process of natural selection must possess a mechanism by which it could engender different types of complexities from a given type of complexity of the ancestors.

However, any such suggestion of natural selection having its own native structuralism is fraught with teleological implications. This fear is indeed justified in view of the centrality of randomness in the Darwinian paradigm. However, this is a category mistake. A causal mechanism need not be a deterministic mechanism. A causal explanation can exist even in a phenomenon characterized by random outcomes. We should think of two similar examples. Firstly, think about Bayesian logic (Press and Clyde 2003). Bayesian logic, like any other logic, has a causal schema. Admittedly, the causality implicit in logic is that of "if so then this" type of causality, but it is causality all the same. However, we never worry while employing Bayesian logic that it represents some design principles. Next, we must think of quantum outcomes. Self-evidently, there cannot be any more random phenomena than quantum outcomes. Even here, given the fact that quantum decoherence occurs in a continuum of spacetime, there exists a causal chain of events (Chhaya 2022b, see Chapter 3). However, because the resulting outcomes are nonlocal that we fail to comprehend the underlying mechanism. The key point is this: A causal mechanism doesn't eliminate randomness, it merely sequesters it into a certain pattern. This is true in the case of Bayesian logic. The probabilities of later outcomes are conditioned by the prior conditionalities. Similarly, in the case of quantum decoherence, while the outcomes are random, they fall within the limited range defined by Bell's inequalities. When viewed from this perspective, it is intuitively clear that natural selection too has given a pattern of morphology. In fact, the topics like punctuated evolution (Gould 2007) or speciation (Coyne and Orr 2004) must be seen as having arisen from these patterns. More importantly, these patterns must have arisen because natural selection must possess its own native structuralism. Therefore, it is intuitively clear that natural selection has given rise to only a few types of complexities. More importantly, if we can deconstruct the types of complexities naturally selected, then we can perhaps deconstruct the native structuralism of natural selection itself.

Apart from the prerequisite of randomness, the reason why we can't visualize natural selection possessing its own native structuralism is that we have all along thought of genomes as consisting of linear sequences of DNA. It is only after the discipline of genomics was established that we have come to accept the possibility of genomes having higher levels of organization. Admittedly, these higher levels of genomic organization still depend on the molecular framework of DNA and chromatin, at least they accommodate three-dimensional stereochemical details into their formal descriptions. If the earlier assumption of linear structuralism of genomes prevented us from contemplating phenomena like chromosome territories, the present stereochemical configurations fail to accommodate long-range influences which are a different type of complexity.

Therefore, the proposed model postulates that just as the linear conception of genomes was replaced by a three-dimensional stereochemical conception of genomes, we need to employ higher dimensional (higher than three-dimensional) conception of genomes. Even in the absence of any prior knowledge of these higher dimensional conceptions of genomes, it is intuitively clear that such a conception would achieve a unitary framework for linear, three-dimensional and higher dimensional complexities. Thus, if the linear perspective of genomes helped us to understand the molecular biology of gene expressions and if the three-dimensional perspective helped us to understand hitherto unexplained phenomenon like epigenetic silencing, it is intuitively clear that any such higher than the four-dimensional perspective would help us to understand hitherto unexplained phenomena like chromosome territories, long-range influences. There is another reason why we should seek higher dimensional configurations of genomes. Purely from the modeling perspective (Laudal 2021), it is intuitively clear that higher the dimensionality of the model, more variety of complexities become discernible. As a logical corollary, it is intuitively clear that as we discern more and more variety of complexities, it would unravel more and more variety of control elements. This is because Life manifests itself in the four-dimensional spacetime. Therefore, if genomes exist in multiple dimensionalities simultaneously, its manifestations would still be at the four-dimensional spacetime. Therefore, each of these higher dimensionalities would need to devolve into the four-dimensional spacetime where genes find their expressions. Therefore, there would be multiple mechanisms, each representing a particular dimensionality. Thus, a conception of higher dimensional configurations of genomes would provide us with insights into newer classes of control mechanisms. Since the four-dimensional spacetime is characterized by the distinction between the time-like and the space-like dimensions, it is axiomatic that these different putative mechanisms (each representing different dimensionalities) would have to manifest in long-range influences of only two types, viz., spatial and temporal controls. Therefore, while there would be a variety of dimensionalities, each influencing the four-dimensional configurations of genomes, they would manifest only in these two types of long-range influences. It would be legitimate to wonder how to differentiate between the influences of different dimensionalities if each of them leads to spatial and temporal controls. The answer lies in the magnitude of the long-range influences. Thus, it is possible that the phenomena like chromosome territories arise from one particular higher dimensionality while the sequence of gene expressions arises from another higher dimensionality.

Thus, it is apparent from the above discussion that irrespective of the exact nature of higher dimensional configurations of genomes, the manifest complexities of genomes at the four-dimensional spacetime would be only of two types, viz., spatial and temporal control elements. However, what would vary would be the magnitude of the influence of these different spatial and temporal control elements. This leads us to the next question of what kind of cancer pathologies can arise from these spatial and temporal long-range control elements. It is important to keep in mind that it is the lack of synchronization between these long-range control elements that causes cancer. At the same time, it is important to keep in mind that there would be some cancer pathologies which arise not from genomic mechanisms, but from different genetic mechanisms. However, it is important to remember that cancer pathologies, as a class, require both genomic and genetic mechanisms to progress and propagate. Thus, even if a particular cancer pathology arises from a genetic mechanism, its metastasis would still require genomic mechanisms to fail to synchronize.

Admittedly, the scenario described above is quite complex, it does offer several good leads to manage the extent of the spread of cancers. For the sake of simplicity, the Schema 7.7 depicts the higher dimensional configurations of genomes without any molecular or genomic details. This abstract representation would help us to deconstruct the nature of intricacies at play in genomics. Schema 7.8 depicts the relationship between complexity and modularity in an abstract manner.

7.11 Complexity According to the Proposed Model

In the previous section, we put forth a schematic abstract representation of higher dimensional configurations of genomes, the long-range influences arising from these higher dimensional configurations. We also looked at the schematic representation of the relationship between complexity and cancer. However, before we establish an exact relationship between complexity and cancer, it is necessary to deconstruct the nature of complexity per se. There are two reasons for this. Firstly, as mentioned above, our conception of genomic architecture evolved from a linear model to a three-dimensional model. This transition revealed some additional types of complexities. Similarly, since now we wish to conceptualize higher dimensional models of genomic architecture, it is necessary to understand how this increase in dimensionality also leads to the discovery of new types of complexities. Secondly, it is necessary to separate out the peculiarity of the proposed model. Let us understand why. Let us assume, albeit temporarily, that genomes exist in multiple dimensionalities simultaneously and some of these dimensionalities are higher than the four-dimensional perspective. Purely from the topological perspective, it is intuitively clear that each dimensionality would have its own degrees of symmetry. Moreover, not all these symmetries are common to all the dimensionalities. Therefore, mere assertion that genomes exist in higher dimensionalities doesn't guarantee what kind of symmetries it could possess. The types of symmetries that genomes could possess depend not just on the dimensionality it is supposed to exist in, but they also depend on the kind of topology that we assign to these higher

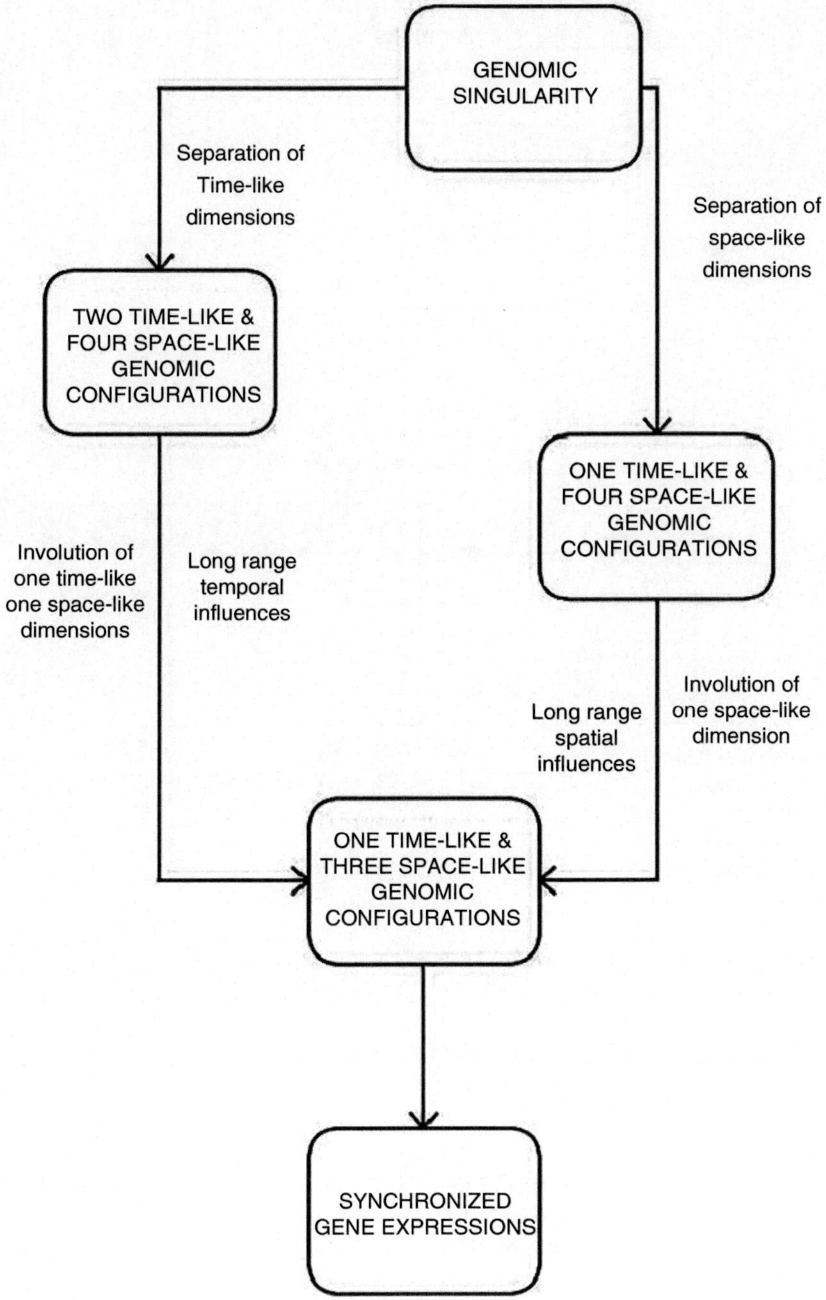

Schema 7.7 Higher dimensional configurations and long-range influences of genomes

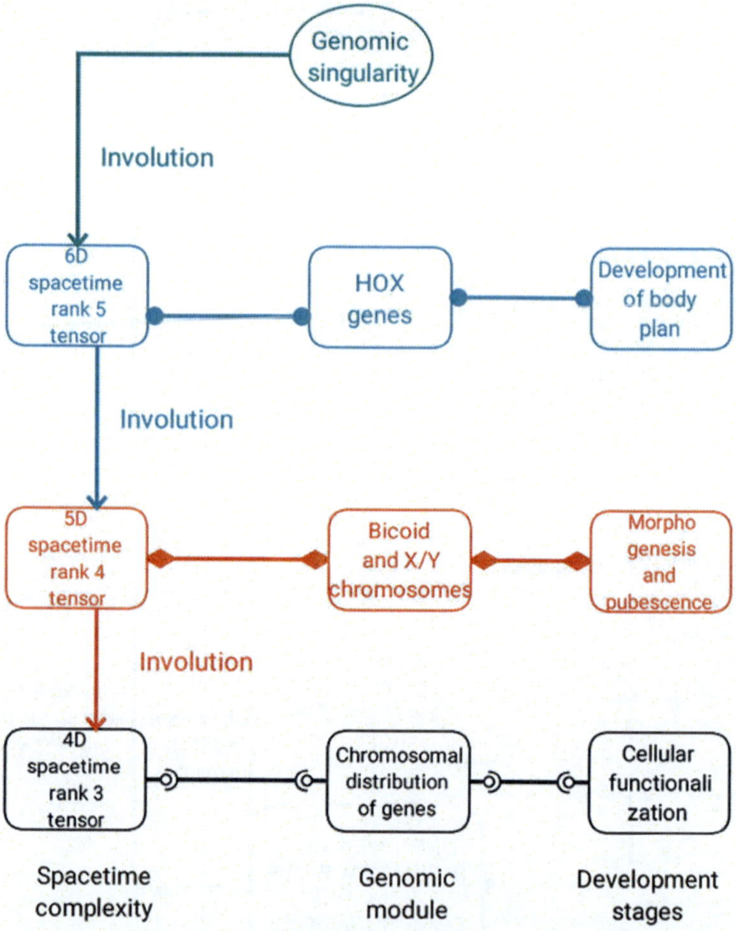

Schema 7.8 Relationship between complexity and modularity

dimensional configurations of genomes. Therefore, it might be the case that genomes exist in higher dimensionalities, but in a different topology than the one provided by the proposed model. Therefore, it is necessary to deconstruct the exact nature of the types of complexities that the proposed model implies. Once we do that, it would give us two important conjectures. Firstly, it would provide us a way to verify the proposed model. Secondly, it would help us to define the relationship between complexity and cancer. For this purpose, we will make one assumption. We will assume that the types of complexities are related to the types of symmetry present in these higher dimensionalities. Admittedly, this is a questionable proposition. However, even if we can't formally prove this proposition, it has a certain intuitive appeal. Therefore, we will provisionally accept this proposition and analyze the

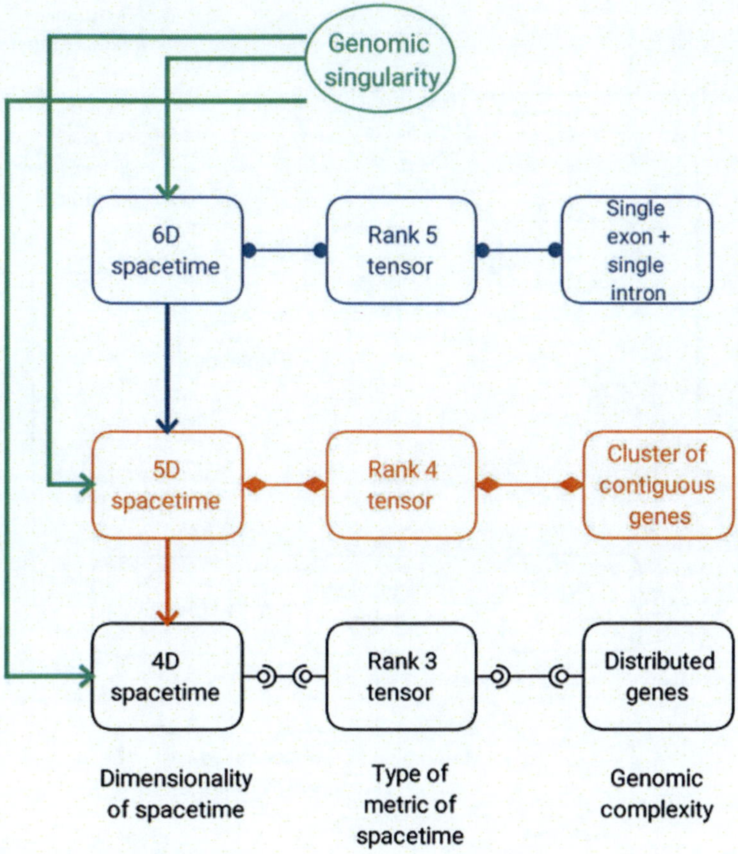

Schema 7.9 Types of complexities in the proposed model of genomes

types of complexities implicit in the proposed model. A schematic representation of the types of complexities implicit in the proposed model is given in Schema 7.9. According to this schema, we can think of genomes as consisting of a pair of single intron and a single exon at a higher dimensionality (tentatively assigned sixth dimensionality). At the fifth dimensionality, there would be clusters of genes which constitute a module in a topological continuity. Finally, in the fourth dimensionality, we would expect the current distribution of genes in different chromosomes.

It is intuitively clear from the Schema 7.9 that in spite of variations in the types of complexities in different dimensionalities, there is a certain degree of commonality. It is also clear that this commonality arises from the topological nature of the manifold postulated in the proposed model. Particularly, it is the inherent feature of the proposed modified operator of involution that gives rise to this commonality

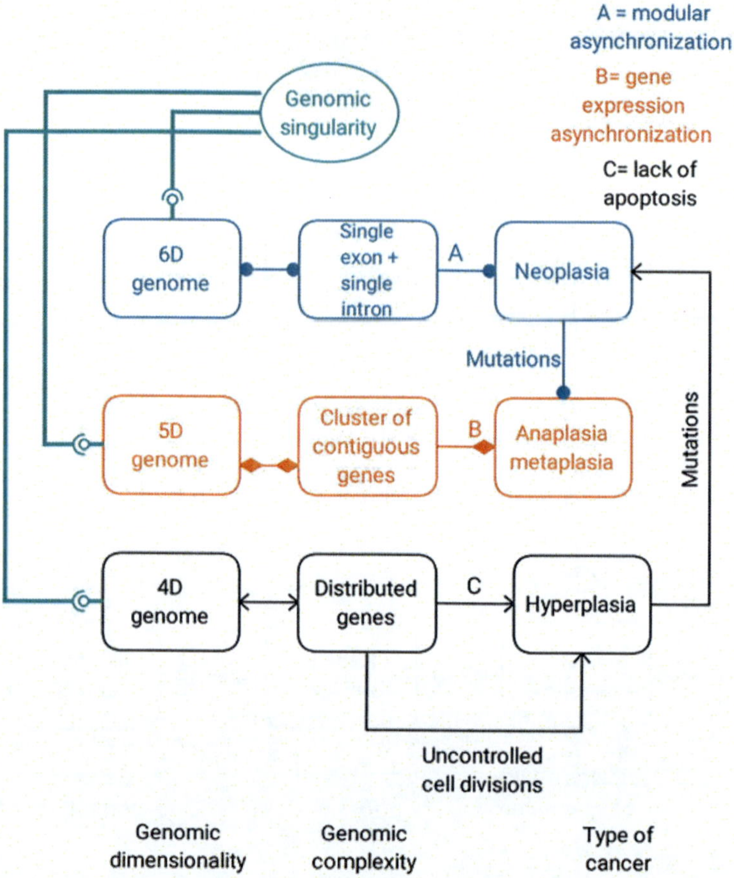

Schema 7.10 Relationship between complexity and cancer

between different types of complexities. Thus, different dimensionalities manifest various types of complexities. However, these variations occur within a limited range.

The next question that arises is this: is it possible to correlate different types of complexity with different types of cancer ? In order to answer this question, let us try to understand the link between complexity and synchronization. Thus, if we can demonstrate that different types of complexities give rise to different types of asynchronization, then it ought to be possible to define the relationship between complexity and cancer. Without going into the mathematical details of complexity, a schematic representation of the types of complexities, the types of asynchronization and the types of cancers is provided in Schema 7.10. Once again, it is intuitively clear that while there is no experimental evidence to support this schema, it has a certain logical appeal.

Having looked at the relationship between complexity and cancer, it is natural to wonder if this highly abstract conception of the relationship between complexity and cancer gets reflected in the genomic architecture. It is one thing to conjure up some fanciful mathematical constructs, but it is quite different from finding its manifestation in a physical entity. The best way to confirm the veracity of this abstract scenario is to search for different types of control elements that would be disturbed whenever an asynchronization sets in. Therefore, in the next section, we will outline different control elements predicted by the proposed model and try to evaluate them with respect to the conventional wisdom.

7.12 Control Elements According to the Proposed Model

In the conventional perspective, we think of control elements in the context of gene expressions. Thus, a control element must be able to influence the initiation and termination of gene expressions by a fixed mechanism. The postulate of a fixed mechanism is necessary to introduce programming. If there exists a fixed mechanism, then its initiation can be controlled. Thus, the conventional perspective accepts the notion and the importance of control elements. This Is exemplified by the conception of operon (Miller and Reznikoff 1980). In addition, the conventional perspective also implicitly accepts the possibility of two types of control elements, viz., the spatial and temporal control elements. Although these are not formalized as such, but the conventional perspective accepts them indirectly. For instance, the phenomenon of chromosome territories (Fritz 2014) refers to spatial control of gene expressions. Similarly, sequence of gene expressions wherein the initiators of gene expressions are synthesized prior to the expression of target genes. This is a type of temporal control of gene expressions. However, the conventional perspective doesn't postulate any genome wide template which subsumes all these different control elements. Similarly, the conventional perspective accepts the conception of operons, but it doesn't provide for any genome wide pattern of such operons. As mentioned above, there are two primary reasons for this lack of genome wide conception of control elements. Firstly, it is the apprehension that such a systemic explanation might introduce some design principles (Fodor and Piattelli-Palmarini 2011) through the back door. Secondly, our conception of genomes so far, has remained confined to three-dimensional stereochemical configurations. We do not define genomic architecture in any higher dimensional configurations. Therefore, our conception of control elements has remained local, based on individual genes and their expressions.

However, the proposed model postulates that genomes exist in multiple dimensionalities which are nested among themselves. This model prompts us to conjure up two tantalizing possibilities. Firstly, it is possible to define a genome wide framework connecting various control elements. Secondly, it is possible to think of more than two types of control elements. Therefore, in this section, we will explore both these possibilities. Let us begin with the first possibility. The key question is this: Is it because the conventional perspective has confined itself to the three-dimensional perspective of genomes that it finds it difficult to formalize a genome wide framework of control elements ? Or to put it differently, if there exists any higher dimensional model of genomes (not necessarily the proposed model), would it enable us to formalize the genome wide framework of control elements ? Leaving aside the semantic compulsions of the conventional perspective in not conceptualizing such a genome wide framework, the problem is essentially what additional dimensions can do ? Our mathematical modeling paradigm (Laudal 2021) suggests that we employ higher dimensional representations only when we wish to include more parameters, each representing some systemic properties.

However, the trouble with this approach is that it fails to explain the origin of these properties. Therefore, in this case, even if we employ such an approach, we will still have to explain the origin of the control elements. This is because while the proposed higher dimensional model might designate different control elements to different higher dimensionalities, the origin of these elements will still have to come from the stereochemical configurations. However, as seen in the conventional perspective, we can't do such a thing unless we introduce some transcendental effects. This is because in such a putative higher dimensional model, the higher dimensionalities represent some unknown information content which can only be transcendental in nature. Thus, mere invocation of a higher dimensional model, by itself, cannot provide a genome wide framework of control elements.

However, the proposed model is different from other higher dimensional models. The proposed model postulates that these higher dimensionalities are not virtual or mathematical abstractions. Rather, according to this model, higher dimensionalities are physical and are synonymous with spacetime dimensionalities. More importantly, according to this model, spacetime is not only a physical entity, but its information content is also physical. Moreover, since according to this model, different dimensionalities of spacetime are not randomly integrated into a single framework. Rather, according to this model, each dimensionality of spacetime is connected to its neighbors by a fixed mathematical operator. This ensures that whatever the nature of the higher dimensionalities may be, they devolve into the four-dimensional (the conventional stereochemical) spacetime in a fixed and predictable manner. Therefore, when we define a genomic framework of control elements, we are providing a fixed mechanism by which these higher dimensional genome frameworks of control elements influence the conventional stereochemical details of genomes. More importantly, since the information content of these higher dimensionalities is also physical in nature, when we define higher dimensional genome wide frameworks of control elements, we are not introducing any transcendental influences. Thus, it is possible to conceptualize a higher dimensional genome wide framework of control elements only if the model subscribes to mathematical realism.

Let us now look at the second question of the number and types of control elements that are implicit in the proposed model. Prima facie, if, as the proposed model postulates, spacetime participates directly in biological evolution, then it is axiomatic that the above mentioned higher dimensional influences would devolve into the four-dimensional spacetime only in the form of spacetime features. Thus, irrespective of the dimensionality from which these long-range influences originate, their eventual manifestation in the four-dimensional spacetime would be perforce in the form of temporal and spatial influences. At first sight, this inference places the proposed model in a redundancy. This is because the conventional perspective already accommodates these two types of long-range influences. Therefore, there is no need for any such complicated and unverifiable model. However, upon a little reflection, it is intuitively clear that this inference is not valid. This is because the conventional perspective, while admitting the possibility of temporal and spatial long-range influences, is not in a position to formalize these influences. Moreover, this inability of the conventional perspective to formalize these two types of long range is not due to lack of mathematical sophistication, but rather due to semantic compulsion. Therefore, the proposed model offers a semantic justification for formalizing these two types of long-range influences. More importantly, this model provides a more detailed account of these long-range influences. Let us see how.

According to this model, genomes exist in multiple dimensionalities simultaneously. Moreover, higher dimensional configurations devolve into the four-dimensional configurations of genomes wherein molecular biological processes operate. While it is true that these higher dimensional configurations of genomes upon devolvement, give rise only to two types of influences, viz., temporal and spatial, these influences need not be identical. Thus, while there are only temporal and spatial long-range influences that manifest in the four-dimensional configurations of genomes, their magnitudes would vary. This is because as mentioned above, the mechanism by which different higher dimensional configurations of genomes devolve into the four-dimensional spacetime is fixed. As a result, depending on the dimensionality from which these long-range influences originate, the magnitude of these long-range influences would be different. Moreover, since the underlying mechanism is the same, different long-range influences would manifest a fixed ratio of their magnitudes. Thus, for instance, if a long-range spatial influence were to arise from the tenth dimensionality and the distance at which this influence would be felt simultaneously (in the form of gene expressions) would be in an exact ratio with another spatial long-range influences arising from say, fifth dimensionality. As a result, if the proposed model is valid then there would be several long-range spatial influences, each manifesting at different distances as measured by genomic distances between two genes expressing simultaneously. Thus, there will be several pairs of genes which are spatially nonlocal but they still synchronize their expressions. The same rationale applies to temporal long-range influences. Thus, the proposed model suggests that there exist multiple long-range temporal and spatial influences. More importantly, these different spatial and temporal influences would manifest a fixed ratio of the magnitudes of their influences among themselves.

Admittedly, since a typical mammalian genome might contain a few thousand genes, it is not possible to determine any such ratio manually. However, with the help of microarray technology and good mathematical modeling, it should be possible to discern any such patterns, if they exist. Thus, the proposed model offers a way to verify itself. At this juncture, when neither genomic architecture is

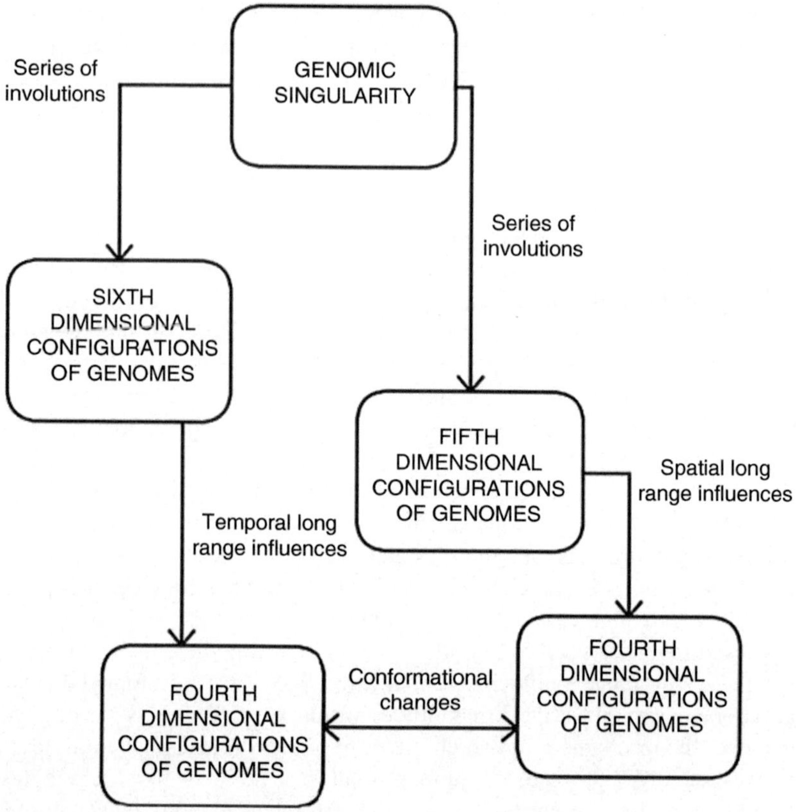

Schema 7.11 Devolvement of long-range influences

formalized and nor the necessary topology of the model elaborated, it is not possible to derive the exact numerical values of these ratios. However, we can conjecture a qualitative scenario. One such scenario is represented in Schema 7.11. It depicts the devolvement from the fifth, sixth, and seventh dimensionalities to the four-dimensional configuration of a putative genome.

7.13 Topological Model of Regulatory Genome

As discussed in the previous section, it is not possible to quantify the long-range influences that arise from the devolvement of the long-range influences onto the four-dimensional spacetime. However, this doesn't prevent us from discerning a general framework for long-range influences. Since according to this model, genomes exist in multiple dimensionalities nested among themselves, it is intuitively clear that the resulting genomic architecture need not manifest itself only in the form of long-range influences. In fact, according to this model, the long-range influences need not manifest themselves only at the four-dimensional configurations of genomes. Admittedly, the four-dimensional configurations of genomes are central to molecular biology of gene expressions. However, genomes could possess a separate framework of these different long-range influences. Therefore, it is intuitively clear that we could think of genomic architecture as consisting of two frameworks, one having a framework for regulating gene expressions and another framework for executing gene expressions. These two frameworks can be named regulatory genome and expressive genome. In this section, we will try to deconstruct the conception of a separate regulatory framework called regulatory genome as implicit in the proposed model.

The fact that the conventional perspective of genomics hasn't been able to formalize a regulatory framework of genomes suggests that if there exists such a template, it must exist in higher dimensionalities of genomes. Had such a framework existed in the four-dimensional configurations of genomes, there would have been some reports in literature. It is just because the conventional perspective doesn't admit the existence of higher dimensional configurations of genomes that a formal description of the regulatory genome is not available. At the same time, it must be admitted that the conception of regulatory elements is well established in the conventional perspective (Pevsner 2015). It is just that the conception of a genome wide framework of regulatory elements is missing. This is important because if the postulate of higher dimensional regulatory genome is valid, it must be congruent with the known regulatory elements tacitly accepted in the conventional perspective. This is because the entire domain of molecular biological perspective of gene expressions manifests itself in the four-dimensional spacetime. Therefore, any such hypothetical entity as a regulatory genome existing in higher dimensionalities, must eventually regulate the molecular biology of gene expressions. Therefore, we will look into two aspects of the regulatory genome in this section, viz., the topological representation of the regulatory genome and its devolvement into the four-dimensional genomic configurations where the molecular biology of gene expressions prevails.

In principle, it is possible that the regulatory genome itself could exist in multiple dimensionalities simultaneously. However, for the sake of simplicity, we will temporarily assign only a couple of dimensionalities to the regulatory genome. Moreover, the numerical value of the dimensionality occupied by the regulatory genome will be assumed to be five and six. Admittedly, the choice of regulatory genome occupying fifth and sixth dimensionalities is arbitrary. However, it is based on the premise that we need to define two types of long-range influences, viz., temporal and spatial. Since these two types of long-range influences are different in nature, according to this model, they must be placed in different dimensionalities. Though the ascription of temporal long-range influences to the sixth-dimensional genomic configurations and the ascription of spatial long-range influences to the fifth-dimensional configurations of genomes are arbitrary, it is consistent with the underlying mathematics of the proposed model.

With these a priori assumptions, let us deconstruct the nature of the regulatory genome. There are two aspects which require immediate clarity. The distribution of different regulatory features in these two dimensionalities and the relationships among these features within the dimensionality. Since according to this model, the higher dimensionalities possess coarser metrics as compared to lower dimensionalities, it is intuitively clear that we must assign sixth dimensionality to all those regulatory elements which are essentially temporal control elements. This is because in the four-dimensional spacetime, the complexity of time-like features is coarser than the corresponding complexities of the space-like features. Thus, it makes sense to assign sixth dimensionality to the temporal regulatory framework because when it devolves into the four-dimensional spacetime, its effects would be coarser. By the same logic, the spatial control elements must occupy the fifth dimensionality.

It is important to note that both these types of control elements must devolve into the four-dimensional spacetime simultaneously. Therefore, when we observe the long-range influences during gene expressions is actually a composite effect of both these types of regulations (and perhaps that is why it has eluded our attention). The next point to consider is the relationship among all the temporal regulatory elements and the relationships among the spatial regulatory elements. This is important for a couple of reasons. Firstly the internal relationships among all the temporal regulatory elements would in any case shape their eventual manifestation in the four-dimensional spacetime, thereby giving rise to a particular type of synchronization between different gene expressions. The same logic applies to the relationship among all the spatial regulatory elements. Secondly, since the proposed model postulates that all the genomic functionalities, including regulatory mechanisms, arise from the primordial entity named here as genomic singularity, it is imperative that these intra dimensionality relationships too must be related to one another. Thus, in reality, even the four-dimensional configuration of genomes carries the echoes of the configurations of genomes at higher dimensionalities. It is just that since there are multiple higher dimensional configurations of genomes, the resulting imprint on the four-dimensional configurations of genomes is beyond our cognitive capabilities of parsing. However, the significance of various higher dimensional configurations of genomes having interchangeable metrics lies in the fact that its existence points toward genomic singularity, thereby confirming the validity of genomic singularity.

Let us begin with different control elements that could occupy sixth dimensionality which upon devolvement into the four-dimensional spacetime give rise to temporal long-range influences in the form of synchronization. In the absence of any prior knowledge of the metric of the sixth dimensionality, we can work backward from our understanding of different temporal controls manifest in functional genomics. Ab initio, we can think of two types of temporal long-range influences, viz., simultaneous gene expressions of otherwise distantly placed genes without any common initiators and a particular sequence of gene expressions within a cluster of genes which do not share a common framework other than that of proximity. Therefore, our task is reduced to formalize how these features are obtained in the proposed model which postulates that all such long-range influences originate when the higher dimensional configurations of genomes devolve into the four-dimensional genomic configurations. Upon a little reflection, it is intuitively clear that there are two independent ways these two features of simultaneous gene expressions and a predefined sequence of gene expressions could arise. The simultaneity of gene expressions would necessarily arise when a control element present in the sixth dimensionality devolves into the four-dimensional spacetime in a symmetric manner. This is because in a topological framework the notion of measure doesn't exist. Therefore, if an element of control were to devolve from the sixth dimensionality into the four-dimensional configuration of genomes, it would do so identically at every point of the four-dimensional configurations. For this to happen, it is necessary that the alignment of this control element in the sixth dimensionality must be identical with the time-like features of the four-dimensional spacetime. In other words, it must possess a bidirectional alignment even in the sixth dimensionality.

As a logical corollary, the control elements that decide the particular sequence of gene expressions must have an alignment in the sixth dimensionality that is nonorthogonal to the direction of time-like features of the four-dimensional spacetime. Therefore, when such a control element devolves into the four-dimensional configurations of genomes, the net effect is the gradual devolvement leading to different gene expressions occurring in a particular sequence. Since the four-dimensional configurations of genomes are dynamic, the order of the sequence of gene expressions must depend on the degree of nonorthogonality of the control elements. Thus, for the observed temporal long-range influences to manifest, according to this model, the sixth dimensionality of spacetime must possess two discrete entities, one orthogonal to the time-like features of the four-dimensional spacetime and another entity which is nonorthogonal to the time-like features of the four-dimensional spacetime. Thus, when the sixth-dimensional configurations of genomes undergo involutions into the four-dimensional configurations of genomes, they give rise to two types of temporal controls of gene expressions, viz., the choice of the sequence of gene expressions and simultaneous gene expressions. This scenario is consistent with the postulate of the proposed model that higher dimensional configurations must have coarser metrics than those of lower dimensional configurations. Thus, at the four-dimensional spacetime we have a metric wherein two degrees of freedom of temporal movement are coupled with three degrees of freedom of spatial movements, while at the six-dimensional spacetime, there are only two degrees of freedom of movement, movement from orthogonal states to nonorthogonal states and vice versa.

This brings us to the spatial control elements present in the fifth dimensionality. It is apparent from the above discussion that the metric of the five-dimensional spacetime must have an intermediate range of degrees of freedom of movement. In other words, the five-dimensional spacetime must possess different control elements whose number must lie between two (as postulated for the six-dimensional spacetime) and four (as witnessed in the four-dimensional spacetime). However, before doing so, it is necessary to understand different spatial long-range influences observed in functional genomics. Ab initio, we can think of spatial long-range influences in the form of bringing together genes which are otherwise widely separated from one another. There are three instances of this type of long-range influences, viz., chromosome territories (Fritz 2014), cis and trans effects (Donaldson 2000). Admittedly, the conventional perspective of these three phenomena rests on conformational changes in the DNA sequences (Weinzierl 1999). Moreover, since conformational changes entail kinetic energy transfers which at ambient temperature are supposed to be random. Therefore, according to the conventional perspective these three phenomena are outcomes of random collisions and therefore, they lack any causal explanation. However, it is important to keep in mind that even in the conventional perspective, it is conceded that these three phenomena do not occur during random conformational changes. They occur only in a limited number of cases. This peculiar feature of spatial long-range influences remains unexplained in the conventional perspective. On the other hand, if we accept the above mentioned scenario wherein a higher dimensional configuration of genomes (fifth-dimensional configurations in this case) causes these spatial long-range influences, it is intuitively clear that these long-range influences would manifest only when the fifth dimensionality configurations of genomes devolve correctly into the four-dimensional spacetime. Moreover, upon a little reflection, it is intuitively clear that each of these spatial long-range influences have different structural requirements. For instance, chromosome territories and its influence on gene expressions require multiple sites at which otherwise distant DNA sequences must converge. On the other hand, in the case of cis and trans effects only DNA sequences must converge onto each other only at one point. Thus, these three long-range influences have different structural requirements of their own. In the case of chromosome territories, more than one chromosome must possess conformations that bring multiple genes in proximity of one another. In the case of trans effects, two chromosomes must possess conformations that bring one DNA each of these chromosomes to be in proximity of one another. In the case of cis effects, one chromosome must possess conformations which bring two DNA sequences in proximity of one another.

This obvious structural perspective finds a different interpretation in the proposed model. According to the proposed model, in the fifth dimensionality, chromosomes need not be noncontiguous. This is because the notion of distances in the four-dimensional spacetime is essentially a geometric notion. In the five-dimensional spacetime, the notion of distances acquires an interpretation of representing a degree of topological proximity. Thus, in fifth-dimensional configurations of genomes, genomes are contiguous objects (not segregated into different chromosomes). However, when the fifth-dimensional configurations of genomes devolve into the four-dimensional spacetime, several new features of genomic architecture manifest. Firstly, we see genomes being segregated into different chromosomes. Secondly, the four-dimensional spacetime manifests two types of dimensions, viz., the time-like and the space-like dimensions. Therefore, these three types of spatial long-range influences arise when a topologically contiguous genomes, upon devolvement into their four-dimensional configurations, become segregated into different chromosomes. Moreover, the kinetic energy transfers are nothing but the energy unleashed by the emergence of two types of dimensions of the four-dimensional spacetime from the undifferentiated dimensions of the five-dimensional spacetime.

Interestingly, we can assign different types of symmetries to these three types of spatial long-range influences. For instance, in the phenomenon of chromosome territories, as mentioned above, it requires more than one chromosome to be in proximity at more than one point. This can be thought to have arisen from a particular structural element of the metric of the fifth dimensionality configurations of genomes. Provisionally, we will call this structural element of the metric of the fifth dimensionality as a three-dimensional element. Similarly, the metric element of the fifth dimensionality giving rise to trans effects can be thought of as a two-dimensional element of the fifth-dimensional spacetime. Finally, the metric element responsible for the cis effects can be thought of as a one-dimensional element of the fifth-dimensional spacetime. Without going into the mathematical details, we can compare these three metric elements of five-dimensional spacetime with their analogues in the four-dimensional spacetime, viz., tensors, vectors, and scalars. Just as tensors, vectors, and scalars have their own internal templates, these three metric elements of the five-dimensional spacetime have their own internal templates. However, these internal templates become manifest as external features of genomic architecture when they devolve into the four-dimensional spacetime. A schematic representation of these temporal and spatial long-range influences is given in Schema 7.12.

Based on these discussions of the temporal and spatial long-range influences, let us deconstruct the exact nature of the regulatory genome that remains embedded in the genomic architecture. If the details given in Schema 7.12 are valid, it is important to define a regulatory genome that is consistent with the details given above. In addition, we need to define two additional categories of control elements. Apparently, every genomic regulation would either be spatial or temporal. Therefore, the new categories of control elements must have different features in addition to the features of spatiality and temporality. Prima facie, we can think of these new categories as being voluntary and nonvoluntary. Before we incorporate these categories into the template of the regulatory genome, let us try to deconstruct them. Voluntary regulatory element must be thought of as a control element which begins its operations only when a certain condition is fulfilled. On the other hand, a nonvoluntary control element must be thought of as a control element that sets in automatically when a particular genomic configuration is arrived at. There is no element of choice in this case. Let us return to molecular biology to distinguish between these new categories. Take the example of initiation of gene expressions. Let us assume that a gene expression has been initiated by the interaction of a messenger molecule with the DNA sequence upstream of the actual gene. Even if this happens, the process of gene expression doesn't begin unless the transcription machinery is already assembled. Interestingly, a similar situation exists when the transcription machinery is assembled but the messenger molecule is missing. It is intuitively clear that the process of a gene expression in this case is conditional. It requires a synchronization between two different processes, viz., the temporal long-range influence in the form of the synthesis of a messenger molecule which acts as an initiator and assembling of the transcription machinery.

When we reflect on this scenario, a few propositions become self-evident. Firstly, this scenario resembles the logic behind the conventional conception of operons, except that the process doesn't employ discrete logic. Secondly, this voluntary or conditional nature of decision making (whether the gene expression should be allowed or not) ensures that the inherent randomness implicit in the Darwinian

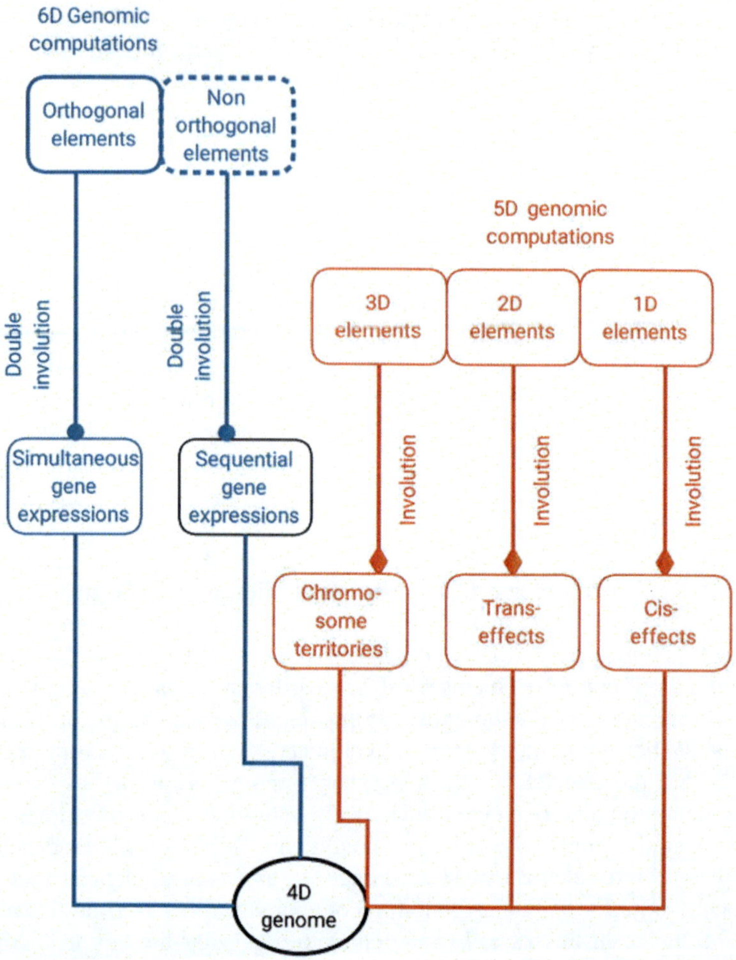

Schema 7.12 Topology of long-range influences

paradigm is retained. Thirdly, since the proposed model postulates that genomes exist in multiple dimensionalities simultaneously, these voluntary controls could manifest between different dimensionalities in parallel. This is because this form of voluntary control would manifest only when a genome undergoes the changes in its dimensionality. Once we accept that these voluntary control elements influence gene expressions through the changes in the dimensionalities of genomic configurations, it is intuitively clear that some of these changes in the dimensionalities would influence gene expressions unconditionally. This brings us to nonvoluntary control elements. These are regulatory control elements which influence gene expressions whenever there is a change in the dimensionalities of genomic configurations.

Thus, now we have four types of control elements, viz., temporal control elements, spatial control elements, voluntary control elements, and nonvoluntary control elements. The key question is this: how these four types of control elements are connected to one another ? The answer is not straightforward because there are overlaps between these four control elements. This is where the significance of multiple dimensionalities comes into the picture. Whenever there are changes in the dimensionalities wherein the numerical value of the changes dimensionality is one, we observe nonvoluntary long-range influences operating. Similarly, whenever the numerical value of the changes in the dimensionalities is more than one, we observe voluntary or conditional long-range influences operating. Similarly, as mentioned above, if the dimensionality change occurs from the sixth dimensionality to the fourth dimensionality, we observe temporal long-range influences operating. Likewise, when the dimensionality change occurs from the fifth dimensionality to the fourth dimensionality, we observe the spatial long-range influences. Interestingly, it is not difficult to realize that some of the temporal long-range influences could also happen through the involvement of the fifth dimensionality. In such a situation, the temporal long-range influences could become a source of the spatial long-range influences because the genomic configuration would occupy the fifth dimensionality on the way to the fourth dimensionality, albeit transiently.

However, upon a little reflection, it is intuitively clear that this would not happen. This is because as depicted in Schema 7.12, the metrics of the sixth and the fifth dimensionalities are different because they possess different degrees of freedom of movement. This explanation raises another interesting inference. The temporal long-range influences involve the changes in the dimensionalities of genomes whose numerical value is more than one. Therefore, it is possible that some of the temporal long-range influences could happen in two steps. In other words, there would be devolvement from the sixth dimensionality into the fifth dimensionality and in the next step, there would be devolvement from the fifth dimensionality to the fourth dimensionality. While as suggested in Schema 7.12, it is unlikely because of the difference between the metrics of the sixth and fifth dimensionalities. However, the possibility cannot be ruled out completely. If these two step transitions were to occur, it throws up an interesting possibility. This is because as mentioned above the direct devolvement from the sixth dimensionality into the fourth dimensionality entails the changes in dimensionalities the value of which is more than one. Therefore, these influences would be voluntary in nature. However, if the same transition were to occur in two steps (via the fifth dimensionality), then both these steps individually, would be involuntary in nature. This is the reason why we have two types of long-range temporal influences. The simultaneously long-range temporal influences possibly represent the direct one step devolvement from the sixth dimensionality into the fourth dimensionality. Similarly, the sequential long-range temporal influences occur through the two step process via the fifth dimensionality. This reasoning makes sense because of the inherent differences between the metrics of the sixth- and fifth-dimensional configurations of spacetime. Because of this difference, during the first step of these two step processes, the devolvement spreads over a range of DNA sequences in the spatial configurations of the fifth-dimensional

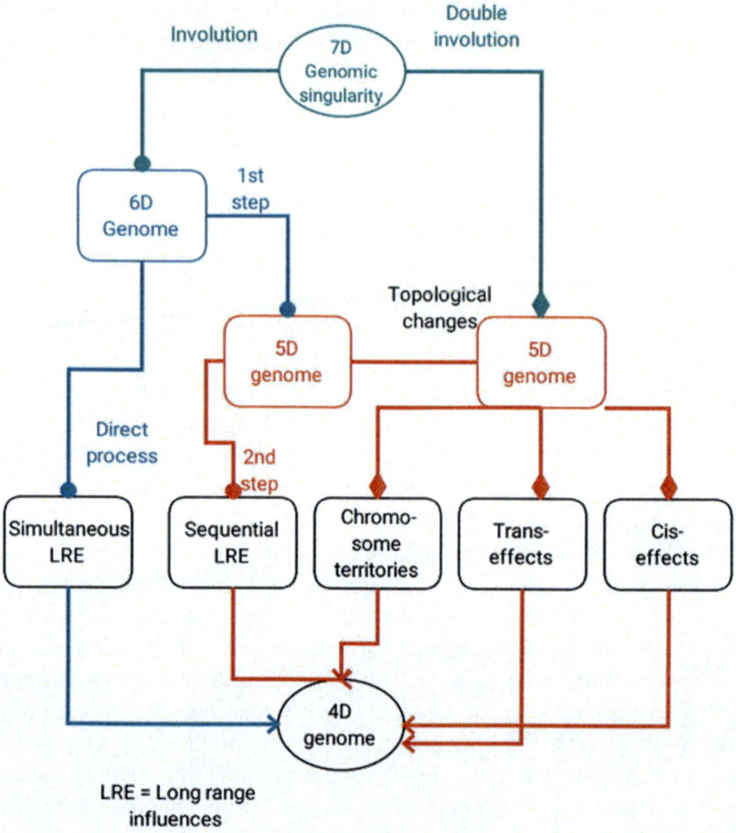

Schema 7.13 Topology of regulatory genome

configurations. Therefore, when the second step is carried out, we get sequential gene expressions. The exact choice of the sequence of gene expressions perhaps reflects spatial orientations of the sequences of different genes in the fifth dimensionality.

Thus, we can now discern a detailed view of the regulatory genome as embedded in the higher dimensionalities of genomes. This scenario is depicted in Schema 7.13.

This model of regulatory genome needs to be analyzed from the perspective of functional genomics. However, to the extent it formalizes different types of long-range influences which play a critical role in synchronization of different gene expressions, the model must be evaluated from the pathologies arising from the lack of synchronization. Since cancer, as a group of pathologies, represents a typical example of asynchronization, it provides an ideal candidate to validate the proposed model of regulatory genome. Therefore, in the next section, we will try to deconstruct cancer using the proposed model of regulatory genome.

7.14 Cancer According to the Proposed Model

In preceding sections, it was pointed out that cancer arises from the lack of synchronization between different gene expressions. It is true that the primary source of cancer lies in mutations of individual genes (Madhusudan and Wilson 2013). However, it is important to keep in mind that while the source of cancer lies in mutations, the source of cancer pathologies lies in the genomes that try to prevent the errant cancer cells from proliferating. This is because mutations are unwelcome but inevitable consequences of transcription. Most of these mutations lead to various corrective measures, failing which, there is always apoptosis. Thus, cancer must be viewed as an outcome of failure to prevent cancer cells from proliferating. This is where genomic architecture comes into play. Genomic architecture essentially decides when and where which gene must express itself. As discussed in Chap. 6 and in the preceding sections, the need for genomic architecture arose only after the emergence of multicellularity. Unicellular organisms do not manifest cancer pathologies. Therefore, the evolution of genomic architecture and its attendant complexity play a critical role in carcinogenesis. It is reasonable to think that since the evolution of genomic architecture (and its complexity) didn't have any premeditated design principles, every additional complexity in genomic architecture would bring about undesirable consequences.

In the context of the present discussion, it is important to note that the conventional perspective of cancer and its pathologies is correct in the sense that ultimately, it is the improper gene expressions or the expressions of harmful genes that cause cancer. However, as discussed in the preceding chapters, functional genomics has eschewed exploring the possibility of genomes having higher levels of organization. This is primarily because there is an apprehension that such a modeling would be antithetical to the randomness implicit in the Darwinian paradigm (Bonner 2013). Admittedly, this belief is further strengthened by the fact that ultimately, it boils down to aberrant genes and their expressions. However, once we concede that cancer pathologies arise from asynchronization of gene expressions, it is intuitively clear that the feature of synchronicity (or the lack of it) cannot arise at the level of individual genes. Thus, genomic architecture becomes *fait accompli*. However, our attempts to formalize genomic architecture haven't succeeded because we don't find any evidence of it in phylogenetic studies. At best, we can discern some modularity of some cluster of genes present in in vivo a fixed order and operating in tandem. Beyond that modular description, there is no genome wide framework of genomic architecture. However, the proposed model offers an opportunity to test our belief that cancer is related to genomic complexity as manifest in genomic architecture.

In the preceding sections, a tentative scenario of various long-range influences and their interactions was outlined. Therefore, it is time to try to correlate this scenario with cancer pathologies. Once again, instead of going into the molecular details of cancer pathologies, we will try to understand whether the implicit topological framework really explains the known cancer pathologies. Therefore, in this section, we will try to deconstruct the relationship between different cancer pathologies and different types of long-range influences. More importantly, we will try to understand whether the topological representation really helps us to define the relationship between complexity and cancer (Dellaire et al. 2014). This is necessary because it can be legitimately argued that the relationship between complexity and cancer is a mainstream idea and that there is no need to interpolate topology in this. Therefore, the focus here will be on the proposition that topological perspective is not an afterthought but the missing link in our understanding of the evolutionary perspective of cancer. Therefore, in this section, we will discuss four specific propositions of the relationship between complexity and cancer, viz., (1) Synchronization of gene expressions cannot be achieved in linear or geometric models of genomes. (2) Synchronization of gene expressions can only be achieved in topological models of genomes, (3) There is a direct relationship between different higher dimensionalities and different types of genomic complexities, and (4) Different types of cancer can arise from different types of complexities which in turn are defined by different dimensionalities.

Let us begin with the first proposition, viz., synchronization of gene expressions cannot be achieved in linear or geometric models of genomes. Prima facie, we know from our experience in computation theory (particularly the Turing machine paradigm; Herken 1995) that we can only formalize sequential operations in any algebraic operations. Any process that operates in higher dimensions can only be approximated. Thus, at best, we can formalize multiple processes only as parallel processes, but never as simultaneous processes. Therefore, even if we discard the proposed model and adhere to the conventional perspective of genomes, it would be very difficult to formalize synchronization of gene expressions for the four-dimensional spacetime (in which stereochemical and conformational configurations of genomes exist), it will be difficult to formalize synchronization of gene expressions. It is possible to argue that even though we can't compute the details of this synchronization between different gene expressions, we can at least obtain a rough outline of such a genomic architecture. However, the fact that we have not been able to formalize long-range influences, by itself, suggests that we need some higher dimensional perspective of these long-range influences to formalize them. The basic thumb rule of modeling (Laudal 2021) suggests that in order to formalize any phenomenon, it is necessary to visualize it from outside. Therefore, in order to obtain such an external perspective, it is imperative that we conceptualize that phenomenon from a higher dimensionality. Therefore, if the long-range influences operate in four-dimensional spacetime, they can't be formalized from the four-dimensional perspective. It is imperative that we obtain a higher dimensional perspective of these long-range phenomena. The trouble with formalizing a higher dimensional formalization is that it can't be formalized as a geometric construct. The Hilbert space is the best example of this. One of the biggest advantages of the Hilbert space is that it helps us to create geometric constructs in higher dimensionalities. However, when we attempt to do so, we end up with a set of solutions, called nonorthogonal states, which are left out of computations. Thus, even if we assume that the long-range influences are essentially electromagnetic in nature, any attempt to formalize the geometric paradigm would be incomplete. Thus, neither linear, nor geometric formalization can give us a correct description of genomic architecture.

Let us now look at the second proposition, viz., synchronization of gene expressions can only be achieved in topological models of genomes. As discussed above, let us concede that any geometric representation of genomes would be unable to formalize synchronization of gene expressions. However, that inference, by itself, cannot justify the assertion that topological approach, per se, is suitable for formalizing synchronization. Therefore, let us understand why a topological framework is a better approach to formalize synchronization. One of the most characteristic features of a topological framework is that it dispenses the notion of distances. In any geometric ensemble, we would require a notion of metrics. The notion of metrics can be thought of as a sum of relationships between neighboring points. Thus, in geometry, we think of a line as a collection of geometric points having fixed relationships among themselves. This rationale connects geometric points with any complex geometric constructs. Thus, a description of metrics is a prerequisite of any geometric model. However, in topology, we don't have a geometric point. Therefore, we don't need the conception of metrics to define any topological framework. Instead, in topology, it is the conception of a surface that is a prerequisite. Topology can be viewed as a set of relationships between topological surfaces just as geometry can be viewed as a set of relationships between geometric points. Thus, it is possible to build a topological framework in any dimensionality. However, as mentioned above, there is a price to be paid for this freedom of formalizing in any dimensionality. It comes in the form of reduced computational power.

However, as discussed in the preceding monograph (Chhaya 2020), a new topological model having comparable computational capabilities is now available. Therefore, it should be possible to define formal description of long-range influences and synchronization between different gene expressions, using this model. Admittedly, since the model of genomic architecture using this model is described in this monograph is in a very nascent stage, we can't formalize the process of synchronization. However, it is possible to arrive at qualitative scenario using this model. This brings us to the third proposition, viz., there is a direct relationship between different higher dimensionalities and different types of genomic complexities. Prima facie, at least from the mathematical perspective, it is intuitively clear that as the dimensionality of a given model increases, the formal representation of the phenomenon under investigation becomes more complex. This is partly because as dimensionality increases, the number of parameters also increases. Therefore, the formal representation develops more complex equations to accommodate different parameters. Secondly, assuming that the phenomenon under investigation is inherently complex, then more and more features of that phenomenon get incorporated into its formal representation as the dimensionality of the model increases.

However, from the perspective of the phenomenon under investigation, the situation is quite different. The complexity of that phenomenon is directly related to the precision of the measurements that we can make. If we can't make more accurate observations, we can't formalize the complexity of such a phenomenon. Quantum phenomenon is a typical example of this situation. We intuitively feel that quantum phenomena must be very complex. However, it is the limitations of accurate measurements that prevent us from conceptualizing higher dimensional or more complex representation of quantum phenomena. Mathematical sophistication, by itself, cannot provide a better understanding of any natural phenomenon. This rationale applies to genomic architecture as well. It is one thing to postulate a complex topological model like the proposed model, but it would amount to nothing more than sophistry if it can't be applied to decode the complexity of phenomena like cancer.

With the preamble, let us look at the basic postulate of the proposed model that each dimensionality represents a different type of complexity. Admittedly, the underlying mathematics justifies this postulate. However, it needs to be reflected in the genomic architecture as well. The most sensible way to defend the proposition that genomes possess different types of complexities (which can manifest in different dimensionalities) is to provide an evolutionary account of such a distribution of different types of complexities. We will discuss it in Sect. 7.15. Presently, let us look at the question whether our present knowledge of functional genomics allows the possibility of different types of complexities coexisting in genomes. Once we discover that, we can justify their manifestations in different dimensionalities. Upon a little reflection, it is self-evident that the very conception of functional genomics rests on a hierarchy of different modules. Once we accept the modular design of genomes, it is axiomatic that different modules would possess different degrees of complexities. Even otherwise, if genomes were to function as a unit, it is imperative that its different modules (even if they possess a uniform level of complexity) must operate in synchronization. Therefore, the very mechanism of synchronization between different modules would create a hierarchy which itself introduces an additional degree of complexity. Thus, to the extent our conception of functional genomics rests on the modular framework, different types of complexities are implicit in that conception of functional genomics. Therefore, it seems reasonable to think that genomes contain different types of complexities embedded in a single framework.

The next question is: Whether different types of complexities of genomes justify their placement into different dimensionalities ? To be honest, the answer to this question is yes, at least from the conventional perspective. This is because the conventional perspective employs mathematical constructs as tokens for different units of genomic architecture. Since the conventional perspective employs mathematics in an abstract sense, there is no problem with assigning different dimensionalities to different types of genomic complexities, say, different modules. The only proviso to this approach is that this assignment of different dimensionalities to different modules must lead to a certain degree of predictive power of how these modules behave in reality. However, this is not the case with the proposed model. The proposed model insists that different dimensionalities do not represent some abstract mathematical constructs. Rather, the proposed model postulates that different dimensionalities are physical in nature and that they represent spatiotemporal dimensionalities. Therefore, if we wish to justify the assignment of different dimensionalities to different modules, we are required to provide a physical explanation of how different types of complexities must manifest in different dimensionalities. Let us see how such an explanation is available from the proposed model.

According to the proposed model, there are two postulates that justifies the assignment of different dimensionalities to different types of complexities (and therefore, to different modules of genomes). Firstly, the proposed model postulates that matter (in the form of molecules, including DNA sequences) and spacetime are isomorphs. Therefore, any molecule including DNA sequence, must be thought of as an entity spread over multiple dimensionalities simultaneously. Therefore, if different dimensionalities of spacetime were to possess different metrics, then naturally each molecule would possess different metrics in each dimensionality. Therefore, each DNA sequence would manifest different structuralism in different dimensionalities. The dimensionality in which each molecule exists depends on the number of electrons present in its highest occupied molecular orbital. The second postulate of the proposed model that justifies the assignment of different dimensionalities to different genomic modules, states that spacetime itself participates directly in biological evolution and natural selection. Thus, once we accept that genomes exist in modular configurations, according to this model, it is intuitively clear that each module must occupy different dimensionalities. Even in the conventional perspective, the very conception of modularity rests on different functionalities. Therefore, it is intuitively clear that even in the conventional perspective, each module would have different types of complexities. The only difference is that in the proposed model, each type of complexity is assigned a different dimensionality. This is because according to this model, spacetime itself participates directly in biological evolution and natural selection.

This brings us to the fourth proposition, viz., different types of cancer arise from different types of complexities. This proposition is not obvious at first sight. However, it arises naturally once we accept the topological representation of genomic architecture. Let us understand how. To begin with, let us distinguish between primary sources of cancer from the emergence of cancer pathologies. Let us accept that it is random mutations that cause cancer. However, as argued above, it is the failure of genomic control elements that allows malignant cells to proliferate and migrate. Thus, it is not the mutations in genes that directly give rise to cancer pathologies. Cancer pathologies arise only when genomic mechanisms fail to execute apoptosis, to prevent cell divisions of malignant cells and to prevent their migration. Once we accept this reasoning, it is intuitively clear that these different types of failure of genomic mechanisms arise from the lack of different types of synchronization. Moreover, as discussed above, different types of synchronization arise from different dimensional configurations of genomes. Therefore, it is axiomatic that if different types of cancer pathologies arise from the lack of different types of synchronizations, then it ought to be possible to categorize different types of cancer pathologies depending on the type of synchronizations. This classification has an added advantage. This advantage arises because according to this method, different types of cancer pathologies can now be placed in a topological hierarchy. Since this topological hierarchy also represents biological evolution of genomic architecture, this classification of cancer pathologies also provides insight into the evolutionary relationship among different types of cancer pathologies. Admittedly, a detailed description of such a classification is beyond the scope of this monograph, but it would be discussed in the forthcoming monograph dealing with the application of the proposed model to genomic therapeutics. While the conjecture that different types of cancer pathologies have an evolutionary perspective is conventionally accepted, the proposed model offers a slightly different perspective on it. We will discuss it in the next section.

7.15 Evolutionary Perspective of Cancer According to the Proposed Model

As discussed in the preceding sections, the emergence of cancer and its pathologies can be linked to the emergence of complexity during biological evolution (Niklas and Newman 2016). This is particularly obvious in the case of complexities arising from emergence of multicellularity during biological evolution. If we think of multicellularity as a primordial type of complexity, then it is axiomatic that different types of morphological features or body plans must have branched away from this primordial complexity (Cabej 2020). Therefore, it seems reasonable to think that these different body plans and their morphological features are shaped by genomic architecture. In that case, it is intuitively clear that different types of complexities must be built into genomic architecture. Of course it is debatable whether every detail of body plans or morphological features is completely determined by genomic architecture. Probably not. However, to the extent that genomic architecture decides these details, complexity per se must be built into the genomic architecture. Moreover, since the Darwinian paradigm doesn't accept any design principles, it is also axiomatically true that different types of complexities must have arisen because of natural selection. Thus, branching out of different types of complexities from the primordial multicellularity must have given rise to several unforeseen consequences of each type of complexities. Therefore, even in the conventional perspective, there is a tacit acceptance of the relationship between complexity and cancer.

In the conventional perspective, there is another explanation for the relationship between complexity and cancer. Natural selection is conceptualized as an ergodic phenomenon. This is best illustrated by the phenomena like punctuated evolution (Gould 2007) and speciation (Coyne and Orr 2004). Appearance of new morphological features has emerged rather randomly. Pre-Cambrian explosion of morphological diversity is a typical example of this ergodicity (Cabej 2020). From the genomic perspective, this situation creates a different set of problems. This is because genomic architecture now has to add additional mechanisms of synchronization. Nature achieves this target by modifying the existing mechanisms of synchronization. This enlargement of scope of existing mechanisms of synchronization or their modifications to include new targets, creates a new set of vulnerability in genomic controls. This scenario can be easily understood if we think of genomic controls as an operating system of modern day computers. As more and more applications are introduced in a computer, the operating system has to be modified to accommodate the additional software. As the number of applications increases, the operating system becomes vulnerable to external hacking or internal systemic glitches. Of course, it is possible for the company manufacturing the operating system to provide periodic patches to secure the operating system. Eventually, the operating system has to be replaced by a newer version. Sadly, neither additional security patches, nor updating to new versions is possible in the case of genomic architecture. Genomic architecture, irrespective of its strengths and weaknesses is *fait accompli*. Neither Nature, nor we can update it. This inability to revise genomic architecture gives another reason why cancer is more likely to manifest as genomic architecture becomes more and more complex.

However, the conventional perspective of cancer on the relationship between complexity and cancer stops here. Different types of cancer arise due diverse causes from aberrant genes to ineffective controls of synchronization. There is no grand scheme in which every type of cancer is linked to one another. Admittedly, the conventional perspective offers a way to classify different types of cancer. However, these methods of classification are based on the individual genes or on the basis of pathologies (Mohammed 2018). There is no way to classify different types of cancer either from the genomic perspective or from the evolutionary perspective. The proposed model however takes a different view on the evolutionary perspective of cancer. To begin with, the proposed model postulates a notional entity called genomic singularity. While the existence of genomic singularity is debatable, its semantic implications are important. According to this model, all the manifest functionalities of genomes are present in incipient forms in the genomic singularity. This scenario is very similar to the assertion that all the details of the manifest spatiotemporal universe are present in the cosmic singularity. Leaving aside how these complexities (functionalities in the case of genomic singularity and a plethora of natural phenomena in the case of the cosmic singularity) could be present in a singularity, let us look at the semantic implications of this scenario. Firstly, the postulate of genomic singularity provides a common ontology. This necessarily means that different functionalities share a common framework. This common framework refers to the mechanisms underlying diverse morphological features. It is important to keep in mind that the implications mentioned so far are integral parts of the conventional wisdom. There could be a plethora of morphological features but their expression patterns are controlled by the same basic mechanisms like enzyme inhibition (Copeland 2013, see Chapter 3) or servo feedback mechanisms (Elsasser 2016). This argument can be extended to higher elements of organization like operons and modules. Effects of different operons could be operational in different cytoplasmic conditions, but the mechanism remains the same. Therefore, the postulate of genomic singularity has always been implicit in the conventional perspective, at least in this sense.

Where the proposed model differs from the conventional perspective is that it postulates that all manifest and potential functionalities are present in genomic singularity in some incipient forms. The reason why the proposed model differs from the conventional perspective is that the proposed model asserts that it is the abstract structural templates that constitute genomic singularity and not the molecular template. On the other hand, the conventional perspective asserts that it is the molecular template that constitutes the genomic architecture. Therefore, it is very difficult (in fact, impossible) in the conventional perspective to conceptualize genomic singularity. Now let us look at the semantic implications of this conception of genomic singularity as a mother template of all abstract templates behind genomic functionalities (and their complexities). Upon a little reflection, it is intuitively clear that not only different functionalities share a common framework, but they also have the potential to replace one another. This is precisely what we observe in phylogenetic studies (Bromham 2008). Think of any exaptation (Gould 2002), it couldn't have happened had it not been for a common abstract structural template. Alternatively, think about neuronal plasticity (Tracy et al. 2015; or synesthesia; Simner 2019). These phenomena manifest because the molecular templates are incidental, but underlying abstract structural templates are essential.

Returning to the present discussion, let us at least temporarily accept that there exists a notional entity called genomic singularity. Now let us look at how the postulate of genomic singularity alters our conception of cancer and its evolutionary perspective. To begin with, genomic singularity occupies the highest dimensionality and it is nonstructural. As biological evolution unfolds, genomic singularity undergoes a series of involutions due to the environment. At each step, the dimensionality of genomic singularity is reduced by one. Thus, we can think of the earliest living organisms as having arisen from genomic singularity after a few involutions. With the passage of time, genomes of these earliest living organisms undergo a series of involutions. There are two parallel processes happening at each stage. Firstly, genomes undergo structural changes due to involutions and more and more genomic functionalities become manifest, with the concomitant increase in genomic complexity and perhaps increased intricacies in genomic architecture. Secondly, in parallel with this process, a second series of involutions unfolds between the environment and the organisms. While the first process entails changes in genomic architecture, the second process results in natural selection. The key point is that while both these processes are essentially involutive in nature, the first process leads to the changes in genomes which are governed by the topological compulsions of the involuted manifold, the second involutive process in which the environment interacts with the living organisms decides the survivability of the changed genomic architecture. This parallel existence of two different but structurally similar involutive processes is at the heart of Darwinian dualities discussed in the preceding chapters.

While the interactions between the environment and living organisms has been central to natural selection, the transformation of genomic singularity into more and more complex genomes has been conventionally overlooked. This explains why Darwinian theory can explain the process of natural selection, but it can't explain biological evolution. Traditionally, we have thought that these two processes are synonymous, if not identical. However, according to this model, both these processes share a common involutive framework, but they possess different semantics. This difference in their semantics arises because the first involutive process entails self reference because genomic architecture reacts within itself. Therefore, it is governed only by the inherent topological compulsions of genomic singularity. On the other hand, the second involutive process doesn't involve direct self reference and therefore obeys Bayesian logic (Press and Clyde 2003), giving rise to randomness implicit in the Darwinian semantics. It is possible to argue why these two processes should occur together. After all, the involutive changes in genomic singularity can happen independent of the second involutive process which is apparently interactive.

Therein lies the secret to the evolution of Life. As mentioned above, the second involutive process doesn't employ direct self reference. However, it employs indirect self reference. This indirect self reference arises because as mentioned in the preceding monographs and in the preceding chapters, spacetime plays an active role in both these involutive processes. In the first process there is a direct self reference because spacetime shares a common structural template with the higher dimensional configurations of genomes. This is because according to this model, matter (in the form of atoms and molecules, including biomolecules) and spacetime are isomorphs. Therefore, spacetime as woven into the higher dimensional configurations of genomes interacts with the higher dimensionalities of spacetime itself. Thus, the direct self reference. In the case of the second involutive process involving the environment and living organisms, the environment is linked to spacetime and so are the living organisms. However, the environment and the living organisms are not directly linked. Therefore, from the perspective of spacetime, the self-reference happens twice, but not simultaneously. Therefore, the indirect self reference. It is important to keep in mind that these two involutive processes occur in conjunction and not separately, otherwise genomes would be independent of the environment. Thus, the parallel existence of two involutive processes explains the Darwinian dualities, viz., phenotype/genotype, units of selection/units of inheritance, functionalities/structuralism, and DNA/RNA.

Returning to the present discussion, let us see how this scenario wherein genomic architecture evolves only under the influence of the topological compulsions. There are two important inferences available from it. Firstly, the changes in genomic architecture are not random. Secondly, different changes in genomic architecture share a common ontology and therefore, form of a pattern which is amenable to classification. Once we place different changes in genomic architecture in an evolutionary tree, a few propositions suggest themselves. Firstly, each variation in genomic architecture would give rise to a different type of complexity. Secondly, the underlying topological compulsions ensure that only a few types of control elements can manifest in each type of complexity of genomes. Therefore, in a modular design of genomes, there would always be some control elements which are incongruent with one another. Therefore, at each level of modularity there would always be control elements which are mutually incompatible. Since each level of modularity is associated with different types of complexities, the relationship between complexity and cancer emerges naturally from the proposed model. The mutual incompatibility of control elements would lead to lack of synchronization between different gene expressions, thereby perpetuating malignancy of individual cells, which in turn lead to cancer pathologies. Admittedly, the sheer size of the human genome and the huge

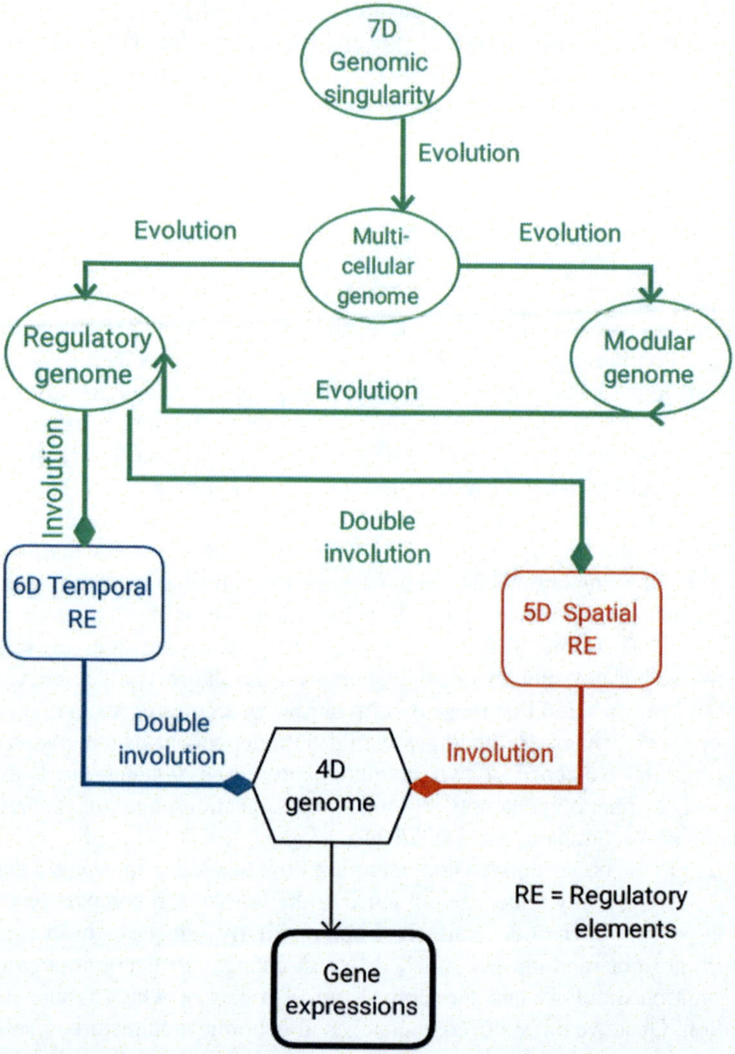

Schema 7.14 General map of genomic architecture

number of identified genes in the human genome precludes us from formalizing a detailed description of genomic architecture as well as that of genomic complexities. However, a tentative scenario of genomic architecture is depicted in Schema 7.14. It is important to keep in mind that the schemas depicted so far, are illustrative representations of biological evolution according to the proposed model. Since these schema do not contain any molecular or morphological details, they must be viewed as abstract representations of the topological perspective of the proposed model. It is legitimate to wonder whether the proposed model has any practical and immediate relevance to cancer research. This is because many of the topics covered in these discussions are not only theoretical, but they are also unverifiable. Therefore, in the next section, we will try to take a pragmatic view and see whether the proposed model has any tangible benefits to offer in therapeutic approaches to cancer.

7.16 Therapeutic Possibilities of the Proposed Model

There is an old dictum: The proof of the pudding lies in eating it. By and large, this is true for scientific theories. Admittedly, there are two notable exceptions to this. The Darwinian paradigm which is at the heart of this monograph and the general theory of relativity. Both these paradigms have certain built-in majestic grandeur and once understood, they are self-evidently true. These two paradigms define our modern scientific thought and have inspired generations of scientists. These paradigms enjoy preeminence not because they are useful, but their preeminence arises from the fact that they are self-evident. However, this is not the case with the rest of the scientific theories (much less can be said about the proposed model). Therefore, if the proposed model has to gain legitimacy, it needs to offer tangible and verifiable predictions. Therefore, in this section, we will try to envisage some realistic strategies that could help scientists to verify (or even falsify) the basic premise of the proposed model. At the outset, it must be mentioned that the therapeutic possibilities mentioned here would not be evaluated in this monograph. Rather, the objective is to articulate what are the practical consequences of the proposed model. For this purpose, we will selectively focus on three areas, viz., (1) classification of cancers, (2) prognostic strategy, and (3) therapeutic strategy.

Let us begin with the classification of cancers. Prima facie, we have at present, fairly accurate classification of cancers based on their molecular imprints. Therefore, the objective here is not to replace it, but to augment it. Since the proposed model postulates higher dimensional configurations of genomes, it is intuitively clear that genes that are otherwise widely separated from one another would be juxtaposed in the higher dimensional configurations. Moreover, since different higher dimensional configurations of genomes arise from genomic singularity, the adjacency of otherwise separated genes would be an outcome of particular evolutionary pathways. This is also manifest in the case of degrees of complexities of different modules. Therefore, to the extent aberrations of control elements are responsible for cancer pathologies, it is intuitively clear that ontological perspectives of different types of cancers would also be witnessed in the higher dimensional configurations of genomes. Thus, if we could create maps of higher dimensional configurations of genomes, a new set of evolutionary linkages between different types of cancers would be discerned. This provides an alternative method of classification.

This brings us to the prognostic perspective of the proposed model. At present, we screen individuals for suspected genes and their mutations to determine their likelihood of developing cancers. The proposed model offers a multidimensional map of individual genomes. Therefore, it is always possible to define more accurate association studies to locate the higher dimensional control elements of synchronization and their failures. Therefore, at least in principle, the proposed model offers a qualitatively and quantitatively improved mapping to identify individuals vulnerable to potential cancer pathologies. However, there is a caveat. We need to develop software capable of generating various higher dimensional configurations of genomes from their DNA sequences.

Finally, let us look at the therapeutic possibilities of the proposed model. Prima facie, as discussed in the preceding chapters and in the preceding sections, cancer is a collateral damage that Nature incurs while incorporating diverse functionalities into genomes. Therefore, to an extent, onset of cancer pathologies is a *fait accompli*. However, there are two possible ways that the proposed model offers to alleviate the problem. Since according to this model, various modules operate from different dimensionalities, it is evident that various control elements that maintain synchronization operate as mechanisms connecting different dimensionalities. More importantly, since the higher dimensional configurations of genomes eventually depend on the length of DNA sequences, it is axiomatic that by altering the length of DNA sequences, it should be possible to alter the higher dimensional configurations. Therefore, by introducing DNA sequences incapable of being transcripted, into target cells, it should be possible to alter the higher dimensional configurations of genomes. These alterations can be manipulated to restore synchronization between different gene expressions. Admittedly, there are lots of nuances of altering DNA sequences to alter genomic architecture. They will be discussed in the forthcoming monograph dealing with the therapeutic approaches arising from the proposed model. This brings us to the end of the discussion on the evolutionary imprints on cancer. In the final section, we will try to summarize the topics discussed in the preceding sections.

7.17 Conclusion

In the preceding sections, we have covered diverse features of the evolutionary perspective of cancer. Therefore, for the sake of simplicity, we will summarize these discussions in a point-wise manner.

1. The conventional perspective of genomics is ambivalent about genomic architecture. On the one hand, it accepts modular design based on the experimental evidence, on the other hand, it is reluctant to accept such a fixed template of genomic architecture for the fear that it might bring back some design principles.
2. If genomic architecture shows legacy argument, it is reasonable to think of genomes themselves being units of selection just as individual genes are. However, if genomes are units of selection, it necessarily implies that natural selection must possess a structural template of its own which selects only a certain type of complexity.
3. The conventional Darwinian paradigm cannot explain the emergence and natural selection of complexities. However, ascription of a structural template to the

process of natural selection, solves the problem of explaining the emergence of complexity during biological evolution and natural selection.

4. Once we accept the native structuralism of natural selection, it is intuitively clear that the evolution of functional modularity in genomes must be reflected in phylogenetic studies. Therefore, there must be a parallel domain of phylogenomics to confirm the native structuralism of natural selection.
5. The functional modularity necessarily involves mechanisms of synchronization. Therefore, phylogenomics must focus on the origin and nature of different processes of synchronization.
6. The proposed model offers a way to formalize the structural template of natural selection and emergence of complexity during biological evolution and natural selection.
7. For this purpose, it postulates a notional entity called genomic singularity in line with the entities like mitochondrial Eve and LUCA (Last Universal Common Ancestor).
8. The model postulates that genomes exist in multiple dimensionalities simultaneously with each dimensionality representing different degrees of complexities. Thus, different functional modules occupy different dimensionalities depending on their degrees of complexity.
9. Thus, the process of natural selection can be represented as the changes in dimensionalities. In this model, genomic singularity occupies the highest dimensionality and the conventional description of genomes occupies the lowest fourth dimensionality.
10. This topological framework reduces the survivability to topological congruence. Therefore, different types of complexities would have different types of congruence and therefore, different potentials for pathologies.
11. Since genomes according to this model exist in multiple dimensionalities simultaneously, higher dimensionalities exert their influences on the fourth-dimensional configurations of genomes through an operator of involution. Thus, different higher dimensionalities simultaneously influence the four-dimensional configurations of genomes. This gives rise to long-range influences.
12. According to this model, while cancer originates from the mutations of individual genes, cancer pathologies arise from the lack of synchronization of these long-range influences.
13. The model defines different temporal and spatial long-range influences using the operator of involution. This scenario gives us a template of the regulatory genome.
14. Since different modules having different degrees of complexities are placed in different dimensionalities, it is intuitively clear that they operate through different types of long-range influences.
15. This provides a way to link different types of cancers to the lack of synchronization of different types of long-range influences. Since different types of long-range influences arise from different dimensionalities and from different types of complexities, it is possible to correlate different types of cancers to different types of complexities.

16. Since the four-dimensional configurations of genomes are linear in nature, it provides a way to alter higher dimensional configurations of genomes (and therefore the nature of long-range influences) by inserting inert nucleotide sequences in genomes. This opens up a new approach to ameliorate cancer pathologies.

References

Alon A (2021) Computational genomics. CRC Press, Boca Raton, FL
Appasani K (2016) (ed.) Genome wide association studies: from polymorphism to personalized medicine. Cambridge University Press, Cambridge
Aubrey BJ, Kelly GL, Janic A, Herold MJ, Strasser A (2018) How does p53 induce apoptosis and how does this relate to p53-mediated tumor suppression. Cell Death Differ 25:104–115
Bard J (2016) Principles of evolution: systems, species and the history of life. Garland Science, New York, NY
Bonner JT (1988) The evolution of complexity by means of natural selection. Princeton University Press, Princeton, NJ
Bonner JT (2013) Randomness in evolution. Princeton University Press, Princeton, NJ
Bromham L (2008) An introduction to molecular evolution and phylogenetics. Oxford University Press, Oxford
Cabej NR (2020) Epigenetic mechanisms of the Cambrian explosion. Academic Press, London
Capuzzo-Dolcetta RA (2019) Classical Newtonian gravity: a comprehensive introduction. Springer, Dordrecht
Chhaya P (2020) The universe within: and a view from that within. Universal Press, Irvine, CA
Chhaya P (2022a) The origin and nature of spacetime
Chhaya P (2022b) The origin of the counterintuitive nature of quantum phenomena
Chowdhury I, Tharakan B, Bhat GK (2006) Current concepts in apoptosis: the physiological suicide program revisited. Cell Mol Biol Lett 1:506–525
Coleman WB, Tsongalis GJ (2010) Essential concepts in molecular pathology. Elsevier, Amsterdam
Copeland RA (2013) Evaluation of enzyme inhibitors in drug discovery. John Wiley & Sons, Hoboken, NJ
Cottingham WN, Greenwood DA (2013) An introduction to standard model of particle physics. Cambridge University Press, Cambridge
Coyne JA, Orr HA (2004) Speciation. Sinauer Publishing, Sunderland, MA
Dellaire G, Berman J, Arceci R (2014) Cancer genomics. Academic, London
Dennett DC (1995) Darwin's dangerous idea: evolution and meaning of life. Simon and Schuster, New York, NY
Dobzanski T (1982) Genetics and origin of species. Columbia University Press, New York, NY
Donaldson KM (2000) Relationships between chromosome structure and long distance regulation of gene expressions. University of California Press, San Diego, CA
Duboule D (1994) Guidebook to homeobox genes. Oxford University Press, Oxford
Elsasser WM (2016) The physiological foundation of biology: an analytical study. Elsevier, Amsterdam
Finn RS, Aleshin A, Slamon DJ (2016) Targeting the cyclin dependent kinases (CDK) 4/6 in estrogen receptor-positive breast cancer. Breast Cancer Res 18:17
Fodor J, Piattelli-Palmarini M (2011) What Darwin got wrong. Profile Books, London

Fritz A (2014) Elucidating chromosome territories organization during the cell cycle and in breast cancer cells. Proquest Publishing, Ann Arbor, MI

Garrett RA, Klenk HP (2007) Archaea: evolution, physiology and molecular biology. Blackwell Science, Oxford

Goss KH, Kahn M (2011) Targeting wnt pathway in cancer. Springer, Dordrecht

Gould SJ (2002) The structure of evolutionary theory. Harvard University Press, Cambridge, MA

Gould SJ (2007) Punctuated evolution. Harvard University Press, Cambridge, MA

Grene M (1986) Dimensions of Darwinism: themes and counter themes in the twentieth century evolutionary theory. Cambridge University Press, Cambridge, MA

Hagen JB (2021) Life out of balance: homeostasis and adaptation in a Darwinian world. University of Alabama Press, Tuscaloosa, AL

Hamilton I (1989) An evaluation of the mitochondrial eve hypothesis: theory, methodology, validity and relationships with the regional-continuity model of human emergence. Simon Fraser University Press, Burnaby, BC

Haseeb M, Ul-Ain Q, Pirzada RH (2019) Wnt signaling in the regulation of immune cell and cancer Therapeutics. Cells 8:1380–1415

Herken K (1995) The universal Turing machine: a half century survey. Springer, Berlin

Laudal OA (2021) Mathematical models in science. World Scientific, Singapore

Lynch M (2007) The origins of genome architecture. Oxford University Press, Oxford

Madhusudan S, Wilson DM (2013) DNA repair and cancer: from bench to clinic. CRC Press, Boca Raton, FL

Miller JH, Reznikoff WS (1980) The operon. Cold Spring Harbour Laboratory Press, New York, NY

Mohammed MMA (2018) A comparison of cancer classification methods on microarray data. University of KwaZulu-Natal, Pietermaritzburg

Morgan DO (2007) The cell cycle: principles of control. Oxford University Press, Oxford

Niklas KJ, Newman SA (2016) Multicellularity: origin and evolution. MIT Press, Cambridge, MA

Pecorino L (2012) Molecular biology of cancer. Oxford University Press, Oxford

Peter IS, Davidson EH (2015) Genome control processes: development and evolution. Academic Press, London

Pevsner J (2015) Bioinformatics and functional genomics. John Wiley & Sons, Hoboken, NJ

Polinski KR (2007) Tumor suppressor genes. Nova Biomedical books, New York, NY

Popper K (1963) Conjectures and refutations: the growth of scientific knowledge. Routledge, London

Press SJ, Clyde CM (2003) Subjective and objective Bayesian statistics: principles, models and applications. John Wiley & Sons, Hoboken, NJ

Ross DW (1998) Introduction to oncogenes and molecular cancer medicine. Springer, Dordrecht

Schutz BF (2009) A first course in general relativity. Cambridge University Press, Cambridge

Simner J (2019) Synaesthesia: a very short introduction. Oxford University Press, Oxford

Sirica AE (1989) The pathobiology of neoplasia. Plenum Press, New York, NY

Spitz DR, Dornfeld KJ, Krishnan K, Gius D (2011) Oxidative stress in cancer biology and therapy. Springer, Dordrecht

Swanton C, Bardelli A, Graham TA, Polyak K (2017) Cancer evolution. Cold Spring Harbour Laboratory Press, New York, NY

Tracy JI, Hampstead BM, Sathian K (2015) Cognitive plasticity in neurological disorders. Oxford University Press, Oxford

Tutar Y, Tutar L (2018) Current understanding of apoptosis: programmed cell death. Intechopen, London

Vlachakis D (2019) Gene expression profiling in cancer. Intechopen, London

Weinzierl ROJ (1999) Mechanism of gene expression: structure, function and evolution of the basal transcriptional machinery. Imperial College Press, London

Welch DR, Fisher PB (2016) Molecular and cellular basis of metastasis: road to therapy. Associated Press, Cambridge

Yarus M (2010) Life from an RNA world: the ancestor within. Harvard University Press, Cambridge, MA
Zambetti GP (2005) p53 Tumor suppressor pathway and cancer. Springer, Dordrecht
Zhang W (2012) Regulation and coordination of homologous pairing and synapsis during Caenorhabditis elegans meiosis. Stanford University Press, Redwood, CA

Nature of Regulatory Genome: The Evolution and Natural Selection of "Genotope"

8

Abstract

In the preceding chapters, several aspects of genomics were deconstructed using the formalism of involuted manifolds. It was demonstrated that several hitherto unknown features of genomic architecture could be explicated and formalized using involutive algebras. As a culmination of these insights into the nature of genomic architecture, we will formalize a topological modular framework of genomes which consists of formal higher dimensional units of genomes to be christened as "*GENOTOPE.*" In this chapter, we will outline an evolutionary and a functional outline of this model. This model necessarily involves a higher dimensional manifold which runs contrary to the four-dimensional spacetime in which the biological evolution has taken place. Therefore, an outline of the reasons why these extra dimensions are required and why they are connected with one another through an operator of involution will be discussed. A topological framework of natural selection using this operator of involution will be presented. We will discuss why genomes should exist both as ecosystems as well as the units of selection.

8.1 Introduction

In the preceding chapters, a higher dimensional topological model of the genome was discussed. In each of these chapters, a different aspect of genomics was discussed separately. The discussion began with the reasons for the lack of a formal model of genomic architecture. Based on the semantic considerations, a higher dimensional topological model using the formalism of the involuted manifold was articulated. It was proposed that the operator of involution can represent the inwardly-directed influence of the larger system on the smaller embedded subsystem. Since the operator of involution is capable of creating submanifolds, the

© The Author(s), under exclusive license to Springer Nature Singapore Pte Ltd. 2023
P. Chhaya, *The Topological Model of Genome and Evolution*,
https://doi.org/10.1007/978-981-99-4318-0_8

generating of nested architecture comes about naturally. Thus, it is possible to visualize how the earliest living organisms like archaebacteria (Garrett and Klenk 2007, see Chapter 3) could have evolved from a "chemical soup" (Oparin 1968). The nested hierarchy of submanifolds suggested by this model fits into our modular conception of genomic architecture. In the case of the earliest living organisms, this modularity can manifest in the form of chemical structuralism and biological functionalities, existing in a hierarchy.

The operator of involution can represent two natural phenomena of the Darwinian paradigm. It can represent the influence of the environment on living organisms. At the same time, it can be used to describe the complexity of genomes since the operator of involution can induct complexity into the resulting submanifold. Thus, the operator of involution can represent two types of processes, viz., the process of a larger system influencing any of its subsystems and the process of induction of complexities into such subsystems through inwardly-directed influences of the larger system. These twin perspectives of involution as an operator makes this operator an ideal candidate for formalizing the Darwinian paradigm. This is because historically, the nature of the influence of the environment in the process of natural selection and justifications for the emergence of complexity in the course of evolution have been problematic features of the Darwinian semantics (Grene 1986; Bonner 2013).

In the preceding chapters, this tentative model was sought to be articulated and corroborated by the deconstruction of various established features of genomics. Thus, the model was applied to deconstruct phenomena like gene expressions, the relationship between the genotype and the phenotype, structural template of developmental processes, and the evolutionary significance of aging. In the previous chapter, the nature of cancer and its evolutionary context were deconstructed using this model. Prima facie, the model proposed here seems to be a serious candidate for the genomic architecture worthy of being considered as a scientific hypothesis amenable to empirical verification. In this chapter, we will outline the formal tenets of this model. Having done that, we will outline here the salient features of the semantics behind this model in the following chapter. There are two structural features that need to be discussed in greater detail. Firstly, we will outline the reinterpreted version of the Darwinian paradigm. Of course, in the preceding monograph (Chhaya 2020), a reinterpretation of the Darwinian paradigm has been discussed. Therefore, here one would point out the structural template of that reinterpretation in the form of involuted manifold. This refers to the nested architecture between a larger system and its smaller subsystem. Therefore, we will look at the ecosystem and a genome as one such pair. Then, we will look at the genome as a larger system and individual genes as smaller subsystems. This will establish a proposition that evolution of life is like any other natural phenomena.

Secondly, since the proposed model has a prerequisite postulate of higher dimensionality, its exact meaning in the model needs to be understood. Generally speaking, we employ a higher dimensional model to accommodate the mathematical variables proposed in the model. This is essentially a hermeneutic strategy (Laudal 2021). We don't think that these postulated higher dimensional configurations are physical entities. However, the postulate of their existence enables us to understand the finer nuances of the phenomenon under investigation. However, occasionally, we employ a higher dimensional model to represent the nature of reality itself. The string theories (Conlon 2016), for instance, assign a certain degree of physicality to these extra dimensions. These theories tacitly accept that the nature of spacetime is indeed higher dimensional. Therefore, we need to understand the underlying epistemology of this higher dimensional model of life, its evolution and its natural selection. It is argued that the higher dimensionality of this model is as "physical" as that is implicit in, say, string theories. It is because our cognitive faculty operates from a particular range of dimensionalities that the higher dimensionalities appear to us as involuted. Moreover, once we ascribe physicality to the higher dimensional configurations of genomes, it is intuitively clear that these higher dimensionalities represent spatiotemporal dimensionalities. It is tempting to think that this points toward a causal linkage between the cosmic singularity and the emergence of life, thereby rendering the phenomenon of life as natural as any other phenomena.

Therefore, in this chapter, we will define a higher dimensional perspective of the genome as the one which supervenes the DNA sequence of a genome. Therefore, the processes that bring about the changes in dimensionality of a genome to its DNA sequence can be formalized as a regulatory genome. This can be defined as an involutive algebra. This formal architecture of a genome, with its higher dimensional spread, its involutive regulatory framework and DNA sequence, as a single unit, is named as "GENOTOPE." It must be seen as a higher dimensional topological object wherein the regulatory functionalities are deemed as surfaces and distribution of genes is deemed as the metric of this topological object. In such a model, higher dimensional surfaces connect with DNA sequence through involutions.

This chapter has been further divided into 22 sections. Section 8.2: The Proposed Model, Sect. 8.3: Significance of Higher Dimensionality, Sect. 8.4: Involuted Model of Ecosystem, Sect. 8.5: Biological Evolution in the Proposed Model, Sect. 8.6: Natural Selection in the Proposed Model, Sect. 8.7: Genome as an Ecosystem, Sect. 8.8: Natural Selection Among Genes, Sect. 8.9: Genomic Architecture, Sect. 8.10: Definition of Genotope, Sect. 8.11: Formal Description of Genotope, Sect. 8.12: Distinction Between Genome and DNA Sequence, Sect. 8.13: Definition of Regulatory Genome in the Proposed Model, Sect. 8.14: Relationship Between Regulatory Genome and Gene Expressions, Sect. 8.15: Operator of Involution as a Genomic Regulator, Sect. 8.16: Operator of Involution as a Source of Complexity, Sect. 8.17: Evolution of Genomic Architecture in the Proposed Model, Sect. 8.18: Significance of Phylogeny in the Proposed Model, Sect. 8.19: Darwinism vs. Lamarckism, Sect. 8.20: Structural Template of Genotope from the Three-Dimensional Perspective, Sect. 8.21: Structural Template of Genotope from Outside, Sect. 8.22: Semantic Implications of the Proposed Model, Sect. 8.23: Conclusion.

8.2 The Proposed Model

The origin of the proposed model, modified involuted manifold model, lies in the epistemology, particularly the epistemology of mathematics (Chhaya 2022a). In epistemology, how our cognitive faculty obtains knowledge has been an enigma since time immemorial (BonJour 2010). In fact, the very conception of epistemology rests on its linguistic root of *episteme* which refers to self. Therefore, the problem of the epistemology of mathematics is that if we have an intuitive understanding of mathematics, then naturally it must arise from *episteme,* the self. However, there has never been any explanation of how and why the self contains mathematics. If mathematics is within self, then naturally, the process of self-reference would axiomatically lead us to the knowledge of mathematics. However, under the Cartesian influence, we have always thought that mathematics was transcendental. Therefore, the origin and knowledge of mathematics had remained an enigma. The proposed model tries to solve this problem. Upon a little reflection, it is intuitively clear that not just our cognitive faculty, but also genomes seem to possess the functionality of self-reference. A fragment of a genome can decide what the other fragment of that genome should do. This is nothing but self-reference. Therefore, this monograph was written to articulate the details of the mechanism by which genomes manifest self-reference.

As discussed in the preceding monograph (Chhaya 2022b), the proposed model of the modified involuted manifold can be applied to spacetime itself. This results in a semantic proposition that what we perceive to be mathematical objects are representations of the fine structure of spacetime. Moreover, as discussed in the preceding monograph, spacetime and matter are one and the same thing. They must be viewed as isomorphs. This development enables us to think of genomes themselves as isomorphs of spacetime which exists in an involuted architecture. This extension from mathematics to spacetime and then to molecules which constitute genomes, provide a semantic justification for employing the proposed model to genomic architecture. As discussed in the preceding chapters, several aspects of genomes seem amenable to translation into the language of the modified involuted manifold model. Therefore, in this section, we will look at the broad framework of genomic architecture that emerges from the proposed model. Some of these details were elaborated in the preceding chapters. Therefore, here we would take a systemic perspective of genomic architecture as suggested by the proposed model. For the sake of simplicity and economy, a point-wise description of this model is presented below.

1. Genomes exist in multiple dimensionalities simultaneously. These dimensionalities are physical spatiotemporal dimensionalities and not abstract dimensionalities necessary for a theoretical model.
2. Genomes switch from one dimensionality to another during their expressions.
3. These changes in the dimensionalities can be formalized as a class of mathematical operators called the operator of involution.

4. Configurations of genomes in each dimensionality are different from the rest of the configurations of genomes in the remaining dimensionalities.
5. The highest dimensionality of genomes is called genomic singularity. It nationally represents all the possible functionalities of genomes which become manifest as and when genomes undergo involutions to give rise to different higher dimensional configurations.
6. Genomic singularity is by definition, a nonstructural entity and contains information present in spacetime at that dimensionality. However, every operation of involution results in inward folding of one of the dimensions into the remaining dimensions of genomic configurations, beginning with genomic singularity.
7. Therefore, after every involution, the resulting genomic architecture occupies a lower dimensionality, but possesses a greater degree of complexity.
8. According to this model, just as genomic singularity represents different genes scattered among different chromosomes, it is possible to formalize all the introns and exons as single entities arising from genomic singularity and then segregating themselves into different chromosomes.
9. Similarly, different genomic modules can be thought of as having arisen from genomic singularity. However, they must be treated as genomic configurations occupying different dimensionalities intermediate between genomic singularity and the four-dimensional DNA sequences.
10. Irrespective of the fact that genomes are spread over multiple dimensionalities simultaneously, gene expressions and the subsequent molecular biological interactions occur only in the four-dimensional configurations of genomes.
11. At the same time, the higher dimensional configurations of genomes do influence the process of gene expression in the four-dimensional spacetime by devolving themselves onto the four-dimensional configurations of genomes through a series of involutions.
12. These devolvements from the higher dimensional configurations of genomes onto the four-dimensional configurations of genomes give rise to different types of long-range influences.
13. Since the higher dimensional spacetime doesn't manifest distinction between the time-like and the space-like dimensions, whenever genomic configurations present in higher dimensionalities devolve into the four-dimensional spacetime, these long-range influences appear in the form of long-range temporal and spatial influences.
14. This happens because the higher dimensional configurations of genomes are isomorphous to spacetime. Since higher dimensional configurations of spacetime do not have any distinction between the time-like and the space-like dimensions, the higher dimensional configurations of genomes do not possess any stereochemical or conformational states. Thus, it is only in the four-dimensional configurations of genomes that we observe stereochemical and conformational changes This is because the four-dimensional spacetime has the distinction between the time-like and the space-like features.
15. Thus, the temporal and spatial long-range influences and the stereochemical and conformational changes arise only from the inherent structural template of

spacetime which manifests the distinction between the time-like and the space-like features only in the four-dimensional spacetime.

This brief description of genomic architecture is adequate for the present. We will elaborate some aspects of this model in the following sections. In the next section, we will discuss the semantic perspective of postulating higher dimensional configurations of genomes.

8.3 Significance of Higher Dimensionality

Conventionally, we employ higher dimensional models in science to accommodate a greater number of parameters that are necessary to formalize a given natural phenomenon (Laudal 2021). However, as mentioned above, the use of a higher dimensional model in this case is necessitated by a different set of compulsions. Firstly, once we accept that information content is a physical entity, it is imperative that any information transfer (which is what the long-range influences are) must occur in the physical dimensions of spacetime. The trouble with this reasoning is that our conception of spacetime is riveted on a four-dimensional model of spacetime. Admittedly, in theoretical physics, particularly in quantum field theory and/or string theory (Conlon 2016), we normally employ higher dimensional representations of spacetime. However, it is debatable whether these higher dimensional representations of spacetime refer to physical entities or they constitute a hermeneutic device. This is because mathematics underlying these theories is essentially abstract in nature. However, as discussed in the preceding monograph (Chhaya 2022a), it is possible to formalize spacetime based on mathematical realism wherein mathematical objects are representations of the fine structure of spacetime. Therefore, in this model every information content and its transfer occurs in a physical dimension of spacetime. However, this model necessitated that spacetime must possess multiple dimensionalities simultaneously.

Therefore, since we wish to formalize long-range influences of genomics as physical information transfers, it is intuitively clear that we must postulate that genomes are spread over multiple dimensionalities of spacetime simultaneously. Moreover, according to this model, spacetime and matter are one and the same, it is intuitively clear why genomes should exist in multiple dimensionalities simultaneously. They exist in multiple dimensionalities simultaneously because the underlying spacetime itself exists in multiple dimensionalities simultaneously. While this argument might appear to be semantically consistent, the problem is that it must be congruent with the known features of functional genomics. Otherwise, it might end up being an example of intellectual sophistry. Therefore, the focus here will be to demonstrate that different features of functional genomics which have defied explanation can be explained by using this model. There are two peculiar features of genomics which have resisted rational explanations. Both these features, viz., higher level modular constructs (Pevsner 2015) and long-range influences (Shmulevich and Dougherty 2014), will be sought to be deconstructed here.

Apart from the physicality of long-range influences of genomes, there is another compulsion why we require higher dimensional configurations of genomes. This refers to inwardly-directed influences in genomics and in the process of natural selection. While in the case of genomes, these inwardly-directed influences are intuitively obvious, the same is not the case with the process of natural selection. Our conventional perspective of natural selection rests on the passive role of the environment wherein different competing species fight out their battles for survival. Therefore, to conceptualize natural selection as inwardly-directed phenomena is difficult. However, if we conceptualize the environment as an ecosystem wherein not just the environment and the coexisting biota, but also spacetime itself, are active ingredients of an ecosystem, it is intuitively clear that an ecosystem ought to be conceptualized as an inwardly-directed system. Thus, it is intuitively clear that any formal description of either an ecosystem or a genome must be based on a mathematical framework wherein inwardly-directed processes can be formalized. Thus, the proposed involuted manifold model is an ideal candidate. In addition, it is not difficult to think of genomes themselves as a type of an ecosystem wherein different genes compete with one another for their survival. Therefore, the proposed model offers a generic framework for representing all inwardly-directed influences wherever they manifest. Moreover, since the basic operation of involution originates from spacetime itself, it is intuitively clear that at least in principle, we can conceptualize the process of natural selection as a system of inwardly-directed influences which can be applied to any natural phenomenon involving multiple units and inwardly-directed influences. Thus, the proposed model offers a way to define the Darwinian paradigm in a domain agnostic manner (Dennett 1995).

In order to understand this domain neutral version of the Darwinian paradigm, it is necessary to deconstruct biological evolution and natural selection using a single framework. It is important to keep in mind that conventionally, the Darwinian paradigm has been successful in explaining the process of natural selection. However, its ability to explain biological evolution per se has been limited. It is tempting to think that this inability is linked to another inability of the Darwinian paradigm. This refers to the problem of explaining the emergence of complexity during natural selection (Bonner 1988). The proposed model offers a new conception of the Darwinian paradigm which explains the emergence of complexity during biological evolution as well as during natural selection. Therefore, in the following sections, we will try to deconstruct natural selection using the proposed model. Before we do that, in the next section, we will try to deconstruct the nature of the ecosystem using the proposed model.

8.4 Involuted Model of Ecosystem

The conception of the ecosystem is of much more recent origins (Raffaelli and Frid 2010). Admittedly, this conception is implicit in the conventional Darwinian perspective (Hodge and Redick 2009; Grene 1986), but it was never articulated until much later. This is possibly because the conventional perspective was preoccupied with creating nonteleological arguments in favor of natural selection. Therefore, any suggestions of Nature acting in concert would compromise the essential Darwinian randomness. In fact, even later, when the conception of the ecosystem as a singular entity emerged, it had a subtext of wholism. This is best exemplified by the popularity of the Gaia hypothesis (Lovelock 2000). The assigning an identity of a willful self to Nature (or to our planet as implicit in the Gaia hypothesis), is actually an act of personification inherent in most of the pantheism doctrines. Leaving aside the validity or justification of this approach, we will look at the conception of the ecosystem in a naturalistic manner which is consistent with the conventional Darwinian semantics. Thus, the invocation of the term ecosystem shouldn't imply any higher level of willfulness, but only a higher level of organization. With these caveats in place, let us deconstruct the ecosystem using the proposed model.

Conventionally, an ecosystem consisting of multiple entities can be naturally represented using a higher dimensional model. However, the underlying semantic compulsion of the proposed model doesn't allow us to use this hermeneutic approach. According to the proposed model, all the dimensionalities are physical in nature and more importantly, they refer to spatiotemporal dimensionalities and not to any theoretical parameters. Therefore, the conception of the ecosystem in the proposed model is radically different from the one conventionally employed. The only thing common between the conventional perspective of an ecosystem and the one according to this model is that it takes a transactional perspective. In other words the ecosystem as a whole, transacts information content among its constituents. However, the nature of transactions differ. For instance, as discussed in the preceding chapters, the role of the environment is passive in the conventional perspective. True, the actions of individual species do change the environment (we know that from our wanton and reckless industrialization); however, the environment remains passive during these changes. Admittedly, the changed environment, in turn, influences the course of natural selection, but it happens as a natural consequence of the changes in the environment. This is best exemplified by the evolution of aerobic organisms which followed the increase in oxygen levels in the atmosphere due to the emergence of blue-green algae (Oparin 1968).

However, according to this model, the role of the environment is not passive. This lack of passivity doesn't arise from any new radical phenomena of the environment. It arises from the inclusion of spacetime in the conception of the environment. Thus, according to this model, an ecosystem also includes spacetime in addition to its conventional constituents. Thus, while the conventional passive role of environment is retained in the proposed model, an additional constituent in the form of spacetime is included in the definition of environment. This inclusion of spacetime results in two features to the conventional perspective of natural selection. Firstly, it shapes the nature and degrees of complexities during natural selection. This happens because spacetime, as formalized in the proposed model, possesses different metrics within itself. Secondly, spacetime influences the long-range influences of genomes, thereby allowing the characteristics of Life to manifest.

However, the most fundamental outcome of inclusion of spacetime in the conception of the environment in the process of natural selection, is that it allows us to formalize the ecosystem as a single entity. According to the conventional perspective, all the constituents of the environment were distinctly different and their influences were independent of one another. Therefore, our conception of the ecosystem was a cognitive artifact because it allowed us to treat these diverse elements of Nature as a unified entity. However, in reality, it is more of an idealization than a reality. However, once we include spacetime itself into the definition of ecosystem, the ecosystem becomes an organically unified entity. There is no need to idealize, the ecosystem becomes a self-evident reality. This is because all the conventional elements of the environment like atmosphere, light, water etc. are conceptualized as isomorphs of spacetime itself. In addition, because spacetime is a continuum, we can define inwardly-directed influences. This provides us with a mathematical framework of ecosystem in which different types of influences arising from the environment can be formalized using a single type of mathematical operators.

In the conventional perspective of the ecosystem (Raffaelli and Frid 2010), the environment and living organisms were separate. Therefore, the influence of the environment on living organisms and the influence of living organisms on the environment were separately defined. However, in this model, living organisms are integral parts of the ecosystem. More importantly, they are on par with the elements of Nature. This parity arises because now we can formalize the ecosystem as a manifold and everything else, including the elements of Nature and living organisms, is defined as submanifolds of the ecosystem. Thus, the proposed model enables us to unify different constituents of the environment with living organisms in a single framework. It allows us to unify the interactions between different elements and different living organisms using a single mathematical operator. Finally, it allows us to formalize the inwardly-directed influences which give rise to characteristic features of Life without invoking any transcendental influences.

This conception of the ecosystem appears to be a reasonably good model. However, this unification is only a formal description. There exists a more fundamental need for unifying the processes of biological evolution and natural selection. As discussed in the preceding chapters, the conventional perspective of Darwin's theory is adequate to deconstruct the nature of natural selection. However, it has nothing much to explain the origin of Life. Biological evolution remains enigmatic from the perspective of the Darwinian paradigm. In fact, it takes biological evolution as a priori. However, as mentioned above, there is a need to extend the Darwinian paradigm to include biological evolution per se. Therefore, in the following two sections, we will try to deconstruct biological evolution and natural selection using this model. This is necessary because the emergence of regulatory elements of genomic architecture (which are unique and definitive features of Life) must be shown to have evolved from a far more general class of influences manifest in Nature. In other words, it is necessary to demonstrate that the regulatory elements of genomes are merely a special class of natural phenomena. There is nothing mysterious or transcendental about regulatory elements of genomes. This can only be achieved when we demonstrate that biological evolution and natural selection have a common framework which is precisely what the proposed model offers.

8.5 Biological Evolution in the Proposed Model

As discussed in the preceding chapters and in the preceding sections, the Darwinian paradigm is not adequate to explain the origin of Life. Even when we try to think of the origin of Life as an independent issue, there is no major explanation available. As a result, our most reasonable explanation of the origin of Life is an amalgamation of known biochemistry and Darwin's theory. We believe that Life evolved from a primordial soup (Oparin 1968) of chemicals that were likely to be present on Earth. The first step must have been some catalysis involving minerals in the form of clay or some autocatalysis of the biomolecules formed in this primordial soup. Once riboproteins were formed, the synthesis of RNA proteins having the functionalities of self-duplication followed in the natural course of events. This "RNA world" hypothesis (Yarus 2010, see Chapter 2) is perhaps the most credible explanation of the origin of Life. The problem with this scenario is not that it is wrong (it is most probably true), but that it hides the correct origin of Life. Rather, it sidesteps the issue. It assumes that the chemical description of Life is the description of Life. It is not. It describes just one aspect of the origin of Life. The RNA world hypothesis represents a category mistake (Actually it is a case of metonymy.).

To be honest, scientists have been aware of this anomaly. Therefore, we have articulated a general principle which is called emergence principle (Smith and Morowitz 2016, see Chapter 4). This principle suggests that as the complexity of a system increases, the system manifests newer functionalities not manifest earlier. Thus, according to this principle, polynucleotides would not have functionalities of riboproteins. Riboproteins would not possess properties of a full-fledged RNA. Similarly, RNA /DNA will not have functionalities of a biological cell. This principle of emergence merely summarizes what is empirically known. It doesn't explain why an increase in the complexity of a biological system should result in the emergence of newer functionalities. To be fair, we can think of the emergence of newer functionalities as having arisen from intricate stereochemical configurations or conformations or even from the shapes of the molecular orbitals of these biomolecules. However, the problem with this conjecture is that it is based on the principles of stereochemistry /quantum chemistry. None of these disciplines have the necessary semantics to justify this conjecture. For instance, the stereochemical paradigm can conceptualize molecular orbitals of these biomolecules spread over long distances (at least on the scale of atoms). However, in order to ascribe any biological functionality to molecular orbitals, stereochemistry would have to invoke quantum mechanical explanation for the transmission of these long-range influences. Quantum chemistry, by definition, is a noncausal domain. Therefore, it cannot offer any mechanistic details of any molecular functionalities (leave aside the biological functionalities). In fact, quantum chemistry (Szabo and Ostlund 1989) rests on a delphic oracle kind of proposition, viz., all the molecular functionalities of a molecule are encoded in its wave function. It precludes any mechanistic details how these incipient functionalities present in a wave function can become manifest. Thus, molecular biological perspective has no answer to the question of the origin of Life. It merely describes a chemical facet of Life. This criticism is not to be taken as a surrogate for some transcendental or deistic origin of Life. On the contrary, this criticism is aimed at the semantic ambiguities that surround the molecular biological perspective (including the RNA world hypothesis) of the origin of Life.

If we wish to formalize the origin of Life in a purely naturalistic description, it is necessary to revisit the conception of natural selection, molecular biology and quantum chemistry. In the preceding monographs, a new topological model of spacetime (Chhaya 2022c) and its relationship with quantum fields is described. It rests on a single postulate that spacetime exists in multiple dimensionalities simultaneously. Our conventional conception of spacetime has been that spacetime is a four-dimensional manifold. It grants that spacetime has multiple dimensions. The proposed model goes one step further and postulates that spacetime consists of multiple dimensionalities with each dimensionality containing multiple dimensions (at least most of the dimensionalities possess multiple dimensions, except for the zero dimensional spacetime). This model has been discussed in the context of the counterintuitive features of quantum phenomena in another monograph (Chhaya 2022c). The same model has been sought to be employed in formalizing genomic architecture in this monograph.

The basic rationale behind employing this model in formalizing biological evolution is simple. Unlike the conventional quantum chemical protocol, the proposed model assigns different dimensionalities to different molecular functionalities. However, these higher dimensional configurations of genomes can express themselves only in the four-dimensional spacetime wherein molecular biological processes operate. There are several advantages of this framework of genomic architecture. Firstly, since the relationship among all the dimensionalities is determined by a singular mechanism consisting of the inward folding, it enables us to define various regulatory elements of genomes in a systemic manner. Secondly, it explains why we can't conceptualize genomic architecture in the conventional perspective of genomics. Thirdly, this framework can be embedded in the framework of the ecosystem described above. This possibility creates a way to unify biological evolution with natural selection. Finally, this framework allows us to think of genomes themselves as a kind of ecosystem wherein, instead of individual species of ecosystem, the individual genes undergo competitive survival, thereby extending natural selection to the individual genes.

However, the biggest advantage of this conceptualization of genomes is that it explains how biological functionalities (which makes Life a unique phenomenon among all the natural phenomena) arise naturally without invoking any transcendental influences. This framework allows genomes to possess memories of its phylogeny. This framework allows individual genomes to execute its gene expressions in an intelligent manner. This framework also allows genomes to copy themselves. Admittedly, this semantic congruence between ecosystem and genomic architecture provides a way to redefine natural selection. Therefore, in the next section, we will deconstruct the process of natural selection as implicit in this framework of hierarchy of dimensionalities.

8.6 Natural Selection in the Proposed Model

In the previous two sections, we discussed how the entire ecosystem and individual genomes can be conceptualized in a single framework. In fact, in this framework, it is possible to argue that genomes themselves can be thought of as ecosystems themselves wherein individual genes compete for survival. In this section, we will try to understand how the process of natural selection can be conceptualized in this framework. There are two reasons why we should redefine natural selection in the language of dimensionalities. Firstly, it allows us to link natural selection and biological evolution in a single framework. This is something that is missing from the conventional Darwinian paradigm. Secondly, this integration of natural selection with the unitary framework for ecosystem and genomes enables us to assign a structural template to the process of natural selection. While this may appear to be inconsistent with the conventional Darwinian paradigm which rests on randomness, the assignment of structuralism to natural selection doesn't lead to any teleology. It helps us to explain phenomena like punctuated evolution and speciation which have defied explanation in the conventional perspective. Secondly, the ascription of structuralism to natural selection provides a semantic foundation for phylogenetic perspective. Lastly, it explains the emergence of complexity during natural selection, something that has remained an unresolved enigma in the conventional perspective.

Let us understand how natural selection fits in this framework of dimensionalities. As discussed above, every influence in the ecosystem (and even in genomes) can be formalized as an inwardly-directed influence. According to this model, these inwardly-directed influences can be formalized as inward folding of one of the dimensions of the ecosystem into the remaining dimensions of the ecosystem. This influence could be in any form. It could be in the form of presence of oxygen in hitherto anaerobic habitats. It could be in the form of a new species in the habitat. The actual details don't matter. The inward folding can be thought of as the addition of information from the dimension undergoing involution (say, chemical reactivity of oxygen in the above mentioned case) to the information content of the recipient dimensions of ecosystem (say the susceptibility of physiological states of the organisms to oxygen in the above mentioned case). This addition of information content from the dimension undergoing involution to the remaining recipient dimensions brings about the changes in the resulting involuted manifold (in the form of oxidative stress in different living organisms in the above mentioned case).

As a result of such an involution, the resulting ecosystem would change. Every species would have their survivability altered because of the changes in the information content of the new habitat (in the form of involuted manifold). Now this altered survivability, in turn, would give rise to the changes in the individual genomes (again through inwardly-directed influences) and genomes would undergo changes to cope with the new habitat. Thus, the entire sequence of influences arising from the global changes in the ecosystem (in the form of emergence of oxygen in the atmosphere) can be defined as inwardly-directed influences. Secondly, at every stage, due to involutions of information content, there are structured changes in the lower dimensionalities starting from the ecosystem, then to individual genomes and then to the individual genes. Moreover, such changes in the information content of different dimensionalities are defined by a single mechanism. Therefore, it is intuitively clear that natural selection can be thought of as the changes in the dimensionalities of the ecosystem. Moreover, it is unitary across the range of dimensionalities. Therefore, the same mechanisms bring about different changes in different dimensionalities. It is also intuitively clear that this description confers a unitary structuralism to the process of natural selection.

Now let us understand what happens during natural selection and how it gives rise to legacy which is manifest in phylogenetics. The key to understanding the legacy argument is the conception of genomic singularity. While the semantics of genomic singularity are discussed in the accompanying chapters, the semantic proposition relevant to the present discussion is that genomic singularity contains all the possible, potential and real, genomic functionalities within itself. Admittedly, such a conception is incongruent with our conventional perspective of genomics. However, this is a legitimate proposition in the present model because it rests on the active participation of spacetime in genomic architecture. Just as different features of different natural phenomena can be traced back to the nature of spacetime, according to this model, genomic features too can be traced back to the nature of spacetime. Therefore, it is a reasonable proposition that all the potential and real genomic functionalities must have originated from genomic singularity. This conception of a notional entity is something similar to the conception of LUCA (Bard 2016, see Chapter 9) and mitochondrial eve (Hamilton 1989).

Natural selection begins when genomic singularity is influenced by the supervening ecosystem. This influence would also be in the form of an inwardly-directed influence. In essence, it amounts to information transfers from the ecosystem to genomic singularity. Therefore, the resulting information changes in the information content of genomic singularity would be determined by the initial information content of genomic singularity. Now, as a result of this inwardly-directed influences of the ecosystem, genomic singularity would start differentiating into different modules. With each inwardly-directed influence (in the form of involution), genomic singularity would keep on differentiating into different degrees of modularity. This scenario establishes that different modularities must share a common ontology. This is essentially reflected in phylogenetics (Bromham 2008, see Chapter 5). However, since according to this model, different modularities occupy different but higher dimensionalities, its ontology would not be traceable in the four-dimensional configurations of genomes. Therefore, what we observe in phylogenetic studies is the legacy of individual genes (Appasani 2016), but not the genomic legacy in the form of its modularity.

It is important to keep in mind that this scenario is consistent with the Bayesian conditional probabilities (Press and Clyde 2003). While genomic singularity can be thought of as omnifunctional, its descendants are not. With each involution, the probability of the unrealized modular plans would diminish. This process continues ad infinitum. Therefore, every evolutionary pathway not taken would be lost. This rationale prevents genomic singularity from being omnipotent and transcendental. Its capacity to encapsulate all the possible genomic functionalities doesn't arise from its Platonic status, but it arises from the underlying structural template of spacetime which contains multiple dimensionalities simultaneously. Moreover, as mentioned above, since the process of involution (and the rules of information transformations implicit in it) ensures that natural selection possesses a unique native structuralism of its own. It is important to keep in mind that neither the postulate of genomic singularity, nor the ascription of structural template to the process of natural selection, undermines the randomness implicit in the conventional Darwinian paradigm. All this model does is to sequester the randomness of outcomes into conditional probabilities. In fact, this model explains phenomena like punctuated evolution (Gould 2007) and speciation (Coyne and Orr 2004) in an intuitive manner. There Is another advantage of this conceptualization of natural selection. It can now be extended to genomes themselves. Prima facie, there is no difficulty in thinking of genomes as ecosystems on their own. Therefore, natural selection can continue to operate among different genes just as it does among different species. We will discuss this topic in the next section.

8.7 Genome as an Ecosystem

As mentioned above, prima facie, it is possible to think of genomes themselves as ecosystems. Keeping aside the romantic allure of this idea, it is necessary to deconstruct it from structural and semantic perspectives. At first sight, it is possible to argue against it by pointing out that genomes have linear structuralism. Therefore, the competitive survival among genes would be compromised by the very fact that they are linearly interconnected. This connectivity would definitely reduce the randomness implicit in the Darwinian paradigm. Moreover, the linear distribution of genes also compromises their autonomy. All these factors raise the question about the nature of competition among genes. In addition, the question arises what source they should compete for? (Incidentally, this logic also applies to the Gaia hypothesis (Lovelock 2000).) After all, one of the motivations behind Darwin's theory was Malthusian economics (Flew 2017, see Part III). Therefore, unless there is a limited quantity of a resource, competitive survival would not set in. Therefore, it is necessary to deconstruct this idea of genomes as ecosystems.

In order to deconstruct this idea, it is necessary to look at the semantic and structural features of the conception of the ecosystem itself. At the heart of the conception of the ecosystem is a belief that Nature is singular in nature. Therefore, even though different elements of Nature appear to be independent (and even autonomous), they are somehow interdependent and interconnected. Therefore, it seems reasonable to think that the conception of an ecosystem accepts that autonomy and interdependence coexist in its constituents. Therefore, by analogy, different genes in a given genome can be interdependent as well as autonomous. Admittedly, our current understanding of ecosystems doesn't tell us how both these mutually contradictory features can be formally represented in a unitary framework. However, if these features can be represented in a common framework, there is no reason why a genome cannot be thought of as an ecosystem manifesting competitive survival among genes. However, the question of limited resources causing competitive survival remains to be answered in the case of genomes acting as ecosystems. Upon a little reflection, it is intuitively clear that different genes depend on genomic signals to express themselves. Therefore, if we think of gene expressions as synonymous with biological survival, then the limited resource could be the genomic signals themselves. Thus, if we accept that the information content (in the form of long-range influences) is a source which enables individual genes to express themselves, then we can conceptualize genomes as ecosystems. Thus, there are two prerequisites for conceptualizing genomes and ecosystems, viz., a single framework to accommodate interdependence and autonomy of genes and the information content as a source of survival. As discussed in the preceding chapters, this is precisely what the proposed model offers. We will return to this topic in Sect. 8.9. Presently, in the next section, we will try to deconstruct how individual genes can be naturally selected within the genome itself.

8.8 Natural Selection Among Genes

There are three historical aspects of natural selection of genes. Firstly, in Darwin's theory, there is a reference to inheritable traits and their survivability (Hodge and Redick 2009). Therefore, there was an implicit suggestion that these traits are discrete entities. This discreteness of inheritable traits was made explicit in Mendel's pioneering experiments (Darden 1991). With the advent of "new synthesis" (Delisle 2021), Darwinian and Mendelian perspectives merged to give us the theory of natural selection as we presently understand. The second aspect became clear with the advent of population genetics which crystallized the inchoate randomness implicit in Darwin's writings into a formal description (Provine 2001, see Chapter 5). Incidentally, it also gave insights into the relationship between discreteness of genes and their interconnectivity. The third aspect of natural selection was the realization that we can think of natural selection taking place between different discrete entities and not necessarily genes. This gave us the semantics of the units of selection (Okasha 2010, see Chapter 2). All these three aspects are extensively discussed in literature.

However, the topic of natural selection among genes (in contrast to earlier topics of selection of genes) is relatively less explored. This is probably because as mentioned above, there was a lack of common framework to define interdependence and autonomy. Therefore, in this section, we will assume, albeit temporarily, that the proposed model correctly represents the relationship between interdependence and autonomy of genes. Our objective is to understand how genes could compete among themselves, without accepting the semantic implications of the proposed model. Admittedly, we will return to the semantics of the proposed model in Sect. 8.22. However, for the present discussion, we will use the proposed model as a proxy to deconstruct the nature selection among genes of a given genome.

In order to map these two parameters of interdependence and autonomy, the proposed model employs a topological framework. According to the proposed model, a genome exists in multiple dimensionalities simultaneously. The dimensionality of a given sequence of DNA is defined by its highest occupied molecular orbital. In turn, the dimensionality of the highest molecular orbital is defined by the number of electrons present in that orbital (Szabo and Ostlund 1989, see Chapters 2 and 3). Since the number of electrons present in the highest molecular orbital roughly depends on the length of DNA sequences, it is possible to assign different dimensionalities to different genes. It must be kept in mind that in the conventional perspective of quantum chemistry, the molecular orbitals remain in four-dimensional configurations. However, as discussed in the preceding monograph (Chhaya 2022c), quantum mechanics can be defined as an involuted manifold using this model. Moreover, in this model, the dimensionality of molecular orbitals vary according to the number of electrons present in them. Thus, the proposed model allows us to assign different dimensionalities to different genes. Once we accept this reasoning, it is intuitively clear that now we have different genes distributed in different dimensionalities.

This gives us a template for defining interdependence and autonomy. Genes occupying different dimensionalities would have greater autonomy than the genes occupying the same dimensionality. It is important to keep in mind that this scenario is different from the conventional perspective because according to conventional perspective interdependence between genes is defined by the length of intergenic distances between them (Pevsner 2015; Weinzierl 1999). The scenario described above is qualitatively and quantitatively different. According to this model, the genes which are conventionally thought to be in close proximity (and therefore having greater interdependence) could turn out to be less interdependent. This can happen if the genes contiguous to one another in a linear model of a given genome possess different lengths of DNA sequences. Thus, according to this model, a pair of contiguous genes would be less interdependent if they possess different lengths. The reverse scenario is also possible. A widely separated pair of genes in a given genome would be thought to be least interdependent according to the conventional perspective. However, if these genes were to possess comparable lengths, then they would occupy the same dimensionality and therefore, according to this model, there would be a greater interdependence between these two genes. Admittedly, the situation is slightly more complicated than the scenarios described here because even according to this model, molecular biology of gene expressions still occurs at the four-dimensional configurations of genomes. Therefore, in practice, we will observe the whole gamut of degrees of interdependence and those of autonomy. As discussed in the preceding chapters, these different degrees of interdependence and autonomy give rise to different types of long-range influences. While these influences were discussed in detail in the preceding chapters, we will return to the systemic framework of regulatory genome in Sect. 8.13.

With this perspective in place, let us see how we can deconstruct natural selection among genes. It is intuitively clear from the above discussion that the proposed model offers a way to accommodate interdependence and autonomy in a framework of dimensionalities. The question is whether the assignment of different dimensionalities to different genes can help us to visualize natural selection among genes. Upon a little reflection, it is intuitively clear that if we define natural selection as a method of changing the dimensionality, then natural selection can operate among different genes. Let us see how. Let us visualize that the overall topology of a given genome is such that the majority of long-range influences are centered around a few dimensionalities. In other words, if the majority of the long-range influences were to arise from only a few dimensionalities of a given genome, then the genes present in these favorable dimensionalities would control a greater number of gene expressions. Thus, some genes would exercise greater influences on the overall gene expressions. Therefore, they would be preferably selected over those genes present in higher dimensionalities but having lesser influences on the gene expressions at the four-dimensional configurations of genomes. Thus, collectively, the fourth-dimensional configurations of genomes would act as an environment which decides which dimensionalities are preferred. In turn, the genes present in these preferred dimensionalities are likely to be naturally selected.

At first sight, this might sound preposterous. However, upon a little reflection, it is intuitively clear that this is precisely what we observe. The only difference being that we have different labels for these phenomena, viz., polygeny (Reavey 2013) and pleiotropy (Lozano 2017). To be honest, the phenomena of polygeny and pleiotropy have remained enigmatic in the conventional perspective. The origin of this enigma is not why they manifest. Rather, the enigma lies in the fact that the conventional perspective can't predict where and when these phenomena would manifest. The proposed model, on the other hand, provides a method of predicting these phenomena. Thus, it is reasonable to think that this model has adequate semantic depth to give a rational perspective of the natural selection among genes. Admittedly, the actual computations would be a humongous task considering the fact that according to the current estimate, the human genome contains thousands of genes. But at least, the proposed model takes a first step toward that goal. We will end this section with one comment. The proposed topological framework offers intuitive insights into the phenomena like gene overlap (Sabath 2009) and open reading frames (Sieber et al. 2018) which have baffled molecular biologists.

8.9 Genomic Architecture

In the preceding chapters and in the preceding sections, we discussed several aspects of genomic architecture. In particular, we discussed several features of genomic architecture that would make sense if a topological framework was formalized. However, in each case, the arguments were put forth to justify one or the other aspects of genomics. In this section, we will discuss the systemic perspective of the topological representation of genomes. In other words, we will assume that there exists enough indirect evidence to justify a topological perspective. Moreover, as discussed in Chap. 1, the proposed model offers an intuitive explanation of some of the ambiguities of the conventional perspective of the Darwinian paradigm. Therefore, combining these two perspectives, viz., the structural implications of genomics and the semantic implications of the Darwinian paradigm, in this section, we will develop a model of genomic architecture. This model would be elaborated in the following sections. For the sake of simplicity, the model will be described in a point-wise manner. Admittedly, some of these propositions are mentioned in the preceding chapters and in the preceding sections. However, even at the cost of being repetitive, these propositions are included here.

1. Genomes exist in multiple dimensionalities simultaneously. These dimensionalities are spatiotemporal dimensionalities and not abstract dimensionalities.
2. According to the proposed model, molecular orbitals occupy different dimensionalities depending on the number of electrons present. Therefore, unlike the conventional quantum chemical perspective, different molecular orbitals can occupy different dimensionalities.
3. Therefore, different DNA sequences would occupy different dimensionalities based on the length of DNA sequences.
4. Therefore, a genome would be spread over multiple dimensionalities simultaneously based on the number of genes and their respective lengths of DNA sequences.
5. Therefore, topological continuity of genes would be different from the stereochemical proximity of the corresponding DNA sequences. As a result, genes which are separated in the DNA sequence of a genome can be topological neighbors. The converse is also true.
6. Different dimensional configurations of genomes possess different structural templates. However, due to the involuted nature of topology, the high dimensionalities possess coarser metrics as compared to the metrics of lower dimensional configurations of genomes.
7. The highest dimensionality of a genome is designated as a genomic singularity. Genomic singularity must be taken as a notional entity representing the source of all the genomic functionalities, both actual and potential.
8. Genomic singularity can also be viewed as a representation of the genome of LUCA (Last Universal Common Ancestor) (Bard 2016, see Chapter 9). In that case biological evolution can be thought of as a series of involutions beginning with the genomic singularity. With each involution, the emerging genome would possess lower dimensionalities and increased complexity.
9. Similar logic applies to the expression of various genomic functionalities. Each higher dimensional configuration of genomes would undergo one or more involution to influence the lower dimensional configurations of genomes. Some of these higher dimensional configurations of genomes would influence the four-dimensional configurations of genomes. Similarly, some of the higher dimensional configurations of genomes would stop involuting before they reach the four-dimensional configurations of genomes.
10. Since these higher dimensional configurations of genomes are spread in spatiotemporal dimensionalities, they don't possess kinetic and thermodynamic energies. This is because the underlying spacetime itself doesn't possess the distinction between the time-like and the space-like features. Therefore, the

changes in these dimensionalities would result in the changes in the information content of higher dimensional configurations of genomes while the corresponding changes involving four-dimensional DNA sequences would result in the stereochemical and conformational changes. .

11. However, when these higher dimensional configurations of genomes manage via a series of involutions to influence the four-dimensional configurations of genomes, these influences would manifest in the form of long-range temporal and spatial influences on the four-dimensional configurations of genomes.
12. According to the proposed model, the entire edifice of molecular biology of gene expressions manifests only in the four-dimensional configurations of genomes.
13. Therefore, according to this model, there would be different types of long-range influences (both temporal and spatial) differing from one another by degrees. This gradation of long-range influences arises because several higher dimensional configurations of genomes influence the four-dimensional configurations of genomes. Thus, it should be possible to figure out the number of higher dimensional configurations of genomes which influence the four-dimensional configurations of genomes by quantifying and sorting out different long-range influences.
14. Since several genes can occupy the same dimensionality, it is intuitively clear that genomic architecture would have modular configuration with each dimensionality being occupied by multiple genes.
15. It is important to keep in mind that this modular design is a product of natural selection. Therefore, the topological continuity of genes enjoys phylogenetic primacy over the stereochemical proximity observed in the DNA sequence of a given genome.
16. Since biological evolution and natural selection share the same topological framework the legacy of modular design ought to manifest in phylogenetics (possibly to be renamed as phylogenomics).
17. Finally, using the analogy of genes, it is possible to define functionally and structurally autonomous units of genomes which are named here as genotopes. Just as a gene is a structural unit of gene expression, genotope can be thought of as a topological unit of a genomic functionality.

With this description in place, in the next section, we will define the topological unit of genomes, viz., Genotope.

8.10 Definition of Genotope

It is possible to take a skeptical view of the model described above. This is partly because it provides an unorthodox perspective of the conventional perspective. Moreover, it appears to have no physical basis. As discussed in the preceding chapters, there are enough semantic justifications for the proposed model. However, it is necessary to endow this model with adequate structural heft to make it a credible scientific hypothesis. Therefore, we will try to employ the general principles given above to create a structural template of the proposed genomic architecture. As mentioned above, according to this model, genomes exist in multiple dimensionalities simultaneously. Moreover, these dimensionalities are spatiotemporal dimensionalities. In other words, these dimensionalities are physical entities and not hermeneutic constructs. In addition, the basic structural element of the underlying topology is the operation of involution. Therefore, it is intuitively clear that any structural unit of this genomic architecture cannot be a simple topological unit. Therefore, in order to understand the details of this structural unit which is named here as "GENOTOPE," we look at the details of this unit in this section. Once again, for the sake of simplicity, we will describe these details in a point-wise manner.

1. The structural unit of this model is named "Genotope". It must be kept in mind that this unit is also a functional unit. This is because all the genomic functionalities arise from the changes in the dimensionalities of genomes. Since this unit straddles multiple dimensionalities simultaneously, it is intuitively clear that it is also a functional unit in addition to being a structural topological unit.
2. Genotope, as mentioned above, is spread over several dimensionalities. Just as it is possible to visualize a higher dimensional geometric object as occupying different dimensions simultaneously, genotope must be visualized as an object occupying different dimensionalities at the same time.
3. In general, a topological object occupying multiple dimensionalities simultaneously can be visualized as a bunch of dimensionalities stacked up. However, just as in geometry, different dimensions are used together and not bunched randomly, in this model, different dimensionalities (implicitly in a sequential order based on the numerical values of these dimensionalities) are also fused together in a predefined manner.
4. According to the proposed model, the sheaves of different dimensionalities are linked to one another by the operator of involution. Therefore, genotope must be visualized not as a heap of dimensionalities ordered in decreasing manner, but as an object wherein these dimensionalities are sewn up together by involutions.
5. In order to understand this better, it is necessary to distinguish between dimensions and dimensionalities. The term dimensionality is used here to denote the maximum number of dimensions that any object can possess in the given dimensionality. For instance, in the three-dimensional space (with which we are familiar), an object can occupy any number of dimensions provided they are three or less than three. Thus, in each dimensionality, there will be built-in degrees of freedom for any object occupying that dimensionality.
6. Thus, Genotope can be occupied by multiple configurations of genomes, and each configuration will have its own degrees of freedom. Therefore, whenever a higher dimensional configuration of a genome devolves into a lower dimensionality, its degrees of freedom would also decrease.
7. This decrease will manifest in the form of a more complex metric of the lower dimensionality.

8. Therefore, when a genomic configuration devolves from the dimensionalities of spacetime wherein there is no distinction between the time-like and the space-like features to the four-dimensional spacetime, this would result in the emergence of thermodynamic and kinetic changes in the form of stereochemical and conformational changes of the DNA sequence. However, from the perspective of the DNA sequences, these changes would appear to be long-range influences without any causal explanations.
9. A genotope per se can exist in any range of dimensionalities. Therefore, it can represent a gene or a genomic module depending on the range of dimensionalities it occupies. This is because according to this model, genomic modules occupy higher dimensionalities of spacetime as compared to the dimensionality occupied by individual genes.

Using this conception of genotope, we will try to deconstruct several features of genomics in the following sections. However, before doing so, in the next section, we will discuss a tentative formal description of a typical genotope.

8.11 Formal Description of Genotope

In the absence of any prior knowledge of the topological framework of genomes, in this section, we will provide a generic description of genotopes. In an ad hoc approach, we will assign seventh dimensionality to genomic singularity. This would undergo involutions to give rise to three different genotopes, viz., the six-dimensional genotope, five-dimensional genotope, and the conventional four-dimensional genome. Some of these aspects are discussed in the previous chapter while discussing the regulatory genome. Therefore, here we will overlook the nature of the long-range influences that arise when a genome changes its dimensionality. Instead, we will discuss the nature of these different genotopes. We will denote genotope with G followed by subscript denoting its highest dimensionality. Thus, G_7 represents a group of genes (or any DNA sequences) occupying seventh dimensionality. As mentioned above, within the seventh dimensionality, different genes (and different DNA sequences) can occupy different dimensions simultaneously, but they are contiguous in the seventh dimensionality, while being noncontiguous in lower dimensions. Therefore, for the present discussion, we will focus on three genotopes. The four-dimensional genotope (G_4) being synonymous with the conventional genome has been omitted from the discussion. A summary of representative dimensionalities is provided in Table 8.1.

Table 8.1 List of representative genotopes

Notation	Dimensionality	Name	Features
G_7	Seven	Genomic singularity	Functional continuum
G_6	Six	Genomic duality	Unified intron and exon
G_5	Five	Genomic trinity	Insertion of intergenic regions

It is intuitively clear that the conventional description of the four-dimensional genotope (G_4), being synonymous with the conventional perspective of genome, would give rise to apparently random distribution of genes in different chromosomes. From the semantic perspective of this assignment of different dimensionalities to different genotopes (although being arbitrary), a few propositions suggest themselves. Firstly, genomic singularity would possess genomic functionalities, both actual and potential, in their latent forms. Therefore, it can be conceptualized as a functional continuum justifying its singular nature. Secondly, genomic duality (G_6) suggests that due to the topological compulsions, the spatial segregation manifests prior to the functional segregation. This is because the distinction between exons and introns rests on functionalities of the transcription machinery. Thirdly, it is possible to explain the origins of the phenomenon like RNAi (Howard 2013), open frame of transcription (Sieber et al. 2018) and RNA splicing (Weinzierl 1999) by assuming that these phenomena manifest when the corresponding biomolecules undergo the inverse of the operator of involution and then revert back to the four-dimensional configurations by succeeding involution. In addition, as mentioned above, by measuring the lengths of all intergenic distances between the known genes, it should be possible to figure out the number of higher dimensionalities and the distribution of different genes in these dimensionalities. Using this primitive description of genotopic conception of genomic architecture, in the following sections, we will try to deconstruct several features of genomics.

8.12 Distinction Between Genome and DNA Sequence

In the context of the discussion presented in Sect. 8.11, it is possible to discuss the relationship between the DNA sequence of a genome and the conception of a genome in terms of the relationship between G_7 and G_4. However, there is a need to broad-base this discussion on a wider semantic foundation of genomes as implicit in genomics and DNA sequences as implicit in molecular biology. This is because when we evaluate this relationship in a broader semantic context, it helps us to understand why we need a topological paradigm to explicate the semantic ambiguities which have been passed on from molecular biology to genomics. In retrospect, it is intuitively clear that it is the molecular paradigm implicit in molecular biology that dominated the semantics of genomics and therefore, it has restricted the semantic depth of genomics. It is possible to argue that if not the molecular paradigm, what else should be the basis of the articulation of semantics of genomics? After all, all the gene expressions and their interconnectivity is governed by molecular biology. To begin with, there is no denying that it is the molecular biology of gene expressions that is at the heart of the conception of genomics. Therefore, even in the proposed model, the gene's expressions have been assigned to G_4 genotope (which is synonymous with the conventional perspective of genomes).

However, the point is that the molecular paradigm is inadequate to deconstruct long-range influences and genomic functionalities in general. Even from the evolutionary perspective, there is no way to deny that genomes themselves could be units of selection (Huges 1999). Therefore, once we accept this reasoning, we need a description of a framework wherein genomes are embedded just as individual species are embedded in an ecosystem. Moreover, as mentioned above, we need to extend the conception of an ecosystem to genomes themselves. In such a scenario, genes become competing species. Both these features, viz., genomes as units of selection and genomes as ecosystems, cannot be represented in a molecular paradigm which defines contemporary genomics.

It is possible to argue that such a hierarchical framework need not be a topological framework. Moreover, the proposed model rests on two additional radical propositions, viz., the physicality of the topological dimensionalities and a peculiar mechanism for connecting different dimensionalities. Obviously, this model would have been more palatable if the hierarchical framework was an abstract schema. Similarly, it would have been easier if there was no unitary mechanism of changing the levels within the hierarchy. Therein lies the semantic core of biological evolution and natural selection. Neither biological evolution, nor natural selection are abstract processes. Therefore, they are not obliged to fit into mathematical abstractions. By insisting on the physicality of dimensionalities, the proposed model offers a naturalistic framework for defining biological evolution and natural selection. Moreover, if different dimensionalities are physical, it is axiomatic that they must be spatiotemporal dimensionalities. Given the universal template of spacetime, it is imperative that any mechanisms that alter the details of spacetime must be unitary. We can't have variable mechanisms that operate in different regions of spacetime. It is as absurd as postulating different types of gravity operating in a cosmological theory (It is important to keep in mind the distinction between the mechanism by which the gravitational field operates and the different consequences arising from it. Be it a supernova explosion or be it a black hole formation, the outcomes are dramatically different, but the underlying mechanism is identical.). The conception of gauge symmetry demands such a unitary conception of spacetime operators (Friedman 1983, see Chapters 6 and 7).

In a similar sense, different dimensionalities coexist, but they must be interconvertible by a singular mechanism. This singular nature of mechanism ensures that semantic integrity of natural selection is maintained. Otherwise, we will end up with different types of natural selection in different ecological niches. Having looked at the semantic considerations for sidestepping the molecular paradigm in defining genomic architecture, let us look at the phenomenological differences between genomes and their DNA sequences. Upon a little reflection, it is intuitively clear that even in the conventional perspective of genomics, there is an implicit suggestion that genomes behave differently as a whole than the underlying ensemble of molecules consisting of DNA sequences and chromatin. The issue is whether the sum of the molecular properties of these constituents can explain the genomic functionalities of not. In the last few decades, genomics has given ample evidence to suggest that there are features of genomes which cannot be reduced to the sum of the molecular properties of the constituent molecules. As discussed in the previous chapter, various types of regulatory elements cannot be formalized using the DNA framework. It is not a DNA sequence per se that can explain these regulatory elements, but their spatiotemporal arrangements possibly can. The trouble with this reasoning is that the conventional perspective is limited by stereochemical and conformational perspectives. These perspectives have failed to formalize the regulatory elements of genomes. Thus, what separates a genome from its constituent molecules is the emergence of regulatory elements. Therefore, when we move from molecular biology of gene expressions to the regulatory elements of genomes, it is imperative that we must employ configurations of genomes which transcend the stereochemical perspective implicit in molecular biology. The proposed model provides one such framework. Leaving aside the implicit metaphysical aspects of the proposed model, as discussed in the preceding chapters, it provides a systemic way to formalize long-range influences of genomes by assigning different dimensionalities to different higher dimensional configurations of genomes. Moreover, by defining a fixed mechanism, this model provides a method of categorizing different types of long-range influences. Therefore, in the next section, we will try to articulate the nature of the regulatory genome which sits upon the molecular description of genomes.

8.13 Definition of Regulatory Genome

As mentioned above, the proposed model offers a way to conceptualize regulatory elements of genomes by placing them in higher dimensionalities. More importantly, it allows us to think of these regulatory elements as a single framework to be named here as the regulatory genome. Once we accept that different regulatory elements are constituents of a single framework, it is necessary to outline its contours. This is precisely what we will attempt in this section. Prima facie, it is intuitively clear that this regulatory genome too must be viewed as a topological construct. It is also obvious that this entity must supervene the molecular framework of genomes. However, these inferences, by themselves, do not tell us anything about either the structural template or its influence on the molecular framework of genomes. Therefore, in this section, we will focus on two aspects of regulatory genome, viz., the relationship between regulatory genome and genotope and the topology of regulatory genome. Let us begin with the relationship between the regulatory genome and genotope. As discussed in Sects. 8.10 and 8.11 a genotope manifests separately in each dimensionality (albeit containing several lower dimensions within itself), with the fourth-dimensional genotope being synonymous with the molecular description of genomes. Purely from the topological perspective, genotopes may be viewed as topological surfaces and accordingly, the entire genomes as manifolds. In this scenario, the regulatory genome must be viewed as a submanifold occupying multiple dimensionalities simultaneously, the only restriction being that it can't occupy four-dimensional spacetime.

However, this description of regulatory genome raises questions about its relationship with genotopes. Prima facie, we can think of the regulatory genome as an ensemble of different genotopes. It is important to keep in mind that there is no restriction on the number of genotopes that a given dimensionality can have. This is because the dimensionality of a genotope would be decided by the number of electrons present in the highest occupied molecular orbital of genes (or any DNA sequence of interest). Therefore, in any given genome, there will be multiple genotopes occupying the same dimensionality. Therefore, regulatory genome must not be viewed as a string of beads wherein each bead consists of a genotope. Rather, we should think of the regulatory genome as a brocade wherein each string represents a dimensionality and each bead in each string represents a genotope in that dimensionality. Albeit, this brocade is made of dimensionalities and not made up of pearls. This scenario hints at another insight into genomic functionalities. As mentioned above, each dimensionality would have its own metric. Therefore, it is intuitively clear that any two genotopes occupying the same dimensionality would have an identical metric. This provides an opportunity to genotopes occupying the same dimensionality to interact with each other. Upon a little reflection, it is intuitively clear that such interactions between different genotopes occupying the same dimensionality are exactly similar to the cis and trans effects of the conventional molecular biology (Donaldson 2000). Or, to put it differently, cis and trans effects are the four-dimensional stereochemical and conformational versions of a broader class of topological long-range influences. This analogy is indicative of the fact that different dimensionalities of genomes are internally more tightly bound than it appears at first sight. This is exactly parallel to our realization that our conception of genomes as being a linear string of genes needed to be abandoned and replaced our current conception of stereochemical details of genomes.

Now, let us look at the second aspect of regulatory genome, viz., its topology. As mentioned above, the regulatory genome must be viewed as a submanifold of the parent manifold of a genome. However, it is necessary to understand how this submanifold is embedded in the framework of the overall genomic architecture. As discussed in the preceding chapters, from the evolutionary perspective, it must arise from genomic singularity. At the same time, as mentioned above, it must be supervening over the molecular description of genomes. However, this reasoning leads to an anomaly. On the one hand, the evolutionary perspective suggests that it must have evolved from genomic singularity. On the other hand, according to the description given above, the regulatory genome arises from the different dimensionalities occupied by different DNA sequences. This is contradictory. This is because according to the evolutionary perspective, the regulatory genome should enjoy ontological primacy over the molecular perspective of genomes. However, it would imply that abstract functionalities of genomes evolved prior to the molecular details of genomes. However, that is the last thing that science can support. How can a functionality emerge prior to manifestation of its structuralism? To admit this possibility amounts to some form of creationism. However, upon a little reflection, this anomaly hides within itself a more fundamental semantic proposition of genomic architecture. Let us see how.

This anomaly can be resolved if we accept that genomes possess a functionality of self-reference. When different dimensionalities of genomes interact with each other, they are manifesting self-reference. This happens because the proposed model postulates an entity named here as genomic singularity. Because genomic singularity is nonstructural in nature, whenever the genomic singularity of a genome interacts with any of the lower dimensionalities of the genome, it provides self-reference. Therefore, when the four-dimensional configurations of genomes (G_4) interact with genomic singularity, it passes through the intermediate dimensionalities. Therefore, whenever a DNA sequence acquires sufficient length and its attendant complexity, it gives rise to genomic singularity. Therefore, genomic singularity doesn't preexist prior to the existence of a sufficiently long DNA sequence (which would qualify it be a genome), it is only when a DNA sequence acquires sufficient complexity that genomic singularity manifests itself. The same logic applies to other higher dimensional configurations like G_6 and G_7. They don't exist prior to the emergence of a sufficiently long and complex DNA sequence. However, once such a DNA sequence is formed, genomic singularity and the intermediate dimensionalities manifest themselves making that DNA sequence a genome.

Because genomic singularity once formed would interact with all the lower dimensionalities starting from G_4, each of these lower dimensionalities would manifest self-reference whenever they interact with genomic singularity. However, the keypoint is that once formed, genomic singularity (and other higher dimensional configurations of genomes) would undergo changes within itself. These changes are not in the form of stereochemical and conformational changes (because there are no time-like features in these dimensionalities), but in the form of information transformations. These information transformations are governed by quantum mechanical fluctuations of the fine structure of spacetime. Therefore, this leads to two types of changes. Firstly, there are changes dictated by mutations of individual genes and their DNA sequences which keep changing the overall topology, including that of genomic singularity. Secondly, due to quantum mechanical fluctuations of the fine structure of spacetime, there are changes within each of these higher dimensional configurations of genomes. While the former gives rise to randomness implicit in the Darwinian paradigm, the latter changes give rise to broad patterns of genomic architecture which get reflected in the phenomena like punctuated evolution (Gould 2007) and speciation (Coyne and Orr 2004). The key point is this: It is the functionality of self-reference that avoids creationism. It is self-reference that allows different higher dimensional configurations of genomes (including genomic singularity) to interact with the four-dimensional configurations of genomes without invoking any creationist argument.

While this scenario makes sense, it gives rise to two problems. Firstly, if this scenario is valid, then it is possible to argue that even the DNA sequences can alter the higher dimensional configurations. Therefore, why should there be a regulatory genome? Alternatively, why can't we think of DNA sequences as regulators of genomic functionalities? The second problem is that if this scenario is valid, it seems reasonable to think maybe Lamarckism wasn't wrong after all. Both these problems require further discussion. While we will discuss the first problem in the next section where we will discuss the mechanism by which the regulatory genome influences gene expressions. This mechanism tells us why reverse changes wherein DNA sequences regulate genomic functionalities, don't take place. The second problem of Lamarckism will be discussed in Sect. 8.19.

8.14 Relationship Between Regulatory Genome and Gene Expressions

In this section, we will discuss two aspects of the regulatory genome. Firstly, we will discuss the topological representation of regulatory elements which arise from the regulatory genome and influence gene expressions. Secondly, we will discuss why this topological mechanism cannot work in the reverse direction whereby DNA sequences modify higher genomic functionalities. While in general, higher dimensional configurations of genomes influence the four-dimensional configurations of genomes and modify molecular biology of gene expressions, our focus in this section will be only on those influences which are regulatory in nature. In other words, we will be looking at only those long-range influences which either activate or suppress gene expressions and those that decide the duration of gene expressions. There will be other long-range influences like modularization and its evolutionary consequences. However, we will exclude them from the present discussion.

Even among the regulatory genome, individual elements of regulation of gene expressions would not be discussed here. This is partly because they have been discussed in Chap. 7 and partly because we wish to develop the importance of the unit of genotope in the regulatory genome. It is apparent from the above discussion that the regulatory genome itself cannot be thought of as a single higher dimensional configuration of genomes. Rather, according to the description given above, the regulatory genome must be viewed as a submanifold of the parent manifold of genomes spread over multiple dimensionalities. As discussed in Chap. 7, this is helpful in formalizing different types of long-range influences and the magnitude of these long-range influences. Given this background, in this section, we will discuss three aspects of the relationship between regulatory genome and gene expressions, viz., (1) regulatory genome as a functional module, (2) regulatory genome as a genotope, and (3) evolution of the regulatory genome.

Let us begin with the first aspect of the regulatory genome itself being a module of genomes. Prima facie, this analogy appears attractive. However, from the semantic perspective, this analogy is incongruent with not only the semantics of the proposed model, but also with the semantics of the Darwinian paradigm. Let us understand why. If the regulatory genome were to be a module by itself, it would imply that regulatory functionalities have arisen later and more importantly, they have arisen in a specialized context. While the possibility of a regulatory genome having arisen later is not inconsistent with the conventional perspective of natural selection, it implies a certain ergodicity in the course of biological evolution. Admittedly, abrupt changes during the course of biological evolution have been observed at several stages like the emergence of eukaryotic organisms (Garrett and Klenk 2007, see Chapter 3) or the morphological diversity during the pre-Cambrian era etc. (Cabej 2020). However, at a molecular level such abrupt changes in the functionality is something that has no precedence. Moreover, from the evolutionary perspective, it is more likely that regulatory elements have evolved after the advent of multicellularity (Niklas and Newman 2016). Therefore, it is reasonable to think that the need for synchronization between the life cycles of different cells of an organism must have the evolution of regulatory elements. However, this scenario leaves one question unanswered. If different regulatory elements evolved independent of one another, how were these elements synchronized? Admittedly, in the conventional perspective, this reasoning acquires a broader context. From the conventional perspective, it is difficult to explain not just the synchronization of different regulatory elements, but it is difficult to explain the conception of modularity per se. Therefore, prima facie, from the conventional perspective, the regulatory genome could have evolved later. However, it leaves several questions unanswered. Before looking at the explanation provided by the proposed model, let us look at the semantic difficulty with the conception of the regulatory genome being a module by itself. For a moment, let us assume that the conventional perspective of modularity is correct. We will also keep aside the ambiguities about modularities per se, about how it could have evolved. In other words, let us assume that the regulatory genome is a module by itself and has evolved due to natural selection just like any other module. The problem with this scenario is that it implies that regulatory functionality has evolved from a very narrow context. This is far from accurate. long-range influences do not manifest only under certain special conditions. These influences are manifest in all types of gene expressions. Therefore, it is a category mistake to think that the regulatory genome is a genomic module. This is corroborated by the conventional perspective as well. Different long-range influences manifest in different genomic modules. Therefore, it makes sense to think of the regulatory genome as a submanifold of the parent manifold of a genome. However, this conception doesn't help us to understand the exact nature of the regulatory genome.

Therefore, let us look at the second aspect of regulatory genome being a genotope. It is legitimate to wonder whether if regulatory genome is not a module per se, how can it be a genotope. This brings us to the distinction between a genomic module and a genotope. As discussed above, the conception of a genomic module rests on multiple functionalities tied together by synchronization. On the other hand, a genotope is conceptualized as a topological unit irrespective of its constituent functionalities. Moreover, in the case of a genomic module, since it synchronizes different gene expressions, it is intuitively clear that different temporal and spatial long-range influences must have comparable magnitudes, if only to effectively synchronize expressions of different genes. However, according to this model, the magnitudes of long-range influences depends on the difference between the numerical values of the dimensionalities devolving into the four-dimensional spacetime (where gene expressions manifest). Therefore, it is intuitively clear that a genomic module must occupy higher but neighboring dimensionalities. However, a genotope by definition, refers to genomic configurations present in the single dimensionality (albeit containing several lower dimensions within itself). Therefore, it is intuitively clear that regulatory genome cannot be conceptualized as a genotope.

Thus, the regulatory genome must be conceptualized as a genomic configuration spread over multiple neighboring dimensionalities. However, the regulatory genome must not be viewed as a unit of genomic architecture, but rather as an architecture itself. Rather, the regulatory genome must be viewed as a governing principle of genomic architecture. This inference is consistent with the earlier suggestion that regulatory framework operates across the different genomic modules and it is not confined to any particular genomic module. Genomic architecture must have acquired its current template because of these elements of the regulatory genome. These regulatory elements owe their origins to the nature of spacetime which manifests itself in its different dimensionalities. This scenario also implies that these types of long-range influences and its influence on the resulting architecture would manifest not just in genomes, but it would manifest in any natural phenomenon occupying multiple dimensionalities simultaneously. Thus, it is the inherent structural template of spacetime (in the form of its fine structure) that builds this involutive architecture irrespective of the domains. If this scenario is valid, then it is axiomatic that even our cognitive faculty, being a natural phenomenon having modularity, too must possess a comparable architecture and multiple long-range influences. This will be discussed in the following monograph.

This brings us to the third aspect of regulatory genome, viz., its evolution. While we will discuss the evolution of genomic architecture in Sect. 8.17, here we will confine ourselves to the evolution of the regulatory genome per se. The distinction between the evolution of genomic architecture and the evolution of the regulatory genome lies in the fact that the former rests on the genomic functionalities whereas the evolution of the regulatory genome rests on the nature of spacetime. This is important for several reasons. Firstly, it suggests that the regulatory framework is a domain agnostic feature of any natural phenomenon involving multiple dimensionalities of spacetime simultaneously. Secondly, while the regulatory framework may be identical in natural phenomena, their functionalities are not domain neutral. Rather, the functionalities of natural phenomena are domain specific. This is because according to this model, the structural details of each dimensionality of spacetime is unique. Therefore, the functionalities of any natural phenomenon would depend on the dimensionality in which they manifest. Thus, if cognitive faculty, as a natural phenomenon, has different functionalities than the functionalities of any other natural phenomena, say, genomes, then it is imperative that these two natural phenomena must manifest in different dimensionalities of spacetime. Thus, the evolution of the regulatory framework is universal, and it would manifest every natural phenomenon existing in multiple dimensionalities of spacetime simultaneously. On the other hand, evolution of functionalities is dependent on the innate functionalities of a natural phenomenon undergoing natural selection. This distinction between the evolution of regulatory genome and the evolution of genomic functionality has been implicit in the conventional perspective. While morphological diversity characterizes natural selection, it is still executed by only a few mechanisms. Before we look at the evolution of genomic architecture in general, it is necessary to deconstruct the mechanism of changing the dimensionality. Therefore, in the next two sections, we will discuss how this mechanism plays a dual role in regulating and giving rise to morphological diversity.

8.15 Operator of Involution as a Genomic Regulator

In the previous section, we discussed the relationship between regulatory genome and gene expressions. It was suggested that the regulatory genome must have evolved within the broader process of the evolution of genomic architecture. At the same time, it was pointed out that the evolution of the regulatory genome and the evolution of genomic architecture are not synonymous with one another. This doesn't seem to be logical at first sight. However, this ambiguity arises from the fact that genomes act as units of selection and ecosystems by themselves at the same time. This dual role of genomes might appear enigmatic. However, this duality owes its origin to the very conception of involution in this model. At the heart of this enigma lies in the fact that the operator of involution is essentially a mechanism for self-reference. Thus, the ecosystem allows biological evolution through self-reference. Moreover, once we accept this self-referential perspective of biological evolution, it is intuitively clear that it can manifest in every case where it is possible to create self-reference. Therefore, genomes must be viewed as products of natural selection as well as the beginning of natural selection. Genomes arise from ecosystems by natural selection and in turn, become ecosystems wherein different genomic functionalities are naturally selected. Thus, it is the self-referential nature of involution which is used here in formalizing natural selection (and by implication, biological evolution) that gives rise to the dual role of genomes. It is important to keep in mind that this ascription of dual roles to genomes is not an artifact of the mathematical formalism employed in this model. This duality of genomes has always been implicit in the conventional perspective. The debate on the unit of selection (Okasha 2010) is a testimony to the underlying semantic ambiguities of the conventional perspective. The proposed model has merely made it explicit and perhaps formalized it.

If this reasoning is correct, then it is imperative that we must deconstruct the self-referential nature of the mechanism by which biological evolution and natural selection are formalized in this model. Therefore, in this section, we will look at the operator of involution as a genomic regulator and in the following section, we will discuss the role of this operator as a source of complexity and diversity. The key point is that in any self-referential system, the system organizes itself using self-reference. However, while doing so, it differentiates its components into well-defined units. In this process, complexity is an inevitable outcome. Thus, the proposed model answers one of the most fundamental questions about the emergence of complexity during natural selection. While the conventional perspective has no answer to this question, the proposed model offers an intuitive answer. During biological evolution and natural selection, complexity arises because both these processes are self-referential.

Returning to the present discussion, let us understand how the operator of involution can act as a regulator. In order to understand this, let us postulate that a given genome has acquired sufficient complexity in the form of several different genomic functionalities. Under the optimum physiological conditions, these different genomic functionalities would naturally begin operating. In the conventional perspective, the focus has been on the competitive expressions of different genes in a given physiological states where some resources are limited (After all, Darwin's own conception of natural selection was influenced by Malthusian economics (Flew 2017)). Therefore, the environment, by changing the availability of different resources, would influence the expressions of different genes, thereby rendering natural selection. However, in the conventional perspective, the resources are conceptualized as the resources available from outside of the organisms. Thus, there are two frames of reference. From the perspective of the individual organisms, the resources come from outside. On the other hand, from the perspective of an ecosystem, the resources come from inside. However, the conventional perspective, perhaps because of these dual frames of reference, has never thought about the internal resources available to different genes waiting for expressions.

However, in this model, there is only one frame of reference. All resources are internal and therefore equally important. Now, according to this model, the resources need not be only in the form of chemical messengers or chemicals providing calories for gene expressions. According to this model, information content itself could be thought of as a resource. Therefore, different genes wait for the information content as a resource necessary for their expressions. This is precisely what the operator of involution provides. The operator of involution provides information to different genes, but in different ways. Depending on the underlying topology, a single operator of involution provides different types of signals to different genes. Therefore, different genes react differently to a single operator of involution. It is this differentiating capacity of the operator of involution that acts as a mechanism for regulation of gene expressions. It is important to keep in mind that this scenario all suggests that the operator of involution can also act as an agent of natural selection. It might sound preposterous at first sight. However, it is important to keep in mind that this operator is actually a physical process involving the changes in the dimensionalities of spacetime itself. Therefore, each operation of involution also brings about the changes in the kinetic energy of the recipient genes and that too in a differential manner. Just as the *lac* operon (Miller and Reznikoff 1980) requires the changes in availability of lactose to activate itself, the operator of involution provides the necessary energies to different genes in different degrees, thereby allowing different activations of different gene expressions. Therefore, this operator, by altering the availability of information content, provides a change in the resource in the form of information to different genes. This naturally affects the timing and duration of expressions of different genes, thereby setting up a race to survive. The only difference being that competitive survival occurs between the products of genes and not between the genes themselves. Since the Darwinian paradigm insists that it is the phenotype that is the unit of selection and not the genotype, the proposed model offers a new perspective on why phenotype must be the unit of selection. It is because phenotype (which is synonymous with the product of gene expressions in this context) needs resources to manifest itself. A genotype, on the other hand, being a part of a larger ensemble of genomes survives whether it can express itself or not. Admittedly, even genotype can perish, but only indirectly if the parent organism fails to survive due to improper phenotype. This is essentially a conventional perspective.

The proposed model simply expands the notion of resources to include the information also as a resource for competitive survival. Thus, the operator of involution, according to this model, acts as a regulatory mechanism and since it allocates information content differentially to different genes, it also acts as an agent of natural selection. In the next section, we will discuss how this mechanism of information transfer becomes a source of complexity.

8.16 Operator of Involution as a Source of Complexity

As discussed in the preceding sections, the operator of involution can represent a variety of aspects of genomic architecture. The reason for this is not the extraordinary nature of this operator. The reason why this operator brings about so many changes in genomes is that it represents self-reference. Any system capable of manifesting self-reference also has a capability to reorganize itself. Moreover, if self-reference is created by information transfers, it is inevitable that reorganization would lead to an increase in complexity. Purely from the epistemological perspective, self-reference must inevitably involve information transfers, it is axiomatic that any system capable of self-reference would become more complex with the passage of time. Since this is a generic feature of self-reference, the emergence of complexity would manifest in a domain agnostic manner. We already know that similar emergence of complexity occurs during the developmental stages of our cognitive faculty. What remains domain specific is the nature of complexity. Thus, while our cognitive faculty and genomes give rise to complexity, the types of complexities would be different in both these cases. In this section, we will discuss how the operator of involution (or rather, its mechanism) introduces complexity. It is important to note that the emergence of complexity occurs within the organism during its developmental stages as well as during natural selection. As discussed in the previous section, since the operator of involution acts as a regulatory mechanism as well as an agent of natural selection, the emergence of complexity also occurs in two different frameworks, individual development and biological evolution.

Let us begin with the framework of individual development of an organism. As discussed above, we can think of the operation of involution as an inward folding of one of the dimensions of genomic configurations into its remaining dimensions. Since genomic architecture exists in multiple dimensionalities simultaneously, each dimension within the genomic configurations in a given dimensionality possesses a certain amount of information. Admittedly, we still don't know the exact nature of information content of different genomic configurations present in different dimensionalities. However, to the extent each genomic configuration has its information content spread over all the dimensions of these multiple dimensionalities simultaneously, what matters is not the exact nature of information content, but that irrespective of its exact nature, this information content present in the dimension undergoing involution would be smeared over the information contents of the remaining dimensions of that genomic configuration. Therefore, it is intuitively clear that this mechanism ensures that at the end of the operation of involution, the resulting genomic architecture would occupy lesser number of dimensionalities. At the same time, since information content of a given genome cannot disappear, the information gets accommodated in the lesser number of dimensions. Therefore, it is inevitable that the resulting configuration of a genome which has undergone involution would contain more complex information content. This is true irrespective of the details by which the information content is transferred. What is peculiar about the proposed mechanism is that it quantifies the information transferred and the type of complexity it gives rise to. Therefore, at least in principle, it is possible to verify the proposed model by constructing an information theoretical model of genomes (Akalin 2021) and measuring the extent of information transfers. While this scenario is intuitive, the scenario of the emergence of complexity during biological evolution and natural selection needs to be deconstructed separately.

Let us try to deconstruct the evolutionary perspective of the operator of involution. The key question is this: if the operator of involution increases the complexity of an organism during its developmental stages, does it also perform the same function during natural selection? Prima facie, if the answer to this question is yes, then the enigma of the emergence of complexity during natural selection can be explained. However, in order to justify this answer, we need to explain how this operator increases complexity even during natural selection. This is not easy. In the case of an individual genome, this operator functions because of the functionality of self-reference of the genome. However, in the case of natural selection, unless we demonstrate that the ecosystem is also capable of manifesting self-reference, we can't justify the proposition that the operator of involution also increases complexity during natural selection. Admittedly, the conception of an ecosystem implicitly accepts that all the components of that ecosystem are bound to one another. However, there is no theoretical rationale behind this romantic notion. At best, we can think of an ecosystem as an interactive ensemble wherein different components interact with each other. To assume that an ecosystem has an identity of its own (the one that is over and above these interactions among its constituents) requires some additional theoretical grounds.

This is where the proposed model has something to offer. One of the key features of this involutive approach is that it is inwardly-directed. Therefore, whenever we try to formalize the interactions between any two or more components of an ecosystem, the involutive paradigm demands that these interactions must be formalized from the perspective of the parent system and not any of these subsystems. Thus, the ecosystem in this description becomes an integrated entity by default. Moreover, once we define an ecosystem in this manner what we obtain is a higher dimensional perspective. This higher dimensional perspective teaches us that our earlier conception of different subsystems interacting with each other was based on the perspective of the lower dimensionality. Therefore, according to this model, just as our conventional lower dimensional perspective (of different subsystems interacting with each other separately) is valid in the lower dimensionalities, the higher dimensional perspective is also valid for that dimensionality. In other words, the proposed model eliminates any preference for a particular dimensionality. All dimensionalities of an ecosystem and all the perspectives available in these dimensionalities are equivalent. There is no preferred dimensionality. This is something very similar to the cosmological principle (Barrow and Tipler 1986) which states that there are no preferred frames of the universe from which we must define the cosmological theories. The universe is the same no matter from where we observe it. The same rationale applies here. There are no preferred dimensionalities. It is just that our cognitive faculty operates from a particular dimensionality that we find an ecosystem as an entity consisting of multiple components. There exists a dimensionality wherein different subsystems of an ecosystem cease to have their independent existence.

Once we accept this reasoning, it is intuitively clear that because of its inwardly-directed influences, the operator of involution can also increase complexity during natural selection just as it does in the case of an individual genome. This explains the mechanism by which the operator of involution operates in both these cases. However, we still need to understand how the genomic architecture could arise through the influence of the operator of involution. Therefore, in the next section, we will try to deconstruct the origin of genomic architecture itself.

8.17 Evolution of Genomic Architecture in the Proposed Model

There are several semantic ambiguities in the Darwinian paradigm about the evolution of complexity during natural selection (Bonner 1988). This is particularly relevant to the inherent randomness implicit in the Darwinian paradigm (Bonner 2013). If natural selection were to be completely random, there would have been no evolution. The point is not whether natural selection is random or not. The point is the degree of randomness. Even in its earliest articulation (Darwin's writings (Hodge and Redick 2009)), the emphasis was on descent with modification. While the modifications can be random, the descent, by its very conception, implies continuity. Therefore, natural selection rests on two mutually incompatible semantic propositions. While descent represents continuity and therefore predictability, modification represents randomness and unpredictability. Given this semantic dichotomy, the Darwinian paradigm has always been at odds with any structural templates that induce rigidity. In fact, it is this reluctance to impose structuralism of the conventional Darwinian paradigm which has prevented formalizing genomic architecture. As discussed in the preceding chapters, the conception of genomic architecture is a *fait accompli*. It is just that we refuse to accept its inevitability lest, it undermines the randomness implicit in the Darwinian paradigm. However, given the large number of genes and their synchronization, we have no option but to look for a suitable genomic architecture. Our current efforts are based on the incremental increase in our knowledge of different genes and their interconnectivity. There are no models of genomic architecture which are based on theoretical considerations. It is in this context that we must evaluate the proposed model and seek explanation for two questions, viz., (1) whether the proposed model can justifiably predict a genomic architecture? (2) whether the proposed model can reconcile Darwinian randomness with the conception of a genomic architecture?

Let us begin with the first question, viz., Whether the proposed model can predict a genomic architecture? As mentioned above, the conventional perspective is consistent with the possibility of a well-defined genomic architecture, but it doesn't endorse such a possibility because it might compromise the essential Darwinian randomness. This reservation about genomes having a fixed architecture arises because such a possibility may lead to some form of creationism or at least some design principles. Both these doctrines are untenable because they seemed to be arbitrarily imposed from outside. However, if it can be demonstrated that genomic architecture arises internally and its fixity is a consequence of natural selection, then the original reservations are annulled. Therefore, the question is whether the proposed model explains how the involuted architecture can arise during biological evolution and can be fixed by natural selection, the proposed model can justifiably predict genomic architecture. The proposed model doesn't employ any ad hoc proposition other than mathematical constructs. Admittedly, it can be argued that even mathematics employed here is ad hoc, and therefore, it brings in some design principles. However, this is not a valid argument. Right from the days of conception of the statistical approach of population genetics (Provine 2001) to the present day information analytics (Akalin 2021), mathematics has been employed in articulating the details of natural selection. Admittedly, there is a far bigger issue of "unreasonable effectiveness" mathematics (Wigner 1960). However, that applies to science in general and not specifically to natural selection. Therefore, mere employment of mathematical constructs in explaining biological evolution and natural selection cannot be taken as an acceptance of some design principles.

In addition, the proposed model employs the notion of self-reference to justify the use of the involutive algebra. However, the belief that genomes manifest self-reference is not an unreasonable belief. It is an established fact that some configurations of genomes (say, stereochemical configurations or conformations) influence the sequence and duration of gene expressions. From the logical perspective, this is nothing but self-reference. Therefore, it seems reasonable to assert that the proposed model offers a way to define genomic architecture from the intrinsic features of genomes and doesn't bring in any ad hoc or transcendental propositions. It is important to keep in mind that this reasoning doesn't, by itself, imply that the proposed genomic architecture is true ab initio. Far from it. The proposed model merely provides a testable and verifiable hypothesis about genomic architecture.

This brings us to the second question, viz., Whether the proposed model can reconcile the Darwinian randomness with the conception of a genomic architecture. As mentioned above, the reservation about genomes possessing a definitive architecture arises from the fear that it may undermine the randomness implicit in the Darwinian paradigm. As discussed in the preceding chapters and in the preceding sections, prima facie, genomic architecture per se, would compromise randomness. However, this restricted randomness doesn't lead to any degree of determinism. However, this restriction is implicit right from the conception of natural selection. Be it Darwin's own conception of modification with descent (Hodge and Redick 2009) or Bayesian conditional probabilities (Press and Clyde 2003), all along the randomness was meant to be restricted. In fact, had it been otherwise, there would be no phylogeny altogether. Therefore, the major obstacle is not the restrictions imposed by a theoretical construct, but whether it leads to any deterministic doctrines or not. As discussed above, this is not the case with the proposed model. This brings us to the end of the discussion on the semantic considerations of the evolution of genomic architecture in natural selection and in the proposed model. In the next section, we will discuss the importance of phylogeny in the proposed model.

8.18 Significance of Phylogeny in the Proposed Model

Phylogenetics (Bromham 2008) has played an important role in refining our intuitive understanding of natural selection. Moreover, it has provided us with a methodology to quantify the degree of evolutionary linkages. However, the most fundamental aspect of phylogenetics is that it crystallized the fundamental semantic proposition of the Darwinian paradigm, viz., descent with modification. When Darwin's theory was published, nothing much was known about genes (let alone molecular biology). Therefore, it was Darwin's prescience to articulate the fundamental idea in this most popular phrase. However, it was left to the pioneers of phylogenetics to transform this simple and evocative phrase into a quantitative discipline. Having said that, it is time to reflect on what is passed on and what is modified during descent with modification. Obviously, Darwin was referring to morphological features and perhaps, by implication, the inheritable traits behind these morphological features. With the advent of population genetics, we realized that what was passed on and modified were the individual genes. Molecular biology taught us that it is the DNA sequence that is passed on and modified.

However, the next transition of what is being passed on and modified when we ushered the discipline of genomics is missing. In view of our current understanding of genomics, it seems reasonable to think that what is being passed on and modified during natural selection is the shape. To be honest, the boundary separating genomics and molecular biology is blurred. However, if we wish to draw such a boundary, it has to be a boundary that separates different long-range influences from the individual gene expressions. Gene expressions are nothing but molecular biological processes. However, what initiates, controls and terminates individual gene expressions is essentially a genomic mechanism. Thus, if a unit of information that was passed on with modification in molecular biology is a DNA sequence, the unit of information that is passed on with modification in genomics must be the shape of different units of genomes. It is the conformations that bring about the phenomenon of chromosome territories (Fritz 2014). Similarly, it is the stereochemical configurations of chromosomes that bring about cis and trans effects (Donaldson 2000). However, it is always the shape that is the unit of information that is passed on with modification.

The proposed model takes this idea one step further. It postulates that the information that is passed on with modification is the topology of genomes. Moreover, the unit of information that is passed on with modification is the dimensionality of genomes. Once we accept this reasoning, it is intuitively clear that time has come to invent the discipline of phylogenomics which instead of measuring evolutionary linkages on the basis of the variations of nucleotides, measures the evolutionary linkages on the bases of variations of topology of genomes. The main objective of this section is to demonstrate that the proposed model is a natural progression from Darwin's theory. The principle of descent with modifications remains unchanged. What changes is the identity of what is passed on with modification. This transition from inheritable traits to genes to DNA sequence to stereochemical shapes to topological dimensionalities provides semantic continuity. This transition is also necessary if we wish to formalize the Darwinian paradigm in a domain agnostic framework for all the natural phenomena. In the next section, we will revisit the classical debate on Darwinism vs. Lamarckism.

8.19 Darwinism Vs. Lamarckism

The debate over Darwinism vs. Lamarckism has several nuances (Steele et al. 1998). Each doctrine has different logical schema, different semantic propositions and an uncanny way to remain topical. With the passage of time, it was felt that the Darwinian paradigm had finally prevailed over the Lamarckian model. However, with the discovery of epigenetic processes we are once again back to square one. The problem with this debate lies in our tendency to take rigid positions. For instance, with the advent of molecular biology, one would have thought that Darwinian randomness would find its correct expression in molecular phenomena like mutations, mistakes in replications of DNA sequences, chromosomal translocations, etc. However, molecular biology gave us another dogma called the central dogma, viz., the information passes from DNA to RNA and never the other way around. It is important to keep in mind that this observation was based on the vast amount of experimental evidence. What was perhaps wrong was its dogmatic rigidity. Therefore, when we eventually discovered epigenetic processes (Robert 2004), we had to abandon that rigidity. The issue is not whether there exists any such inviolable propositions. The issue is whether our tendency to take rigid positions on such issues hampers our scientific research or not.

Does a universal theory point toward some Platonic absolutes (Panza and Sereni 2013) which justify our dogmatic belief in that theory? Not really. Firstly from the Godelian perspective (Smullyan 1992), the existence of such a universal theory devoid of any exception is ruled out. There cannot be any such theory. Secondly, from the Popperian perspective (Popper 1963), a scientific theory must be in principle, falsifiable. In other words a scientific theory can never be a dogma. No scientific theory is right or wrong in any absolute sense. A scientific theory is right or wrong in a given context. Admittedly, this agnostic conception of truth is as dangerous as dogmatic perception of truth. However, this is what our understanding of knowledge suggests. There is only one way to resolve this dilemma. We must resort to Hegelian dialectics (Deng 2022). There must be some framework in which the apparently paradoxical propositions can be reconciled. Therefore, instead of going into the highly nuanced debate on this topic of Darwinism vs. Lamarckism which is available in literature, we will try to understand whether the proposed model can offer any reconciliation between these two doctrines.

For this purpose, we will take a simplistic description of both these approaches and see whether the proposed model can accommodate them in its topological framework. In retrospect, it is intuitively clear that the difference between these two approaches lies in the role of the environment. In the Darwinian paradigm, the environment doesn't play any active role in natural selection. Its role is limited to altering resources necessary for survival. The resulting competition has its own logic. Driven by the inexorable force of the logic of survival, the outcomes are dramatically different. There is a certain degree of inevitability in Darwinian competitive survival (Though there is no inevitability about who would survive). In contrast, the Lamarckian model assigns an active role to the environment. Rather, it assigns an active role to the adaptability of different species to respond to the changes in the environment and their ability to pass on adapted features to the next generation. According to the Darwinian paradigm, adaptability of species is passed on to the next generation but not the adapted features. On the other hand, according to the Lamarckian model, it is the adapted features that are passed onto the next generation. Thus, both these doctrines accept that the competing species pass on the necessary information to the next generation. What they differ from one another is what is passed on to the next generation. According to the Darwinian theory, it is the adaptability that is passed on, and according to the Lamarckian theory, what is passed on is the adapted features themselves.

Apparently, we know that the Darwinian paradigm is nearer to the truth than the Lamarckian theory is. However, the actual truth is somewhere in between. Therefore, let us try to reframe both these theories in the language of information transfers. This is justified because both these theories agree that something is indeed passed on to the next generation. The choice of information theoretical framework is also justified because the proposed model deals with information transfers (albeit from one dimensionality to another). If we employ molecular biology to define information transfers, the resulting comparison is intuitively clear. The Darwinian information transfers can be exactly stated as the above mentioned central dogma, viz., the information content is passed on from DNA to RNA and never the other way around. Similarly, the Lamarckian information content can be thought of as epigenetic silencing of DNA. This is intuitively clear because what the DNA passes on is the adaptability in the form of RNA sequences. It is the subcellular cytoplasmic environment that tailors RNA via splicing (Weinzierl 1999), RNA interference (Howard 2013) into the required traits to facilitate the cell to survive. Similarly, in the case of epigenetic silencing, it is the adapted feature in the form of secondary messengers that is passed on to the next generation via epigenetic silencing.

However, we will take a broader perspective of information transfers. Therefore, let us summarize the nature of information transfers as postulated by the proposed model and then try to deconstruct both these doctrines in that scenario. According to the proposed model, information transformations can occur between two dimensionalities and information transfers within the dimensionality. Therefore, let us understand natural selection using this model. Albeit, this has been discussed in greater detail in the preceding chapters, but here we will look at a specific model in which we can compare Darwinism and Lamarckism. According to this model, the Darwinian concept can be summarized thus: The information content of genotypes would be passed on via duplication accompanied by occasional changes due to translocations, imperfect copying, etc. According to this model, all these instances are examples of information transfers, and therefore, they occur in the same dimensionality. However, according to this model, information changes that occur during translation to RNA and then to proteins, are information transformations. As explained in the preceding chapters, this is because in the case of biomolecules, the dimensionality is defined by the number of electrons present in the highest occupied molecular orbital of biomolecules. Therefore, the dimensionalities of a given DNA sequence, its RNA counterpart and the resulting proteins are different.

Let us now look at the corresponding scenario of epigenetic silencing. In this process, there are no information transformations, but only information transfers. This is because the primary mode of silencing is methylation or acetylation. None of these modifications alter the highest occupied molecular orbital of the DNA sequence being modified. Surprisingly, the process of natural selection also constitutes information transfers because it doesn't alter either the genotype or phenotype. In contrast, biological evolution requires information transformations. With these observations in place, let us see how Darwinism and Lamarckism can be accommodated in a single framework. On the one hand, there are transformations from genotype to phenotype and the more generalized biological evolution wherein information transformations take place. According to the proposed model, this involves the changes in the dimensionalities. On the other hand, we have the role of the environment in Darwinism and Lamarckism wherein only information transfers take place. According to the proposed model, this doesn't involve any changes in the dimensionalities. Thus, according to this model, prima facie, natural selection either by Darwinian or by Lamarckian processes are equivalent. In other words, there is nothing to distinguish between these two types of processes. However, there is one fine difference between these two processes which ensures that the Darwinian model of natural selection prevails. This refers to the feature of self-reference. The Darwinian model of natural selection doesn't require the environment and the species to be a unified entity. Therefore, there is no manifestation of self-reference. However, in the case of Lamarckian model, the reverse information transfers can occur only if the highest dimensionality is involved in the information transfer. In other words, it must manifest self-reference.

Now, this distinction explains why the Lamarckian model works in epigenetic processes but not in natural selection. In epigenetic processes the system consisting of genomes, surrounding cytoplasm (or the biological cell itself) is capable of manifesting self-reference. Therefore, the reverse information transfers can be brought about. However, in the ecosystem, there is no way to ascribe and justify the self-reference to an ecosystem. This might sound contradictory to what was discussed above in Sect. 8.15. However, this is not the case. Let us see why. It was proposed above that ecosystems can manifest self-reference at the higher dimensionalities. However, as mentioned in Sect. 8.15, this is true only for the higher dimensionalities in which an ecosystem is formalized. Epigenetic processes occur only in the four-dimensional configurations. Therefore, if we try to formalize ecosystems from the four-dimensional perspective, it would not manifest self-reference Therefore, the Lamarckian model wouldn't be operational in natural selection which operates in an ecosystem, but it would be operational as epigenetic processes at a cellular level where self-reference manifests through genomes.. This deconstruction also provides a sobering effect to the romantic allure of Gaia hypothesis. Maybe, the conception of Gaia hypothesis (Lovelock 2000) is a cognitive artifact and not a reality. More importantly, we must continue treating an ecosystem as an ensemble and not as an independent singular entity at least from the four-dimensional perspective. However, this scenario also complicates the definition of Life. The biological cell may be similar to an ecosystem, but it is not. The biological cell has a feature of self-reference, and therefore, it is not an ensemble of a genome, cytoplasm, and lipid bilayer. It is an independent entity having a capacity of self-reference. Fortunately, the proposed model offers a way to formalize this distinction between an ecosystem and a biological cell. An ecosystem can be formalized as a topological manifold. However, a biological cell can only be formalized as an involuted manifold. This functionality of self-reference that separates living from nonliving also raises a question about the internal and external perspectives of such a system capable of manifesting self-reference. Therefore, in the next two sections, we will discuss the internal and external perspectives of genomes. This is important because our conventional perspective has been that of the internal perspective of genomes. This is an internal perspective because our cognitive model and genomes cohabit the three-dimensional space. However, it doesn't necessarily imply that genomes would retain the same structuralism when viewed from an external perspective of a higher dimensionality.

8.20 Structural Template of Genotope from Three-Dimensional Perspective

It was suggested that in this and the next section, we will discuss the internal and external perspectives of genomes. However, due to the sheer magnitude of the genome, it is not easy to construct such perspectives, particularly when we know very little about genomic architecture. Therefore, in order to simplify this discussion, we will confine ourselves to the conception of genotope per se. Therefore, in this section, we will discuss how a genotope may appear to be from the three-dimensional space. In other words, why did we miss observing genotopes. In the next section, we will venture out and try to conceptualize the perspective of a genotope from any of the higher dimensionalities. Another reason for choosing a genotope rather than genomes for this deconstruction is that it would give us chance to find out whether the conception of a genotope is topologically consistent or not. This in a way, can provide additional reasons for believing in the validity of the conception of a genotope. As discussed in Sect. 8.11, for the present discussion we will arbitrarily think of genotopes as a topological object occupying seventh, sixth, and fifth dimensionalities and hovering over the fourth-dimensional configurations of genomes as we presently understand. It is important to keep in mind that these fourth-dimensional configurations consist of three-dimensional space and the four dimensions of time are treated separately. In principle, whatever is projected from the higher dimensionalities of genotope (G_7–G_5) would be projected onto the time dimension as well. However, for the present discussion, we will confine ourselves to three space dimensions.

With these restrictions in place, let us see how genotope would appear from the three-dimensional space. In principle, according to this model, biomolecules themselves must be seen as projections from G_7 to G_5 because at these higher dimensionalities the distinction between spacetime and matter ceases to exist. However, in addition to this molecular framework, the remaining information transfers from these higher dimensionalities onto the three-dimensional space would also be manifest. Therefore, instead of focusing on the molecular framework (which we know in detail), we try to understand how the remaining information contents from higher dimensionalities manifests themselves. It is intuitively clear that these informed contents must be present in the form of different stereochemical and conformational orientations. Usually, we associate different stereochemical and conformational configurations as possessing different energy levels. However, these energy differences would be in the form of projections from G_7 to G_5 to the time dimension which eventually manifest in the form of thermodynamic and kinetic changes of these biomolecules.

Confining ourselves to three space dimensions, since each of these dimensionalities G_7–G_5 would be projected differently onto the three-dimensional space, it would result in spatial segregation of these biomolecules. Thus, according to this model, the separation of different genes between different chromosomes and within each chromosome must be decided by the details of genotope in G_7–G_5. This opens up a possibility of verifying the proposed model by mapping the intergenic distances between every pair of genes and creating a hyper dimensional model of genomes. It is legitimate to wonder that if this scenario is as simple as it is made out to be, surely, we would have succeeded in decoding the higher dimensional configurations of genomes. However, there are a few problems with this scenario. Firstly, we don't know exactly how many genes could be present in any given genome. For instance, even today, we aren't sure how genes are actually present in, say, the human genome. Our current estimate of a few thousand genes makes it difficult to compute all the intergenic distances. Moreover, we are saddled with phenomena like open reading frames (Sieber et al. 2018) which complicate the computation of all the intergenic distances. In fact, according to this model, if a genome possesses fractal dimensionalities, they would give rise to phenomena like open reading frames and RNA interference (Howard 2013). Thus, we can reasonably conclude that while the proposed model can't predict any new features of genomes by looking at the projections from genotope to the three-dimensional space of biomolecules, the conception of genotope is at least able to explain some of the hitherto unexplained phenomena of genomics. In the next section, we will try to visualize genotope from outside.

8.21 Structural Template of Genotope from Outside

Before we look at the external perspective of genotope, let us look at the importance of this external perspective. Firstly, as discussed in the previous section, the internal perspective of genotope didn't help us to visualize overall topology of genomes. Admittedly, it did tell us how the hitherto unexplained phenomena of open reading frames and RNA interference could arise. However, it didn't help us to visualize the overall topology of genomes. Secondly, as discussed in Sect. 8.19. The functionality of self-reference is the key to understanding Life. However, as discussed in that section, simply having multiple dimensionalities simultaneously, by itself, doesn't give rise to self-reference as is the case with ecosystems. According to this model, the ecosystem remains an ensemble but without the functionality of self-reference. Therefore, it is necessary to deconstruct the relationship among all the dimensionalities to find out why self-reference is critical to the emergence of Life. Admittedly, we will discuss the semantics of dimensionalities and their changes in the next section. Presently, let us look at the external perspective of genotope as it would help us to develop the semantics of dimensionalities. Before we look at the external perspective of genotope, it is necessary to keep in mind that this perspective is created by our cognitive faculty. Therefore, essentially, it is a perspective developed by a projection and not by a direct observation. Therefore, this perspective would still be anthropic in nature. However, since we wish to formalize anthropic semantics of the dimensionalities, this bias is acceptable and perhaps inevitable.

When viewed from outside, genotope would not appear to be a block in which different dimensionalities would appear in a nested hierarchy. Rather, genotope would appear to be sliver and a genome as a bundle of such slivers entwined with one another. It is tempting to compare this organization of multiple genotopes as some kind of spiral helix (very similar to the one conventionally employed to describe DNA sequences). However, this is not an exact analogy because in higher dimensionalities the description of a helix would be different than the one available in three-dimensional perspective. The reason why it should be congruent with the higher dimensional spiral lies in two features of mathematics underlying this model. Firstly, the modified operator of involution demands that one of the dimensions gets involuted into the remaining dimensions. Therefore, at every dimensionality, the recipient dimensions would be curved. Similarly, different genotopes are also connected with one another through involutions. Therefore, this curvature would manifest within each genotope and among different genotopes within a genome. Secondly, each submanifold would be nested within the parent manifold is a similar curved manner. Thus, a genome would appear as a collection of slivers of genotopes entwined in a curved manner and in turn, each sliver representing a genotope would also be curved. The similarity between these curved slivers of genotopes and the double helix is not accidental. The semantics of this similarity will be discussed in the next section.

8.22 Semantic Implications of the Proposed Model

The proposed model employs a topological framework to formalize genomic architecture. However, it is necessary to justify the change of framework from three-dimensional stereochemical perspective to higher dimensional topological perspective. While the overall semantic foundation of the proposed model is discussed in various chapters of this monograph, in this section, we will discuss semantic implications of the conception of genotope. Even here, there are several semantic implications that need to be justified. However, for the sake of brevity, we will discuss only three semantic propositions that are implicit in the discussion presented above, viz., (1) semantics of the bundling of several dimensionalities in a genotope, (2) semantics of involution as a bundling mechanism in a genotope, and (3) semantics of spatiotemporal dimensionalities in a genotope.

Let us begin with the first semantic proposition, viz., semantics of the bundling of several dimensionalities in a genotope. It is legitimate to wonder why we should replace the three-dimensional stereochemical perspective of genomics with a higher dimensional model? The answer to this question lies in the nature of spacetime. For instance, when we employ a three-dimensional stereochemical perspective of molecules, we assume that three dimensions of space are bundled together in a predefined manner. We don't question this bundling. This Is because spacetime appears to be bundled in this manner. It is a *fait accompli*. Therefore, our cognitive faculty has evolved around this bundled nature of spacetime. Therefore, to our cognitive faculty, this bundling of three space dimensions is a priori. However, from the epistemological perspective, we ought to question this as well. This is because this particular way of bundling of three space dimensions reflects on the type of information content of spacetime. We often don't realize that our employment of vectors and tensors works because it is congruent with the nature of information content present in spacetime in which three space dimensions are bundled together in a particular format (Wieder 2012, see Chapter 3). Similarly, the conception of a genotope requires multiple dimensionalities bundled together by involutions, because genomes contain that kind of information content which is essentially topological in nature. In addition, it is important to keep in mind that unlike the other scientific theories, in the proposed model, these dimensionalities are spatio-temporal dimensionalities having a certain degree of physicality. These dimensionalities are not theoretical constructs to satisfy the mathematical variables. Rather, these are the dimensionalities of spacetime itself. It is possible to argue why spacetime should contain information that genomes require to manifest various functionalities? The answer is simple. We never question why spacetime contains information that gives rise to the properties of fundamental particles. We must accept that all the types of phenomenology, be it a quantum phenomenology or be it a biological phenomenology, owe their origins to the fine structure of spacetime. As to the question how a single set of information content of spacetime can give rise to different types of phenomenology, there is no answer in the conventional perspective. However, the proposed model offers a way to explain this plurality of types of phenomenology. Each dimensionality of spacetime possesses a unique metric.

Therefore, each dimensionality of spacetime would manifest a unique type of phenomenology. Thus, the conception of a genotope requires multiple dimensionalities bundled together because it represents a unique type of information content.

Now, let us look at the second semantic proposition, viz., semantics of involution as a bundling mechanism in a genotope. It is important to keep in mind that this model postulates that all the natural phenomena that are manifest, arise from the bundling of the information content of spacetime. Because there are multiple ways of bundling the information content of spacetime, we obtain multiple natural phenomena. Moreover, according to this model, there is only one way to bundle the information content of spacetime, viz., by bundling different regions of spacetime. Since the proposed model offers an ontological perspective of all the natural phenomena starting from the cosmic singularity, it can't postulate any external constructs to define the bundling mechanism. Therefore, the only logical mechanism for bundling different regions of spacetime itself is to allow these different regions of spacetime to mingle with one another. Since the proposed model is conceptualized in the language of topology, it is inevitable that different regions of spacetime are defined as different dimensionalities. Moreover, the conception of dimensionalities as representations of different regions of spacetime is justified because spacetime can be thought of possessing different regions only if we assume that spacetime contains different types of information contents. Once we accept this reasoning, given the physicality of information content, it is inevitable that different types of information contents would require different dimensionalities to represent them. This is necessary because according to the cosmological principle (Barrow and Tipler 1986), spacetime must be uniform no matter from where we observe it. The proposed model employs a different version of that principle. According to the proposed model, spacetime is uniform in each dimensionality no matter from where we observe it within that dimensionality. Even in the conventional perspective, we represent vectors and tensors as having different dimensionalities.

The key point is why spacetime should be bundled using involution? The answer to this question is that the proposed model begins with the cosmic singularity. In the conception of the cosmic singularity (Chhaya 2022b, see Chapter 1), there is nothing outside of the cosmic singularity. Therefore, any mechanisms that alter the cosmic singularity must be inwardly directed. Therefore, involution is the most obvious choice. The reasons for choosing this operator of involution and its application to the cosmic singularity are discussed in the preceding monograph. Returning to the need for the involution as a bundling mechanism for a genotope, it is intuitively clear from the conception of a genotope that it is a unit which is created by involution from the larger framework of genomic architecture and more importantly, a genotope passes on its information content (in the form of long-range influences) to different DNA sequences through involutions. Therefore, it is intuitively clear that the operation of involution must be the right choice for the mechanism of bundling of multiple dimensionalities of a genotope. It is important to keep in mind that as discussed in the previous section, genomes possess the functionality of self-reference and the operator of involution provides a way to formalize self-reference.

This brings us to the third semantic proposition, viz., semantics of spatiotemporal dimensionalities in a genotope. One of the main objectives of this monograph is to formalize Life in a completely naturalistic manner. As discussed in the preceding chapters, it is necessary to eliminate any reference to transcendentalism. Therefore, it is intuitively clear that all the details of genomic architecture and each of its functionality must be viewed as having arisen from the cosmic singularity, just like any other natural phenomena. Therefore, it is imperative that dimensionalities defined for representing genomic architecture must not be some abstract entities, but they must be physical entities. This can happen only if we assign spatiotemporal dimensionalities to genomic architecture. This completes the discussion on the semantic implications of the proposed model. In the last section, we will summarize various aspects discussed in the preceding sections.

8.23 Conclusion

In the preceding sections, we discussed several aspects of genomic architecture. Some of these aspects were thematically connected with one another, while the rest of the aspects were reflective of the semantics behind this model. Therefore, in this section, we will summarize these discussions in a point-wise manner.

1. The key difficulty in defining genomic architecture is that our molecular perspective of genomes cannot formalize the regulatory elements of genomes in a systematic manner.
2. Therefore, a topological model of genomes based on higher dimensionalities has been discussed in the preceding chapters.
3. Using the proposed model, a topological unit of natural selection of genomes has been defined. This has been named as a genotope.
4. In order to understand the significance of genotope, a topological model of natural selection was discussed. Using genotope as a unit of selection, this topological model of natural selection was presented.
5. In this model of natural selection, a higher dimensional model of regulatory genome emerges naturally. This regulatory genome consists of multiple dimensionalities of genomes which supervenes the DNA sequence of genomes.
6. Genotope can also be thought of as a module, thereby providing a modular design of genomic architecture.
7. Using the topological perspective, it is possible to formalize ecosystems also as a topological framework. Similarly, it is possible to formalize genomes also as ecosystems.
8. Thus, a single framework is capable of formalizing natural selection from the level of ecosystem to the level of genomes. Using this reasoning, it is possible to define genes as competing species.
9. Using this model, it is possible to find intuitive explanations for hitherto unexplained phenomena like open reading frames, RNA interference, gene overlap, punctuated evolution, and speciation.

References

Akalin A (2021) Computational genomics. CRC, Boca Raton
Appasani K (ed) (2016) Genome wide association studies: from polymorphism to personalized medicine. Cambridge University Press, Cambridge
Bard J (2016) Principles of evolution: systems, species and the history of life. Garland Science, New York
Barrow JD, Tipler FJ (1986) The anthropic cosmological principle. Oxford University Press, Oxford
BonJour L (2010) Epistemology: classic problems and contemporary responses. Rowman and Littlefield, Lanham
Bonner JT (1988) The evolution of complexity by means of natural selection. Princeton University Press, Princeton
Bonner JT (2013) Randomness in evolution. Princeton University Press, Princeton
Bromham L (2008) An introduction to molecular evolution and phylogenetics. Oxford University Press, Oxford
Cabej NR (2020) Epigenetic mechanisms of the Cambrian explosion. Academic Press, London
Chhaya P (2020) The universe within: and a view from that within. Universal Press, Irvine, CA
Chhaya P (2022a) The origin and nature of mathematics. Submitted
Chhaya P (2022b) The origin and nature of spacetime. Submitted
Chhaya P (2022c) The origin of the counterintuitive nature of quantum phenomena. Submitted
Conlon J (2016) Why string theory? CRC, Boca Raton
Coyne JA, Orr HA (2004) Speciation. Sinauer, Sunderland
Darden L (1991) Theory change in science: strategies from Mendelian genetics. Oxford University Press, Oxford
Delisle RG (ed) (2021) Natural selection: revisiting explanatory role in evolutionary biology. Springer, Dordrecht
Deng X (2022) A new exploration of Hegel's dialectics: the tensions of speculation. Routledge, London
Dennett DC (1995) Darwin's dangerous idea: evolution and meaning of life. Simon and Schuster, New York
Donaldson KM (2000) Relationships between chromosome structure and long distance regulation of gene expressions. University of California Press, San Diego
Flew A (2017) Darwinian evolution. Routledge, London
Friedman M (1983) Foundation of Spacetime Theories: relativistic physics and philosophy of science. Princeton University Press, Princeton
Fritz A (2014) Elucidating chromosome territories organization during the cell cycle and in breast cancer cells. Proquest, Ann Arbor
Garrett RA, Klenk HP (eds) (2007) Archaea: evolution, physiology and molecular biology. Blackwell Science, Oxford
Gould SJ (2007) Punctuated evolution. Harvard University Press, Cambridge
Grene M (1986) Dimensions of Darwinism: themes and counter themes in the twentieth century evolutionary theory. Cambridge University Press, Cambridge
Hamilton I (1989) An evaluation of the mitochondrial eve hypothesis: theory, methodology, validity and relationships with the regional-continuity model of human emergence. Simon Fraser University Press, Burnaby
Hodge J, Redick G (eds) (2009) Cambridge companion to Darwin. Cambridge University Press, Cambridge
Howard K (ed) (2013) RNA interference: from biology to therapeutics. Springer, Dordrecht
Huges AL (1999) Adaptive evolution of genes and genomes. Oxford University Press, Oxford
Laudal OA (2021) Mathematical models in science. World Scientific, Singapore
Lovelock J (2000) Gaia: a new look at life on earth. Oxford University Press, Oxford

Lozano NA (2017) The influence of epistasis, epistatic mutability and pleiotropy on evolvability. California State University Press, Long Beach
Miller JH, Reznikoff WS (1980) The operon. Cold Spring Harbour Laboratory Press, New York
Niklas KJ, Newman SA (eds) (2016) Multicellularity: origin and evolution. MIT Press, Cambridge
Okasha S (2010) Evolution and the levels of selection. Oxford University Press, Oxford
Oparin AI (1968) Genesis and development of life. Academic Press, New York
Panza M, Sereni A (2013) Plato's problem: an introduction to mathematical Platonism. Springer, Dordrecht
Pevsner J (2015) Bioinformatics and functional genomics. Wiley, Hoboken
Popper K (1963) Conjectures and refutations: the growth of scientific knowledge, vol 16. Routledge, London, p 80
Press SJ, Clyde CM (2003) Subjective and objective Bayesian statistics: principles, models and applications. Wiley, Hoboken
Provine WB (2001) The origins of theoretical population genetics. University of Chicago Press, Chicago
Raffaelli DG, Frid CLJ (eds) (2010) Ecosystem ecology: a new synthesis. Cambridge University Press, Cambridge
Reavey C (2013) Analysis of transcription activating distance as a polygenic trait in saccharomyces cerevisiae. Harvard University Press, Cambridge
Robert JS (2004) Embryology, epigenesis and evolution: taking development seriously. Cambridge University Press, Cambridge
Sabath N (2009) Molecular evolution of overlapping genes. University of Houston, Houston
Shmulevich I, Dougherty ER (2014) Genomic signal processing. Princeton University Press, Princeton
Sieber P, Platzer M, Schuster S (2018) The definition of open reading frames revisited. Trends Genet 34(3):167–170
Smith E, Morowitz HJ (2016) The origin and nature of life on earth: the emergence of the fourth geosphere. Cambridge University Press, Cambridge
Smullyan RM (1992) Godel's incompleteness theorems. Oxford University Press, Oxford
Steele EJ, Lindley RA, Blanden RV (1998) Lamarck's signature: how retrogenes are changing Darwin's natural selection paradigm. Allen & Unwin, New South Wales
Szabo A, Ostlund NS (1989) Modern quantum chemistry: an introduction to advanced electronic structure theory. Dover, Mineola
Weinzierl ROJ (1999) Mechanism of gene expression: structure, function and evolution of the basal transcriptional machinery. Imperial College Press, London
Wieder S (2012) The foundations of quantum theory. Elsevier, Amsterdam
Wigner E (1960) The unreasonable effectiveness of mathematics in natural science. Commun Pure Appl Math 13:1–14
Yarus M (2010) Life from an RNA world: the ancestor within. Harvard University Press, Cambridge

Principles of Genomic Evolution

9

Abstract

In the preceding chapters, we looked at a possible topological model of genomic architecture. Even if we were to accept the proposed model to be true on the basis of the arguments presented in these chapters, the litmus test for this proposed model lies in its scrutiny in the context of the Darwinian paradigm. This is a historical necessity. Every paradigm shift in biology has come about only after it has survived the semantic purging by the Darwinian paradigm. In this chapter, we will look at the specific aspects of the proposed architecture vis a vis the Darwinian paradigm. This includes the units of selection, the need for the distinction between the units of selection, and the units of inheritance and the semantics of information storage as the architectural details. Admittedly, there are several fundamental semantic propositions of the Darwinian paradigm which need to be revisited. Some of these propositions and the reasons why they require reinterpretation will be discussed.

9.1 Introduction

In the preceding chapters, a topological model of genomic architecture has been outlined. This model was sought to be justified by demonstrating its congruence with various facets of conventional perspective on the major topics of genetics and genomics. It is possible to take diverse views on the validity of the proposed model. It is only when the insights provided by the proposed model are verified by experiments that we can pass a judgment on the merits of the proposed model. However, it is possible to evaluate any such speculative proposal even prior to any such experimental verification. This method of evaluation consists of its scrutiny in the context of the Darwinian paradigm. As Dobzanski famously concluded that nothing in biology makes sense unless it is viewed from the perspective of

© The Author(s), under exclusive license to Springer Nature Singapore Pte Ltd. 2023

P. Chhaya, *The Topological Model of Genome and Evolution*,
https://doi.org/10.1007/978-981-99-4318-0_9

Darwinian theory (Dobzanski 1982). Therefore, in this chapter, we will seek to make sense of the proposed model by looking for its Darwinian interpretation. There are certain fundamental semantic ambiguities of the conventional interpretation of the Darwinian paradigm. Therefore, the best way to verify any new model of biological evolution and natural selection is to find out how it deals with these semantic ambiguities. However, since we will be dealing with the postulate that the genome is also subject to natural selection, we will confine ourselves, in this chapter, only to those semantic ambiguities which are relevant to this postulate.

The postulate that the genome itself is subject to natural selection, raises certain questions. Historically, the identity of the units of selection has been a matter of debate (Okasha 2010, see Chapter 2). On one extreme, we have the Gaia hypothesis which places the entire ecosystem of our planet as being subject to natural selection (Lovelock 2000). On the other extreme, we have a hypothesis that it is the individual genes (Dawkins 1999) and not the members of individual species that are subject to natural selection. In the middle of this spectrum of hypotheses, we have a theory of group selection which asserts that it is the group of individuals that are collectively taken as a unit of selection (Borrello 2010). In that sense, the postulate that the genome itself is a unit of selection lies in the middle of this spectrum. However, this postulate leads to several logical inferences which are debatable, and therefore, they need to be deconstructed. Therefore, in this chapter, we will deconstruct these inferences using the proposed model and find out whether the proposed model offers any significant semantic resolution of these inferences. This is important because the conventional perspective is agnostic about this postulate of the genome being a unit of selection.

In addition to the possibility of the genome being a unit of selection, there is a fundamental problem of the mechanism by which natural selection comes about. The classical definition of natural selection is based on the competitive consumption of limited resources resulting in the elimination of the competing species (Flew 2017, see Part III). However, such a scenario cannot be applied to the postulate of the genome being a unit of selection because whatever the resources required for genomic functionalities are provided within the genome itself. Therefore, the nature of competition must be different. More importantly, once we accept that the genome is subject to natural selection, we cannot deny the lower level natural selection among the constituent genes. Therefore, whatever the mechanisms by which the genome undergoes natural selection, must be generic enough to include the competition among its constituent genes. This necessarily leads to different conceptions of natural selection, competitive survival and more importantly that of biological evolution itself. This is because once we concede that individual genes also compete among themselves, there is no reason why we should reject the possibility of genes themselves being products of natural selection. This necessarily implies that biological evolution itself must be governed by the same model that governs natural selection. It must be kept in mind that conventional wisdom behind the Darwinian paradigm excludes the evolution of Life from its scope (Hodge and Redick 2009). Darwin's theory is conventionally defined only for the natural selection only after the competing species were available for natural selection. Darwin's theory is remarkably silent on the origin of Life, it successfully explains how different forms of Life compete and survive.

Thus, the postulate that the genome is also subject to natural selection, carries, within itself, a theory of biological evolution. Admittedly, these issues are intricately interwoven. Therefore, for the sake of brevity and simplicity, we will discuss them in a point-wise manner. Accordingly, this chapter is further divided into 19 sections. Section 9.2: Debate on the Units of Selection, Sect. 9.3: Conventional Perspective of the Mechanisms of Natural Selection, Sect. 9.4: Does the Mechanism Change with the Changes in the Units of Selection? Sect. 9.5: Does the Changes in the Mechanism Change the Semantics of Natural Selection? Sect. 9.6: New Definition of Units of Selection, Sect. 9.7: New Definition of Resources for Competitive Survival, Sect. 9.8: New Definition of Natural Selection, Sect. 9.9: New Model of Natural Selection, Sect. 9.10: New Model of Biological Evolution, Sect. 9.11: Unifying Different Units of Selection, Sect. 9.12: Why We Need the Duality of Units of Selection and the Units of Inheritance, Sect. 9.13: Genome as a Duality of Units Personified, Sect. 9.14: Regulatory Genome Versus Expressive Genome, Sect. 9.15: The Concept of "Genotope," Sect. 9.16: Mechanism for Genotopic Natural Selection, Sect. 9.17: Information Theoretical Perspective of the New Model, Sect. 9.18: Principles of Genomic Evolution, Sect. 9.19: Revisiting the Darwinian Paradigm, Sect. 9.20: Conclusion.

9.2 Debate on the Units of Selection

There are two facets to the conception of the unit of selection. Firstly, it is important to decide what is the exact nature of the unit of selection. Secondly, it is equally important to decide the identity of the unit of selection. It is intuitively clear both these facets are semantically linked to one another. If the conception of the unit of selection is based on the exact nature of what is being selected, then it is axiomatic that that definition would also decide the class of such entities which are qualified by that definition. Conversely, if the conception of the unit of selection is based on the definition of entities eligible for the selection, then it is axiomatic that that conception would decide the nature of what is being selected. However, the real problem with the Darwinian paradigm is that there exists a semantic ambiguity about the nature and identity of the unit of selection. This semantic ambiguity has been passed on during several paradigm shifts that the Darwinian paradigm has undergone. Thus, originally, the morphological features implicit in Darwin's writings were thought to be a unit of selection (Hodge and Redick 2009). With the advent of genetics, we broadened the definition of the unit of selection from morphological features to a broader notion of phenotypes (Delisle 2021, see Part II). With the advent of population genetics, we realized that it is the individual genes that are the units of selection (Provine 2001). Molecular biology taught us that it is the DNA sequences that must be viewed as the units of selection (Nei 2013, see Chapter 4). Finally, with the advent of genomics, it seems that it is the genomic functionalities that must be viewed as the units of selection (Pevsner 2015). Moreover, as a natural consequence of the development of population genetics, we also found that the nature of competing species need not be confined to biological species. It could be even a group of species that collectively can be thought of as the units of selection (Borrello 2010). Finally, by analogy, it seems natural to think that the idea of group acting cohesively can be extended to include a group of genes as well. Thus, genomes become logical units of selection.

Upon a little reflection, it is intuitively clear that behind these smooth transitions of the conception of the units of selection lies a fundamental semantic ambiguity in the conception of the unit of selection. As mentioned above, if we knew what is being selected, then we can certainly define the unit of selection unambiguously. This argument leads us to another semantic ambiguity. In order to determine what is being selected, it is imperative that we must know how natural selection operates. However, the conventional perspective of natural selection is also plagued by semantic ambiguities. As discussed in the preceding chapters, the Darwinian paradigm rests on randomness. It is decidedly against any design principles. Therefore, if natural selection were to possess a structural template, it would lead to two unacceptable consequences. Firstly, it would impact on the degree of randomness implicit in the Darwinian paradigm. Secondly, if natural selection were to possess a structural template, it would implicitly point toward some design principles (Fodor and Piattelli-Palmarini 2011). Moreover, the origin of this structuralism of natural selection would bring back transcendentalism, if not theism. However, the irony lies in the fact that if to recall Darwin's evocative phrase, descent with modification, is correct, descent implies certain continuity of forms. This continuity cannot be guaranteed unless the process of natural selection could recognize this continuity. Therefore, in order to recognize this continuity, natural selection must possess structuralism, at least in principle capable of recognizing the continuity of forms. It is important to keep in mind that this inherent structuralism of natural selection need not be deliberate or preconceived, but like any other natural processes, it must possess a native structuralism nonetheless. However, such an ascription of structuralism to natural selection is an anathema to the conventional Darwinian perspective. Thus, irony or self-contradiction is built into the very conception of the Darwinian paradigm. To be honest, the conventional perspective of natural selection doesn't deny the possibility of natural selection having its own structuralism per se. It is just that this perspective is also semantically ambiguous. We will discuss this topic in the next section.

9.3 Conventional Perspective of the Mechanisms of Natural Selection

It was suggested in the previous section that natural selection ought to possess a native structuralism of its own. It was also mentioned that the conventional perspective of natural selection is ambiguous. In this section, we will try to deconstruct this ambiguity in the conventional perspective. It was mentioned that the conventional perspective can't ascribe any structuralism to natural selection because it might compromise the essential Darwinian randomness and may bring in some design principles through backdoor. However, the conventional perspective has a clear idea about the mechanism by which natural selection operates. Therefore, it appears unwarranted to claim that natural selection has no structuralism of its own. If natural selection operates through a certain fixed mechanism, then it should be axiomatic that it possesses a native structuralism. This is precisely where the semantic ambiguity of the conventional perspective manifests. It is clear from the above discussion that there is some difference between structuralism and a mechanism of natural selection. However, we tend to think of both these terms as being synonymous. Therefore, it doesn't make sense to assert that natural selection operates through a certain mechanism and still it doesn't possess a structuralism of its own. In this section, we will try to understand the distinction between a mechanism and a structuralism of natural selection. The objective of this section is to demonstrate that the conventional perspective implicitly accepts that natural selection possesses a structuralism of its own, but refuses to acknowledge it, lest it undermines the randomness implicit in itself.

In order to understand this argument, let us look at the mechanism of natural selection as conventionally understood. The conception of natural selection rests on the principle of competitive survival in an environment having finite resources (Flew 2017). In an idealized scenario, initially there is an abundance of resources and only few biological species thriving by consuming these resources. This abundance of resources allows biological species to multiply at a rapid rate. This rapid growth also ensures divergent inheritable traits to manifest. However, as the number of biological species increases and the resources start depleting, the competitive survival sets in. Admittedly, this is an idealized scenario, but it is adequate for the present discussion. There are two key features of this scenario. Firstly, during the phase of abundant resources, there is a corresponding increase in the number and range of inheritable traits. However, as the resources start getting depleted, this plurality of inheritable traits helps biological species to survive through competitive survival. Secondly, according to this scenario, both these phases of natural selection, viz., the abundance of resources leading to the proliferation of inheritable traits and the scarcity of resources leading to the competitive survival, are necessary for natural selection. In the absence of abundant resources, there would be no emergence of divergent inheritable traits, and in the absence of any scarcity of resources, there would be no competitive survival and therefore no natural selection.

From the conventional perspective, this idealized scenario has a definite mechanism. If we think of the environment as a collection of a certain number of resources, then it is intuitively clear that the above-mentioned scenario would replicate itself in the context of each of its resources, sometimes, even simultaneously. Thus, there is a mechanism in the form of consumption cycles of different resources and it results in the cycle of abundance of inheritable traits followed by competitive survival among these inheritable traits. Moreover, in accordance with the conventional perspective, this scenario, in spite of having a fixed mechanism, produces randomness in the form of selected inheritable traits. Since there are multiple types of resources and there are multiple inheritable traits, the outcomes of this scenario are random. Thus, according to the conventional perspective, natural selection operates through a single mechanism but due to the plurality of resources and the plurality of inheritable traits, the process of natural selection cannot have a structural template of its own.

However, there is a lacuna in this rationale. It assumes that neither the inheritable traits, nor their methods of consumption of resources of the environment, have their own structuralism. In fact, the above mentioned scenario is an idealization of a statistical analysis often used in population genetics (Provine 2001). It is important to keep in mind that the fault doesn't lie in this statistical approach of population genetics. This is because it is possible to create a more detailed scenario wherein different rates of consumption of different resources by different inheritable traits can be formalized. This approach would accommodate the inherently different methods of consumption into the analysis, thereby accounting for different structural templates of consumption. Similarly, it is possible to assign different survival values to different inheritable traits while analyzing a given population of inheritable traits. This would account for different structural templates of inheritable traits in the outcomes of natural selection.

However, the problem lies in assuming that different resources and different inheritable traits are autonomous. While in the case of the environment, it is understandable that we don't normally think of the environment as an ecosystem manifesting a considerable degree of integration, it is surprising that we think of inheritable traits as autonomous units. This is because in the case of the environment, our conception of an ecosystem is more of a hermeneutic device than a reality. However, in the case of inheritable traits, it is surprising that we still treat them as autonomous units in population genetics when we already know that inheritable traits are not really discrete or autonomous. Here again, it must be clarified that it is possible to formalize different degrees of interdependence between these inheritable traits and construct suitable models for predicting outcomes of natural selection (Provine 2001). It is just that with every additional level of parameters, we are relegating the structural foundation of natural selection. This process of parameterization helps our cognitive faculty to circumvent its inability to discern the structural template of natural selection. In fact, we must view the success of the parametric approach as a confirmation of the underlying structural template of natural selection.

It is legitimate to wonder why we should search for a putative but indiscernible structural template of natural selection when the above mentioned statistical methodology works reasonably well. After all, the putative structural template of natural selection, if formalized, would improve our ability to predict the course of natural selection which in any case we manage with the conventional statistical analysis. This reasoning, while being pragmatic, doesn't preclude us from having a better understanding of the Darwinian semantics. As discussed in the preceding chapters, there are several fundamental semantic ambiguities which cannot be resolved unless we deconstruct the nature of natural selection. Apart from this semantic imperative, there is an additional reason why we should seek to formalize the mechanism by which natural selection operates. This refers to practical implications of natural selection having its own structuralism. With the advent of genomics, it is possible to address several widely spread pathologies arising from genetic disorders. If we knew the evolutionary trajectory of individual genes and the structuralism by which they were selected, it would help us to treat these pathologies in a proactive manner without waiting for the pathologies to manifest. Before we look at one such model of natural selection, it is necessary to deconstruct the conventional perspective of natural selection with respect to the ambiguity about the unit of selection. Therefore, in the next section, we will try to understand what the conventional perspective implicitly suggests about the changes in the process of natural selection when the unit of selection is changed.

9.4 Does the Mechanism Change with the Changes in the Units of Selection?

It was suggested in the previous section that the conventional perspective accepts a mechanism of natural selection, but it is reluctant to accept that this mechanism represents an underlying structuralism of natural selection. Of course, it is debatable whether a specific mechanism necessarily implies a definitive structural template of the process itself. We will return to this topic in the following sections. Presently, we will provisionally accept the conventional perspective that our current understanding of natural selection allows us to concede the existence of a fixed mechanism of natural selection without forcing us to concede the existence of the native structuralism of natural selection per se. If this view is correct, it necessarily means that the details of this mechanism would vary if natural selection operates at multiple levels simultaneously. In other words, if there are different units of selection on which natural selection operates, then the details of the mechanism of natural selection would vary from unit to unit of selection. Therefore, it is necessary to deconstruct the debate on the unit of selection (Okasha 2010) in the context of the mechanistic details of natural selection. It must be admitted at the outset that there exists a vast amount of literature on the topic of the units of selection. For the present discussion, we will assume this literature as having been read. Instead of revisiting literature, we will pick up the underlying semantics and deconstruct it.

Prima facie, the fact that the units of selection have been a subject matter of research suggests that there is something more fundamental about the process of natural selection than the details of each individual unit of selection. The possibility of having a generic framework for accommodating different units of selection in a single framework of natural selection suggests that the individual details of each unit of selection play a less important role in natural selection than a systemic requirement. In retrospect, this reasoning explains the conception of an ecosystem. It is the framework of an ecosystem that defines the nature of natural selection while the contexts of this process change depending on the identity of the unit of selection. This raises several questions about the nature of natural selection, viz., (1) Are there different mechanisms of natural selection for different units of selection?, (2) If so, are these different mechanisms derived from one common framework, say the structural template of natural selection per se?, and (3) Does this mean that we need to redefine the semantics of natural selection? In this section, we will try to answer the first two questions while we will discuss the third question in the next section.

We looked at an idealized version of natural selection as implicit in the conventional perspective in the previous section. Therefore, it is necessary to find out whether the conventional perspective of natural selection accepts the possibility of different mechanisms for different units of selection. Prima facie, it appears the conventional perspective views different details of natural selection as instantiations of an abstract schema of natural selection. Therefore, by implication, the conventional perspective doesn't accept that there are different mechanisms of natural selection for different units of selection. This might sound contradictory to what was discussed in the previous section. This is because if there is a unitary mechanism of natural selection, it necessarily implies a structural template of natural selection. However, as discussed in the previous section, the conventional doesn't concede the possibility of natural selection having its own structuralism. This contradiction can be resolved if we realize that the conventional perspective thinks that a unitary mechanism of natural selection (as described in the idealized version above) is an abstraction and not a physical entity. The proof of the assertion that the conventional perspective perceives the mechanism of natural selection as an abstraction lies in the fact that it doesn't provide for different structural templates for different types of environmental resources. As mentioned above, in the idealized version, the exact nature of resources or the exact nature of the consumption of these resources by competing biological species or even the exact nature of biological species don't play any role in the process of natural selection.

It is possible to argue that this is because we chose to conceptualize natural selection in an idealized scenario. In reality, the details matter. However, upon a little reflection, this argument is not valid. Let us look at different units of selection and how they have been fitted into a unitary mechanism by the conventional perspective. For this purpose, we will discuss 3 units of selection, viz., (1) classical inheritable traits like morphological features, (2) biological species, and (3) individual genes. Ideally, we could have included a group of species as well. However, there are inherent semantic ambiguities about the conception of group selection (Borrello 2010). Therefore, we will omit it from the list. In the case of morphological features (Incidentally, this was implicit in Darwin's own writings (Hodge and Redick 2009).), the focus is on the survivability of species possessing these morphological features. Admittedly, the underlying genotypes are implicit in this perspective, there is no explicit reference to them in Darwin's writings. However, we will assume that Darwin meant the underlying genotypes while referring to the survivability of morphological features. Surprisingly, in this conception of the unit of selection and the corresponding natural selection, there is no reference to any abstraction.

The units of selection and the corresponding natural selection are thought of as physical entities. However, there is no reference to the exact nature of either the morphological features or to the exact nature of the role of the environment on natural selection of these morphological features. This is because not much was known about the chemical and genetic details of living organisms. Therefore, the mechanism by which natural selection operates at the level of morphological features conforms to the idealized scenario described above. However, there is one aspect which remains ambiguous. A given morphological feature may have a different degree of contribution to the survivability of the species possessing it. Therefore, in a situation where in a given biological species possessing several morphological features with each morphological feature contributing to the survivability of that species, it would be difficult to disentangle different contributions by different morphological features. As discussed above, it is possible to employ statistical analysis to factorize the contribution of each morphological feature in the survivability of the species possessing multiple morphological features. However, this is possible only because we resort to abstraction of details about different morphological features and choose to ignore the physical details of these morphological features.

Of course, after the advent of genetics and molecular biology, it is now possible to include the physical details of various morphological features. However, as discussed later in this section, it leads to its own set of problems. As far as the mechanism by which natural selection operates on morphological features is concerned, it is reasonable to think that it conforms to the idealized version described above if we conceptualize it in an abstract sense. However, the introduction of the physical details of these morphological features isn't possible unless we change the units of selection from the morphological features to the underlying genotypes.

Now let us look at individual species as units of selection (Delisle 2021, see Part II). This is perhaps the most intuitive example of a unit of selection. Apart from its intuitiveness, this example of a unit of selection also embodies the semantics of competitive survival. Competitive survival requires a certain degree of sentience to perceive the nature of resources and the identity of rivals. Living organisms manifest this type of sentience in varying degrees. This is not the case with any other units of selection. For instance, morphological features *suo moto,* can't plan a strategy to compete for limited resources. The same is true for individual genes. In these two cases, there is no volition on the part of these units of selection to participate in competitive survival. If at all they appear to be participating in competitive survival, it is only because of our anthropic bias that they perceive competition in their struggle to survive. Once again, in the case of individual species as units of selection, the resources and the ensuing competition are physical in nature. The abstraction of a schema, which Darwin's own writings articulate, is an afterthought and not an anthropic projection.

This emphasis on the physicality of the process of natural selection in contrast to its abstraction is necessary to deconstruct the semantics of the units of selection. However, before we discuss why it is necessary to deconstruct this semantics, let us look at the third example of the unit of selection, viz., individual genes. The idea that genes compete among themselves is of much recent origins. It was suggested by Dawkins (1999). In fact, from the chronological order, it appeared after the idea of group selection was propounded (which incidentally would not be discussed here). The reason why this ascription of the units of selection to individual genes is important is that it eliminates the notion of volition of the participants in competitive survival. It brings about a certain inevitability to competitive survival that elevates it to a law of Nature. Prima facie, it is difficult to imagine that individual genes can have a degree of autonomy to participate in competitive survival. However, with the advent of molecular biology, it was apparent that while we may debate on what constitutes a gene (Mukherjee 2016) (or even how many genes are present in a given genome (Makalowski 2001)), gene expressions give us discrete outputs. Therefore, to the extent these outputs of gene expressions are discrete and possess chemical interactivity, they would be participating in competitive survival.

This realization broadened our conception of competition, and therefore, that of natural selection. Now, it is possible to think of the environment not only as an ensemble of different natural elements of Nature, but also as an ensemble of different consumable resources. What appeared to be volition and sentience in the classical Darwinian paradigm can now be replaced by thermodynamic imperative that forces competitive survival. It is only when we think of genes as units of selection that we realize that resources for which competition occurs need not be calorific resources. Individual genes don't compete for calories (though their gene expressions do). They compete for information in the form of signals from genomic functionalities. Theoretically speaking, even this wouldn't have been justified but for the fact that as we realized later that information too is a physical entity.

This brings us to the point made earlier about the distinction between physical aspects of natural selection and its abstraction. When we made a transition from the physical aspects of natural selection to its abstraction, we didn't accommodate the physicality of information content. Our attempts to abstract the details of natural selection from the physical details of natural selection as implicit in Darwin's own writings were based on the belief that information coded in the abstract representation of natural selection were virtual. However, now we know from our understanding of quantum mechanics (Nielsen and Chuang 2010) that information content per se is a physical entity, we need to modify the abstract conception of natural selection that we have formalized in population genetics. The classical Darwinian paradigm didn't incorporate information theoretical interpretation into its semantics because the physicality of information content wasn't established at that time. However, in an era of functional genomics and system biology, we need to modify the conception of available resources and the nature of competitive survival by including information content because it is a physical resource like any other natural resources. If we can succeed in formalizing information theoretical models of natural selection, it is possible to recast the Darwinian paradigm into a law of Nature. Returning to the present discussion, let us summarize the nature of different mechanisms of natural selection for different units of selection and whether they can be shown to arise from a native structuralism of natural selection.

In the case of morphological features or biological species being units of selection, there is not much need to modify the conception of natural selection. Accordingly, the mechanism by which natural selection operates is essentially a Malthusian logic (Flew 2017). However, there are two implicit structural features that we need to keep in mind. Firstly, the relationship between the environment and different competing species is not defined as a unitary framework. Just as the population of different competing species would vary not only due to natural selection, but also due to mutations, the environment also changes in two different ways. Firstly, the environment would be changed due to consumption of its constituent elements by the competing species. Secondly, the composition of the environment would also change due to inherent fluctuations. Thus, the relationship between the environment and the competing species is loosely defined and not defined as an integrated system, something that we now want the conception of an ecosystem to manifest. It is important to keep in mind that this scenario also explains the origin of randomness in natural selection in the classical Darwinian paradigm. In the present day notion of an ecosystem (Raffaelli and Frid 2010), because of its integrated framework, imposes a certain degree of predictiveness to the outcomes of natural selection.

However, as far as the mechanism of natural selection is concerned, in the case of morphological features and the biological species, it has no directionality and like any thermodynamic system fluctuates. There are two parallel thermodynamic equilibria in this mechanism, one in the fluctuations of the environment and another one in the near-neutral genetic drift (Kimura 1983). It is important to keep in mind that these fluctuations arise because there is an absence of any systemic influences. However, when we introduce a systemic perspective, say, in the form of an ecosystem in the case of the environment and in the form of genomic architecture in the case of genetic drift, the mechanism changes. It is no longer like thermodynamic fluctuations. It is now directed by its internal compulsions.

It is legitimate to wonder that if this scenario is valid, the introduction of the conception of an ecosystem or a genome as unified entities ought to have changed the semantics of natural selection. However, we know that this is not the case. This paradox can be resolved if we realize that our conventional conception of ecosystems (Raffaelli and Frid 2010) and genomes (Roy and Kundu 2021, see Part III) is based on the abstract nature of information. When we formalize either an ecosystem or a genome, we don't think that the information transfers are physical processes. In our minds we create an abstraction of an ecosystem or a genome as abstract models. Therefore, the information transfers are abstract processes. However, in reality, neither an ecosystem nor a genome is an abstract entity. They are as physical as any other phenomena. Similarly, the information transfers in an ecosystem and in a genome are as physical as any natural processes. Once we accept this physicality of information content and their transfers, we have to accept that each type of information content would have its own structuralism, and therefore, its transfer from one part of the system to another part will be defined by its internal structuralism. Therefore, the mechanism of information transfers would vary from case to case. The reason why the introduction of the concepts of an ecosystem or a genome didn't alter our conception of the mechanism of natural selection is that we assumed that these concepts are mental constructs and not physical entities. Once we introduce the physicality of these concepts and the physicality of their information contents, the mechanism of natural selection would be different in each case and the mechanism would be defined by the kind of information transfer that happens during natural selection. This is best exemplified by our misconception of natural selection in which the units of selection are individual genes.

Let us look at the mechanism by which individual genes compete with one another. It is intuitively clear that genes do not compete with one another for their survival. They compete for their expressions. Therefore, the nature of natural selection must be different from the one operational in the case of wherein individual biological species are competing with one another for their survival. Upon a little reflection, it is intuitively clear that the resource for which individual genes compete must be the signals which initiate gene expressions. Therefore, in this case, it is intuitively clear that the genome should be thought of as the environment from which these signals are made available to individual genes. Therefore, the resource must be in the form of long-range influences (Shmulevich and Dougherty 2014). In the context of the earlier discussion on the physicality of resources versus their abstract representation, it is important to note that in this case, the resources (in the form of long-range influences) must be physical in nature. The conventional perspective is inadequate to define the nature of long-range influences in physical terms. Admittedly, the conventional perspective can explain some of these long-range influences in terms of stereochemical and conformational changes of the DNA sequences. This is exemplified by the phenomena like chromosome territories (Fritz 2014) and cis and trans effects (Donaldson 2000). However, as discussed in the preceding chapters, there are long-range temporal and spatial influences which cannot be defined as physical forces in the conventional perspective. Of course, as discussed in the preceding chapters, the proposed model offers a physical description of these long-range influences.

In the context of the present discussion, the key point is this: It is possible to think of individual genes as units of selection provided we accept that long-range influences are physical entities. However, there is one subtle difference between the physicality of resources in the case of biological species as units of selection and the physicality of resources in the case of individual genes as units of selection. In the case of biological species, the resources are available in a particular geographical distribution. However, otherwise, these resources are available to all the competing species. However, in the case of individual genes, the resources are not available uniformly to all the genes. Even if we draw analogies between the geographical distribution of resources and the configurations of genomes themselves as a distributed form of resources, there is still one difference between these two processes. The resources in the case of biological species are nonspecific. Once a biological species gets access to these resources, it can consume it irrespective of its identity. A resource is not differentially available to different biological species. The variation exists in terms of degree of availability but not in the terms of which species gets the access. This is not the case with the individual genes. The long-range influences are not equally available to all the genes. Apart from topological proximity (which is analogous to geographical distribution), the long-range influences are specific for each of these competing genes. Thus, in the case of long-range influences as resources, they are inherently asymmetric. This is because the physical nature of long-range influences has its own structuralism which may or may not be congruent with the recipient genes. Therefore, the mechanism by which natural selection operates now depends on the nature of resources.

Table 9.1 Units of selection and natural selection

Unit of selection	Types of resources	Nature of resources	Types of mechanisms
Morphological features	Elements of nature	Systemic	Structural alterations
Biological species	Caloric sources	Geographically distributed	Intelligent search
Individual genes	Long-range influences	Topologically distributed	Information transfers

This reasoning forces us to accept that there cannot be a universal mechanism for natural selection. However, as evident from the discussion presented above, these different mechanisms are derived from an indiscernible structural template of natural selection. Thus, it is possible to apply Malthusian logic (Flew 2017) to information transfers provided we accept the physicality of information content. Similarly, it is possible to employ abstract representation of natural selection provided we allow this abstract representation to possess a degree of asymmetry that arises from the physical nature of representation. For the ease of comprehension, these details are summarized in Table 9.1. In the next section, we will discuss whether having different mechanisms of natural selection for different units of selection can change the semantics of natural selection. This will be followed by sections discussing new conceptions of units of selection, resources and natural selection itself. The objective of these sections is to demonstrate that genomes too can act as the environment for natural selection of genes.

9.5 Does the Changes in Mechanism Change the Semantics of Natural Selection?

As discussed in the previous section, it is possible to distinguish between different mechanisms of natural selection for different units of selection. We will return to this topic in the following sections. However, before doing so, in this section, we will discuss whether this postulate of the relationship between mechanisms and the units of selection alters the basic semantics of natural selection. This is necessary because we don't want to either reduce the randomness implicit in the conventional perspective of natural selection or to introduce some form of design principles. Therefore, in this section, we will discuss whether having different mechanisms of natural selection alters the semantics of natural selection. For this purpose, we will discuss three aspects of the semantics of natural selection, viz., (1) does Bayesian approach to natural selection permit different mechanisms of natural selection?, (2) does having a mechanism per se reduce the randomness of outcomes?, and (3) does having a structural template of natural selection introduce any predictable complexity?

Let us begin with the first question, viz., does Bayesian approach to natural selection permit different mechanisms of natural selection? Prima facie, any logic has a definite structural template which can be likened to a mechanism. For instance, deductive logic has a certain fixed template. Therefore, whenever a premise fulfills the requirements of that template, deduction follows axiomatically. However, Bayesian logic is different (Press and Clyde 2003). It defines the degree of probability of a given outcome on the basis of the previous outcome. Thus, in a chain of events, the probability of each succeeding outcome is decided by the probability of the previous outcome in the given chain of events. In the case of a chain having multiple outcomes at every step, the probabilities of each outcome at each step is affected by the algebraic sum of the probabilities of all the predecessors of a given outcome. It is intuitively clear that the sum of all the probabilities would be one. Given this simplistic scenario, it is obvious that when the number of outcomes and number of steps in a given chain of events are large (as is the case in biological evolution and natural selection), the outcomes are essentially random. This justifies the use of Bayesian statistics in population genetics.

The question that is germane to the present discussion is this: Does this scenario allow multiple mechanisms? Let us imagine that in a given chain of events, say, evolutionary trajectory of an individual species, there are several parallel processes of natural selection, say, one each for morphological feature, a biological species under investigation and the individual genes present in the genome of that species. Naturally, as discussed in the previous section, natural selection of morphological features would be shaped by the elements of Nature. Similarly, natural selection of that species would be shaped by other competing species present in that niche of the ecosystem. Finally, the natural selection of the individual genes present in the genome would be shaped by the genomic architecture of that genome. Let us further assume that each of these processes of natural selection are brought about by different mechanisms (see Table 9.1). Upon a little reflection, it is intuitively clear that whether these three processes of natural selection occur in parallel or in a sequence, the outcomes would remain within the same range of randomness.

Therefore, as far as Darwinian randomness is concerned, having multiple mechanisms of natural selection for multiple units of selection do not alter the outcomes of natural selection. If at all, these mechanisms give rise to discontinuities in the distribution of probabilities of the possible outcomes. As discussed in the preceding chapters, this is precisely what is observed. In fact, there are phenomena like punctuated evolution (Gould 2007) and speciation (Coyne and Orr 2004) which have defied explanation in the conventional perspective. It is reasonable to think that these phenomena manifest because of multiple mechanisms of natural selection operating on different units of selection. It is important to keep in mind that the multiplicity of the mechanisms of natural selections, per se, does not introduce any teleology, nor do they represent any design principles. Thus, it is reasonable to conclude that the possibility of having multiple mechanisms of natural selection doesn't undermine the Darwinian semantics. In fact, it throws light on the origin of several phenomena like punctuated evolution and speciation. However, it must be admitted that mere semantic compatibility, by itself, doesn't justify the belief in multiple mechanisms of natural selection. This is because according to this model, these multiple mechanisms must be shown to have arisen from a common structural template of natural selection. Therefore, we need to understand the exact nature of this elusive structural template. In the next few sections, we will discuss a new paradigm of natural selection, its linkage with biological evolution.

9.6 New Definition of Units of Selection

As discussed in the preceding sections, the conventional perspective admits that there are different units of selection. However, the conventional perspective doesn't provide for any mechanistic differences in natural selection of different units of selection. In addition, the conventional perspective doesn't explain the origin and the reasoning behind this multiplicity of units of selection. This is surprising because the conventional perspective has accepted the unifying paradigm of ecosystems (Raffaelli and Frid 2010). If Nature behaves like a unified entity, there is no reason why any such plurality cannot have any ontological perspective. Therefore, if there are multiple units of selection, it is reasonable to expect a common ontology of this multiplicity. However, the proposed model accepts that Nature is a singular entity. More importantly, it accepts that all natural phenomena share a common structural template. Therefore, just as different types of quantum phenomenology arise from a unified quantum field due to a singular structural template (Lawrie 2013, see Chapter 5), biological phenomenology, in the form of biological diversity, must arise from the singular Nature. Therefore, it is intuitively clear that what appears to be multiple units of selection is also a phenomenological plurality having a common underlying unitary structuralism. Therefore, in this section, we will try to conceptualize a universal conception of the unit of selection and demonstrate that different units of selection are in fact, different instantiations of this universal conception of the unit of selection. In the following sections, we will extend this rationale to redefine natural resources and the process of natural selection itself.

For this purpose, we will employ two features of natural selection. We will try to define a unit of selection from the perspective of information content and its topological features. The choice of these two features is based on the following semantic considerations. Firstly, the proposed model postulates that information content is a physical entity. Therefore, it can be thought of as a resource for which competing species can fight. This ensures that all the three resources named in Table 9.1 can be unified. Secondly, once we accept the physicality of information as a resource for which competition occurs, it is intuitively clear that this physical resource must have a structural template. Therefore, it makes sense to think of this structuralism of information content in the language of topology. While it is intuitive to think that a resource would occupy space, the choice of topology over geometry (which is what geographical distribution actually is) needs some justification. The choice of topological framework is based on the fact that if we wish to formalize any systemic properties, say, an ecosystem or a genome, topological framework offers inherent advantages. For instance, if we think of sunlight as a resource, different organisms compete for it for different reasons. Some organisms use sunlight for photosynthesis. While some organisms use it for navigation, cf., Phototaxis, (Hader and Lebert 2001), still others use it to maintain their body temperature (cf. Thermoregulation (Blatteis 1998). If we employ a geometric framework, we need to add additional parameters for formalizing the efficiencies of different conversion processes that harness the energy of sunlight for their respective objectives. However, if we employ a topological framework, it is possible to include these different parameters as a single parameter of topological proximity. This argument is also applicable to long-range influences in genomes because there are different types of influences propagating through different mechanisms. However, if we define a parameter of topological proximity which incorporates these different mechanisms, it is possible to formalize a universal template for different types of resources.

Now, let us define the unit of selection based on the topological framework of information content as a resource for competitive survival. Apparently, there are two yardsticks that must be considered while defining the unit of selection, viz., the degree of complexity of information content and congruence between the ecosystem and the unit of selection. In the proposed model, the degree of complexity of the information content has been linked to the dimensionality. Therefore, the definition of the unit of selection must be some type of topological unit. Secondly, the congruence between the ecosystem and the unit of selection can be measured by the changes in the dimensionalities during natural selection. This also leads us to a topological description of the unit of selection wherein natural selection can be conceptualized in the language of the changes in the dimensionalities. Let us now define a topological unit of selection. It is intuitively clear that this unit of selection must be a tightly integrated information system. This is because it is only when different bits of information content are integrated into a single unit that it would possess different degrees of complexities. Therefore, we require two prerequisites for conceptualizing the unit of selection, viz., discrete collection of information content and their variable degrees of complexities. These prerequisites provide us with another perspective of natural selection. According To the proposed model, the degree of discreteness changes with the changes in the dimensionalities. Thus, according to this model, an entity might appear to be discrete in one dimensionality and nondiscrete in another dimensionality. Therefore, according to this model, natural selection can operate at multiple levels simultaneously. What would be subjected to natural selection would depend on the dimensionality in which natural selection operates. Since at each dimensionality, different entities possess discreteness, it is axiomatic that the identity of the unit of selection would vary from dimensionality to dimensionality. This is precisely what we observe. Thus, genomes could be the units of selection in dimensionality, and individual genes could be the units of selection in another dimensionality. Thus, as discussed in the preceding chapters, genomes occupy fifth, sixth and seventh dimensionalities. Therefore, natural selection can operate on genomes at each of these dimensionalities. For instance, if genomic modules occupy sixth dimensionality, they would be participating in competitive survival at that dimensionality. Similarly, individual genes occupy fourth dimensionality. Therefore, they would also be subjected to natural selection. However, in their case the nature of the environment consists not only of the conventional environment but in addition, it also consists of the higher dimensionalities of genomes as well. Thus, we must define a unit of selection as a topological object occupying a single dimensionality. It is important to keep in mind that within every single dimensionality, an object can occupy any number of dimensions provided they are less than the numerical value of that dimensionality. Thus, a unit of selection occupying the fifth dimensionality can occupy any number of dimensions which are equal to or less than five. This provision allows different units of selection occupy the same dimensionality to possess different degrees of complexities. This is necessary because the environment would act differentiably on the units of selection occupying the same dimensionality only if these units have different degrees of complexities. Thus, the unit of selection occupying a given dimensionality must possess at least one element of information content that occupies that dimensionality while the rest of elements of the information content may occupy lower dimensions, albeit integrated with the element in other dimensions. This explains the need for integration. All the different elements of information content of a given unit of selection must be tightly integrated into a discrete unit. Within those units, different elements of information can occupy different dimensions provided at least one element of information occupies the highest permissible dimension of that dimensionality. In Sect. 9.17, we will formally define this unit of selection as a genotope. In the next section, we will redefine the resources necessary for competitive survival.

9.7 New Definition of Resources for Competitive Survival

When we try to think of natural selection in the language of information theoretical perspective, we are required to think of information content as a source that can be consumed and utilized for replication. Admittedly, this scenario is intuitive to comprehend in the case of calories or food as a resource. However, a similar conception of information as a source is not easy to visualize. Of course, it can be argued that even calories are also a type of information units in the form of photons. However, the key difference between calories as information content and informational content as a resource for biological consumption is that calories (in the form of photons) are a nonstructural entity and therefore universally acceptable by all biological species. However, the information content in the present context has a distinct structural template, and therefore, its transfer depends on the structural template of the recipient (in this case, units of selection). Therefore, the nature of information content, particularly its structuralism, must be included in the definition of resources in this information theoretical description of natural selection. Therefore, in this section, we will try to deconstruct the nature of resources that are relevant to natural selection.

In continuation with our attempt to define the units of selection in the previous section, we will employ a topological framework to define the nature of resources for natural selection. As mentioned above, the environment is normally represented as a topological manifold. However, from the perspective of natural selection, it is intuitively clear that resources must be thought of as the information available from the higher dimensionalities with respect to the units of selection. This is important because according to this model, the information transfers must ideally occur via the operator of involution from a higher dimensionality to a lower dimensionality. The information transfers occurring in the same dimensionality are essentially in the form of competition. However, it is the information content present in the dimensionalities higher than the dimensionality of the unit of selection that acts as the environment. Moreover, if we think of different units of selection occupying the same dimensionality, it is intuitively clear that the environment (in the form of information content present in the dimensionalities higher than that of the units of selection) uniformly devolves into the dimensionality of the units of selection. This is not the case for the information content being transferred either from the same dimensionality or even from the dimensionalities lower than the dimensionality of the units of selection. Before we try to understand why only the higher dimensionalities must be thought of as the environment, it is necessary to understand how this information content of the dimensionalities higher than the dimensionality of the units of selection acts differentially on different units of selection.

It is here that the above-mentioned structural template of information content comes into play. While the information content made available through the operator of involution devolves uniformly into the dimensionality of the units of selection, its transfer to different units of selection would depend on the structural template of different units of selection. This scenario can be likened to the transport of chemicals across the double-layered cell membrane. The receptors present in cellular membranes must be stereochemically congruent with the molecules being transported across the membrane (Stein 1967). In serum, there will be a number of different molecules; however, it is only the congruence between these molecules and the receptors present in the cellular membrane that decides which molecules would be allowed to enter a cell. Similarly, the operator of involution enables all the information content of the higher dimensionalities to devolve into the dimensionality of the unit of selection. However, how much of this information content would be ingested by different units of selection (and become a resource for competitive survival) would depend on the specific structural template of different units of selection. Therefore, while the information content of the environment is equally available to different units of selection, it is the inherent structural template of each unit of selection that decides how much of this information content is absorbed into the unit of selection and used for replication. Therefore, it is imperative that we define different resources in the language of their topological framework just as we have done in the case of the units of selection in the previous section. Before we formally define resources in this manner, it is necessary to understand why the information content transferred from the same dimensionality and from the lower dimensionalities cannot act as the environment in the strict sense.

Let us begin with the information transfers from the same dimensionality. It was mentioned above that these information transfers must be viewed as competition rather than natural selection. Let us understand why. According to this model, the information content of each dimensionality has a unique metric. Therefore, the information transfers, say, from one unit of selection to another, would result in no change in the total information content because of the indestructible nature of information content. Therefore, if one unit of selection transfers information content to another, it would result in the loss of the information content for the donor unit of selection and the corresponding gain for the recipient unit of selection. Thus, irrespective of the identity of the units of selection, these information transfers would result in gain or loss from the perspective of the individual unit of selection. This is precisely what competitive survival is. The unit of selection that gains from the information transfers would be better equipped to receive information content made available from the higher dimensionalities. The converse is true for the unit of selection that loses the information content via information transfers in the same dimensionality.

Now, let us look at the information transfers occurring from the dimensionalities lower than the dimensionality of the units of selection. Admittedly, these information transfers must occur via an inverse of the operator of involution. However, our concern here is whether it would shape natural selection or not. This is where the conception of the unit of selection plays an important role. Since the conception of the unit of selection in this model is essentially a topological construct, the information transfers from the lower dimensionalities would be confined within these units of selection. However, since these information transfers would alter the details of the unit of selection, it would influence its competitive survival. Most obvious example of this upward information transfers is that of the relationship between genotype and phenotype. According to this model, genotypes exist in lower dimensionalities as compared to their respective phenotypes. Therefore, genotypes give rise to phenotypes through upward information transfers. However, this information transfer influences only the phenotype and not the environment. Therefore, the influence of upward information transfers would be local and its impact on natural selection will be indirect. It is important to keep in mind that in this model, unit of selection is a topological construct including both genotype and phenotype.

Since, such information transfers would not contribute to the environment as such. Therefore, they would not shape the course of natural selection, at least not directly. Thus, we can think of information content as resources because the information transfers result in the changes in the ecosystem. However, from the perspective of natural selection, it is only the information transfers from the dimensionalities higher than the dimensionality of the units of selection that alter the course of natural selection and therefore act as the environment in the conventional sense. The information transfers occurring from the same dimensionality as that of the units of selection result in competitive survival. Similarly, the information transfers from the dimensionalities lower than the dimensionality of the units of selection results in the changes in the nature of the units of selection which alters their own survivability.

It is important to note that this scenario also unravels several semantic propositions of the Darwinian paradigm which are not often acknowledged. Firstly, it suggests that there is a certain asymmetry in the information transfers. A larger chunk of information transfer occurs in the downward direction. Some information transfers occur horizontally and only a limited number of information transfers occur upwardly. This suggests that the ecosystem determines the extent and the direction of natural selection. The process of competitive survival merely shifts the equilibrium between different competing species without any direct influence on the ecosystem. Similarly, the upward information transfers normally are local influences that alter the survivability of the individuals of the competing species. Secondly, since the downward information transfers obey topological compulsions, they spread in the lower dimensionalities in a noncontiguous manner giving rise discontinuities. This perhaps explains the emergence of punctuated evolution (Gould 2007). Similarly, since the upward information transfers are local influences, they lead to the phenomenon of speciation (Coyne and Orr 2004) which is otherwise difficult to explain.

However, there is one aspect which needs to be deconstructed here. This refers to the influence of the competing species on the environment. Prima facie, if these species were to transfer information content upward, as mentioned above, it would lead to local changes. This is because as discussed in the preceding chapters, in any given genome, the functionalities occupy higher dimensionalities than the dimensionality of the DNA sequences from which they emerge. Therefore, it makes sense to think that these upward information transfers end up altering the functionalities of the organisms and thereby altering their survivability. However, biological evolution has witnessed several turning points in which the ecosystem has changed dramatically because of the individual competing species. The increase in oxygen levels in the atmosphere due to blue-green algae in the prime example of this (Decker and Van Holde 2011). This observation flies in the face of our earlier inference that upward information transfers are local in nature. Let us try to resolve this obvious contradiction.

This contradiction arises from the conception of systems. We define an ecosystem as an integrated system consisting of multiple constituents, including biological species. However, this system is essentially a closed system. For instance, in the case of Earth as an ecosystem, it is essentially a closed system (By referring to Earth as an essentially closed system, we are ignoring the fact that it receives energy from the Sun. Therefore, in principle, Earth is an open system. However, the argument given here can be applied to the solar system as a whole. Therefore, what matters is that close systems have certain properties. It is this distinction between an open system and a close system that is at the heart of the argument presented here). Therefore, in this case the inference that the information transfers from the lower dimensionalities are restricted to the unit of selection is valid. However, in the case of living organisms (and in case individual genomes as well as individual genes), they are thermodynamically open systems. Therefore, information transfers (and even the material transfers) are dissipative, leading to an increase in entropy. More importantly, they predominate over internal information transfers mentioned above. In fact, historically, we couldn't formalize life as a formal system because we didn't have a formal theory of thermodynamics of open systems (Prigorgin 1968). It is important to keep in mind that it is this distinction between Earth as a closed system and living organisms as open systems that prevents us from subscribing to the Gaia hypothesis (Lovelock 2000). For the Gaia hypothesis to be valid, it is necessary to demonstrate that Earth is an open system competing with other planets for some common source. This is not possible, at least not presently. This brings us to the end of this section. In view of these revised conceptions of units of selection and resources for competitive survival, in the next section, we will discuss a new conception of natural selection.

9.8 New Definition of Natural Selection

In the preceding sections, we discussed that there are perhaps different mechanisms for different units of selection. However, it was suggested that these different mechanisms of natural selection share the same structural template of natural selection. The proposition that natural selection can possess a native structuralism of its own might appear to contradict the conventional Darwinian randomness. However, it was argued that this fear of abandoning the inherent randomness of natural selection by this proposition is misplaced. Using a topological framework, it can be seen that the postulated native structuralism of natural selection doesn't introduce any teleology or design principles. Using this topological model, new conceptions of units of selection and resources for competitive survival were defined in the preceding sections. It is intuitively clear that once we accept this rationale of using information content as a resource for which individual units of selection compete and that these information transfers are inherently topological in nature, it is necessary to redefine natural selection using this topological model. Therefore, in this section, we will try to articulate a definition of natural selection based on the topological model of information transfers. Prima facie, even without going into the details of topology of such a model, it is intuitively clear that natural selection, in this model, must be in the form of congruence between the source and the target of information content. To put it differently, natural selection must be thought of as a process wherein the information transfers make the recipient of the information more congruent with the environment. Upon a little reflection, it is obviously true that this is precisely what the conventional perspective of natural selection suggests.

In the conventional perspective, competing species try to achieve two objectives. Firstly, they try to survive the dangers that the environment imposes on them. Secondly, the competing species exploit the resources to multiply. In both these activities, the notion of congruence is implicit. In order to survive, the competing species need to conform to the exigencies of the environment. The competing species do so by employing their morphological and physiological features. This process of adaptation is nothing but congruence. Similarly, in order to multiply, the competing species have to consume resources available in the environment. This consumption would not occur unless the competing species don't have the capacity to consume a given resource. This capacity to consume is also nothing but a metabolic congruence between the resource and physiology / biochemistry of the competing species. Thus, the notion of congruence has always been implicit in the conventional perspective.

What is different in the proposed model is that it codifies the notion of congruence in the language of topology. What is significant is that in this process, the proposed model offers a way to define a domain neutral description of natural selection. More importantly, it redefines natural selection. Let us see how. When we think of information content as a physical entity, it is axiomatic that it would possess some structural template that occupies spacetime. This property of "occupancy" necessarily bestows a certain degree of symmetry (or asymmetry as we will discuss later) to the information content of both the environment and the competing species. Therefore, any information transfer between the environment and the competing species (and vice versa) must be governed by symmetry principles. The key point in this perspective of natural selection is this: The information transfers would take place unhindered so far as they preserve the symmetry between the environment and the competing species. In other words, after the information transfers, neither the environment, nor the competing species would lose symmetry. Normally, we think of depletion of resources of the environment during natural selection (This is essentially the Malthusian argument implicit in Darwin's theory (Flew 2017)). However, from the perspective of information theory, the information content of an ecosystem remains constant during natural selection. It is just transferred from the environment to the competing species. What appears to be the depletion of the resources of the environment is actually the inability of the competing species to reverse the information transfers. The competing species transfer information content back to the environment, but in different forms. It is this inability to reverse the same information content back to the environment by the competing species that is asymmetry between two types of information transfers between the environment and the competing species.

From the information theoretical perspective, the information transfers occur in both the directions between the environment and the competing species. It is just that both these information transfers are asymmetric. What is transferred from the environment to the competing species is not symmetric to what is transferred from the competing species to the environment. Thus, whenever the asymmetry between these two types of information transfers, viz., from the environment to the competing species and vice versa increases, natural selection sets in. It was mentioned above that we normally think of natural selection causing depletion of the resources of the environment. Similarly, we normally think of extinction of competing species as the products of natural selection. Both these perceptions arise because we look at natural selection from two different frames of reference. If we choose to view natural selection from the perspective of the environment, we perceive depletion of the resources. Similarly, when we choose to view from the perspective of the competing species, we perceive extinction or survival as the outcomes. However, from the perspective of natural selection, what transpires during the process of natural selection is the elimination of asymmetry. Let us understand why. Let us imagine that the environment prior to biological evolution was a mixture of various elements. For the sake of simplicity, let us think of the environment as a primordial soup from which Life evolved (Oparin 1968). In that scenario, the environment was perfectly symmetric in the sense that various ingredients were randomly distributed in that soup. The evolution of Life started organizing the molecules into an organized ensemble of biomolecules like RNA-proteins. As natural selection began working on these molecules, more and more complex biological organisms emerged. At every stage, there was an increase in asymmetry from the information theoretical perspective.

This asymmetry manifested in two different forms. Firstly, it was manifest in the entropic sense. Out of the chaotic distribution of nutrient molecules in the primordial soup, biological evolution gave rise to the more and more organized ensembles. This gave an appearance of loss of entropy as implicit in the second law of thermodynamics (Haynie 2008). (Historically, this was one reason why we couldn't claim that Life was like any other natural phenomena. It was only after the formalization of thermodynamics of open systems (Prigorgin 1968) that this anomaly was resolved and we could justifiably claim that Life was like any other natural phenomena because it also led to increase in entropy.) It is this entropic connotation of asymmetry that gives rise to the perception of depletion of resources from the perspective of the environment. However, there is another form of asymmetry that arises during biological evolution. This refers to the arrangements of the information vis a vis the environment. To use a simplistic analogy, the environment can be thought of as a lock and biological organisms (as highly organized ensembles of information) as keys. Since the information content, say, in the form of genomic architecture, is organized by the rules of stereochemistry. Therefore, it is not necessary that these ensembles should fit into the environment just as a key would fit into a lock. Therefore, there is an asymmetry of information organization between biological organisms and the environment.

When viewed from this perspective, it is intuitively clear that survival must be thought of as the removal of asymmetry between the biological species and the environment. It is important to keep in mind that this process of reducing asymmetry is not one sided. Just as the biological species change in order to fit into the environment, the environment also changes. It is just that the capacity to change is greater in the biological species than in the environment. This is because the environment is inherently less organized than the biological species. The key point is that survival is essentially a process of reducing asymmetry between the environment and the biological species. Now, we can understand the information theoretical conception of natural selection. Natural selection tries to restore symmetry. There is an exact parallel between the second law of thermodynamics and natural selection. The second law of thermodynamics strives to reduce asymmetry by increasing entropy. Similarly, natural selection tries to reduce asymmetry by eliminating unfit biological species. Therefore, from the thermodynamic perspective, we observe depletion of resources, and from the information theoretical perspective, we see extinction or survival of biological species.

Thus, natural selection aims to restore symmetry in the information theoretical perspective, and therefore, it must be regarded as a law of Nature and not just a biological rule. At first sight, it might appear that natural selection works as an opposite of the second law of thermodynamics because natural selection yields more and more complex ensembles, whereas the second law of thermodynamics leads to the least organized ensemble. The reason for this anomaly lies in the frame of reference in which these laws are conceptualized. The second law of thermodynamics is framed in the language of mass, energy and their distributions. Natural selection is framed in the language of information content. It is important to keep in mind that survival is congruence between the information content of the environment and that of biological species. As discussed in the preceding monographs (Chhaya 2020, 2022b) and in the preceding chapters, the conception of thermodynamics is conceptualized on the basis of the structural template of the four-dimensional spacetime. On the other hand, the information content present in molecules (including the biomolecules that constitute genomes) are constituted on the basis of the structural template of the higher dimensional spacetime. Therefore, the apparent contradiction between the second law of thermodynamics and natural selection reflects the mismatch between the structuralism of spacetime among its different dimensionalities. This brings us to the end of this section. In the next section, we will try to articulate a new model of natural selection without any reference to thermodynamic perspective.

9.9 New Model of Natural Selection

In the previous section, it was suggested that natural selection is a process of reducing asymmetry or of restoring symmetry. It was also suggested that in that sense, natural selection was more like the second law of thermodynamics. However, if natural selection were to be thought of as a law of Nature, it is imperative that it must have a structural template which is universal. This structuralism must manifest wherever natural selection begins to operate. As discussed in the preceding monograph (Chhaya 2020, see Chapter 8) and in the preceding chapters, conventionally, the possibility of natural selection having its own structuralism has not been considered seriously because it was thought that such a scenario may introduce a certain degree of predictiveness which might undermine the randomness implicit in Darwin's theory. However, as discussed in the preceding chapters, this is not the case. Admittedly, if natural selection were to possess a structural template of its own, it does reduce the degree of randomness. However, this reduction in the degree of randomness is already implicit in Darwin's central argument of descent with modification. More importantly, it is at the heart of our phylogenetic studies (Bromham 2008, see Chapter 5). In fact, it can be argued that the postulate that natural selection possesses a structuralism of its own explains the central premise of phylogenetics.

In this section, we will try to outline a new model of natural selection that is domain independent and universal. In the next section, we will demonstrate that biological evolution itself is an instantiation of this generic model. This is important because historically speaking, the Darwinian paradigm (Grene 1986) has focused on natural selection and has taken biological evolution as a priori. It is only recently that we have sought to deconstruct biological evolution as a precursor to natural selection. The RNA world hypothesis is the best example of reverse engineering of the Darwinian paradigm (Yarus 2010, see Chapter 2). Therefore, this model provides the necessary semantic foundation for unifying natural selection with biological evolution.

Let us begin with the topological perspective of natural selection. As discussed above, according to this model, the environment seems to influence the competing biological species by downward transfers of its information content. Similarly, competitive survival, according to this model, occurs within the same dimensionality in the form of horizontal information transfers. Finally, upward information transfers result in changes in the units of selection, as in the case of a genotype influencing its phenotype. In view of this perspective, it seems logical to define natural selection in the language of dimensionalities and information transfers. Accordingly, we can visualize natural selection as a combination of upward, downward and horizontal information transfers. While natural selection results from restoration of symmetry (or the reduction of asymmetry), competitive survival consists of this process of restoration of symmetry (or the reduction of asymmetry). It is important to note that in this scenario, while the changes in genotype are a part of horizontal information transfers (which is integral to competitive survival, the relationship between genotype and phenotype constitutes upward information transfers. Incidentally, according to this model, phenotypes can influence genotypes through downward information transfers. This is something largely neglected in the conventional perspective of Darwinian theory, but it is acknowledged in the form of epigenesis in the modern perspective of Darwinian theory. The role of epigenesis and its connection with the Darwinian paradigm and Lamarckism are discussed in the preceding chapters. As mentioned above, this topological model of natural selection has three advantages. Firstly, it enables us to define a domain neutral model of natural selection. Secondly, it allows us to deconstruct the relationship between genotype and phenotype. Finally, because it is conceptualized in topological terms, it is amenable to formalization, something that has eluded natural selection so far. For the sake of clarity, the process of natural selection is presented as a Schema 9.1.

9.10 New Model of Biological Evolution

As discussed in the preceding monograph (Chhaya 2020, see Chapter 8) and in the preceding chapters, the Darwinian paradigm mainly deals with natural selection. Biological evolution has remained outside the Darwinian logic of differential survivability. Admittedly, there is an increasing awareness of including biological evolution within the ambit of the Darwinian paradigm. In Darwin's theory biological evolution of life is taken as a priori. It was much later that attempts were made to extend the Darwinian logic to biological evolution itself. As mentioned above, the RNA world (Yarus 2010) is an attempt in that direction. The motivation behind including biological evolution within the Darwinian paradigm is straightforward and imperative. If natural selection is manifest in all living organisms, there is no reason why Life itself should be excluded from this perspective. More importantly, if we accept that biological evolution is exempt from the principles of natural selection, we are tacitly conceding some kind of creationism. However, the problem with this

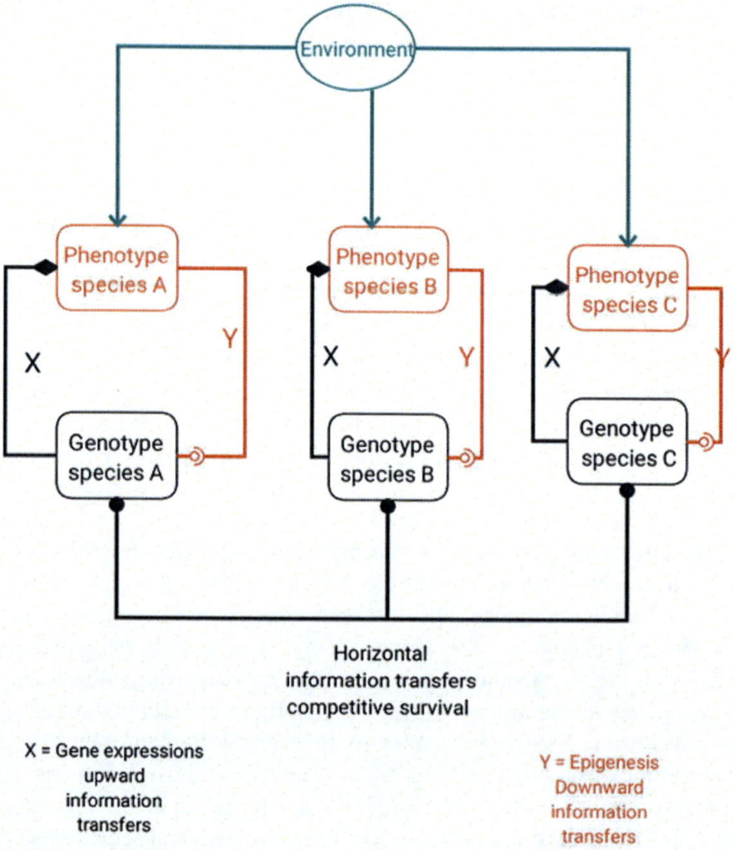

Schema 9.1 A natural selection

approach is not the lack of justification, but a lack of a common framework to accommodate biological evolution and natural selection. As discussed above, there are two problems why the conventional perspective of the Darwinian paradigm can't explain biological evolution. Firstly, for competitive survival to occur, there ought to be some resources for which these biomolecules must compete. However, in current understanding of the origin of life, there is nothing that these biomolecules like RNA-proteins can compete for their survival. These molecules undergo changes out of stereochemical or kinetic compulsions and not because they compete for some external resources. Neither the micronutrients, nor the catalytic clay (which could have catalyzed the syntheses of these biomolecules) could have been in short supply (Oparin 1968). Therefore, it is difficult to define competitive survival for these biomolecules.

Secondly, in order for these biomolecules to acquire autocatalysis or self-replicative memories, they need to be sufficiently complex. However, as discussed in the preceding chapters, the emergence of complexity during natural selection has always been problematic. Therefore, if we wish to define biological evolution from the perspective of the Darwinian paradigm, we need to address the problem of the emergence of complexity in natural selection. Thus, there are two difficulties with the conventional perspective of natural selection which prevent us from including biological evolution within the framework of the Darwinian paradigm, viz., the lack of limited resources for which competition occurs and the prerequisite of complexity prior to natural selection. The proposed model offers a way to redress these issues. According to this model, the stereochemical or kinetic compulsions arise from the information content present in the higher dimensional configurations of these biomolecules. Therefore, the competition is for the information content available from the higher dimensionalities of these molecules and the information content is the limiting resource for competitive survival of these biomolecules. We have missed this idea because we have confined ourselves to the four-dimensional configurations of molecules (which give rise to stereochemical or kinetic forces). However, as discussed in the preceding monograph (Chhaya 2022b, see Chapter 2), spacetime itself consists of multiple dimensionalities simultaneously. More importantly, molecules are spread over multiple dimensionalities of spacetime simultaneously. Therefore, different competing biomolecules have different higher dimensional configurations from which different types of information transfers take place depending on the stereochemical orientations of the biomolecules. Thus, information transfers, both quantitatively and qualitatively, act differentiating factors, thereby setting up competitive survival among these biomolecules. Thus, with the help of the information content as a resource, it is possible to expand the scope of natural selection to biological evolution.

Similarly, the emergence of complexity can be explained within the Darwinian semantics. According to this model, information transfers are essentially topological processes. Moreover, since according to this model, the information content is a physical entity. Therefore, the information transfers are determined by the congruence between the information content of the information being transferred and the information content of the recipient of these information transfers. It is intuitively clear that due to the inherent topology of the information content, some types of complexities would be preferred. Therefore, during biological evolution, complexity emerges naturally because of the topological prerequisite of the information content being transferred. As discussed in the preceding chapters, this reasoning explains why RNA was replaced by DNA as a repository of information content of genomes. Thus, by including spacetime itself in the conception of the environment and by accepting the information content of spacetime as a resource for natural selection, it is possible to include biological evolution within the Darwinian paradigm. In the following section, we will define a unitary conception of the unit of selection and demonstrate that different types of units of selection are different instantiations of this unitary conception.

9.11 Unifying Different Units of Selection

In Sects. 9.2 and 9.6, we discussed the plurality of the units of selection. One way to interpret this plurality is to think of it as evidence of the universality of natural selection. In other words, we can say that there are multiple examples of the units of selection because natural selection operates at more than one level. However, as discussed above, if natural selection operates using a single structural template, then it is imperative that these different units of selection too must be different instantiations of a universal conception of the unit of selection. Therefore, in this section, we will try to unify different units of selection. Sincc, as discussed above, the proposed model employs the topological nature of information and its transfer, we will try to define a unit of selection based on its information content and its vulnerability to natural selection. Even prior to any articulation of the universal conception of the unit of selection, it is possible to think of several informational prerequisites that the unit of selection must possess. This is reasonable because from the information theoretical perspective, every natural phenomenon can be thought of as information transfer and information transformations. Therefore, there must be something definitive about the unit of selection for it to be preserved and passed on to the next generation. It is this selectivity of natural selection to leave the unit of selection unchanged during the universal flux that characterizes all natural phenomena. The unit of selection must be viewed as an island of stability amidst the ocean of information transfers.

Having outlined the motivations for conceptualizing a unitary conception of the unit of selection, let us look at what the proposed model has to offer. To begin with, according to this model, information content, being physical entities, must possess structural templates. Therefore, every information transfer must be treated as a physical process having its own structuralism. Therefore, the fact that all natural phenomena involve constant information transfers and transformations, would suggest that all natural phenomena are structurally agnostic to structuralism of the process of information transfers. The only exception to this situation is that all natural phenomena retain some part of information content unchanged. Therefore, while all natural phenomena are in a flux of information transfers, there exists a small part in each natural phenomenon that retains its identity even amidst constant changes. It is intuitively clear that these unvarying elements of every natural phenomenon must be potential candidates for being called the units of selection. Therefore, any definition of the universal unit of selection must be based on the information content that remains unchanged. More importantly, it must define the reasons for this constancy.

In view of the fact that according to this model, the process of natural selection possesses a structuralism of its own, it is necessary that we define this constancy in the unit of selection in terms of the structural template of natural selection. As discussed in Sect. 9.8, natural selection tries to restore symmetry. Since different competing species possess different types of organized information content (in the form of genomic architecture), they possess different degrees of information asymmetries. Therefore, natural selection restores symmetry by eliminating those competing species which possess higher degrees of asymmetries vis a vis the structuralism of natural selection itself. By implication, what survives must be congruent with the structuralism of natural selection. Thus, the most intuitive way to define the universal unit of selection must be in the language of congruence between the unit of selection and the structuralism of natural selection. While this rationale sounds attractive, the problem is how does one define the structuralism of natural selection. This is where the proposed model has something to offer. There are four features of the proposed model that help us in defining the structuralism of natural selection. Firstly, according to this model, natural selection occurs via downward information transfer. Therefore, according to this model, we should be able to define natural selection as an involutive algebra as defined in this model (Chhaya 2022a, see Chapter 3). Secondly, according to this model, every involution connecting higher dimensionality to a lower dimensionality allows only a certain type of information content to be transferred. This limits the types of information content that can be transferred downward. Thirdly, since the operators of involutions are essentially a mathematical group, the entire evolutionary trajectory is governed by these restrictions. This permits us to define different units of selection in different dimensionalities. Finally, at the end of every involution, the information is transferred to different submanifolds (which represent different competing species) by fixed algebraic rules of associations. Therefore, while the information being transferred downward is uniform, its assimilation into different species would be different, depending on the degree of asymmetry of each species.

With this scenario in place, we can define units of selection as those submanifolds that assimilate the maximum information content being transferred downward. Thus, the unit of selection must possess two prerequisites. Firstly, it must maintain a certain degree of congruence with the information being transferred downward, and due to this congruence, the unit of selection must absorb and assimilate the maximum amount of this downwardly transferred information. The key point is that it must assimilate this information into its permanent structural template of information. If the information received were to alter the structure template of the unit of selection completely, it would cease to be a unit of selection and it would become a nondescript information content like the rest of the natural phenomena. This, prima facie, sounds self-contradictory. However, in this contradiction lies the secret to natural selection and biological evolution. In order to deconstruct this apparent contradictory demand that information theoretical interpretation of the unit of selection imposes, in the next section, we will discuss the need for the dualities of the Darwinian paradigm mentioned in the preceding chapters, viz., the duality of the units of selection and inheritance.

9.12 Why We Need Duality of the Units of Selection and the Units of Inheritance

In the preceding chapters, three forms of dualities were mentioned, viz., the duality of DNA and RNA; the duality of genotype and phenotype; and the duality of structuralism and functionality. In this section, we will add another duality of the unit of selection and the unit of inheritance. To be honest, the conventional perspective of the Darwinian paradigm has no explanation for either its origin or its necessity. This is true not just for the classical Darwinian theory, but also for later paradigms like genetics (Provine 2001) and molecular biology (Weinzierl 1999). It is possible to take a metaphysical view and suggest that this duality is inherent in every natural phenomenon. This phenomenological view of dividing reality into reality as it appears to be and reality as it is, without asking for any explanation, is something that science can't accept. To the extent scientific inquiry consists of seeking causality, these dualities must be sought to be explained in terms of causality. Moreover, unlike the other natural phenomena, Life is an exception. While the distinction between the reality as it appears to be and the reality as it is in natural phenomena is implicit, in the case of Life, this distinction is explicit. Our biological theories are characterized by these dualities. For instance, in quantum field theory (Lawrie 2013), it is possible to overlook the underlying nature of the quantum field and instead focus on its phenomenology. We don't need to know what constitutes a quantum field in order to formalize unified quantum field theory. However, this is not the case with Life. Genomic functionalities and genomic architecture are both palpably manifest. We can't articulate one without the other (Though it is a different matter that we know so little about either of them). With this perspective in mind, in this section, we will try to deconstruct the semantics of having dual units.

In the previous section, it was mentioned that the unit of selection, from the information theoretical perspective, is required to possess two contradictory features. Firstly, the unit of selection must maximally assimilate the downward information transfers during natural selection to reduce the inherent asymmetry between itself and its environment. Secondly, while assimilating, it must retain its own characteristic identity. Upon a little reflection, it is intuitively clear that the identity of the unit of selection rests on how asymmetric it is with respect to its environment. Thus, we can see that this conception of the unit of selection is self-contradictory. This contradiction can be resolved if we postulate that there are two different units serving two different purposes. Thus, the unit that maximally assimilates the information content from the environment to reduce the inherent asymmetry must be separated from the unit that tries to retain its inherent informational asymmetry that characterizes it. This explains the need for having a separate unit of selection and a unit of inheritance.

It is tempting to think that this simple hermeneutic device explains the enigma of Life. The enigma lies not in having two types of units. Rather, it lies in having these two units bound to one another. The key point is that Life not only has two different units of selection and inheritance, but that these two types of units are bound to one another by a definitive relationship. A Genotype and its phenotype are not autonomous entities, but they are causally interconnected ensembles. It is this relationship between two types of units that defines Life. The proposed model merely formalizes this relationship between this and other types of dualities in the framework of topological dimensionalities using involutive algebra. More importantly, it postulates a unitary operator to define the relationship between the twins which occupy different dimensionalities. It is this unitary nature of the relationship between different types of units of selection and inheritance explains the mystery of Life. There are no units of selection only, nor are there any pure units of inheritance. It is the dimensionality from which we try to understand Life that defines the unit of selection and the unit of inheritance. Does that mean that genotype can behave like a phenotype? Or vice versa? The answer is yes, but with a proviso. From the perspective of genomes, individual genes are phenotypes of the lower dimensional DNA sequences. Similarly, genomic functionalities are normally thought of as phenotypes are actually genotypes or structural elements of genomic architecture and therefore, behave like genotype while defining genomic architecture. It was mentioned above that Life is unlike any other natural phenomena. The reason for this uniqueness lies in the fact that in the case of Life, different dimensionalities representing the units of selection and inheritance are rather too close to one another. Therefore, units of selection and the units of inheritance interact with one in a perceptible manner. As a result, we can observe these interactions. This is not the case with other natural phenomena. For instance, we can, by analogy, think of quantum fields and the fundamental particles as the corresponding genotype and phenotype. However, since the topological separation between them is so large that we only observe phenotypic polymorphism in the form of different types of particles and their properties, but rarely, if ever, we observe the underlying genotypic quantum field. As discussed in the preceding chapters, the relationship between genotypes and phenotypes is also subtle and is no more a one-way relationship. Phenotypes too can influence genotypes through epigenesis. However, it is more intriguing to think of a scenario wherein a single entity can act both as a unit of selection and a unit of inheritance. Upon a little reflection, it is intuitively clear that a genome is one such entity. Therefore, in the next couple of sections, we will try to deconstruct the dual nature of genomes.

9.13 Genome as a Duality of Units Personified

It was suggested in the previous section that the units of selection as currently understood, requires it to possess two mutually contradictory features of an ability to reduce asymmetry by assimilating the information from the environment and an ability to retain the inherent asymmetry to retain its identity. It was suggested that this contradiction can be resolved if employ a topological device of dimensionalities to define the unit of selection and the unit of inheritance. The advantage of this approach is twofold. Firstly, it explains the hitherto unexplained Darwinian dualities, viz., the duality of DNA and RNA; the duality of structuralism and functionality; and the duality of genotype and phenotype. Secondly, this approach suggests that there is only one unit. It is the dimensionality from which we choose to view that decides whether the unit behaves as a unit of selection or as a unit of inheritance. Such a contextual definition enables us to unify these above mentioned Darwinian dualities. However, this approach also throws open a possibility that there could be a unit which acts simultaneously as a unit of selection and a unit of inheritance. Therefore, in this section, we try to deconstruct genomes as dual units of selection and inheritance.

The reason why it is possible to think of genomes both as units of selection and the units of inheritance lies in the inherent dual nature of genomes. A genome seems to possess a set of functionalities which cannot be inferred from the functionalities of its constituent genes. At the same time, we can only define structuralism of a genome which is nothing more than the sum of the structures of its constituent genes. As discussed in the preceding chapters, conventionally, we have focused on the structural template of genomes. The irony lies in the fact that we have tried to conceptualize structuralism of genomes based on their functionalities, while denying a degree of autonomy to these genomic functionalities. It is this mismatch between our conception of the functionalities and the structuralism of genomes that has obscured the dual role of genomes as the units of selection and the units of inheritance. It is the genomic functionalities that are selected, but it is the structuralism of genomes that is inherited.

It is possible to argue that a similar situation exists in the case of individual genes. After all, the conception of genotype and phenotype rests on the distinction between the DNA structure of individual genes and their physiological effects. However, there is one aspect which separates genetics from genomics. In genetics, genotype and phenotype are both manifest and can be subjected to measurements. In the case of genomics, genomic functionalities (which may be thought of as synonymous with phenotypes) are not amenable to direct measurements. To put it differently, in genetics, both genotype and phenotype exist in the four-dimensional spacetime. On the other hand, in genomics, only genomic structural templates exist in the four-dimensional spacetime. Genomic functionalities are essentially nonexistent in the four-dimensional spacetime. If at all, genomic functionalities are evanescent or ephemeral entities in the four-dimensional spacetime.

It is possible to argue that while this reasoning might appear alluring, there is no hard evidence in favor of any higher dimensional perspective of genomes. However, there are hosts of genomic features which a four-dimensional configuration of genomes can never explain. The long-range influences, polygeny, and pleiotropy would remain inscrutable until we concede higher dimensional configurations of genomes. In addition, there is another aspect of genomics that forces us to concede that genomics must include such higher dimensional configurations of genomes. This refers to functional genomics (Roy and Kundu 2021). While, prima facie, functional genomics takes a reductionist perspective of studying only those aspects of genomes that directly deal with individual genes and their expressions, a realistic model of functional genomics has to provide for formalization of long-range influences. This requirement cannot be fulfilled by restricting our modeling to molecular ensembles of biomolecules.

It is only when we accept that genomes behave like the units of selection and the units of inheritance simultaneously that we can think of creating a reasonably good model of functional genomics. This is where the proposed model comes into play. According to this model, genomic functionalities are assigned a separate higher dimensionality and the structural template of genomes is assigned four-dimensional configurations. More importantly, this model offers a way to define genomic functionalities from the underlying genomic architecture. As discussed in the preceding chapters, it is possible to think of genomic functionalities as units of selection by postulating that there exists a notional entity named genomic singularity which shapes different genomic functionalities. Similarly, the conventional perspective of the environment acts upon the DNA sequences to give rise to necessary modifications. Thus, there are two parallel processes occurring. Genomic singularity shapes the genomic functionalities and the conventional environment shapes the genomic architecture in the form of DNA sequences. However, both these parallel selection processes are intertwined because of the definitive relationship between genomic architecture and the genomic functionalities. Finally, according to this model, it is possible to unify these parallel selection processes by including the structuralism of spacetime into our definition of the environment. Thus, by enlarging our conception of the environment to include spacetime, we can define the dual nature of genomes wherein genomic functionalities are the units of selection and the genomic architecture is the unit of inheritance. In order to clarify this duality, in the next section, we will deconstruct two faces of genomes, viz., the regulatory genome and the expressive genome.

9.14 Regulatory Genome Versus Expressive Genome

Genomics faces a peculiar problem. It is required to explain a host of functionalities without any structural foundations. Admittedly, the earlier belief that a large part of genomes (about 90%) constitutes "junk" DNA (Carey 2015). With the discovery of RNA interference (Howard 2013), open reading frames (Sieber et al. 2018) and antisense transcription (Mostovoy 2014), genomics has moved on to a more nuanced conception of genomic functionalities. However, the basic problem of locating the structural basis of genomic functionalities remains unaddressed. However, in accordance with our current understanding of functional genomics, it seems reasonable to think that genomes possess a dual framework of gene expressions and their regulations. Therefore, for the present discussion, we will postulate that there exists a notional entity called regulatory genome. Similarly, we will postulate that there exists a notional entity called expressive genome epitomizing as a framework of all the gene expressions. As discussed in the preceding chapters, the conventional perspective has shied away from formalizing genomic architecture for the fear that it might undermine the randomness that is emblematic of Darwin's theory. Therefore, without committing ourselves to the physical status of regulatory and expressive genomes, we will treat them as abstractions. However, it must be kept in mind that the proposed model asserts that the regulatory and expressive genomes are physical entities, or rather physical frameworks, existing in different dimensionalities and yet entwined with one another. If this physicality of these two frameworks is not insisted upon in this section, it is because this physicality is incidental to the present discussion. This discussion is aimed at convincing everyone that even in the conventional perspective, this duality of frameworks is implicit. The proposed model merely formalizes this implicit duality using a topological framework to accommodate both these frameworks. As for the putative physicality of dual frameworks, we will have to wait until the technology for making measurements in different dimensionalities of spacetime is developed.

Let us begin with a proposition that the regulatory genome is actually a higher dimensional configuration of the molecular ensemble of genomes. Similarly, we will adhere to the proposition that the expressive genome is actually a topological surface which is contiguous even though the underlying DNA sequences appear to be segregated into different chromosomes. In the absence of any prior knowledge of these higher dimensional configurations of genomes, we will arbitrarily assign six-dimensional configurations to the regulatory genome and five-dimensional configurations to the expressive genome. Obviously, in this schema, the conventional description of genomes would be assigned to four-dimensional configurations of genomes. With this simplistic scenario in place, let us try to visualize how genomes act as dual units of selection and inheritance. A more detailed description of different genomic configurations and their interactions is given in Chap. 8. Therefore, we will not go into the details here. Instead, we will discuss how genomes can behave like a unit of selection and a unit of inheritance.

To further simplify the discussion, a graphical representation is given in Schema 9.2. According to this model, the regulatory genome occupies sixth dimensionality, and the expressive genome occupies the fifth dimensionality. Thus, according to this model, neither the regulatory framework, nor the expressive framework is directly visible in the fourth-dimensional configurations of genomes. Since our conventional perspective of genomes is confined to the four-dimensional spacetime, we have not been able to formalize either the regulatory or the expressive framework of genomes. As a result, our conception of natural selection has been confined to just the four-dimensional configurations of genomes. Therefore, we have been able to articulate natural selection of genotypes. Moreover, since gene expressions and the resulting physiological changes occur only in the four-dimensional spacetime, we can observe them, but we can't perceive the guiding forces which operate from the sixth- and fifth-dimensional configurations of genomes. What is germane to the present discussion is that natural selection also occurs when the sixth-dimensional configurations of genomes influence the fifth-dimensional expressive genome, inducing natural selection. Similarly, the sixth-dimensional configurations of genomes undergo natural selection under the influence of the higher dimensional genomic singularity. Thus, in each case, it is the higher dimensionalities that act as the environment and bring about natural selection in the lower dimensional configurations of genomes. More importantly, according to this scenario, the units of selection and the units of

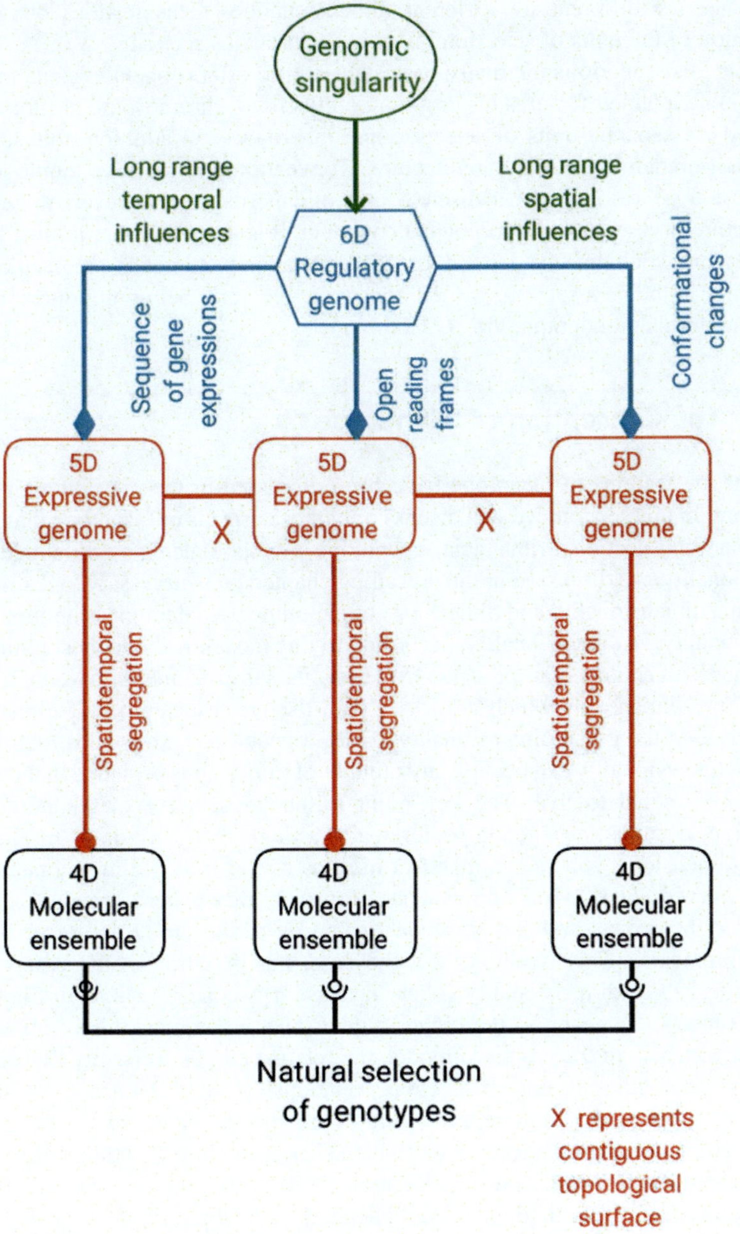

Schema 9.2 Topology of natural selection

inheritance are different for different dimensionalities. This justifies our earlier conception of the units of selection given in the previous section.

While this scenario is internally consistent, it also offers cogent explanations for hitherto unexplained features of long-range influences, chromosome territories and the need for separate units of selection and inheritance. In addition, this scenario helps us to redefine genomic architecture. Conventionally, we have employed the notion of a gene to denote a discrete unit of inheritance. However, if genomic architecture is essentially a topological construct, it is necessary that it must have a discrete unit which must be defined as a topological unit and not as a molecular unit. Therefore, in the next section, we will discuss a new term being introduced for the first time in this monograph, viz., Genotope.

9.15 The Conception of "Genotope "

Some of the features of Genotope have been discussed in the preceding chapters. Therefore, in this section, we will discuss only those aspects of Genotope that play a role in natural selection. Prima facie, a genotope is a topological object defined by its dimensionality. As discussed in the preceding chapters, it is necessary to distinguish between a dimensionality and dimensions of genotopes. A genotope is defined by its dimensionality. A dimensionality, according to this model, is the highest number of dimensions a genotope can possess. However, its substructures of a genotope can occupy any dimensions which are lesser than the greatest number of dimensions permissible in the given dimensionality. Thus, a genotope and its substructures, in the fifth dimensionality can occupy any number of dimensions provided that they are less than or equal to five. The key point is that irrespective of the number of dimensions occupied by a genotope, it remains a single entity. To draw the analogy, a three-dimensional sphere can manifest itself as circles in two-dimensional planes and as curved lines in one dimension. However, this sphere remains a three-dimensionality object and can be reconstructed by adding up all its lower dimensional fragments. (This is always not the case, but it is true in the case of this hypothetical three-dimensionality sphere because it is conceptualized in Euclidean solid geometry. However, in the higher dimensional topologies, this is not always true. For instance, in the proposed model, the relationship between any two successive dimensionalities is not an algebraic relationship. It is defined by a special involutive algebra.) A genotope, according to this model is a discrete entity having functional and structural autonomous status. Therefore, during natural selection, it acts as a unit. Once again, as discussed above, whether a genotope acts as a unit of selection or as a unit of inheritance would depend on the dimensionality in which we wish to formalize natural selection. In the next section, we will discuss genotopic natural selection.

9.16 Mechanism of Genotopic Natural Selection

In the previous section, it was mentioned that a genotope may act either as a unit of selection or as a unit of inheritance depending on the dimensionality from which we define natural selection. Therefore, in this section, we will discuss both these scenarios wherein a genotope acts as a unit of selection and as a unit of inheritance. Let us look at the scenario wherein a genotope acts as a unit of selection. Apparently, in this scenario, the role of the environment is played by the higher dimensional configurations of genomes which are above the dimensionality of the genotope under investigation. Therefore, there will be downward information transfers from the higher dimensional configurations of genomes which would force the genotope to absorb these downward information transfers and to alter its own information content to become more congruent with the information content being transferred downward. When the genotope absorbs the information content of the information transfers, it must alter its own information content. This would lead to adaptations. Since the extent of adaptation decides the survivability, it is clear that the genotope would behave like a unit of selection. Secondly, if the genotope fails to absorb the information content available through the downward information transfers, it would lead to an increased information asymmetry. This is because if the information was absorbed by the genotope, due to inherent mathematical compulsions, the information content of the genotope would have been altered. It is only when the information content inherent to the genotope is asymmetric in reference to the information content available from the downward information transfers that the genotope would fail to absorb the information content. This increased asymmetry would decrease the survivability of the genotope, leading to natural selection of only those genotopes that have successfully absorbed the information content available through the downward information transfers. Thus, in this scenario, a genotope acts as a unit of selection.

Now let us look at the second scenario wherein a genotope acts as a unit of inheritance. Apparently, in this scenario, a genotope must be able to transfer its own information content to its descendants. Therefore, these information transfers must occur horizontally. This is because according to this model, each dimensionality has its own metric. Therefore, if the information content of a genotope has to be transferred intact, it can happen only when it is transferred in the same dimensionality. Upon a little reflection, it is intuitively clear that this is precisely what the conventional perspective of genetics suggests. During transcription, genes are duplicated and passed on to the descendants. Since the information content of a gene and its replicas are structurally identical, they would be represented in the proposed model as occupying the same dimensionality. Therefore, such information transfers would always be horizontal.

It is natural to wonder what would happen if the information transfers were upwardly directed. According to this model, upward information transfers can occur only within a genotope. This is possible because as discussed above, a genotope in a given dimensionality can occupy any number of dimensions provided they are less than the numerical value of its dimensionality. Let us say that we have a genotope in sixth dimensionality. Apparently, some substructures of this genotope would also manifest itself in five and four dimensions as well. In such a scenario, it is possible to transfer information within the genotope, say from the fourth dimensionality to the fifth dimensionality. The key point is that these information transfers must occur within the genotope and between the two genotopes. Upon a little reflection, it is intuitively clear that this is precisely what is implicit in the conventional perspective of the relationship between a genotype and phenotype. The only difference being that in the conventional perspective, we don't assign different dimensionalities to phenotypes. Thus, this scenario allows us to formalize the relationship between a genotype and its phenotype. It is important to note that according to this model, the downward information transfers can occur even within a genotope. Therefore, this model provides an explanation of the origin of epigenesis (Robert 2004). When a higher dimensional phenotype within a genotope influences its lower dimensionality genotypic counterpart through downward information transfers, we observe epigenetic phenomena, something that has defied the conventional Darwinian perspective of natural selection. It is apparent from this discussion that the proposed model offers a new conception of the unit of selection and the unit of inheritance. The proposed model offers a way to formalize the relationship between genotype and phenotype. More importantly, it binds genotype and phenotype into a single entity which can act either as a unit of selection or as a unit of inheritance. It is important to note that the ascription of different dimensionalities to different types of information contents is not ad hoc, but it is semantically consistent with the Darwinian semantics of natural selection. In the next section, we will discuss the information theoretical perspective of natural selection.

9.17 Information Theoretical Perspective of the New Model

In the preceding sections, we tried to reconceptualize natural selection from the perspective of information content and its topological features. While this perspective offers us new insights into the nature of natural selection, it is necessary to understand why we need to employ information theoretical perspective to deconstruct natural selection. After all, there exists a humongous amount of literature on this topic (Grene 1986). Therefore, unless the employment of information theoretical perspective can give us a better interpretation or an improved predictive power, such an exercise would be a mere pedagogy. Therefore, in this section, we will discuss the reasons why we need to employ information theoretical perspective of natural selection. One of the drawbacks of employing mathematical modeling (Laudal 2021) or information theoretical modeling is that it is based on an inchoate belief that such models are fortuitously congruent with the natural phenomena under investigation. Therefore, these exercises are more adventurous attempts. As discussed in the preceding monograph (Chhaya 2020), this ambiguity about the reason why mathematical formalisms are effective in explaining Nature, arises from our incomplete understanding of the origin and nature of mathematics.

However, the proposed model is based on a school of thought that is based on mathematical realism. It postulates that what we perceive to be mathematical objects are abstractions of the material details of Nature. As discussed in the preceding monographs (Chhaya 2020, 2022b), our belief in the indestructibility of the information content arises from this belief in mathematical realism. In addition, we have found it difficult to formalize Life without invoking any transcendental proposition. The Darwinian paradigm in that sense, is critical because it seeks to define biological evolution and natural selection in a purely naturalistic perspective. Therefore, it seems worthwhile to reinforce the naturalism implicit in the Darwinian paradigm with the mathematical realism implicit in the proposed model. As discussed in the preceding chapters, the Darwinian paradigm, for poorly understood reasons, seems to define a law of Nature rather than a scientific theory of biology. This domain neutral conception of the Darwinian paradigm requires that it needs to be recast in a domain neutral framework. Once we accept this rationale, the information theoretical perspective becomes a natural choice. Thus, an information theoretical conception of natural selection can be useful for several reasons. Firstly, it can provide us with an expression of natural selection that elevates it to the status of being a law of Nature. Therefore, it can be applied to various scientific domains. Secondly, it can help us to disambiguate some of the semantic ambiguities of the conventional Darwinian perspective. Several such ambiguities like the emergence of complexity in natural selection, the need to have duality of genotype and phenotype etc., have been successfully resolved in this and the preceding chapters. However, the most fundamental outcome of this reinterpretation is that it provides a structural template of natural selection. This was something that was not explored before. As discussed in the preceding chapters, the Darwinian paradigm, as a scientific theory, is unique. Unlike other theories, it lacks predictive powers, even while possessing excellent explanatory power. Therefore, this information theoretical approach can fulfill this lacuna. Once we are able to find quantitative approaches using this formalism, it ought to be possible to enhance predictiveness of natural selection without compromising its semantic core of randomness. It is important to keep in mind that this is possible only because the proposed model is founded on the physicality of information content and its attendant topological framework. The purists might object to this insistence on mathematical sophistication as it might rob the natural simplicity and charm of Darwin's theory. However, if the introduction of mathematics can help us to resolve the semantic ambiguities of the Darwinian paradigm and if it adds predictiveness to the original Darwinian semantics, this approach is worth its while. Having digressed from the main topic of this chapter, we will return to it in the next section which discusses a new conception of genomic evolution.

9.18 Principles of Genomic Evolution

As discussed in the preceding chapters, while we accept the possibility of genomes being a unit of natural selection, we have not pursued this possibility seriously. This is because we are reluctant to accept that genomes possess a definitive structural template. This reluctance arises from the fear that any definitive genomic architecture would undermine the randomness that is implicit in Darwin's theory. However, as discussed in the preceding chapters, this need not be true. It is possible to reconcile structural complexity with the essentially random nature of natural selection. The proposed model allows us to reconcile this dichotomy between randomness and complexity by using a topological framework wherein different dimensionalities possess different degrees of complexities and randomness. Thus, natural selection can be defined as the changes in the dimensionalities in this model. This helps us to explain how complexity could emerge from natural selection. With this strategy in place, it is necessary that we look at genomic architecture as an instantiation of the coexistence of multiple dimensionalities simultaneously within genomes. Therefore, as discussed above, we can conceptualize genomes both as units of selection and as the environment in which different genes are subject to natural selection. Moreover, as discussed below, genomes can also act as units of inheritance. This makes it imperative that we must discuss genomic evolution. Therefore, in this section, we will discuss the principles of genomic evolution with reference to the conventional perspective of the principles of genetic evolution.

As discussed above, we can think of genomic architecture itself as the environment for natural selection of genes. It is important to keep in mind that this natural selection is in addition to the natural selection implicit in the conventional perspective. This is because in the conventional perspective, natural selection occurs when a phenotype generated from a given genotype competes for the natural resources present in the environment. However, in this case, it is the genotype itself that competes for the information content available through long-range influences from genomic architecture. The role of phenotypes in this case is not that of an interface between genotype and the environment. If anything, phenotypes are part of genomic architecture and therefore influence gene expressions (via epigenetic processes). Thus, to extend Dawkins' metaphor now (Dawkins 1999), genes behave selfishly but for the information content available through the downward information transfers from the higher dimensionalities of genomes. Survivability, in this scenario, consists of gene expressions. However, this is only one side of the story.

Genomes also act as units of selection as well as the units of inheritance. Let us understand how. As discussed in the preceding chapters, this model postulates an entity named here as genomic singularity. This entity acts as a repository of all the structural and functional features of genomes. According to this model, genomic singularity must have arisen when the biomolecules capable of replicative memory were synthesized from the primordial soup. However, once created, genomic singularity acquired autonomy. Since according to this model, all the molecules, including biomolecules, are isomorphs of spacetime itself, the inherent structuralism of spacetime is woven into these biomolecules. Therefore, genomic singularity, once formed, would have its own structuralism and functionality. As the name suggests, both the structuralism and functionality of genomic singularity are also singular and therefore, undifferentiated. Therefore, during natural selection, in parallel with the conventional natural selection, genomic singularity also evolved via a series of involutions to give rise to increasingly complex genomic architectures with differentiated structural and functional features. In such a scenario, it is intuitively clear that different genomes, having different genomic architectures (and therefore, different structuralisms and functionalities) would compete with one another to optimize the information content in different topological frameworks. Thus, in this case, genomic singularity acts as the environment and different genomes as units of selection.

It is legitimate to doubt this scenario because unlike the conventional description of natural selection, this genomic natural selection doesn't have any physical entity which acts as the environment. After all, even if genomic singularity were to exist in the distant past, it surely can't still shape the current genomes. However, the earlier assertion that genomic singularity, once formed, would be an autonomous entity needs to be elaborated in order to understand how a past entity can influence the present genomes. As mentioned above, the structuralism (and by implication, functionalities) of genomic singularity is a manifestation of the structural template of spacetime itself. Therefore, genomic singularity continues to exist in the higher dimensionalities of spacetime in which the present genomes continue to evolve. Thus, what we perceive to be genomes is actually a lower dimensional manifestation of genomic singularity. In other words, genomic singularity is still manifest, but only in the higher dimensionalities.

This brings us to the role of genomes as the units of inheritance. In comparison to the abstract conception of genomic singularity and its influence on genomes as the units of selection, the role of genomes as the units of inheritance is implicit in the conventional perspective of natural selection. Our phylogenetic studies (Bromham 2008) embody this concept, albeit tacitly. According to this model, genomes exist in multiple dimensionalities simultaneously. Moreover, each dimensionality of a genome has its own structuralism. Therefore, during the process of gene expressions, it is this structural information that is passed on to the lower dimensionalities of genomes. As discussed in the preceding chapters, this happens in the form of long-range influences. Therefore, each lower dimensionality of genomes (except for the highest dimensionality occupied by genomic singularity) receives information content from multiple dimensionalities which are higher than itself. This is precisely what the units of inheritance are. Each downward information transfer constitutes a unit of inheritance. It is just that there are so many units of inheritance present in each genome that we collectively call them long-range influences. Thus, each dimensionality of genomes can be thought of as a unit of inheritance so far as it influences dimensionalities lower than itself.

Upon a little reflection, it becomes apparent that this discussion has been shifting frames of reference while describing different roles of genomes, viz., as the environment, as the units of selection, and as the units of inheritance. Therefore, purists might object to this frame shifting as an intellectual trick to mislead. However, the key point is this: Natural selection occurs at multiple levels simultaneously. Therefore, it is legitimate to deploy different frames of reference to define various roles of genomes. This brings us to the end of this section. In the next section, we will discuss, albeit very briefly, the need to revisit the Darwinian paradigm.

9.19 Revisiting the Darwinian Paradigm

Progress of science is characterized by a series of paradigm shifts (Kuhn 2021). Therefore, it is axiomatic that every paradigm reaches a point where there is an inherent need to replace it with a new paradigm. It is important to keep in mind that this need to replace doesn't invalidate the established paradigm. It is just that the underlying semantic imperative demands such a paradigm shift. Thus, every paradigm shift in science has provided new insights that such transitions bring about. It enriches even the outgoing scientific paradigm. Perhaps, it is more appropriate to think that old and new paradigms remain in a symbiotic relationship. A classic example of this is the transition from the Newtonian paradigm (Capuzzo-Dolcetta 2019, see Chapter 2) to the relativistic paradigm (Schutz 2009, see Chapters 7 and 8). Both these paradigms have thrived due to such a symbiosis. However, in the case of the Darwinian paradigm, the situation is quite different. As discussed in the preceding chapters, the Darwinian paradigm has already gone through such semantic transitions. With the advent of genetics (Delisle 2021), the Darwinian paradigm was enriched by the clarity on genotypes and phenotypes. Similarly, with the advent of population genetics (Provine 2001, see Chapter 5), the conception of natural selection found a formal expression of its inherent randomness. Similarly, molecular biology (80) offered a structural template of the mechanisms by which genotype gives rise to phenotypic polymorphism. Finally, with the advent of genomics (Roy and Kundu 2021), natural selection has acquired its semantic heft.

The key point is that, unlike the Newtonian paradigm, the Darwinian paradigm can never be replaced. All along these transitions, the Darwinian paradigm has found itself rejuvenated and reinforced, albeit with subtle semantic shifts. The Darwinian paradigm, for some unknown reasons, seems to be endowed with prescience. It is as if the Darwinian paradigm embodies within itself a law of Nature which transcends biology. Therefore, it would always be difficult to contemplate its replacement with some alternative paradigm. Having conceded the semantic primacy of the Darwinian paradigm in modern scientific thought, it is necessary to deconstruct its semantic ambiguities, if only to resolve them. This monograph is meant to do just that. There are three key semantic ambiguities that require us to revisit (and possibly reinterpret) the Darwinian paradigm. In this section, we will merely mention these three semantic ambiguities, viz., (1) the relationship between randomness and complexity in natural selection, (2) plurality of levels of natural selection, and (3) natural selection as a law of Nature.

Let us begin with the first aspect of the relationship between randomness and complexity in natural selection. As discussed in the preceding chapters, the main objection to any explanation of the emergence of complexity is the fear that it might bring back some kind of teleology or design principles into what is essentially a random process of natural selection. However, this apprehension of undermining the essential randomness is actually a category mistake. It rests on the assumption that the earliest living organisms were simple. Admittedly, from the morphological perspective, these organisms were simple. However, from the molecular perspective, every organism is incredibly complex. Therefore, we must accept that what appears to be an emergence of complexity is in fact incremental changes in complexity. Once we accept this rationale, it is intuitively clear that the changes in the degree and nature of complexity would always follow certain structural compulsions. Therefore, there is a need to redefine biological evolution and natural selection in the language of information content and the formal description of its complexity. It is possible to argue that this kind of formal description is already available in the form of the models of population genetics (Provine 2001). However, as discussed in the preceding monographs, the employment of mathematics in formalizing scientific theories remains enigmatic because of the lack of ontological perspective. To quote Wigner's famous phrase (Wigner 1960), the "unreasonable effectiveness of mathematics" in natural science is indeed enigmatic because we can't explain why mathematics should be so effective, particularly when we think of mathematics as a transcendental entity. However, the proposed model rests on the premise that mathematics is an abstraction of the fine structure of spacetime and therefore, it must be congruent with all the natural phenomena manifest in the spatiotemporal universe, including Life. Secondly, the proposed model goes one step further and instead of employing mathematics, it employs the information content of genomes and postulates that this information, like mathematical objects, is a physical entity and therefore has certain shapes. Therefore, it employs a topological framework to organize the information content of genomes into multiple dimensionalities. Admittedly, the success of such an approach would be subject to its ability to predict genomic functionalities. However, the underlying semantics, particularly its ability to formalize complexity within the Darwinian randomness merits a serious reinterpretation of the Darwinian paradigm.

Let us now look at the second aspect of the plurality of levels of natural selection. While it is generally accepted that there are several candidates for units of selection, there is no precise definition of units of selection. Admittedly, there is a varying degree of consensus on how many units of selection exist. For instance, the notion of group selection (Borrello 2010) is not widely accepted as the idea of individual genes being the units of selection. Admittedly, this situation arises because of the lack of semantic clarity. However, it is preferable if we had a structural definition of the units of selection. However, as discussed above, it is possible to think of a structural template of the units of selection, only if we accept that natural selection possesses a structuralism of its own. This is something that is apparently antithetical to Darwinian randomness. However, as discussed above, if we employ an information theoretical conception of natural selection, it is possible to formalize the structuralism of natural selection. This ought to help us in defining the structuralism of the units of selection. Therefore, we need to redefine the conception of natural selection and its underlying mechanisms.

This brings us to the third ambiguity in the conventional perspective of natural selection, viz., the status of natural selection as a law of Nature. As discussed in the preceding chapters, the Darwinian paradigm has an uncanny ability to rejuvenate itself during every paradigm shift. This points toward the possibility that biological natural selection somehow encapsulates a natural law. However, the problem with this inference is that unlike other laws of Nature, natural selection does not have any predictive powers. Admittedly, as discussed in the preceding chapters, the very notion that natural selection should lead to predictive outcomes is antithetical to the core of randomness implicit in the Darwinian paradigm. On the other hand, there are several domains which are unlike biology, where we can discern the logic of natural selection operating itself. Therefore, there exists a legitimate claim to consider natural selection as a law of Nature. The proposed model, as discussed in the preceding chapters, offers a way to accommodate a limited degree of predictiveness within what is essentially a random process of natural selection. Therefore, there is a semantic imperative that we must redefine the Darwinian paradigm using an information theoretical framework which accommodates the conflicting demands of randomness and predictivity. Upon a little reflection, it is intuitively clear that such a framework must be a topological framework, something that the proposed model fulfills. In the concluding section, we will summarize various topics discussed above.

9.20 Conclusion

In the preceding sections, we have covered a wide range of topics. Therefore, it is necessary to summarize them. Therefore, in this section, we will present a point-wise summary.

1. The conventional perspective of natural selection accepts that there are multiple units of selection and that natural selection operates at more than one level.
2. The conventional perspective of natural selection also accepts that natural selection operates on competitive survival in the face of limited resources. This implicitly suggests that there exists a universal mechanism by which natural selection operates.
3. However, the conventional perspective is ambiguous about the possibility that natural selection possesses a structuralism of its own. This ambiguity arises from the conflicting demands of accepting a common mechanism of natural selection and the inherent randomness of natural selection.
4. However, plurality of the units of selection points toward a common structural template of natural selection which operates at more than one level.
5. The proposed model offers a way to unify different units of selection by postulating that different types of resources can be thought of as information content. Therefore, natural selection can be formalized using information transfers from the environment to different competing species.
6. The proposed model postulates that information content, as a resource, is a physical entity, and therefore, it occupies spacetime. Therefore, in order to unify different resources, the proposed model includes spacetime itself as an integral part of the environment.
7. Accordingly, the proposed model formalizes a topological framework wherein different types of information content (or different types of resources) are assigned different dimensionalities.

8. The role of the environment is to shape natural selection through information transfers. Therefore, natural selection can be formalized in the language of information transfers and survival can be formalized as the preservation of symmetry between the information content of the environment and the information content of the competing species.
9. Thus, the units of selection can be formalized as packets of information content that try to retain their information content (and its organization) in the face of constant information flux that represents an ecosystem.
10. Using this definition, it is possible to unify different units of selection and define a structural template of natural selection. The information packets that retain their identity are the units of selection and the structural template of natural selection consists of inducing symmetry between the information content of the environment and the information content of the competing species. Due to inherent structuralism of the topological model proposed here, natural selection can be formalized as downward information transfers.
11. In order to retain the integrity of the information content, the units of selection are required to pass on the unchanged information content to their descendants. Therefore, natural selection requires a separate unit of inheritance.
12. Therefore, the proposed model defines a topological unit called genotope. The same unit can act as the unit of selection as well as the units of inheritance depending on the dimensionality in which we define natural selection. Thus, the duality of the units of selection and the units of inheritance exists in the conventional perspective because it is conceptualized in the four-dimensional spacetime. In contrast, the proposed model defines a single unit which acts either as a unit of selection or as a unit of inheritance in different dimensionalities. Thus, topological framework unifies different types of dualities that characterize the Darwinian paradigm.
13. Due to the inherent topology of the proposed model, the units of selection occupy higher dimensionalities than the corresponding units of inheritance.
14. When the units of inheritance are genotypes and the units of selection are phenotypes, genotypes exist in the four-dimensional spacetime. Therefore, the changes in genotypes including mutations must be thought of as horizontal information transfers.
15. Since gene expressions occur only in the four-dimensional spacetime (because the higher dimensional spacetime does not have the distinction between the time-like and the space-like dimensions, thereby excluding any thermodynamic consequences), phenotypes do not manifest themselves because they occupy higher dimensionalities. Therefore, we have failed to formalize phenotypic structuralism. However, according to this model, the higher dimensional phenotypes can transfer information content to genotypes through downward information transfers. This perhaps explains epigenesis.
16. Once we accept this topological framework of natural selection, it is intuitively clear that natural selection occurs in any system that manifests itself in multiple dimensionalities simultaneously. This provides us with a possibility of developing domain neutral conception of natural selection.

17. Interestingly, genomes, as discussed in the preceding chapters, also manifest multiple dimensionalities simultaneously. Therefore, they manifest a unique feature of being a unit of selection, a unit of inheritance and the environment.
18. Just as the environment shapes the course of natural selection of morphological features, spacetime (because of its inherent structuralism as proposed model implies) shapes genomic architecture. Therefore, genomes can be thought of as units of selection.
19. Similarly, we can think of genomes as playing the role of the environment for the individual genes. In this case, the information content in the form of long-range influences acts as a limited resource for which individual genes compete. In this scenario, gene expressions can be thought of as being synonymous with survivability.
20. Finally, genomes also act as the units of inheritance. As we know from phylogenetics, genomic architecture is also passed on to the succeeding generations.
21. The proposed model of natural selection unifies different types of units of selection and inheritance. It resolves several semantic ambiguities. It also provides a naturalistic foundation of the domain neutral model of natural selection. More importantly, it offers a platform for formalizing genomic evolution as distinct from conventional genetic evolution.

References

Blatteis CM (ed) (1998) Physiology and pathophysiology of temperature regulation. World Scientific, Singapore

Borrello ME (2010) Evolutionary restraints: the contentious history of group selection. University of Chicago Press, Chicago

Bromham L (2008) An introduction to molecular evolution and Phylogenetics. Oxford University Press, Oxford

Capuzzo-Dolcetta RA (2019) Classical Newtonian gravity: a comprehensive introduction. Springer, Dordrecht

Carey N (2015) Junk DNA: a journey through the dark matter of the genome. Icon Books, London

Chhaya P (2020) The universe within: and a view from that within. Universal Press, Irvine, CA

Chhaya P (2022a) The origin and nature of mathematics. Submitted

Chhaya P (2022b) The origin and nature of spacetime. Submitted

Coyne JA, Orr HA (2004) Speciation. Sinauer Publishing, Sunderland

Dawkins R (1999) The selfish genes. Oxford University Press, Oxford

Decker H, Van Holde KE (2011) Oxygen and the evolution of life. Springer, Dordrecht

Delisle RG (ed) (2021) Natural selection: revisiting explanatory role in evolutionary biology. Springer, Dordrecht

Dobzanski T (1982) Genetics and origin of species. Columbia University Press, New York

Donaldson KM (2000) Relationships between chromosome structure and long distance regulation of gene expressions. University of California Press, San Diego

Flew A (2017) Darwinian evolution. Routledge, London

Fodor J, Piattelli-Palmarini M (2011) What Darwin got wrong? Profile Books, London

Fritz A (2014) Elucidating chromosome territories organization during the cell cycle and in breast cancer cells. Proquest, Ann Arbor
Gould SJ (2007) Punctuated evolution. Harvard University Press, Cambridge
Grene M (1986) Dimensions of Darwinism: themes and counter themes in the twentieth century evolutionary theory. Cambridge University Press, Cambridge
Hader DP, Lebert M (2001) Photo movement. Elsevier, Amsterdam
Haynie DT (2008) Biological thermodynamics. Cambridge University Press, Cambridge
Hodge J, Redick G (eds) (2009) Cambridge companion to Darwin. Cambridge University Press, Cambridge
Howard K (ed) (2013) RNA interference: from biology to therapeutics. Springer, Dordrecht
Kimura M (1983) The neutral theory of molecular evolution. Cambridge University Press, Cambridge
Kuhn S (2021) The structure of scientific revolutions. University of Chicago Press, Chicago
Laudal OA (2021) Mathematical models in science. World Scientific, Singapore
Lawrie ID (2013) A unified grand tour of theoretical physics. CRC, Boca Raton
Lovelock J (2000) Gaia: a new look at life on earth. Oxford University Press, Oxford
Makalowski W (2001) The human genome: structure and organization. Acta Biochim Pol 48(3): 587–598
Mostovoy Y (2014) Regulation and function of antisense transcription in saccharomyces cerevisiae yeast. University of California Press, Berkeley
Mukherjee S (2016) The gene: an intimate story. Penguin Press, London
Nei M (2013) Mutation driven evolution. Oxford University Press, Oxford
Nielsen MA, Chuang IL (2010) Quantum computations and quantum information. Cambridge University Press, Cambridge
Okasha S (2010) Evolution and the levels of selection. Oxford University Press, Oxford
Oparin AI (1968) Genesis and development of life. Academic Press, New York
Pevsner J (2015) Bioinformatics and functional genomics. Wiley, Hoboken
Press SJ, Clyde CM (2003) Subjective and objective Bayesian statistics: principles, models and applications. Wiley, Hoboken
Prigorgin I (1968) Introduction to thermodynamics of irreversible processes. Wiley, Hoboken
Provine WB (2001) The origins of theoretical population genetics. University of Chicago Press, Chicago
Raffaelli DG, Frid CLJ (eds) (2010) Ecosystem ecology: a new synthesis. Cambridge University Press, Cambridge
Robert JS (2004) Embryology, Epigenesis and evolution: taking development seriously. Cambridge University Press, Cambridge
Roy S, Kundu T (2021) Chemical biology of the genome. Academic Press, London
Schutz BF (2009) A first course in general relativity. Cambridge University Press, Cambridge
Shmulevich, I. , , Dougherty, E. R. , 2014, Genomic signal processing. Princeton University Press, Princeton
Sieber P, Platzer M, Schuster S (2018) The definition of open reading frames revisited. Trends Genet 34(3):167–170
Stein WD (1967) The movement of molecules across cell membrane. Academic Press, London
Weinzierl ROJ (1999) Mechanism of gene expression: structure, function and evolution of the basal transcriptional machinery. Imperial College Press, London
Wigner E (1960) The unreasonable effectiveness of mathematics in natural science. Commun Pure Appl Math 13:1–14
Yarus M (2010) Life from an RNA world: the ancestor within. Harvard University Press, Cambridge

Epilogue

In days of technological sophistication and therapeutic contingencies, to write a book on the philosophical perspective of genomics may appear to be an anachronism. All I can say is mea culpa. However, upon a little reflection, it is intuitively clear that if genomics were to fulfill these expectations of being the ultimate answer to the illnesses that plague us, it is imperative that we deconstruct genomics ab initio. It was this conviction that has prompted me to venture.

When I began this book, I was motivated by two unresolved aspects of the biology of Life. Firstly, the formal description of the biology of Life must be unitary in nature. Therefore, if we accept that Darwin's theory represents this description, it must also encompass genomes as competing species. However, we do not have any theoretical framework of genomic architecture. Therefore, I wanted to develop a model of genomic architecture that is consistent with the Darwinian paradigm. Secondly, it is necessary to formalize Life in a completely naturalistic framework. Therefore, it is imperative that the underlying formalism of genomic architecture too must be defined naturalistically. However, our understanding of the origin and nature of mathematics is not completely naturalistic. Therefore, it is necessary to employ mathematical constructs that are based on the model of mathematics that is based on mathematical realism. Thus, my objective behind this book is to enlarge the scope of the Darwinian paradigm to include biological evolution per se and to demonstrate that genomes too can act as the competing species in natural selection. Incidentally, this rationale has led to the inference that genomes act not just as the units of selection but also as units of inheritance. Moreover, they also act as the environment in which individual genes compete with one another.

The mathematical framework employed here is based on mathematical realism. Therefore, it has different semantics and metaphysics. These aspects are discussed in my previous books. However, it is pertinent to note that the genomic model outlined here enables us to define Life as natural computation, just like any other natural phenomena. This completes the process of naturalization of Life and eliminates any transcendental arguments. However, it is important to keep in mind that the proposed model does not discard any transcendentalism. Rather, it suggests that what we perceive to be transcendental influences are in fact spatiotemporal influences which

© The Author(s), under exclusive license to Springer Nature Singapore Pte Ltd. 2023

P. Chhaya, *The Topological Model of Genome and Evolution*,
https://doi.org/10.1007/978-981-99-4318-0

are topologically inaccessible to our cognitive faculty. If Life appears to be transcendental, so is knowledge and our quest for it.

Finally, while unraveling the details of this model of genomic architecture, it became clear that this model is capable of resolving several fundamental semantic ambiguities of the Darwinian paradigm. These have been discussed in various chapters of this book. However, any such reinterpretation of the Darwinian paradigm needs a separate monograph of its own. Life permitting, I would deal with it. However, before doing that, I hope to write about the therapeutic possibilities of the proposed model of genomic architecture.